€ 72,-
48
Physik

Light Cone 2016

Alfred Stadler · Teresa Peña
Editors

Light Cone 2016

Challenges for Theory and Experiment in Hadron and Nuclear Physics on the Light Front

The articles have previously been published with Springer in the journal *Few-Body Systems*, Volumes 57 & 58 (2016-2017)

Springer

Editors
Alfred Stadler
Department of Physics
University of Évora
Évora
Portugal

Teresa Peña
Instituto Superior Técnico, CFTP
University of Lisbon
Lisbon
Portugal

ISBN 978-3-319-65731-8

Library of Congress Control Number: 2017951187

Printed on acid-free paper

This Springer imprint is published by Springer Nature
The registered company is Springer International Publishing AG
The registered company address is: Gewerbestrasse 11, 6330 Cham, Switzerland

Contents

LIGHT-CONE 2016: Challenges for Theory and Experiment in Hadron and Nuclear Physics on the Light Front 1
M.T. Peña and Alfred Stadler

Trends and Progress in Nuclear and Hadron Physics: A Straight or Winding Road 3
James P. Vary, Lekha Adhikari, Guangyao Chen, Meijian Li, Yang Li, Pieter Maris, Wenyang Qian, John R. Spence, Shuo Tang, Kirill Tuchin and Xingbo Zhao

Study of Generalized Parton Distributions at Jefferson Lab 15
Andrey Kim

GPDs and Orbital Angular Momentum 21
Matthias Burkardt

Transverse Densities of Octet Baryons from Chiral Effective Field Theory 29
Jose Manuel Alarcón, Astrid N. Hiller Blin and Christian Weiss

An Algebraic Model for the Pion's Valence-Quark GPD: A Probe for a Consistent Extension Beyond DGLAP Region Via Radon Transform Inversion 35
Nabil Chouika, Cédric Mezrag, Hervé Moutarde and José Rodríguez-Quintero

Twist-3 Distribution Amplitudes of Pion in the Light-Front Quark Model 43
Ho-Meoyng Choi and Chueng-Ryong Ji

The GLUEX Experiment: First Results 49
Cristiano Fanelli

Meson Spectroscopy from Lattice QCD 55
Christopher E. Thomas

Nucleon Structure and the Neutron Electric Dipole Moment from Twisted Mass Lattice QCD 63
Giannis Koutsou

Gluon and Ghost Dynamics from Lattice QCD 71
O. Oliveira, A.G. Duarte, D. Dudal and P.J. Silva

Lattice Calculations and the Muon Anomalous Magnetic Moment 79
Marina Krstić Marinković

Confinement at Finite Temperature 87
Nuno Cardoso, Pedro Bicudo and Marco Cardoso

Gluons at Finite Temperature 95
P.J. Silva, O. Oliveira, D. Dudal, P. Bicudo and N. Cardoso

Parametrization of the Transverse Momentum Dependent Light-Front Correlator for Gluons 103
Sabrina Cotogno

The Gauge-Field Propagator in Light-Cone Gauge: Which is the Correct One? 109
L. Mantovani, A. Bacchetta, B. Pasquini and X. Xiong

Heavy Flavour Production as Probe of Gluon Sivers Function 115
Rohini M. Godbole, Abhiram Kaushik, Anuradha Misra, Vaibhav Rawoot and Bipin Sonawane

Interpolation Between the Instant Form and the Front Form of Relativistic Dynamics 123
Chueng-Ryong Ji

The CST: Its Achievements and Its Connection to the Light Cone 131
Franz Gross

Quarkonium Wave Functions on the Light Front 141
Yang Li

Application of the Covariant Spectator Theory to the Study of Heavy and Heavy-Light Mesons 149
Sofia Leitão, Alfred Stadler, M.T. Peña and Elmar P. Biernat

More About the Light Baryon Spectrum 157
Gernot Eichmann

Two-Fermion Bethe–Salpeter Equation in Minkowski Space: The Nakanishi Way 167
Giovanni Salmè, Wayne de Paula, Tobias Frederico and Michele Viviani

Quantum Field Theory in Two Dimensions: Light-front Versus Space-like Solutions 177
L'ubomír Martinovič

Euclidean Relativistic Quantum Mechanics: Scattering Asymptotic Conditions 185
W.N. Polyzou and Gordon J. Aiello

Relativistic Model of Hamiltonian Renormalization for Bound States and Scattering Amplitudes 191
Kamil Serafin

Effective Particles in Quantum Field Theory 199
Stanisław D. Głazek and Arkadiusz P. Trawiński

From Asymptotic Freedom Toward Heavy Quarkonia Within the Renormalization-Group Procedure for Effective Particles 209
María Gómez-Rocha

Advances in Light-Front QCD: Supersymmetric Properties of Hadron Physics from Light-Front Holography and Superconformal Algebra 217
Stanley J. Brodsky

Nonperturbative Strange Sea in Proton Using Wave Functions Inspired by Light Front Holography 233
Alfredo Vega, Ivan Schmidt, Thomas Gutsche and Valery E. Lyubovitskij

Wigner Distributions of Quarks for Different Polarizations 241
Jai More, Asmita Mukherjee and Sreeraj Nair

Symmetry Breaking in Light-Front ϕ^4 Theory 247
J.R. Hiller

Physical and Nonphysical Modes in the Light Front Formulation for the LC Gauge and the Lorentz Gauge Conditions 253
Elżbieta Dzimida-Chmielewska and Jerzy A. Przeszowski

Light-Front Versus Equal-Time Quantization in ϕ^4 Theory 261
S.S. Chabysheva

Alternate Light Front Quantization Procedure for Scalar Fields 267
Jerzy A. Przeszowski

The Mirror Nuclei ^{3}H and ^{3}He Program at JLab 275
Javier Gomez

A Perturbative QCD Approach to $\pi^- p \rightarrow D^- \Lambda_c^+$ 285
Stefan Kofler and Wolfgang Schweiger

Baryon-to-Meson Transition Distribution Amplitudes: Formalism and Models 291
B. Pire, K. Semenov-Tian-Shansky and L. Szymanowski

In-Medium Parton Branching Beyond Eikonal Approximation 301
Liliana Apolinário

In-Medium Pion Valence Distribution Amplitude 309
K. Tsushima and J.P.B.C. de Melo

In-Medium ρ-Meson Properties in a Light-Front Approach 319
J.P.B.C. de Melo and K. Tsushima

Phase Transition in Topologically Massive QED 327
Yuichi Hoshino

Born Level Bound States 355
Paul Hoyer

Few-Body Syst (2017) 58:145
DOI 10.1007/s00601-017-1309-z

M. T. Peña · Alfred Stadler

LIGHT-CONE 2016: Challenges for Theory and Experiment in Hadron and Nuclear Physics on the Light Front

Received: 10 June 2017 / Accepted: 15 June 2017 / Published online: 29 June 2017

The LIGHT-CONE 2016 conference belongs to the series of Light-Cone meetings founded by the International Light Cone Advisory Committee (ILCAC), Inc. (http://www.ilcacinc.org). The 2016 conference, as the ones before it, followed the objectives defined by ILCAC, Inc., namely "to advance research in quantum field theory, in particularly light-cone quantization methods to the solution of physical problems", and "to assist in the development of crucial experimental tests at hadron facilities". The presentations and lively discussions focussed on the most recent advances obtained both within the light-front approach as well as with other relativistic frameworks that deal with non-perturbative QCD, such as Lattice QCD and Bethe-Salpeter/Dyson-Schwinger type functional methods.

Light-front theory was first introduced by Dirac as an alternative to the usual instant form of relativistic dynamics. It provides a theoretical framework for non-perturbative studies of the evolution of composite systems, such as hadrons and nuclear systems, in light-front time $\tau = t + z/c$. In simple terms, the light-front form description is the analogue of a flash snapshot of a relativistic system.

The 2016 edition of the Light-Cone meeting took place at Instituto Superior Técnico, University of Lisbon, Portugal, from the 5th to the 8th of September. The members of the local organizing committee were Teresa Peña (co-chair, CFTP-IST U. Lisboa), Alfred Stadler (co-chair, U. Évora), Orlando Oliveira (U. Coimbra), João Seixas (LIP U. Lisboa), Pedro Bicudo (CFTP-IST U. Lisboa), Elmar Biernat (CFTP-IST U. Lisboa), and Sofia Leitão (CFTP-IST U. Lisboa).

The conference was supported in part by generous contributions from the Jefferson Science Association (JSA) and from Jefferson Laboratory. This financial support allowed ILCAC to award this year's McCartor Fellowships to two young physicists, María Gómez-Rocha (ECT* Trento) and Luca Mantovani (U. Pavia), enabling them to attend the conference and to present the results of their research. We also gratefully acknowledge additional funding by the Portuguese Funding Agency, Fundação para a Ciência e Tecnologia (FCT) and logistic support from the Portuguese Physical Society (SPF).

The LIGHT-CONE 2016 conference brought together 57 registered participants, 20% of which were female scientists. The program included 12 plenary sessions (with a total of 33 invited talks) and 4 parallel sessions

This article belongs to the Topical Collection "Light Cone 2016".

M. T. Peña · A. Stadler
CFTP, Departamento de Física, Instituto Superior Técnico, Universidade de Lisboa, Av. Rovisco Pais, 1049-001 Lisbon, Portugal
E-mail: alfred.stadler@tecnico.ulisboa.pt

M. T. Peña
E-mail: teresa.pena@tecnico.ulisboa.pt

A. Stadler (✉)
Departamento de Física, Universidade de Évora, 7000-671 Évora, Portugal
E-mail: stadler@uevora.pt

(with a total of 17 contributed talks). 20 of the participants were young scientists, who delivered 30% of the plenary talks and 60% of the contributed talks of the program.

The conference addressed new frontiers and challenges in hadron physics, both in experiment and in theory. Examples are talks on the connection between methods for nuclear and hadron-physics calculations (J. Vary), on high-performance computing in light-cone physics (P. Maris), on the GlueX (C. Fanelli) and mirror nuclei ^{3}H and ^{3}He (J. Gomez) programs at JLab, and on lattice QCD calculations of nucleon structure (G. Koutsou), meson spectroscopy, and meson-meson scattering (C. Thomas). Other talks on results of lattice QCD calculations covered the string tension as function of the temperature (N. Cardoso/P. Bicudo), the anomalous magnetic moment of the muon (M. Marinkovic), and gluon and ghost dynamics (O. Oliveira).

A variety of other methods to solve problems in QCD and QED were presented, such as the Dyson-Schwinger approach (G. Eichmann, P. Tandy), solutions of the Bethe-Salpeter equation in Minkowski and Euclidean space (T. Frederico, G. Salmè), the Bethe-Salpeter-type Covariant Spectator Theory (F. Gross) and its results for the meson spectrum (S. Leitão), light-front quantization in QED and QCD calculations (C.-R. Ji, X. Zhao, Y. Li), scattering in Euclidean relativistic quantum mechanics (W. Polyzou), and effective particles in quantum field theory (S. Glazek, M. Gómez-Rocha). A number of talks discussed theoretical and experimental results for hadron structure in terms of GPDs (A. Kim, F. Steffens, M. Burkardt, J. Rodriguez-Quintero, W. Schweiger) and TDAs (L. Szymanowski, E. Atomssa), and various in-medium effects (K. Tsushima, L. Apolinário, J. Cabral de Melo).

In the closing session, S. Brodsky gave a captivating overview of a great variety of recent results on spectroscopic and dynamical properties of QCD from light-front holography and AdS/QCD, with remarkable supersymmetric relations between the masses of mesons and baryons with equal parity.

One of the important take-away messages of the conference is that the comparison of the large variety of existing theoretical methods to analyze multi-quark systems is very valuable for assessing and quantifying the uncertainties in the theoretical predictions of hadron properties, which are to be confronted with experimental data especially when new hadronic states are found.

The conference's half-day excursion began with a visit of the Jerónimos Monastery in Belém and a bus trip along and even in the Tagus river—the bus was of the amphibian kind, we hasten to add. The excursion continued at the XVIIIth century palace of the Marquis of Pombal in Oeiras near Lisbon, where the Marquis' wife herself, unaware of the restrictions relativity imposes on time travel, received the international visitors and gracefully guided them through the palace and its gardens. The visit concluded in the palace's wine cellar with a wine tasting and the conference dinner, which also featured a performance of Portuguese Fado by local artists. During the dinner, the chair of ILCAC, Chueng-Ryong Ji, together with the conference chairs presented the awards to this year's two McCartor Fellowship recipients, and James Vary gave a brief address about the history and objectives of the McCartor Fellowships awards.

All articles published in this Topical Collection of Few-Body Systems were peer-reviewed. Details about the conference, including the presentations, can be found on its webpage, http://www.lc2016.net.

The co-chairs of LIGHT-CONE 2016 wish to express their gratitude to the members of the scientific advisory committee and of the local organizing committee and secretarial staff, for their valuable contributions to the preparation of the conference. We are especially grateful for the commitment and teamwork of the youngest members of the organizing committee, Elmar Biernat and Sofia Leitão. Last but not least, we thank all participants for their contributions to making this conference a success.

Few-Body Syst (2017) 58:56
DOI 10.1007/s00601-016-1210-1

James P. Vary · Lekha Adhikari · Guangyao Chen · Meijian Li · Yang Li · Pieter Maris · Wenyang Qian · John R. Spence · Shuo Tang · Kirill Tuchin · Xingbo Zhao

Trends and Progress in Nuclear and Hadron Physics: A Straight or Winding Road

Received: 17 December 2016 / Accepted: 28 December 2016 / Published online: 23 January 2017

Abstract Quantitative calculations of the properties of hadrons and nuclei, with assessed uncertainties, have emerged as competitive with experimental measurements in a number of major cases. We may well be entering an era where theoretical predictions are critical for experimental progress. Cross-fertilization between the fields of relativistic hadronic structure and non-relativistic nuclear structure is readily apparent. Non-perturbative renormalization methods such as similarity renormalization group and Okubo–Lee–Suzuki schemes as well as many-body methods such as coupled cluster, configuration interaction and lattice simulation methods are now employed and advancing in both major areas of physics. New algorithms to apply these approaches on supercomputers are shared among these areas of physics. The roads to success have intertwined with each community taking the lead at various times in the recent past. We briefly sketch these fascinating paths and comment on some symbiotic relationships. We also overview some recent results from the Hamiltonian basis light-front quantization approach.

1 Introduction

Nuclear and hadronic physics share methods for non-perturbatively solving the quantum many-body problem. Lattice, wave-equation and Hamiltonian eigenvalue methods represent major many-body approaches where developments in one subfield have impacts on the other. Renormalization and effective field theory techniques are also under development in both subfields. For these reasons, researchers often enjoy "dual citizenship" in these subfields. Advances in computational physics and high-performance computing have spurred great progress in nuclear and hadronic physics by enabling solutions with unprecedented accuracy and quantified uncertainties.

In nuclear physics, the development of effective interactions has a long history. The pioneering works of Brueckner, Bethe and Goldstone (see Ref. [1] for a review) led to decades of successful applications by combining non-perturbative and perturbative approaches. Eventually, more robust non-perturbative methods such as Okubo–Lee–Suzuki (OLS) (see Ref. [2] for a review) and the unitary correlation operator method (UCOM) (see Ref. [3] for a review) paved another road to extensive successes.

These methods were developed to enable many-body Hamiltonian approaches to achieve improved convergence in nuclear physics within available computational methods and resources. The methods improved

This article belongs to the Topical Collection "Light Cone 2016".

Presented at LightCone 2016, Lisbon, Portugal.

J. P. Vary (✉) · L. Adhikari · G. Chen · M. Li · Y. Li · P. Maris · W. Qian · J. R. Spence · S. Tang · K. Tuchin
Department of Physics and Astronomy, Iowa State University, Ames, IA 50011, USA
E-mail: jvary@iastate.edu

X. Zhao
Institute of Modern Physics, Chinese Academy of Sciences, Lanzhou 730000, China

and the resources grew simultaneously so that increasingly precise solutions have been attained both for a growing number of nuclei and for a wider array of experimental observables [2].

In the hardron arena, we have witnessed parallel rapid developments of the Hamiltonian approach and some developments have been intertwined with those in nuclear physics. Based upon Dirac's front form of relativistic Hamiltonian dynamics [4], hadronic physics witnessed the development and application of the practical discretized light cone quantization (DLCQ) approach (see Ref. [5] for a review). On the renormalization frontier, we saw the introduction and application of the similarity renormalization group (SRG) technique [6,7] which has enjoyed remarkable success in nuclear physics as well (see Ref. [8] and references therein).

A major example of a many-body method that has enjoyed success in an array of scientific fields is the coupled-cluster method of Coester and Kuemmel [9,10]. From its genesis within nuclear physics, it has been actively developed and applied in quantum chemistry (see Ref. [11] for a review), nuclear physics (see Ref. [12] for a review) and it has also been introduced into light-front field theory with promising results [13].

2 Basis Function Approach

Given these rapid advances in many-body methods, renormalization approaches, algorithms for simulations and high-performance computers sketched above, it is appropriate to survey a few applications that have emerged from these advances. Due to space limitations, I will select a few highlights of recent results from development and applications of basis light-front quantization (BLFQ) [14–16] and its time-dependent generalization tBLFQ [17,18]. For a more complete overview of current and planned research in light-front Hamiltonian theory, see Ref. [19] and references therein.

Simply stated, our goal is to solve the Hamiltonian eigenvalue problem expressed as,

$$\hat{P}^+\hat{P}^-|\psi_h\rangle = M_h^2|\psi_h\rangle \tag{1}$$

where $\hat{P}^{\pm} = \hat{P}^0 \pm \hat{P}^3$ are the longitudinal momentum (+) and the light-front quantized Hamiltonian operator (−). The invariant-mass spectrum results from the product of their eigenvalues. The eigenstates $|\psi_h\rangle$, or light-front wavefunctions (LFWFs) in either coordinate or momentum space, yield predictions for hadron structures in deep inelastic scattering (DIS) [5], deeply virtual Compton scattering (DVCS) [20] and diffractive hadron production in ultra-peripheral heavy ion collisions [21,22], among many other applications.

The Fock space expansion for $|\psi_h\rangle$ produces a sparse-matrix many-body eigenvalue problem. BLFQ was introduced to implement azimuthal rotational symmetry and to accelerate convergence in bound state applications, especially with confining interactions as in QCD. By choosing the two-dimensional (2D) harmonic-oscillator (HO) for the transverse modes, one benefits from the developments of the no-core shell model (NCSM) used successfully in nuclear many-body theory [2,23–25] while retaining a fully covariant framework. The ability to factorize transverse center-of-mass motion in the LFWFs in order to preserve transverse boost invariance is one of the appealing features of BLFQ [14,26,27]. For applications to bound state problems in QCD, we note that the choice of the 2D-HO for the transverse basis space is harmonious with the phenomenologically successful light-front AdS/QCD soft-wall Hamiltonian for the hadrons [28,29].

BLFQ has been used to solve QED problems at strong coupling such as the electron anomalous magnetic moment [15,30], non-linear Compton scattering [17,18] and the positronium spectrum [31]. In this paper, we briefly summarize recent BLFQ applications to bound-state and scattering problems: electron form factors [32] in Sect. 3, Yukawa model [33] in Sect. 4, heavy quarkonium [34] in Sect. 5, vector meson production in Sect. 6, and the extension to tBLFQ [17,18,35] in Sect. 7, each of which points pathways to future developments and applications. A promising approach to non-perturbative renormalization is also presented at this meeting [36,37].

3 Electron Form Factors in BLFQ

Following earlier applications to QED where we evaluated the electron anomalous magnetic moment in BLFQ and compared the results with perturbation theory [15,30], we now evaluate the electron's generalized parton distribution (GPD) $E(x, q^2)$, Pauli form factor $F_2(q^2)$, and anomalous gravitomagnetic form factor $B(q^2) = B_f + B_b$. These are displayed in Fig. 1 with values for the regulators provided in the legends. Here, λ_k (Λ_k) signifies the infrared (ultraviolet) cutoff matched to the BLFQ regulators which are defined in terms of b_2 and $N_{\max}$. The quantity $b_2 = b/\sqrt{2}$ represents the 2D-HO parameter of relative motion and $b = M$ is the

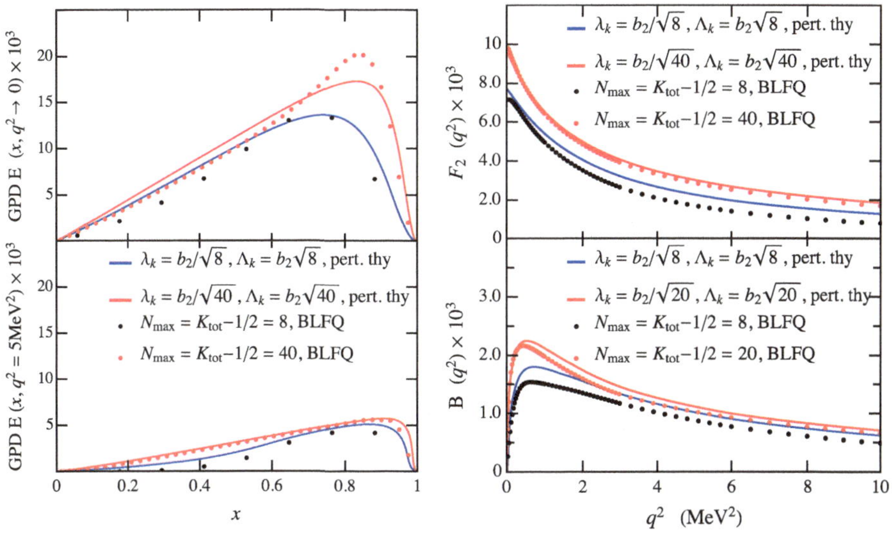

Fig. 1 (Color online) Electron form factors obtained from LFPT (*solid lines*) and BLFQ (*dotted curves*) at infrared (*ultraviolet*) regulator λ_k (Λ_k) matched to BLFQ regulators N_{max} and K_{tot}. *Left* comparison of GPD E(x, q^2), for $q^2 = 0$ (*top panel*) and $q^2 = 5$ MeV^2 (*bottom panel*). The integral of GPD E(x, q^2) over the momentum fraction x contributes to F_2. *Right* form factors as functions of q^2. *Top panel* Pauli form factor $F_2(q^2)$, *bottom panel* gravitomagnetic form factor $B(q^2) = B_f + B_b$

2D-HO parameter in the one-particle sector. The regulator N_{max} defines the upper limit of the sum of 2D-HO quanta for the transverse basis states and the regulator $K = K_{tot}$ defines the upper limit of the longitudinal plane wave modes for where we adopt antiperiodic (periodic) boundary conditions for fermions (bosons); i.e. $K = K_{tot}$ is the sum of the electron's (half-odd-integer) and the photon's (integer) longitudinal quanta. In addition to showing the expected agreement between light-front perturbation theory (LFPT) and DLCQ, we re-confirm the vanishing of the electron anomalous gravitomoment, i.e. $B(q^2 = 0) = 0$ [38].

4 Yukawa Model in BLFQ

We follow the scheme we employed in our application of BLFQ to positronium [31] to solve the Yukawa model with a scalar boson exchange [33]. To be specific, we solve for the mass eigenstates (in units of the fermion mass m_f) as a function of two basis space regulators, N_{max} and K in a Fock space consisting of a fermion f and an antifermion $\bar{f}$ exchanging a scalar boson of mass $m_b = 0.15m_f$ with coupling $\alpha = 0.3$ and HO length parameter $b = \sqrt{(M\Omega)} = 0.16m_f$. Our $f\bar{f}$ effective interaction follows the one employed in Ref. [31] except that a scalar boson replaces the vector exchange of QED. We neglect fermion self-energy and annihilation processes. Our mass eigenstates are presented in units of the fermion mass in Fig. 2 for $N_{\text{max}} = K = 19$ (left panel) and, for the singlet ground state, as a function of $1/N_{\text{max}}$ at various values of K. The spectra in the left panel is a representative case at fixed N_{max} and K. It shows degeneracy with respect to M_j whose multiplicity implies the total angular momentum for each mass according to $J = \max(|M_j|)$. Making a simple extrapolation in $1/K$ of the extrapolants obtained in the right panel with respect to $1/N_{\text{max}}$, produces an estimated ground state mass of 1.99933 m_f in the continuum limit which is close to the non-relativistic result (solution of the Schroedinger equation with the Yukawa potential) of 1.99954 m_f. Current work focuses on extending the results in the right hand panel, obtaining corresponding extrapolations for excited states and quantifying the uncertainty of the extrapolated masses.

5 Heavy Quarkonium in BLFQ

We begin here with an effective Hamiltonian H^1_{eff}, where the superscript indicates our first approximate form, based on the holographic QCD Hamiltonian [28]

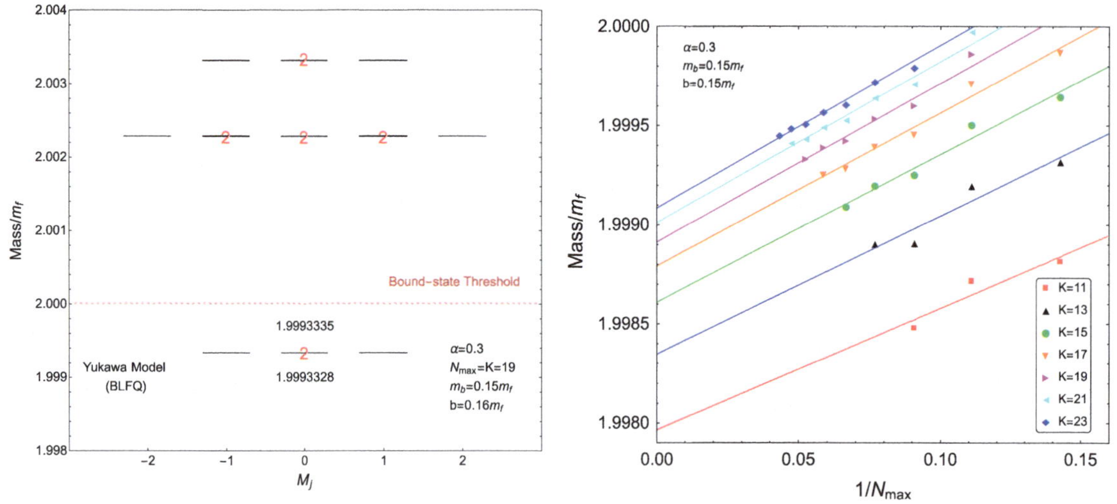

Fig. 2 (Color online) Mass spectrum of the Yukawa model for an $f\bar{f}$ system interacting by a scalar boson exchange as a function of total magnetic projection M_j at fixed $N_{\max} = K=19$ (*left panel*) and, for the singlet ground state, as a function of $1/N_{\max}$ at various values of K (*right panel*). Masses are quoted in units of the fermion mass m_f. Degenerate states are indicated with the "2" to indicate two states nearly coincide

$$H^1_{\text{eff}} \equiv P^+\hat{P}^-_{\text{eff}} - \boldsymbol{P}^2_\perp = \frac{\boldsymbol{k}^2_\perp}{x(1-x)} + \kappa^4 x(1-x)\boldsymbol{r}^2_\perp \tag{2}$$

where $x = p^+_q/P^+$ is the longitudinal momentum fraction of the quark, $\boldsymbol{k}_\perp = \boldsymbol{p}_{q\perp} - x\boldsymbol{P}_\perp$ is the intrinsic transverse momentum, $\boldsymbol{r}_\perp = \boldsymbol{r}_{q\perp} - \boldsymbol{r}_{\bar{q}\perp}$ is the transverse separation of the quarks. κ is the strength of the confining potential.

The holographic QCD Hamiltonian is only 2-dimensional and is defined for massless quarks without interactions. For the heavy quarkonium systems and other applications we incorporate the mass of the quarks m_q, add one-gluon exchange V_g, and add longitudinal dynamics to arrive at our H_{eff} [34],

$$H_{\text{eff}} = \frac{\boldsymbol{k}^2_\perp + m^2_q}{x} + \frac{\boldsymbol{k}^2_\perp + m^2_{\bar{q}}}{1-x} + \kappa^4 x(1-x)\boldsymbol{r}^2_\perp + V_g + V_L(x) \tag{3}$$

Our longitudinal confining potential V_L has some similarities with other proposals [42–44]:

$$V_L(x) = -\frac{\kappa^4}{(m_q + m_{\bar{q}})^2}\partial_x\big(x(1-x)\partial_x\big). \tag{4}$$

Among V_L's appealing features are that it gives rise to longitudinal basis functions which are analytic solutions of the resulting 1D wave equation, which introduces a new quantum number ℓ, and resemble the known asymptotic parton distribution $\sim x^\alpha(1-x)^\beta$. In addition, in the non-relativistic limit $m_f \gg \kappa$, our longitudinal confinement combines with transverse confinement to form a 3D-HO potential, $V_{\text{con}} = \frac{m_q m_{\bar{q}}}{(m_q+m_{\bar{q}})^2}\kappa^4 \boldsymbol{r}^2$, and rotational symmetry is preserved. This non-relativistic reduction also provides us with simple estimates of the model parameters. On the other hand, in the massless limit $m_f \ll \kappa$, the longitudinal mode stays in the ground state and the longitudinal wavefunction is a constant, thereby restoring the massless model of Brodsky and de Téramond [29].

Our main results for heavy quarkonium are presented in more detail by Li [45] at this meeting so we will use this opportunity to provide comparisons in Fig. 3 with a sample of approaches that incorporate massive quarks and quark-antiquark interactions. Each approach involves parameter fits but they differ widely on the choice of experimental data to incorporate in the fit and on the results presented as predictions. The details the BLFQ approach highlighted here are presented in Ref. [34] with more recent improvements presented in this meeting [45]. A variational approach leading to a wave equation resembling a Bethe–Salpeter equation provides the results quoted from Ref. [40]. In this approach, confinement arises from an ansatz for the non-perturbative scale and a variational treatment of the multi-gluon dynamics that provides the shape and magnitude of the confining interaction. The two-particle Dirac equation with one gluon-exchange and confinement has been

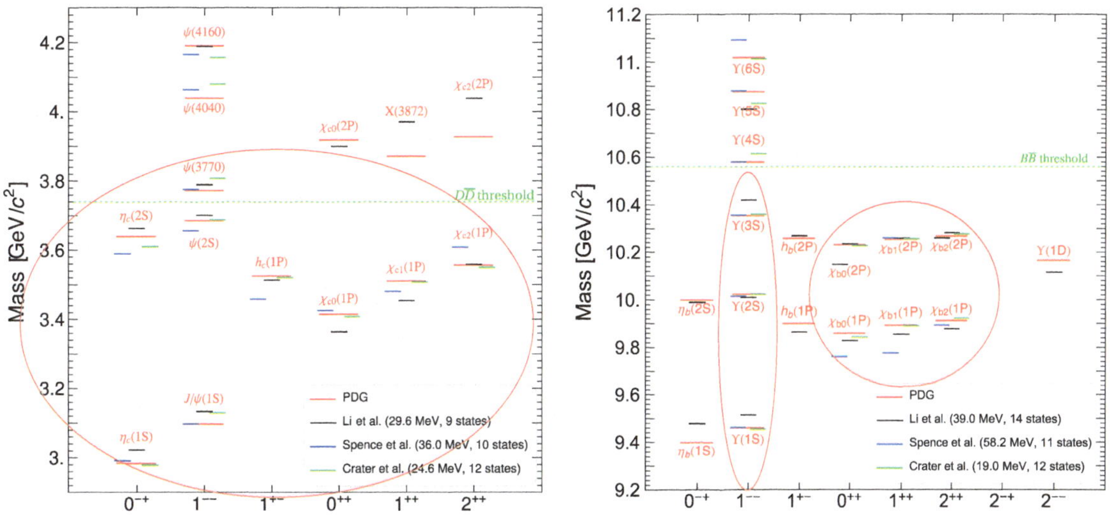

Fig. 3 (Color online) *Left* (*Right*) charmonium (bottomonium) spectrum in GeV$/c^2$; known states are labeled by their PDG symbols [39] while others are labeled by non-relativistic symbols. Different theoretical approaches (Refs. [34] for"Li et al.", [40] for "Spence et al." and [41] for "Crater et al."), as described in the text, are compared with each other and with experiment (PDG). *Red* ovals encircle states that are quoted in all three theoretical approaches

intensively investigated for all the known mesons and has led to the results shown from Ref. [41] that exhibit the lowest rms deviation between theory and experiment.

We have highlighted the states that have been produced with all the cited methods by encircling them in Fig. 3. Some of these states are also involved in the respective fitting procedures. The comparisons outside the ovals provide an impression of similarities and differences in the predicted mass eigenstates where available.

Before closing this section, we present an example of the range of observables accessible with our LFWFs developed in Ref. [34]. For brevity, we present only the helicity-non-flip GPD $H(x, \xi = 0, t = -\mathbf{\Delta}_\perp^2)$ for a selection of two bound states of charmonium in Fig. 4. The helicity-non-flip GPDs can be written, for the case where the photon couples only to the quark, as overlap integrals between LFWFs [20,46–48,50]

$$H\left(x, \xi = 0, t = -\mathbf{\Delta}_\perp^2\right) = \sum_{\lambda_q, \lambda_{\bar{q}}} \frac{1}{16\pi^3 x(1-x)} \int d^2\mathbf{k}_\perp \, \psi^*\left(\mathbf{k}'_\perp, x, \lambda_q, \lambda_{\bar{q}}\right) \psi\left(\mathbf{k}_\perp, x, \lambda_q, \lambda_{\bar{q}}\right). \quad (5)$$

Here, $\mathbf{k}_\perp$ and $\mathbf{k}'_\perp$ are the respective relative transverse momenta of the quark before and after being struck by the virtual photon, Δ is the momentum transfer (we choose the Drell–Yan frame $\Delta^+ = 0, t \equiv \Delta^2 = -\mathbf{\Delta}_\perp^2$, ξ is the skewness parameter and in the Drell–Yan frame, $\xi = 0$, and $\lambda_e(\lambda_{\bar{e}})$ is the spin of the quark (antiquark). This is a demonstration case illustrating hadronic GPDs in BLFQ similar to a positronium application [49,50].

Now, referring to Ref. [51], the impact-parameter dependent GPDs are defined as the Fourier transform of the GPDs with respect to the momentum transfer $\mathbf{\Delta}_\perp$

$$q(x, \mathbf{b}_\perp) = \int \frac{d^2\mathbf{\Delta}_\perp}{(2\pi)^2} e^{-i\mathbf{\Delta}_\perp \cdot \mathbf{b}_\perp} H\left(x, 0, -\mathbf{\Delta}_\perp^2\right). \quad (6)$$

Here, the impact parameter $\mathbf{b}_\perp$ corresponds to the displacement of the quark (q) from the transverse center of momentum of the entire system $(q\bar{q})$. We present the the impact-parameter dependent GPDs for the same two states of charmonium in the lower panels of Fig. 4 to provide a visual impression of these coordinate space distributions. Note, especially, the appearance of secondary peaks in the η'_c.

6 Vector Meson Production

With our BLFQ LFWFs discussed above, we compute the production of charmonium states in diffractive deep inelastic scattering and ultra-peripheral heavy ion collisions within the dipole picture [52]. The assumed process is depicted schematically in Fig. 5 where a virtual photon dissociates into a massive $q\bar{q}$ pair that subsequently interacts with the gluon field of the hadronic system producing a vector meson in the final state.

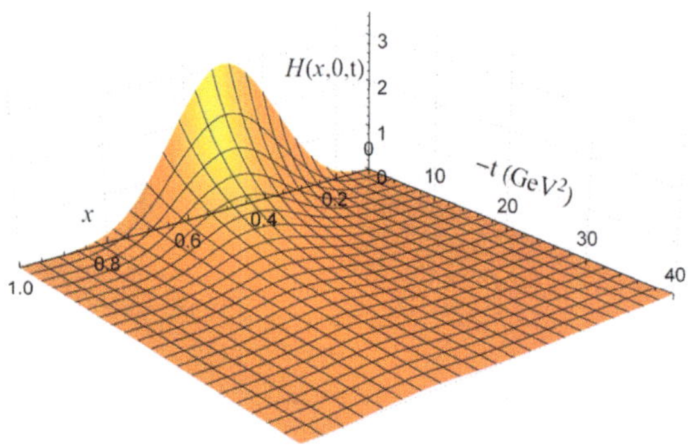

(a) $J/\psi : 1^3S_1(1^{--})$

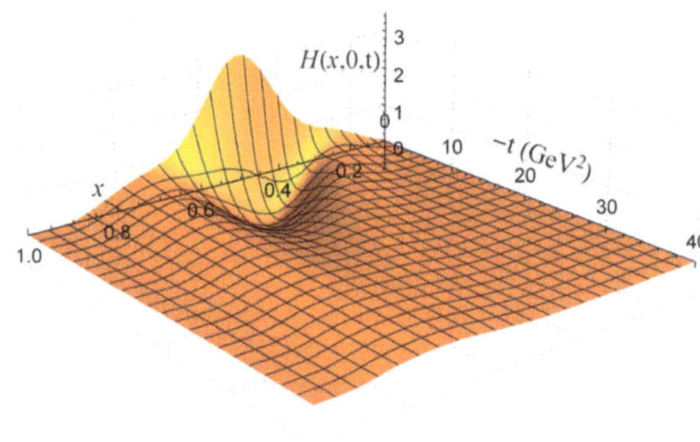

(c) $\eta_c' : 2^1S_0(0^{-+})$

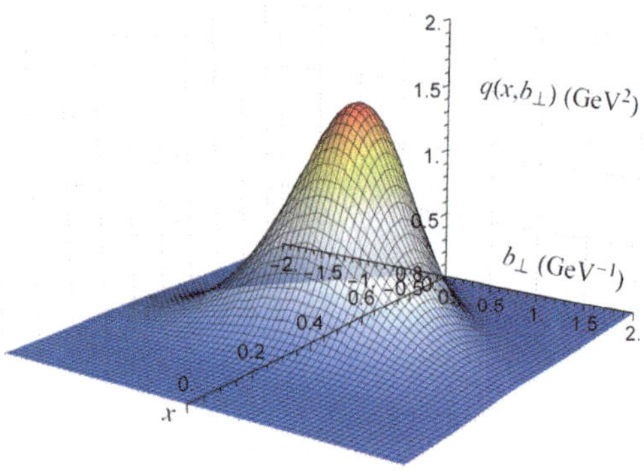

(b) $J/\psi : 1^3S_1(1^{--})$

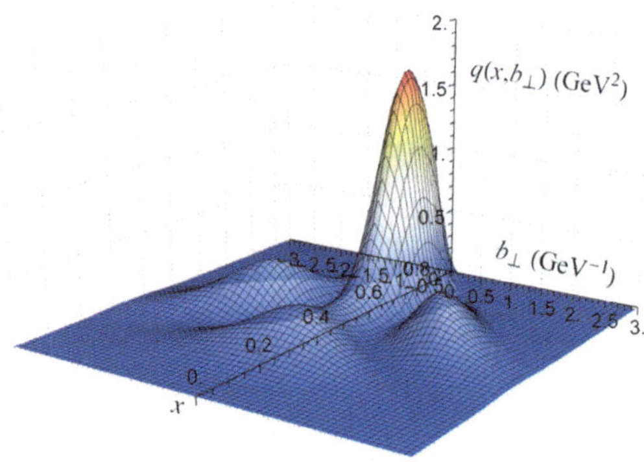

(d) $\eta_c' : 2^1S_0(0^{-+})$

Fig. 4 (Color online) Top plots: 3D plot of helicity non-flip GPDs $H(x, \xi = 0, t = -\Delta_\perp^2)$ (Eq. 5) and bottom plots: impact-parameter dependent GPDs $q\,(x, b_\perp)$ (Eq. 6) for the two bound states of charmonium ($c\bar{c}$) with $N_{\max} = 8$, $L_{\max} = 8$, $m_J = m_J' = 0$, coupling constant $\alpha = 0.3595$, the confining strength $\kappa = 0.963$ GeV, the 2D-HO basis scale $b = \kappa$, the quark (charmonium) mass $m_q = 1.492$ GeV, and gluon mass $\mu_g = 0.02$ GeV. Note states are identified with their non-relativistic quantum numbers (relativistic quantum numbers) $n^{2S+1}L_J\,(J^{PC})$, where n is the radial quantum number, the relation between N, the principal quantum number, and n, is $N = n + L$, L is the total orbital angular momentum, S is the total intrinsic spin, J is the total angular momentum, P is the parity and C is the charge conjugation. **a** $J/\psi : 1^3S_1(1^{--})$, **b** $J/\psi : 1^3S_1(1^{--})$, **c** $\eta_c' : 2^1S_0(0^{-+})$, **d** $\eta_c' : 2^1S_0(0^{-+})$

In the dipole model, the amplitude for exclusive heavy quarkonium production in DIS can be calculated as [53]

$$\mathcal{A}_{T,L}^{\gamma^* p \to Ep}(x, Q, \Delta) = \mathrm{i} \int d^2\boldsymbol{r} \int_0^1 \frac{dz}{4\pi} \int d^2\boldsymbol{b}\, \left(\Psi_E^* \Psi\right)_{T,L}(r, z, Q)\, \mathrm{e}^{-\mathrm{i}[\boldsymbol{b}-(1-z)\boldsymbol{r}]\cdot\boldsymbol{\Delta}}\, \frac{d\sigma_{q\bar{q}}}{d^2\boldsymbol{b}}(x, r), \tag{7}$$

where T and L denote the transverse and longitudinal polarization of the virtual photon (with virtuality Q^2) and the produced quarkonium, and $t = -\boldsymbol{\Delta}^2$ denotes the momentum transfer. On the right-hand side, $\boldsymbol{r}$ is the transverse size of the color dipole, z is the LF longitudinal momentum fraction of the quark, $\boldsymbol{b}$ is the impact parameter of the dipole relative to the proton and x is the Bjorken variable. Ψ and Ψ_E^* are LFWFs of the virtual photon, treated perturbatively, and the exclusively produced quarkonium respectively. The cross section is written in terms of the amplitude as

$$\frac{d\sigma_{T,L}^{\gamma^* p \to Ep}}{dt} = \frac{1}{16\pi} |\mathcal{A}_{T,L}^{\gamma^* p \to Ep}(x, Q, \Delta)|^2. \tag{8}$$

For the results presented in Fig. 5, we employ the parametrization in the impact-parameter-dependent saturation model provided in Ref. [53], which is referred to as "b-Sat I" in Ref. [52]. The calculations using the boosted Gaussian LFWF can also be found in Ref. [53]. The results with the BLFQ LFWF (solid curves)

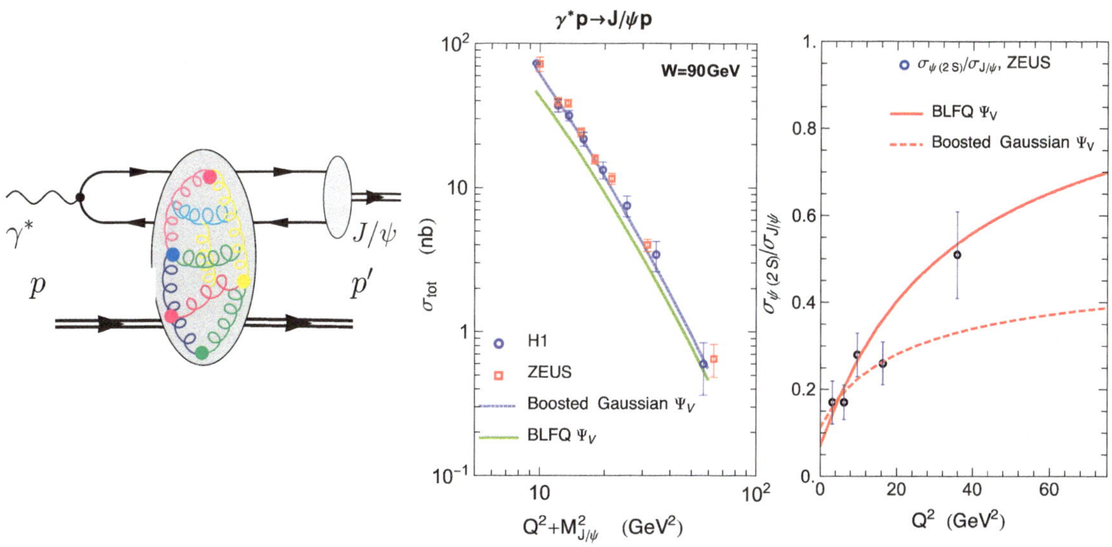

Fig. 5 (Color online) *Left* schematic of vector meson production in the dipole model that begins with virtual photon dissociation to a $q\bar{q}$ pair, interaction with the gluon field of a hadron with 4-momentum p and hadronization of the vector meson in the final state. *Middle* total J/Ψ cross section for different values of Q^2 and W [54,55]. *Right* cross-section ratio $\sigma_{\Psi(2s)}/\sigma_{J/\Psi}$ for different Q^2. *Error bars* indicate the statistical uncertainties only [56]

and the phenomenological boosted Gaussian LFWF (dashed curves) both provide reasonable descriptions of the HERA experimental data.

We obtain additional charmonium production cross sections and find reasonable agreement with experimental data at HERA, RHIC and LHC. We observe that the cross-section ratio $\sigma_{\Psi(2s)}/\sigma_{J/\Psi}$ reveals significant independence of the gluon distribution parameters while showing sensitivity to the charmonium LFWFs as seen in the right panel of Fig. 5. Additional results are found in Ref. [52].

One of the most important uses of vector meson production is to probe the non-perturbative low-x region of hadronic systems. To this end, we investigate the time-dependent evolution of the $q\bar{q}$ system within the color glass condensate (CGC) model [57] of hadrons at low-x. As an initial step, we obtained the gluon field by solving the Yang–Mills equation of stochastic color sources on a 2D lattice. The left panel of Fig. 6 shows one color component of the gluon field generated by a randomly sampled color source distribution. To test whether the stochastic solutions represent the CGC, we examine the Wilson line correlator $C(x_\perp - y_\perp) = \langle TrU^\dagger(x_\perp)U(y_\perp)\rangle$ whose Fourier transform (times k^2) is plotted in the right hand panel of Fig. 6. From this we extract the saturation scale defined as the maximum of the curves in the right hand panel of Fig. 6 as $Q^2/g^2\mu = 0.32$ and observe good convergence with respect to the number of grid points N as well as good agreement with the established CGC saturation scale [57,58].

7 Electron Motion in External Field with tBLFQ

We adopt the tBLFQ framework [17,18] to investigate the effects of electromagnetic (EM) fields generated by ultra-relativistic heavy ions on charged particles [35]. This study is motivated by the fact that strong EM fields are generated during heavy ion collisions [59], and a quantitative study of their effects on charged particles is essential for extracting properties of the quark gluon plasma (QGP) as well as the properties of the gluon distributions in nuclei as discussed above. Here, we investigate the real-time evolution of a quantized electron field under the influence of a strong external time-dependent EM field of a relativistic heavy ion.

When the coupling between the electron and the external field is small such as the field generated by an ultra-relativistic proton with coupling $\alpha = \alpha_{\text{em}} \approx 1/137$, the transition rate between two kinetic energy eigenstates calculated within tBLFQ approach shows agreement with LFPT. On the other hand, for EM fields generated by an ultra-relativistic gold nucleus, the coupling between an electron and the fields is $\alpha = Z_{\text{Au}}\alpha_{\text{em}} \approx 79/137$. The transition rate between the same two kinetic energy eigenstates calculated within tBLFQ approach deviates

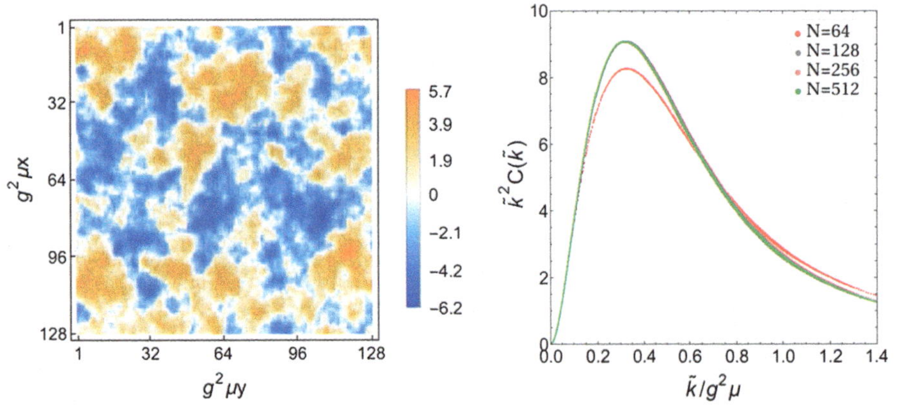

Fig. 6 (Color online) *Left*: an $x-y$ slice of the amplitude of one color component of the color glass condensate in covariant gauge with lattice size $g^2\mu = 1.0$ GeV, $g^2\mu L = 100$ and $N = 128$ grid points on an edge. *Right* Fourier transform of the Wilson correlator $C(k)$ times k^2. *Each curve* represents an average over 500 color configurations with lattice spacing $g^2\mu a = 0.5$, lattice size $L = Na$

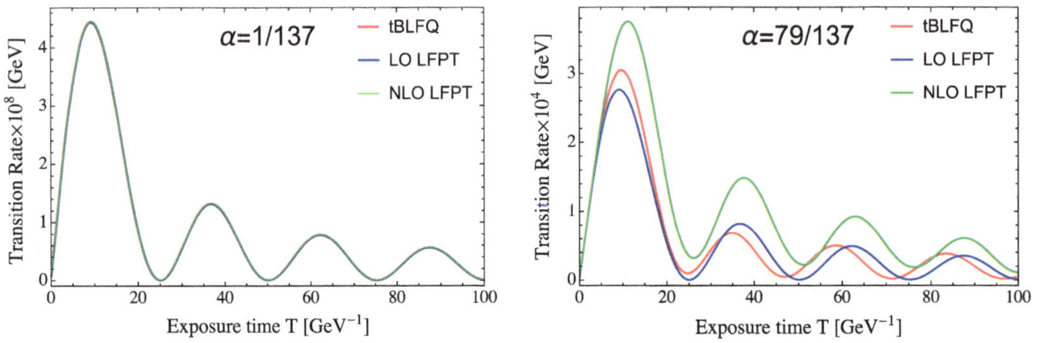

Fig. 7 (Color online) Transition rate of an electron between a specific initial and a specific final state induced by the potential which is generated by nuclei with atomic number $Z = 1$ (*left panel*) and $Z = 79$ (*right panel*) with $\alpha \equiv Z\alpha_{\text{em}}$ with $\alpha_{\text{em}} = 1/137$. The nucleus is moving along positive z-axis with $y = 5.3$. The initial and final states are kinetic energy eigenstates in BLFQ with energies $P^-_{\beta,i} = 0.455$ and $P^-_{\beta,f} = 0.955$ GeV. The calculation is performed using $N_{\max} = 32$, $K = 32$, $L = 10$ GeV^{-1} and $b = 1000m_e$

from LFPT calculations [both leading-order (LO) and next-to-leading order (NLO)], and the differences among these calculations is a clear indication that higher order effects are significant. See Fig. 7 for an illustration.[1]

We next demonstrate strong time-dependent field effects with the real-time evolution of the transverse and longitudinal momentum distributions of an electron evolving in the electromagnetic fields generated by a gold nucleus moving along positive z-axis with rapidity $y = 5.3$. Figure 8 shows a snapshot of the transverse and longitudinal momentum distributions of the electron after evolving 50 GeV^{-1} (roughly 10 fm/c) inside such strong EM fields. We observe that the tBLFQ simulation, which reconciles the higher order effects, predicts that the peak value of the transverse momentum will increase, accompanied by a decreased width, as compared with the LFPT calculations predictions. See the left panel of Fig. 8 for detailed comparisons. The probability that the electron has been excited to higher longitudinal momentum is about 35% according to the tBLFQ simulation.

Good prospects for applying the tBLFQ formalism to heavy ion collisions and electron ion collisions impel us to carry out further applications. For instance, as discussed briefly in the previous section, we will adopt the classical description of gluon fields in high energy nuclear collisions from the color glass condensate (CGC) effective theory [57], and study the real-time evolution of colored objects interacting with these classical gluon fields. Within the tBLFQ framework, we can study the effects of initial gluon fields generated in the initial stage of relativistic heavy-ion collisions on heavy quarks and jets. The advantages of tBLFQ framework

[1] The kinetic energy eigenstates are obtained through diagonalization of kinetic energy matrix in the BLFQ basis. We then evolve the electron amplitude in light-front time using the external field in the interaction picture. Figure 7 shows transition rates between two states with energy 0.0648 and 1.886 GeV.

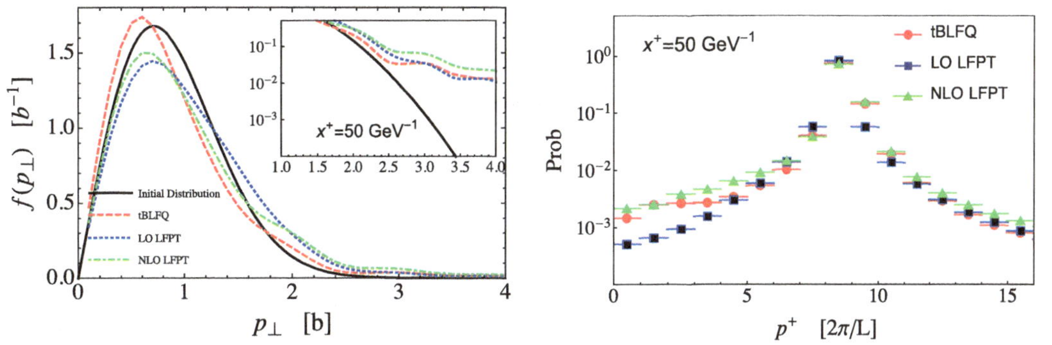

Fig. 8 (Color online) Comparisons between LFPT and tBLFQ of the transverse (left) and longitudnal (right) momentum distribution of an electron after evolving in the EM fields generated by a gold nucleus moving along positive z-axis with rapidity $y = 5.3$ for $x^+ = 50\ \text{GeV}^{-1}$. The initial state of the electron is the BLFQ basis state with $k = \frac{17}{2}$, $n = 0$, $m = 0$ and $\lambda = \frac{1}{2}$ at $x^+ = 0$. The calculation is performed using $N_{\max} = 32$, $K = 32$, $L = 10\ \text{GeV}^{-1}$ and $b = 1000m_e$. The inset in the left panel presents a semi log plot to highlight the significant differences in the large transverse momentum region. *Bars* in the plot of the longitudinal momentum distribution indicate the bin size for discretized plane waves in the longitudinal direction

are distinctive: the calculation is both relativistic and at the amplitude level thereby incorporating quantum interference effects. In addition, we can naturally extend our calculation to higher Fock sectors and go beyond the Eikonal approximation.

8 Summary and Outlook

It is reasonable to expect that the interplay between nuclear and hadronic physics to continue to flourish owing to the common challenges of strong interactions, the complexities of many-particle dynamics and the need to exploit the disruptive technologies of high-performance computers. Likewise, both subfields will benefit greatly from collaborations with applied mathematicians and computer scientists.

The BLFQ applications summarized here represent potentially fruitful research paths. Beyond those outlined above, we would like to mention the extension of BLFQ to other meson sectors (mixed flavor and light mesons), to the baryons and to multi-baryon systems which are underway.

Acknowledgements This work was supported by the Department of Energy under Grant Nos. DE-FG02-87ER40371 and DESC0008485 (SciDAC-3/NUCLEI). Computational resources were provided by NERSC, which is supported by the Office of Science of the U.S. DOE under Contract No. DE-AC02-05CH11231 and by an INCITE award, "Nuclear Structure and Nuclear Reactions", from the US DOE Office of Advanced Scientific Computing. This research also used resources of the Oak Ridge Leadership Computing Facility, which is supported by the US DOE Office of Science under Contract DE-AC05-00OR22725. X. Zhao is supported by new faculty startup funding by the Institute of Modern Physics, Chinese Academy of Sciences.

References

1. B.H. Brandow, Linked-cluster expansions for the nuclear many-body problem. Rev. Mod. Phys. **39**, 771 (1967)
2. B.R. Barrett, P. Navratil, J.P. Vary, Ab initio no core shell model. Prog. Part. Nucl. Phys. **69**, 131 (2013)
3. R. Roth, T. Ne, H. Feldmeier, Nuclear structure in the framework of the unitary correlation operator method. Prog. Part. Nucl. Phys. **65**, 50 (2010)
4. P.A.M. Dirac, Forms of relativistic dynamics. Rev. Mod. Phys. **21**, 392 (1949)
5. S.J. Brodsky, H.-C. Pauli, S. Pinsky, Quantum chromodynamics and other field theories on the light cone. Phys. Rept. **301**, 299 (1998)
6. S.D. Glazek, K.G. Wilson, Renormalization of hamiltonians. Phys. Rev. D **48**, 5863 (1993)
7. F. Wegner, Flow-equations for hamiltonians. Ann. Phys. **506**, 7791 (1994)
8. S.K. Bogner, R.J. Furnstahl, P. Maris, R.J. Perry, A. Schwenk, J.P. Vary, Convergence in the no-core shell model with low-momentum two-nucleon interactions. Nucl. Phys. A **801**, 21 (2008)
9. F. Coester, Bound states of a many-particle system. Nucl. Phys. **7**, 421 (1958)
10. F. Coester, H. Kuemmel, Short-range correlations in nuclear wave functions. Nucl. Phys. **17**, 477 (1960)
11. H. Kuemmel, A biography of the coupled cluster method. Int. J. Mod. Phys. B **17**, 5311 (2003)
12. G. Hagen, T. Papenbrock, M. Hjorth-Jensen, D.J. Dean, Coupled-cluster computations of atomic nuclei. Rept. Prog. Phys. **77**(9), 096302 (2014)
13. J.R. Hiller, Nonperturbative light-front hamiltonian methods. Prog. Part. Nucl. Phys. **90**, 75 (2016)

14. J.P. Vary, H. Honkanen, J. Li, P. Maris, S.J. Brodsky, A. Harindranath, G.F. de Teramond, P. Stern-berg, E.G. Ng, C. Yang, Hamiltonian light-front eld theory in a basis function approach. Phys. Rev. C **81**, 035205 (2010)
15. H. Honkanen, P. Maris, J.P. Vary, S.J. Brodsky, Electron in a transverse harmonic cavity. Phys. Rev. Lett. **106**, 061603 (2011)
16. J.P. Vary, Hamiltonian light-front field theory: recent progress and tantalizing prospects. Few Body Syst. **52**, 331 (2012)
17. X. Zhao, A. Ilderton, P. Maris, J.P. Vary, Non-perturbative quantum time evolution on the light-front. Phys. Lett. B **726**, 856 (2013)
18. X. Zhao, A. Ilderton, P. Maris, J.P. Vary, Scattering in time-dependent basis light-front quantization. Phys. Rev. D **88**, 065014 (2013)
19. B.L.G. Bakker et al., Light-front quantum chromodynamics: a framework for the analysis of hadron physics. Nucl. Phys. Proc. Suppl. **165**, 251–252 (2014)
20. S.J. Brodsky, M. Diehl, D.S. Hwang, Light-cone wavefunction representation of deeply virtual compton scattering. Nucl. Phys. B **596**, 99–124 (2001)
21. S. Afanasiev et al., Photoproduction of J/psi and of high mass e^+e^- in ultra-peripheral $Au+Au$ collisions at $s**(1/2) = 200$-GeV-PHENIX Collaboration. Phys. Lett. B **679**, 321–329 (2009)
22. E. Abbas et al., Charmonium and e^+e^- pair photoproduction at mid-rapidity in ultra-peripheral Pb-Pb collisions at $\sqrt{s_{\rm NN}}$=2.76 TeV [ALICE Collaboration]. Eur. Phys. J. C **73**(11), 2617 (2013)
23. P. Navratil, J.P. Vary, B.R. Barrett, Properties of ^{12}C in the ab initio nuclear shell model. Phys. Rev. Lett. **84**, 5728 (2000)
24. P. Navratil, J.P. Vary, B.R. Barrett, Large basis ab initio no-core shell model and its application to ^{12}C. Phys. Rev. C **62**, 054311 (2000)
25. P. Maris, J.P. Vary, A.M. Shirokov, Ab initio no-core full configuration calculations of light nuclei. Phys. Rev. C **79**, 014308 (2009)
26. Y. Li, P.W. Wiecki, X. Zhao, P. Maris, J.P. Vary, in Introduction to Basis Light-Front Quantization Approach to QCD Bound State Problems, ed. by A.M. Shirokov, A.I. Mazur. Proceedings of International Conference Nuclear Theory in the Supercomputing Era (NTSE-2013), Ames, IA, USA, May 13-17, 2013 (Pacic National University, Khabarovsk, Russia, 2014), p. 136
27. P. Maris, P. Wiecki, Y. Li, X. Zhao, J.P. Vary, Bound state calculations in QED and QCD using basis light-front quantization. Acta Phys. Polon. Supp. **6**, 321 (2013)
28. G.F. de Teramond, S.J. Brodsky, Light-front holography: a first approximation to QCD. Phys. Rev. Lett. **102**, 081601 (2009)
29. S.J. Brodsky, G.F. de Teramond, H.G. Dosch, J. Erlich, Light-front holographic QCD and emerging confinement. Phys. Rept. **584**, 1 (2015)
30. X. Zhao, H. Honkanen, P. Maris, J.P. Vary, S.J. Brodsky, Electron g-2 in light-front quantization. Phys. Lett. B **737**, 65 (2014)
31. P. Wiecki, Y. Li, X. Zhao, P. Maris, J.P. Vary, Basis light-front quantization approach to positronium. Phys. Rev. D **91**(10), 105009 (2015)
32. S. Tang, et al., Form factors of the electron (in preparation)
33. W. Qian, et al., Bound states of the Yukawa model (in preparation)
34. Y. Li, P. Maris, X. Zhao, J.P. Vary, Heavy quarkonium in a holographic basis. Phys. Letts. B **758**, 118 (2016)
35. G. Chen, X. Zhao, Y. Li, P. Maris, K. Tuchin, J.P. Vary, Light-front time evolution in intense fields (in preparation)
36. X. Zhao, Advances in basis light-front quantization. Few Body Syst. **56**(6–9), 257 (2015)
37. X. Zhao, Positronium in basis light-front quantization. (invited talk at this meeting)
38. S.J. Brodsky, D.S. Hwang, B.Q. Ma, I. Schmidt, Light cone representation of the spin and orbital angular momentum of relativistic composite systems. Nucl. Phys. B **593**, 311 (2001)
39. K.A. Olive et al., (Particle Data Group), Review of particle physics. Chin. Phys. C 38, 090001 (2014). [http://pdg.lbl.gov]
40. J.R. Spence, J.P. Vary, Variational tamm-danco treatment of quantum chromodynamics III: a QCD-motivated treatment of meson spectroscopy (in preparation)
41. H. Crater, P. Van Alstine, Relativistic calculation of the meson spectrum: a fully covariant treatment versus standard treatments. Phys. Rev. D **70**, 034026 (2004)
42. S.D. Głazek, Similarity flow of a neutral scalar coupled to a fixed source. Acta Phys. Polon. B **42**, 1933 (2011)
43. A.P. Trawiński, S.D. Głazek, S.J. Brodsky, G.F. de Teramond, H.G. Dosch, Eective confining potentials for QCD. Phys. Rev. D **90**, 074017 (2014)
44. S.S. Chabysheva, J.R. Hiller, Dynamical model for longitudinal wave functions in light-front holo-graphic QCD. Ann. Phys. **337**, 143–152 (2013)
45. Y. Li, Quarkonium wave functions on the light-front. [arXiv:1612.01259 [Nucl-th]]; paper in these proceedings
46. T. Frederico, E. Pace, B. Pasquini, G. Salme, Pion generalized parton distributions with covariant and light-front constituent quark models. Phys. Rev. D **80**, 054021 (2009)
47. M. Diehl, Generalized parton distributions. Phys. Rept. **388**, 41 (2003)
48. S.J. Brodsky, G.F. de Teramond, Light-front dynamics and AdS/QCD correspondence: the pion form factor in the space- and time-like regions. Phys. Rev. D **77**, 056007 (2008)
49. J.P. Vary, L. Adhikari, G. Chen, Y. Li, P. Maris, X. Zhao, Basis light-front quantization: recent progress and future prospects. Few Body Syst. **57**(8), 695 (2016)
50. L. Adhikari, Y. Li, X. Zhao, P. Maris, J.P. Vary, A.A. El-Hady, Form factors and generalized parton distributions in basis light-front quantization. Phys. Rev. C **93**, 055202 (2016)
51. M. Burkardt, Impact parameter dependent parton distributions and off forward parton distributions for $\zeta \rightarrow 0$. Phys. Rev. D **62**, 071503 (2000); *Erratum:ibid.* **66**, 119903(E) (2002)
52. G. Chen, Y. Li, P. Maris, K. Tuchin, J.P. Vary, Diffractive charmonium spectrum in high energy collisions in the basis light-front quantization approach. [arXiv:1610.04945 [nucl-th]]
53. H. Kowalski, L. Motyka, G. Watt, Exclusive diffractive processes at HERA within the dipole picture. Phys. Rev. D **74**, 074016 (2006)
54. S. Chekanov et al., [ZEUS Collaboration], Exclusive electroproduction of J/psi mesons at HERA. Nucl. Phys. B **695**, 3 (2004)

55. A. Aktas et al., [H1 Collaboration], Elastic J/ψ production at HERA. Eur. Phys. J. C **46**, 585 (2006)
56. H. Abramowicz et al., [ZEUS Collaboration], Measurement of the cross-section ratio $\sigma_{\psi(2S)}/\sigma_{J/\psi(1S)}$ in deep inelastic exclusive ep scattering at HERA. PoS DIS **2015**, 078 (2015)
57. L.D. McLerran, R. Venugopalan, Computing quark and gluon distribution functions for very large nuclei. Phys. Rev. D **49**, 2233 (1994)
58. T. Lappi, Wilson line correlator in the MV model: relating the glasma to deep inelastic scattering. Eur. Phys. J. C **55**, 285 (2008). [arXiv:0711.3039 [hep-ph]]
59. K. Tuchin, Particle production in strong electromagnetic fields in relativistic heavy-ion collisions. Adv. High Energy Phys. **2013**, 490495 (2013)

Few-Body Syst (2017) 58:142
DOI 10.1007/s00601-017-1306-2

Andrey Kim · for CLAS Collaboration

Study of Generalized Parton Distributions at Jefferson Lab

Received: 20 January 2017 / Accepted: 18 May 2017 / Published online: 6 June 2017

Abstract The quark-gluon dynamics manifests itself in a set of non-perturbative functions describing all possible spin-spin and spin-orbit correlations. The Generalized Parton Distributions (GPDs) carry information not only on the longitudinal momentum but also on the transverse position of partons, providing rich and direct information on the orbital motion of quarks. The hard exclusive production of photons and pions provide a variety of spin and azimuthal angle dependent observables, sensitive to the dynamics of quark-gluon interactions. The study of the GPDs is one of the main goals of Jefferson Lab 12 GeV upgrade. In this talk, we present an overview of the current status and some future measurements of hard exclusive processes and extraction of underlying GPDs at Jefferson Lab.

1 Introduction

In the last years Generalized Parton Distributions (GPDs) [1–3] were recognized as a versatile theoretical framework to describe the nucleon internal structure and provide an important piece of information that is missing in one-dimensional parton densities, in particular, the distribution of partons in the plane transverse to the direction of motion. The GPD formalism demonstrated the ability to construct a multi-dimensional image of the internal structure of the nucleon.

At twist-2 of the GPD framework, there are eight GPDs. Four of them, denoted by H, $\tilde{H}$, E, $\tilde{E}$, correspond to the parton helicity-conserving (chiral-even) processes, and the remaining four GPDs, denoted by H_T, $\tilde{H}_T$, E_T, $\tilde{E}_T$, correspond to parton helicity-flip (chiral-odd) processes. They depend on three kinematic variables: x, ξ and t, where x is the average parton longitudinal momentum fraction and ξ (skewness) is half of the longitudinal momentum fraction transferred to the struck parton. The skewness can be expressed in terms of the Bjorken variable x_B as $\xi \simeq x_B/(2-x_B)$, in which $x_B = Q^2/(2pq)$, q is the four-momentum of the virtual photon and $Q^2 = -q^2$. The momentum transfer to the nucleon is $t = (p - p')^2$, where p and p' are the initial and final four momenta of the nucleon.

Deeply virtual exclusive processes, such as Deeply Virtual Compton Scattering (DVCS) and Deeply Virtual Meson Production (DVMP), allow us to probe the hadronic structure and create a transverse spatial image of partons as a function of their longitudinal momentum fraction. In deep inelastic scattering (DIS) regime the QCD factorization theorem states that the measured amplitudes of these processes can be expressed as a

This article belongs to the Topical Collection "Light Cone 2016".

A. Kim (✉)
University of Connecticut, Storrs, CT, USA
E-mail: kenjo@jlab.org

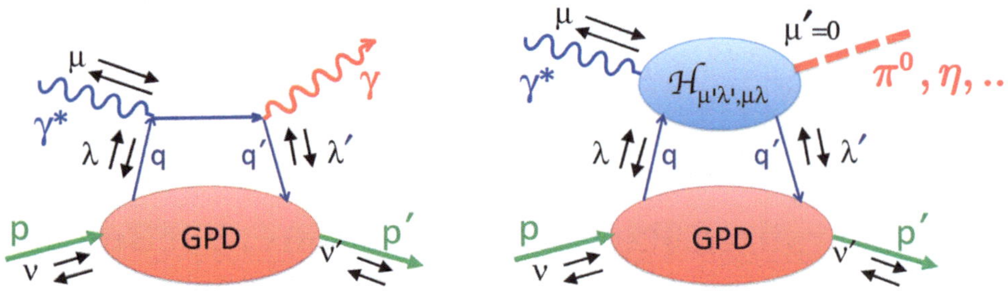

Fig. 1 Schematic DVCS (*left*) and DVMP (*right*) in the handbag mechanism framework. The helicities of initial (final) nucleon and quark are denoted as ν (ν') and λ (λ') respectively. The incident virtual photon and produced photon (meson) have μ and μ' helicities

convolution of a perturbatively calculable hard scattering with Generalized Parton Distributions. This provides an opportunity of using deeply virtual exclusive electroproduction processes to systematically explore the three-dimensional structure of hadrons at a quark-gluon level. Schematic illustrations for DVCS and DVMP in the handbag framework are given on Fig. 1.

DVCS is one of the cleanest and most suitable processes to extract GPDs [4–6] from experimental data at moderate energies and in the valence quark regime [7–11]. However, DVCS observables are primarily sensitive to the chiral-even GPDs which describe unpolarized and longitudinally polarized quarks. The pseudoscalar DVMP channels, on the other hand, contain significant contributions from chiral-odd GPDs. The recent measurements of exclusive π^0 electroproduction structure functions by CLAS collaboration [12,13] showed the strong sensitivity of this process to chiral-odd GPDs [14–16].

2 Deeply Virtual Compton Scattering

The first measurements of DVCS cross sections in the valence quark region were presented by Hall A collaboration [8]. The experimental data were collected with a 5.75 GeV longitudinally polarized electron beam and a 15-cm-liquid H_2 target in Hall A [17] at Jefferson Lab.

The helicity-dependent and helicity-independent cross sections for $ep \rightarrow ep\gamma$ process were measured for three kinematic points at $Q^2 = 1.5$, 1.9 and 2.3 GeV^2. The unpolarized cross section exhibits a significant deviation from the Bethe–Heitler (BH) process in measured kinematics, compatible with a large contribution from the leading twist DVCS term. The helicity-dependent cross section is dominated by the twist-2 interference term of the DVCS amplitude with BH amplitude. DVCS amplitude is proportional to sums of convolutions of specific GPDs and hard scattering amplitudes called Compton Form Factors (CFF). The effective Compton Form Factors used to describe the kinematical dependence of cross sections were extracted. Twist-2 and twist-3 contributions in the interference terms were separated analyzing the azimuthal dependence of the helicity-dependent cross sections. The twist-2 interference term is dominant, and twist-3 contribution is small with large systematic errors. The extracted CFFs show no Q^2 dependence over accessible Q^2 range (see Fig. 2). The absence of Q^2 dependence supports the twist-2 dominance of the DVCS amplitude even at modest Q^2.

The DVCS beam spin asymmetry (BSA) [20], target spin asymmetry (TSA) [21–23] and cross sections [24] were measured over wide kinematic range using CLAS spectrometer [25] at Hall B, Jefferson Lab. Each polarization observable is sensitive to a particular combination of GPDs. For instance, the unpolarized cross section is dominated by the contribution from real part of CFF $\mathcal{H}$, and the beam spin asymmetry is dominated by $Im\mathcal{H}$. Conversely, the longitudinally polarized target spin asymmetry (shown on Fig. 3) is sensitive to the combination of $Im\mathcal{H}$ and $Im\tilde{\mathcal{H}}$. Therefore, combined together, these experimental observables provide strong constraints for the global analysis and extraction of Generalized Parton Distributions.

3 Deeply Virtual π^0 Production

In the past decade, the pseudoscalar meson electroproduction was identified as one of a few sources of information on chiral-odd GPDs [15,29–31]. Early efforts to calculate π^0 electroproduction focused on chiral-even GPDs to parametrize the contributions from longitudinally polarized virtual photons but failed to describe the large cross section and beam spin asymmetry measurements of exclusive π^0 electroproduction [32,33]. The

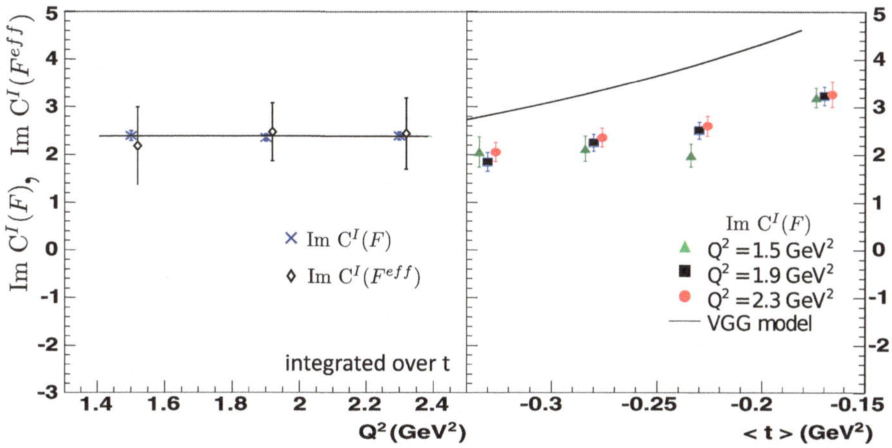

Fig. 2 *Left* Q^2 dependence of imaginary parts of (twist-2) $C^I(F)$ and (twist-3) $C^I(F^{eff})$ angular harmonics, averaged over t. The *horizontal line* is the fitted average of Im$[C^I(F)]$). *Right* Extracted imaginary parts of the twist-2 angular harmonics as a function of t. The theoretical predictions are from the Ref. [6,18,19]

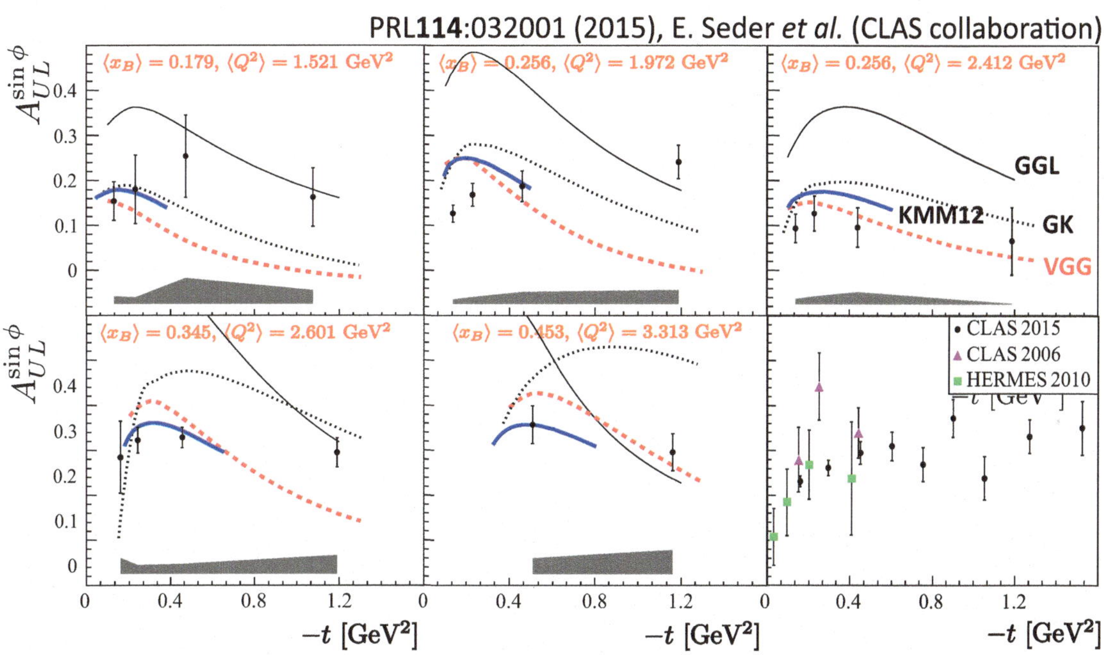

Fig. 3 (color online) First 5 plots: $-t$ dependence of the $\sin\phi$ amplitude of target spin asymmetry for each $Q^2 - x_B$ bin. The *shaded bands* represent the systematic uncertainties. The curves show the predictions of four GPD models: VGG [19] (*red dashed line*), GK [26] (*black dotted line*), KMM12 [27] (*blue thick solid line*) and GGL [28] (*black solid line*). Sixth plot: Comparison of the $\sin\phi$ amplitude of target spin asymmetry as a function of $-t$

large experimental values imply that both transversely and longitudinally polarized virtual photons participate in the process.

Two parallel theoretical approaches have been developed to calculate pseudoscalar meson electroproduction with substantial contribution from transverse virtual photon amplitudes: the Goloskokov-Kroll (GK) model [15,29,30] and Goldstein-Liuti (GL) model [31]. Although different in details they both utilize chiral-odd GPDs accompanied by twist-3 meson wave functions to calculate the dominant contribution from transversely polarized photons in the Q^2 range accessible in the present-day experiments. Inclusion of the chiral-odd components to the hard exclusive amplitude brings theoretical calculations into fair agreement with the experimentally measured cross sections [34]. The extracted structure functions for different kinematics are shown on Fig. 4. They show small σ_{LT} and substantial σ_{TT}, which is what one would expect for a process dominated by transverse photon amplitudes.

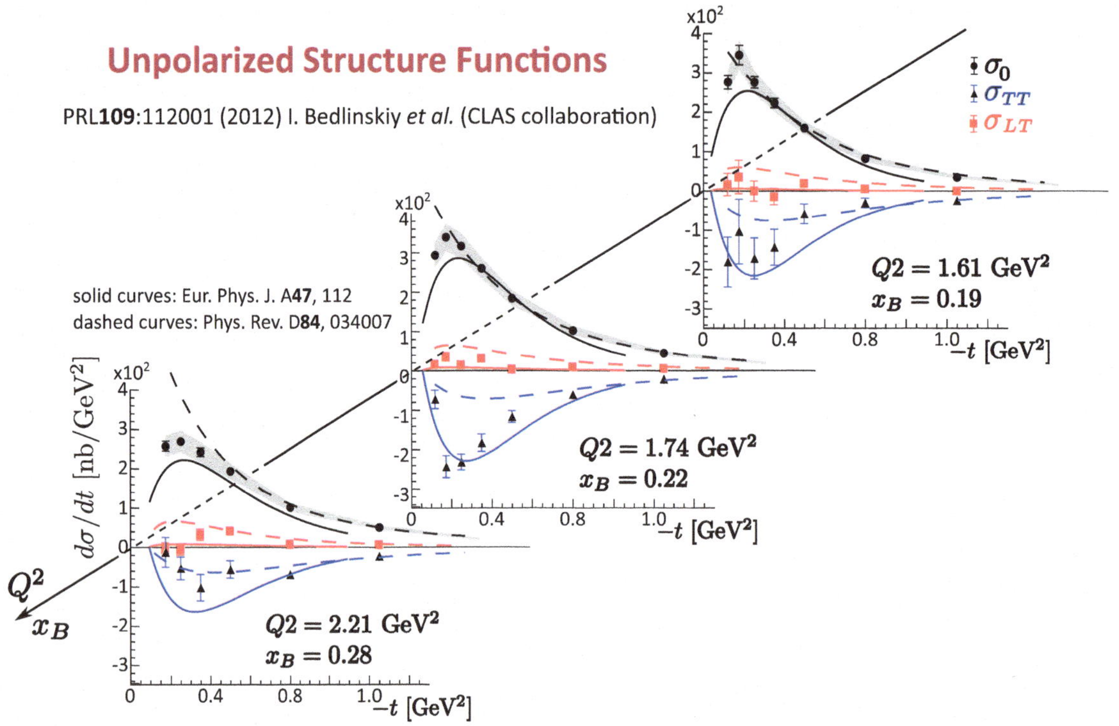

Fig. 4 The extracted structure functions for exclusive π^0 electroproduction as a function of t for the bins with best kinematic coverage. The curves are theoretical predictions and explained on the graph. The *shaded bands* represent experimental systematic uncertainties

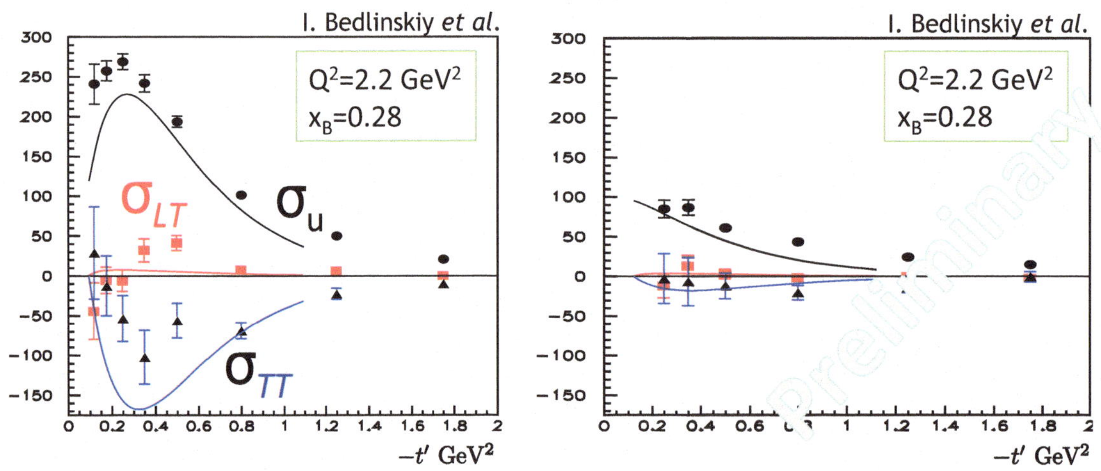

Fig. 5 Structure function $\sigma_T + \epsilon\sigma_L$ (*black*), σ_{TT} (*blue*) and σ_{LT} (*red*) as a function of $-t$ for π^0 (*left*) and η (*right*) exclusive electroproduction for kinematic point (Q^2 =2.2 GeV^2, x_B =0.28). Data points: CLAS, preliminary. Curves: theoretical predictions produced with GK handbag model

Recent Rosenbluth L/T separation of the π^0 electroproduction cross section by Hall A Collaboration at Jefferson Lab [35] also demonstrated the dominance of transverse component and supported the exciting idea that chiral-odd GPDs can be accessed via pseudoscalar meson electroproduction in the available Q^2 regime.

In parallel, cross sections and structure functions for $ep \to ep\eta$ were also obtained. The comparison of the π^0 and preliminary η structure functions is shown on Fig. 5. σ_U drops by a factor of 2.5 for η-mesons in comparison with π^0 and σ_{TT} drops by a factor of 10. The GK GPD model (curves) follows the experimental data. The inclusion of η data into consideration strengthens the statement about the transversity

GPD dominance in the pseudoscalar electroproduction process. Combined π^0 and η data open the way for the flavor decomposition of the underlying transversity GPDs.

The extraction of longitudinally polarized target spin asymmetry is underway using the experimental data collected in Hall B, Jefferson Lab [36]. The unpolarized and polarized structure functions can be expressed as a combination of the partonic subprocess amplitudes and GPDs convolutions. Therefore measurements of beam, target and double spin asymmetries and extraction of azimuthal moments provide several observables, combined analysis of which may allow separation of contributions from different underlying GPDs.

4 Conclusion

The 6 GeV program at Jefferson Lab provided a large set of experimental observables over a wide kinematic range for DVCS and DVMP processes. The DVCS measurements supported the dominance of the leading-order handbag mechanism. The successful description of unpolarized cross sections for pseudoscalar meson production supported an exciting opportunity to access chiral-odd GPDs via DVMP channels. Combined together, these experimental observables can be used to access chiral even and chiral odd GPDs. The variety of DVMP channels combined with precision of DVCS measurements provide a set of tight constraints for the parametrization of the underlying GPDs. The first interpretations of deep exclusive processes in Generalized Parton Distributions framework are already showing remarkable results.

The work presented here leads directly to the program of the Jefferson Lab 12 GeV upgrade. The increased energy and luminosity will allow higher accuracy measurements over significantly wider kinematic range providing us with the opportunity to extend our analysis at higher Q^2 and test the mechanism of exclusive photon and meson electroproduction.

Acknowledgements We acknowledge the outstanding efforts of the staff of the Accelerator and Physics Divisions at JLab. This work was supported in part by the U.S. Department of Energy and National Science Foundation, the French Centre National de la Recherche Scientifique and Commissariat à l'Energie Atomique, the Italian Istituto Nazionale di Fisica Nucleare, the National Research Foundation of Korea and the U.K. Engineering and Physical Science Research Council. Jefferson Science Associates (JSA) operates the Thomas Jefferson National Accelerator Facility for the United States Department of Energy under contract DE-AC05-06OR23177.

References

1. X.D. Ji, Phys. Rev. Lett. **78**, 610 (1997)
2. A.V. Radyushkin, Phys. Lett. B **380**, 417 (1996)
3. A.V. Belitsky, D. Mueller, A. Kirchner, Nucl. Phys. B **629**, 323 (2002)
4. A. Radyushkin, Phys. Rev. D **56**, 5524 (1997)
5. D. Mueller et al., Fortsch. Phys. **42**, 101 (1994)
6. K. Goeke, M.V. Polyakov, M. Vanderhaeghen, Prog. Part. Nucl. Phys. **47**, 401 (2001)
7. S. Stepanyan et al., Phys. Rev. Lett. **87**, 182001 (2001)
8. C. Munoz Camacho et al., Phys. Rev. Lett. **97**, 262002 (2006)
9. F. Girod et al., Phys. Rev. Lett. **100**, 162002 (2008)
10. G. Gavalian et al., Phys. Rev. C **80**, 035206 (2009)
11. M. Defurne et al., Phys. Rev. C **92**, 055202 (2006)
12. R. De Masi et al., Phys. Rev. C **77**, 042201 (2008)
13. I. Bedlinskiy et al., Phys. Rev. Lett. **109**, 112001 (2012)
14. S. Ahmad, G.R. Goldstein, S. Liuti, Phys. Rev. D **79**, 054014 (2009)
15. S. Goloskokov, P. Kroll, Eur. Phys. J. A **47**, 112 (2011)
16. G.R. Goldstein, J.O.G. Hernandez, S. Liuti, Phys. Rev. D **84**, 034007 (2011)
17. J. Alcorn, B. Anderson, K. Aniol, J. Annand, L. Auerbach et al., Nucl. Instrum. Meth. **A522**, 294 (2004)
18. M. Vanderhaeghen, P.A.M. Guichon, M. Guidal, Phys. Rev. D **60**, 094017 (1999)
19. M. Guidal, M.V. Polyakov, A.V. Radyushkin, M. Vanderhaeghen, Phys. Rev. D **72**, 054013 (2005)
20. F.X. Girod et al., Phys. Rev. Lett. **100**, 162002 (2008)
21. S. Chen, H. Avakian, V. Burkert et al., Phys. Rev. Lett. **97**, 072002 (2006)
22. E. Seder et al., Phys. Rev. Lett. **114**, 032001 (2015)
23. S. Pisano et al., Phys. Rev. D **91**, 052014 (2015)
24. H.S. Jo et al., Phys. Rev. Lett. **115**, 212003 (2015)
25. B. Mecking et al., Nucl. Instrum. Meth. **A503**, 513 (2003)
26. P. Kroll, H. Moutarde, F. Sabatié, The European Physical Journal C **73**(1), 2278 (2013)
27. K. Kumerički, D. Müller, M. Murray, Physics of Particles and Nuclei **45**(4), 723 (2014)
28. J.O. Gonzalez-Hernandez, S. Liuti, G.R. Goldstein, K. Kathuria, Phys. Rev. C **88**, 065206 (2013)
29. S.V. Goloskokov, P. Kroll, Eur. Phys. J. C **53**, 367 (2008)

30. S.V. Goloskokov, P. Kroll, Eur. Phys. J. C **65**, 137 (2010)
31. G.R. Goldstein, J.O.G. Hernandez, S. Liuti, Phys. Rev. D **91**, 114013 (2015)
32. R. De Masi et al., Phys. Rev. C **77**, 042201 (2008)
33. E. Fuchey, A. Camsonne, C. Munoz Camacho, M. Mazouz, G. Gavalian et al., Phys. Rev. C **83**, 025201 (2011)
34. I. Bedlinskiy et al., Phys. Rev. Lett. **109**, 112001 (2012)
35. M. Defurne et al., Phys. Rev. Lett. **117**, 262001 (2016)
36. A. Kim et al., Phys. Lett. B **768**, 168 (2017)

Few-Body Syst (2017) 58:137
DOI 10.1007/s00601-017-1296-0

Matthias Burkardt

GPDs and Orbital Angular Momentum

Received: 22 December 2016 / Accepted: 3 April 2017 / Published online: 29 April 2017

Abstract Generalized Parton Distributions provide information on the distribution of quarks in impact parameter space. For transversely polarized nucleons, these impact parameter distributions are transversely distorted and this deviation from axial symmetry leads on average to a net transverse force from the spectators on the active quark in a DIS experiment. This force when acting along the whole trajectory of the active quark leads to transverse single-spin asymmetries. For a longitudinally polarized nucleon target, the transverse force implies a torque acting on the quark Orbital Angular Momentum (OAM). The resulting change in OAM as the quark leaves the target equals the difference between the Jaffe–Manohar and Ji OAMs. In the context of the scalar diquark model we demonstrate that the effect from that torque contributes at the same order as single-spin asymmetries.

1 Impact Parameter Dependent Parton Distributions

Generalized Parton Distributions (GPDs) can be obtained from the same light-cone wave function overlap integrals that yield form factors, except that the momentum fraction x of the active quark is not integrated over, i.e. GPDs can be understood as an x decomposition of form factors. The 2-dimensional Fourier transform of the GPD $H_q(x, 0, t)$ yields the distribution $q(x, \mathbf{b}_\perp)$ in impact parameter spacefor unpolarized quarks and target [1]

$$q(x, \mathbf{b}_\perp) = \int \frac{d^2\Delta_\perp}{(2\pi)^2} H_q(x, 0, -\Delta_\perp^2)\, e^{-i\mathbf{b}_\perp \cdot \Delta_\perp}, \tag{1}$$

with $\Delta_\perp = \mathbf{p}'_\perp - \mathbf{p}_\perp$. For a transversely polarized target (e.g. polarized in the $+\hat{x}$-direction) the impact parameter dependent PDF $q_{+\hat{x}}(x, \mathbf{b}_\perp)$ is no longer axially symmetric and the transverse deformation is described by the gradient of the Fourier transform of the GPD $E_q(x, 0, t)$ [2]

$$q_{+\hat{x}}(x,\mathbf{b}_\perp) = q(x,\mathbf{b}_\perp) - \frac{1}{2M}\frac{\partial}{\partial b_y}\int \frac{d^2\Delta_\perp}{(2\pi)^2} E_q(x, 0, -\Delta_\perp^2)\, e^{-i\mathbf{b}_\perp \cdot \Delta_\perp} \tag{2}$$

This article belongs to the Topical Collection "Light Cone 2016".

This work was supported by the DOE under grant number DE-FG03-95ER40965.

M. Burkardt (✉)
Department of Physics, New Mexico State University, Las Cruces, NM 88003, USA
E-mail: burkardt@nmsu.edu

$E_q(x, 0, t)$ and hence the details of this deformation are not very well known, but its x-integral, the Pauli form factor F_2, is Eq. (2) allows to relate the average transverse deformation

$$d_y^q \equiv \int dx \int d^2\mathbf{b}_\perp q(x, \mathbf{b}_\perp) b_y = \frac{1}{2M} \int dx E_q(x, 0, 0) = \frac{\kappa_q^p}{2M} \tag{3}$$

to the contribution from the corresponding quark flavor to the anomalous magnetic moment. Neglecting the contribution from strange quarks—which should be fine for a rough estimate—one finds $\kappa_u^p = 2\kappa_p + \kappa_n = 2 * 1.793 - 1.913 = 1.673$ and $\kappa_d^p = 2\kappa_n + \kappa_p = 2 * (-1.913) + 1.793 = -2.033$. Since $\frac{1}{2M} \approx 0.1\ fm$ this implies a very significant deformation $|d_q^y| = \mathcal{O}(0.2\ fm)$ for both u and d quarks and in opposite directions. More specifically, since u quarks in a proton contribute with a positive anomalous magnetic moment and d quarks (after factoring out the negative d quark charge) with a negative value. Equation (2) thus implies that for a nucleon target polarized in the $+\hat{x}$ direction, the leading twist distribution of u quarks is shifted in the $+\hat{y}$ direction while that of d quarks is shifted in the $-\hat{y}$ direction.

GPDs provide information about the longitudinal momentum and the transverse position of partons. This is reminiscent of angular momentum, which requires position and momentum information in orthogonal directions. In a state that is an eigenstate under rotations about the $\hat{x}$ axis, the expectation value of the momentum density in the $\hat{y}$ direction weighted by the position in the $\hat{z}$ direction is up to a sign numerically equal to the expectation value of the momentum density in the $\hat{z}$ direction weighted by the position in the $\hat{y}$ direction. This is why the transverse deformations of quark distributions described by GPDs can tell us about the contribution from that quark flavor to the nucleon spin. Given that $E(x, 0, t)$ describes the transverse deformation of quark distributions in a transversely polarized target it is thus not surprising the the x-moment of $E(x, 0, t)$ plays a role in the angular momentum sum rule. However, $E(x, 0, t)$ describes only the deformation relative to the center of momentum of the entire nucleon. It turns out that a delocalized nucleon wave packed for a transversely polarized nucleon that is centered at the origin when boosted to the infinite momentum frame is has a center of momentum that is shifted sideways by half a Compton wavelength. As a result there is an additional contribution involving the momentum fraction carried by that quark flavor [3]

$$J_q = \frac{1}{2} \int_0^1 dx\, x \left[H^q(x, 0, 0) + E^q(x, 0, 0)\right], \tag{4}$$

which was first derived in Ref. [4].

2 Angular Momentum Decompositions

Since the famous EMC experiments revealed that only a small fraction of the nucleon spin is due to quark spins [5], there has been a great interest in 'solving the spin puzzle', i.e. in decomposing the nucleon spin into contributions from quark/gluon spin and orbital degrees of freedom. In this effort, the Ji decomposition [4]

$$\frac{1}{2} = \frac{1}{2} \sum_q \Delta q + \sum_q L_q^z + J_g^z \tag{5}$$

appears to be very useful: through GPDs, not only the quark spin contributions Δq but also the quark total angular momenta $J_q \equiv \frac{1}{2}\Delta q + L_q^z$ (and by subtracting the spin piece also the the quark orbital angular momenta L_q^z) entering this decomposition can be accessed experimentally. In the Ji decomposition (5) the quark OAM is defined as the expectation value

$$L_q^z = \int d^3r \langle PS| q^\dagger \left(\mathbf{r} \times \frac{1}{i}\mathbf{D}\right)^z q |PS\rangle / \langle PS|PS\rangle \tag{6}$$

in a nucleon state polarized in the $+\hat{z}$ direction. Here $\mathbf{D} = \boldsymbol{\partial} - ig\mathbf{A}$ is the gauge-covariant derivative. The main advantages of this decomposition are that each term can be expressed as the expectation value of a manifestly gauge invariant local operator and that the quark total angular momentum $J^q = \frac{1}{2}\Delta q + L^q$ can be related to GPDs [4] and is thus accessible in deeply virtual Compton scattering and deeply virtual meson production and can also be calculated straightforwardly in lattice gauge theory.

Jaffe and Manohar have proposed an alternative decomposition of the nucleon spin, which does have a partonic interpretation, and in which also two terms, $\frac{1}{2}\Delta q$ and ΔG, are experimentally accessible [6]

$$\frac{1}{2} = \frac{1}{2}\sum_q \Delta q + \sum_q \mathcal{L}^q + \Delta G + \mathcal{L}^g. \tag{7}$$

In this decomposition the quark OAM is defined as

$$\mathcal{L}^q \equiv \int d^3r \langle PS|q_+^\dagger \left(\mathbf{r} \times \frac{1}{i}\boldsymbol{\partial}\right)^z q_+|PS\rangle / \langle PS|PS\rangle, \tag{8}$$

where light-cone gauge $A^+ = 0$ is implied. Although Eq. (8) is not manifestly gauge invariant as written, gauge invariant extensions can be defined [11–13]. Indeed, manifestly gauge invariant definitions for each of the terms in Eq. (7) exist, which, with the exception of Δq, involve matrix elements of nonlocal operators. In light-cone gauge those nonlocal operators reduce to a local operator, such as Eq. (8).

To illustrate that there may be issues with how OAM is defined, let us start from a definition of photon total angular momentum based on the Poynting vector. First one writes the magnetic field in terms of the vector potential (9)

$$\mathbf{J}_\gamma = \int d^3r\, \mathbf{r} \times (\mathbf{E} \times \mathbf{B}) = \int d^3r\, \mathbf{r} \times [\mathbf{E} \times (\boldsymbol{\nabla} \times \mathbf{A})] \tag{9}$$

$$= \int d^3r \left[E^j \left(\mathbf{r} \times \boldsymbol{\nabla}\right) A^j - \mathbf{r} \times (\mathbf{E} \cdot \boldsymbol{\nabla})\mathbf{A}\right] \tag{10}$$

$$= \int d^3r \left[E^j \left(\mathbf{r} \times \boldsymbol{\nabla}\right) A^j + (\mathbf{r} \times \mathbf{A})\, \boldsymbol{\nabla} \cdot \mathbf{E} + \mathbf{E} \times \mathbf{A}\right] \tag{11}$$

$$= \int d^3r \left[E^j \left(\mathbf{r} \times \boldsymbol{\nabla}\right) A^j + (\mathbf{r} \times \mathbf{A})\, e\psi^\dagger\psi + \mathbf{E} \times \mathbf{A}\right]. \tag{12}$$

Upon simplifying the double cross product inside the square bracket, one finds a term that involves the photon fields (A^j) as well as their canonical momenta (E^j) sandwiching $\mathbf{r} \times \boldsymbol{\nabla}$ (9). That term has a physical interpretation as orbital angular momentum for the photons. In the next step, one can integrate by parts (drop surface terms). This results in two contributions: one where the gradient acts on the position vector, resulting in a term $\mathbf{E} \times \mathbf{A}$ that acts only on the internal degrees of freedom and thus has an interpretation as the photon spin (11). The other term that arises from integration by parts involves the divergence of the electric field which, using the equations of motion, can be replaced by the charge density. This results in a term that is up to a sign identical to the term involving the vector potential in the local manifestly gauge invariant definition of the quark OAM (12) and thus cancels that term in the total angular momentum. In summary, what has happened is that integrating by parts and using the equations of motion we also shuffled angular momentum from photons to electrons.

3 OAM from Wigner Distributions

Given the fact that different spin decompositions and corresponding definitions for quark OAM are possible raises the question about the physical interpretation of those differences. In this effort significant progress has been made based on 5-dimensional Wigner distributions[1]

Wigner distributions can be defined as off forward matrix elements of non-local correlation functions [7–9,11] with $P^+ = P^{+\prime}$, $P_\perp = -P_\perp^\prime = \frac{q_\perp}{2}$

$$W^{\mathcal{U}}(x, \mathbf{b}_\perp, \mathbf{k}_\perp) \equiv \int \frac{d^2\mathbf{q}_\perp}{(2\pi)^2} \int \frac{d^2\xi_\perp d\xi^-}{(2\pi)^3} e^{-i\mathbf{q}_\perp \mathbf{b}_\perp} e^{i(xP^+\xi^- - \mathbf{k}_\perp \xi_\perp)} \langle P'S'|\bar{q}(0)\Gamma \mathcal{U}_{0\xi} q(\xi)|PS\rangle. \tag{13}$$

Throughout this paper, we will chose $\mathbf{S} = \mathbf{S}' = \hat{\mathbf{z}}$. Furthermore, we will focus on the 'good' component by selecting $\Gamma = \gamma^+$. To ensure manifest gauge invariance, a Wilson line gauge link $\mathcal{U}_{0\xi}$ connecting the quark

[1] For a recent overview, see Ref. [10].

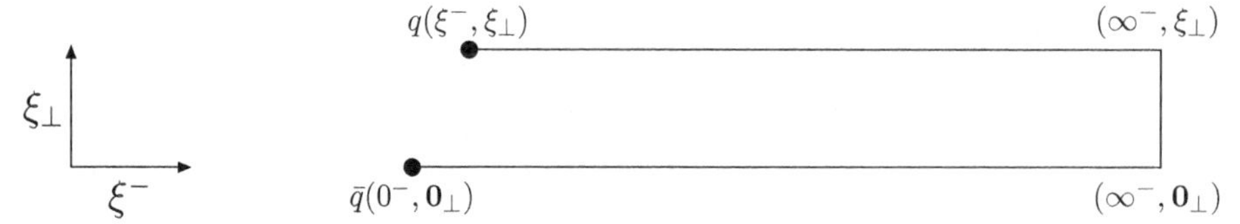

Fig. 1 Illustration of the path for the Wilson line gauge link $\mathcal{U}_{0\xi}^{+LC}$ entering W^{+LC}

field operators at position 0 and ξ is included. The issue of choice of path for the Wilson line will be addressed below.

In terms of Wigner distributions, TMDs and OAM can be defined as [11]

$$f(x, \mathbf{k}_\perp) = \int dx d^2\mathbf{b}_\perp d^2\mathbf{k}_\perp \mathbf{k}_\perp W^{\mathcal{U}}(x, \mathbf{b}_\perp, \mathbf{k}_\perp) \tag{14}$$

$$L_{\mathcal{U}} = \int dx d^2\mathbf{b}_\perp d^2\mathbf{k}_\perp \left(\mathbf{b}_\perp \times \mathbf{k}_\perp\right)^z W^{\mathcal{U}}(x, \mathbf{b}_\perp, \mathbf{k}_\perp). \tag{15}$$

No issues with Heisenberg's uncertainty principle arise here: only perpendicular combinations of position $\mathbf{b}_\perp$ and momentum $\mathbf{k}_\perp$ are needed simultaneously in order to evaluate the integral for $L_{\mathcal{U}}$.

A straight line connecting 0 and ξ for the Wilson line in $\mathcal{U}_{0\xi}$ results in [7]

$$L^q_{straight} = L^q_{Ji}. \tag{16}$$

However, depending on the context, other choices for the path in the Wilson link $\mathcal{U}$ should be made. Indeed for TMDs probed in SIDIS the path should be taken to be a straight line to $x^- = \infty$ along (or, for regularization purposes, very close to) the light-cone. This particular choice ensures proper inclusion of the FSI experienced by the struck quark as it leaves the nucleon along a nearly light-like trajectory in the Bjorken limit. However, a Wilson line to $\xi^- = \infty$, for fixed $\boldsymbol{\xi}_\perp$ is not yet sufficient to render Wigner distributions manifestly gauge invariant, but a link at $\xi^- = \infty$ must be included to ensure manifest gauge invariance. While the latter may be unimportant in some gauges, it is crucial in light-cone gauge for the description of TMDs relevant for SIDIS.

Let $\mathcal{U}_{0\xi}^{+LC}$ be the Wilson path ordered exponential obtained by first taking a Wilson line from $(0^-, \mathbf{0}_\perp)$ to $(\infty, \mathbf{0}_\perp)$, then to $(\infty, \boldsymbol{\xi}_\perp)$, and then to $(\xi^-, \boldsymbol{\xi}_\perp)$, with each segment being a straight line (Fig. 1) [12,13].

The shape of the segment at ∞ is irrelevant as the gauge field is pure gauge there, but it is still necessary to include a connection at ∞ and for simplicity we pick a straight line. Likewise, with a similar 'staple' to $-\infty$ we define the Wilson path ordered exponential $\mathcal{U}_{0\xi}^{-LC}$, and using those light-like gauge links, we define

$$W^{\pm LC}(x,\mathbf{b}_\perp,\mathbf{k}_\perp) \equiv \int \frac{d^2\mathbf{q}_\perp}{(2\pi)^2} \int \frac{d^2\xi_\perp d\xi^-}{(2\pi)^3} e^{-i\mathbf{q}_\perp \cdot \mathbf{b}_\perp} e^{i(xP^+\xi^- - \mathbf{k}_\perp \cdot \boldsymbol{\xi}_\perp)} \langle P'S'|\bar{q}(0)\Gamma\mathcal{U}_{0\xi}^{\pm LC} q(\xi)|PS\rangle. \tag{17}$$

This definition for W^{+LC} is the same as that in Ref. [12,13] and similar to that of W_{LC} in Ref. [7] (the link segment at $\xi^- = \infty$ was not included in the definition of W_{LC}).

4 Bashinsky–Jaffe Orbital Angular Momentum

Imposing the condition $A^+ = 0$ does not completely fix the gauge as one can still apply residual gauge transformations $A^\mu(x) \longrightarrow A^\mu(x) + \partial^\mu \phi(x_\perp)$. In the context of quark OAM there are (at least) two options to deal with this situation. One option is to attach light-cone staple-shaped Wilson line gauge links to render the definition of $\mathcal{L}^q$ manifestly gauge invariant. In the following we will denote this option as $\mathcal{L}_\pm$ for future and past pointing lightcone staples respectively. PT invariance implies that $\mathcal{L}_- = \mathcal{L}_+$ [12,13]. In light-cone gauge $A^+ = 0$, only the segment at $\xi^- = \pm\infty$ contributes and the OAM looks similar to the local manifestly gauge invariant expression, except

$$\mathbf{r} \times \mathbf{A}(\mathbf{r}) \longrightarrow \mathbf{r} \times \mathbf{A}(r^- = \pm\infty, \mathbf{r}_\perp). \tag{18}$$

Bashinsky and Jaffe addressed the issue of residual gauge invariance by introducing a gauge-covariant derivative that contains the 'zero-mode' of $\mathbf{A}_\perp$

$$\mathcal{A}_\perp(\mathbf{x}_\perp) \equiv \frac{\int dx^- \mathbf{A}_\perp(x^-, \mathbf{x}_\perp)}{\int dx^-} \tag{19}$$

into the definition of quark OAM, which renders $\mathcal{L}^q_{JB}$ invariant under residual gauge transformations [15]. Since the gluon field strength tensor should vanish at $x^- = \pm\infty$, $\mathbf{A}_\perp(x^-, \mathbf{x}_\perp)$ should approach a limiting value at $x^- = \pm\infty$ and hence $\mathcal{A}_\perp(\mathbf{x}_\perp)$ should be dominated by these values, which results in

$$\mathcal{L}^q_{JB} = \frac{1}{2}\left(\mathcal{L}^q_+ + \mathcal{L}^q_-\right) = \mathcal{L}^q_+ = \mathcal{L}^q_-. \tag{20}$$

In the last steps PT invariance was used. This demonstrates that $\mathcal{L}^q_{JB}$ is identical to definitions using light-cone staples.

From PT invariance one finds that $\mathcal{L}^q_+ = \mathcal{L}^q_-$ [12,13]. In the Bashinsky–Jaffe definition of OAM $\mathcal{L}^q_{BJ}$ [15], the vector potential in the gauge covariant derivative is replaced by

$$\frac{\int_{-\infty}^{\infty} dx^- A_\perp(r^-, \mathbf{r}_\perp)}{\int_{-\infty}^{\infty} dx^-} = \frac{1}{2}\left[A_\perp(r^- = \infty, \mathbf{r}_\perp) + A_\perp(r^- = \infty, \mathbf{r}_\perp)\right], \tag{21}$$

and is thus equivalent to the light-cone-staple definition

$$\mathcal{L}^q_{BJ} = \frac{1}{2}\left(\mathcal{L}^q_+ + \mathcal{L}^q_-\right) = \mathcal{L}^q_+ = \mathcal{L}^q_-. \tag{22}$$

This observation is crucial for understanding the difference between the Ji versus Jaffe–Manohar OAM, which in light-cone gauge[2] involves only the replacement $A^i_\perp(\mathbf{r}) \longrightarrow A^i_\perp(r^- = \pm\infty, \mathbf{r}_\perp)$. Using

$$A^i_\perp(r^- = \infty, \mathbf{r}_\perp) - A^i_\perp(r^-, \mathbf{r}_\perp) = \int_{r^-}^{\infty} dz^- \partial_- A^i_\perp(z^-, \mathbf{r}_\perp) = \int_{r^-}^{\infty} dz^- G^{+i}(z^-, \mathbf{r}_\perp) \tag{23}$$

where $G^{+\perp} = \partial_- A^\perp$ is the gluon field strength tensor in $A^+ = 0$ gauge. Note that [16]

$$-\sqrt{2} g G^{+y} \equiv -g G^{0y} - g G^{zy} = g\left(E^y - B^x\right) = g\left(\mathbf{E} + \mathbf{v} \times \mathbf{B}\right)^y \tag{24}$$

yields the $\hat{y}$ component of the color Lorentz force acting on a particle that moves with the velocity of light in the $-\hat{z}$ direction ($\mathbf{v} = (0, 0, -1)$)—which is the direction of the momentum transfer in DIS. Thus the difference between the Jaffe-Manohar and Ji[3] OAMs

$$\mathcal{L}^q - L^q = -g\int d^3x \langle P, S|\bar{q}(\mathbf{x})\gamma^+\left[\mathbf{x} \times \int_{x^-}^{\infty} dr^- F^{+\perp}(r^-, \mathbf{x}_\perp)\right]^z q(\mathbf{x})|P, S\rangle / \langle PS|PS\rangle \tag{25}$$

has the semiclassical interpretation of the change in OAM due to the torque from the FSI as the quark leaves the target [14]: while L^q represents the local and manifestly gauge invariant OAM of the quark *before* it has been struck by the γ^*, $\mathcal{L}^q$ represents the gauge invariant OAM *after* it has left the nucleon and moved to $r^- = \infty$.

5 Intuitive Picture for the Torque from Final State Interactions

In order to estimate the effect from the final state interactions on the quark OAM we first consider the effect on a positron moving through the magnetic dipole field of an electron, which is polarized in the $+\hat{z}$ direction. This should be the most simple analogy to a proton polarized in the $+\hat{z}$ direction because more quarks are polarized in the same direction as the nucleon spin and the color-electric force between the active quark and the spectators is attractive.

[2] As L^q involves a manifestly gauge invariant local operator, it can be evaluated in any gauge.

[3] Here we replaced $\gamma^0 \to \gamma^+$ in L^q as discussed in Ref. [17].

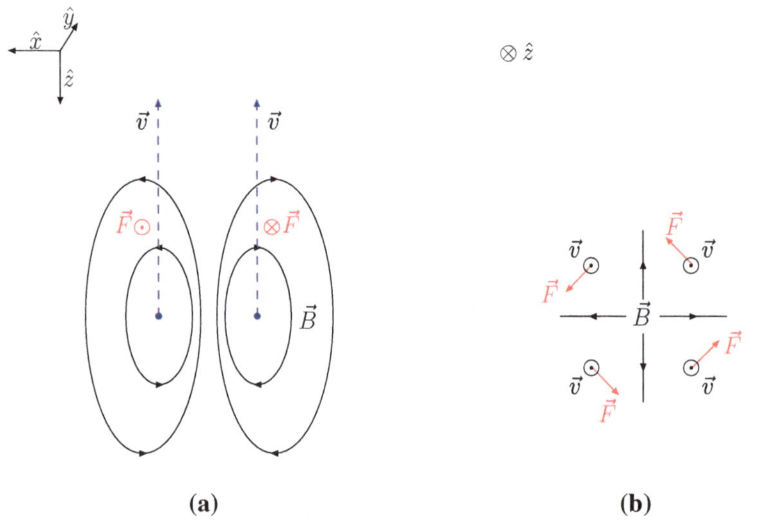

Fig. 2 Illustration of the torque acting on a positron moving in the $-\hat{z}$ direction through a magnetic dipole field caused by the magnetic moment of an electron polarized in the $+\hat{z}$ direction. **a** side view; **b** top view. In this example the $\hat{z}$ component of the torque is negative as the positron leaves the bound state

As illustrated in Fig. 2 the magnetic FSI leads to a negative torque. Since the example was chosen such that the signs of polarization and forces are the same as in the nucleon this implies that the color-magnetic torque acting on quarks ejected from the proton are negative as well.

The dipole example also illustrates why the matrix element of the torque vanishes immediately after the absorption of the virtual photon:

$$\langle P, S|\bar{q}(0)\left[xF^{+y}(0) - yF^{+x}(0)\right]q(0)|P, S\rangle = 0. \tag{26}$$

When comparing the $\perp$ magnetix field (for fixed $\perp$ position) at positions at the 'front' and 'back' side of the nucleon are equal and opposite, and therefore the torque at the original position of the active quark averages to zero Note that this cancellation is a consequence of averaging over all possible initial positions of the active quark. For a specific initial position the initial torque will in general be nonzero. While we used here an intuitive picture to motivate this result, it can also be shown to be a rigorous consequence of PT invariance.

The same example also illustrates the nonzero net effect from the final state interaction torque: although there is some partial cancelation from the 'front' and 'back' of the nucleon, after integrating along the trajectory of the ejected quark, the torque from the 'back' impacts more quarks than that from the 'front' resulting in a nonzero ensemble average.

6 Torque in Spectator Models

[4]Given the fact that the Jaffe–Manohar and Ji definitions of quark orbital angular momentum are differ by the potential angular momentum raises the question as how significant that difference actually is. In Ref. [18] it was shown that to $\mathcal{O}(\alpha)$ in QED $L_{Ji} = \mathcal{L}_{JM}$. This corrects an earlier result [17] where the contribution from states with longitudinally polarized Pauli–Villars photons had been omitted.

In order to assess the significance of FSI effects for quark OAM we thus considered the effects of the vector potential in the scalar diquark model. While we do not consider this model a good approximation for QCD, it has been very useful in several respects:

Most importantly, the model allows for a fully Lorentz invariant calculation of 'nucleon' matrix elements—which is not the case for almost all other models for nucleon structure. Since Lorentz invariance is heavily used for example in the derivation of Eq. 4, this is a crucial aspect of the model. We also note that this was the first model that clearly illustrated the role of FSI and the Sivers in SIDIS and DY [19].

In this model we directly evaluated the potential angular momentum by calculating nonforward matrix elements of $\bar{q}\gamma^{+}A^{y}q$ with momentum transfer $\Delta_\perp$. Upon taking $\frac{d}{d\Delta^x}\big|_{\Delta=0}$ of these matrix elements one obtains the

[4] In collaboration with C. Lorcé.

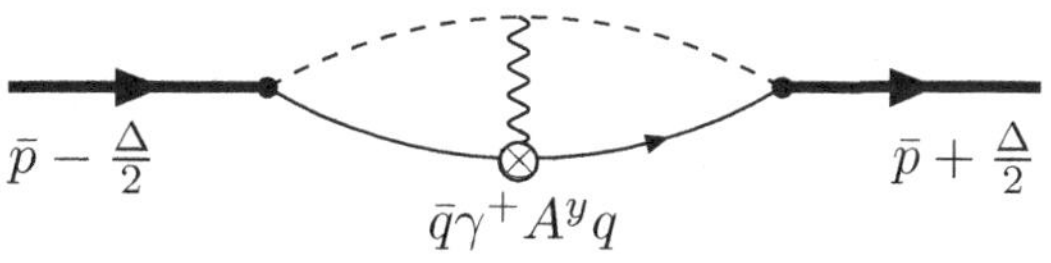

Fig. 3 Diagram contributing to the potential angular momentum in spectator models

forward matrix element of $x\bar{q}\gamma^+ A^y q$. The vector potential A^y in this matrix element is evaluated perturbatively using Feynman rules. This results in an 'exchange' diagram where the photon originates from the spectators (Fig. 3) as well as a diagram where the photon originates from the active quark line.

We found that a nonzero potential angular momentum arises at the same order as transverse single-spin asymmetries [19], i.e. one photon/gluon exchange. Numerically, one finds at lowest nontrivial order that the average transverse momentum and the (dimensionless) potential angular momentum differ only by a 'quark'/'nucleon' mass factor times a factor of $\mathcal{O}(1)$

$$\langle k_\perp^q \rangle = \frac{3m_q + M}{12} \pi \langle \bar{q}\gamma^+ (\mathbf{r} \times \mathbf{A})\tilde{q} \rangle \tag{27}$$

which confirms that the potential angular momentum is as important as transverse single-spin asymmetries.

For the example of the scalar diquark model we found that $L_{pot} \equiv L_{JM} - L_{Ji}$ is positive. This result can be understood as follows: in the scalar diquark model both the OAM of the 'quark' and that of the diquark are positively correlated with the nucleon spin. Moreover, when the active quark and the diquark are orbiting their common center of momentum their transverse momenta are always pointing in opposite directions. For example, when the 'nucleon' spin is pointing up, the magnetic field from the orbital motion of the diquark (negative charge) at the position of the active quark (positive charge) is always pointing 'out' above the plane of orbit and 'in' below it. Thus regardless whether the active quark is ejected 'upward' or 'downward', the Lorentz force due to the transverse component of the magnetic field of the spectator will be directed in the direction of motion before being ejected, i.e. the change in OAM is positive.

7 Summary

Final state interactions manifest themselves in a variety of parton correlation functions. In transverse momentum dependent PDFs the color Lorentz force from the spectators on the active quark causes transverse single-spin asymmetries. The difference between the Jaffe–Manohar–Bashinsky definition of quark orbital angular momentum and that of Ji is equal to the change in orbital angular momentum due to the torque from the final state interactions. While Ji's definition of OAM corresponds to the intrinsic OAM, the Jaffe–Manohar–Bashinsky definition corresponds to the OAM after the active quark has left the target in a DIS experiment. Intuitive pictures as well as explicit calculations in the scalar diquark model confirm that the effect on the OAM is is significant as transverse single-spin asymmetries.

Acknowledgements Funding was provided by U.S. Department of Energy (Grant No. DE-FG03-95ER40965).

Conflict of interest The author declares that he has no competing interest.

References

1. M. Burkardt, Impact parameter dependent parton distributions and off forward parton distributions for $\zeta \to 0$. Phys. Rev. **D62**, 071503 (2000) Erratum: Phys. Rev. **D66**(2002), 119903
2. M. Burkardt, Impact parameter space interpretation for generalized parton distributions. Int. J. Mod. Phys. A **18**, 173–208 (2003)
3. M. Burkardt, Generalized parton distributions and distribution of partons in the transverse plane. in Proceedings for workshop on Lepton Scattering, Hadrons and QCD, eds. by W. Melnitchouk , A.W. Schreiber, A.W. Thomas, P.C. Tandy (World Scientific Publishing, Singapore, 2001), p. 45
4. X. Ji, Gauge-invariant decomposition of nucleon spin. Phys. Rev. Lett. **78**, 610 (1997)
5. J. Ashman et al., A measurement of the spin asymmetry and determination of the structure function g_1 in deep inelastic muon–proton scattering. Phys. Lett. B **206**, 364 (1988)
6. R.L. Jaffe, A. Manohar, The g_1 Problem: fact and fantasy on the spin of the proton. Nucl. Phys. B **337**, 509 (1990)

7. A.V. Belitsky, X. Ji, F. Yuan, Quark imaging in the proton via quantum phase space distributions. Phys. Rev. D **69**, 074014 (2004)
8. S. Meissner, A. Metz, M. Schlegel, Properties and applications of GTMDs. Generalized parton correlation functions for a spin-1/2 hadron. AIP Conf. Proc. **1149**, 539 (2009)
9. S. Meissner, A. Metz, M. Schlegel, JHEP **0908**, 056 (2009)
10. M. Burkardt, B. Pasquini, Modelling the nucleon structure. Eur. Phys. J. A **52**, 161 (2016)
11. C. Lorcé, B. Pasquini, Quark wigner distributions and orbital angular momentum. Phys. Rev. D **84**, 014015 (2011)
12. Y. Hatta, Gluon polarization in the nucleon demystified. Phys. Rev. D **84**, 041701 (2011)
13. Y. Hatta, Notes on the orbital angular momentum of quarks in the nucleon. Phys. Lett. B **708**, 186 (2012)
14. M. Burkardt, Parton orbital angular momentum and final state interactions. Phys. Rev. D **88**, 1 (2013)
15. S. Bashinsky, R.L. Jaffe, Quark and Gluon orbital angular momentum and spin in hard processes. Nucl. Phys. B **536**, 303 (1998)
16. M. Burkardt, Transverse force on quarks in DIS. Phys. Rev. D **88**, 114502 (2013)
17. M. Burkardt, H. BC, Angular Momentum decomposition for an electron. Phys. Rev. D **79**, 071501 (2009)
18. X. Ji et al., Spin decomposition of the electron in QED. Phys. Rev. D **93**, 054013 (2016)
19. S.J. Brodsky, D.-S. Hwang, I. Schmidt, Final state interactions and single spin asymmetries in semiinclusive deep inelastic scattering. Phys. Lett. B **530**, 99 (2002)

Few-Body Syst (2017) 58:121
DOI 10.1007/s00601-017-1283-5

Jose Manuel Alarcón · Astrid N. Hiller Blin · Christian Weiss

Transverse Densities of Octet Baryons from Chiral Effective Field Theory

Received: 19 January 2017 / Accepted: 8 March 2017 / Published online: 24 March 2017

Abstract Transverse densities describe the distribution of charge and current at fixed light-front time and provide a frame-independent spatial representation of hadrons as relativistic systems. We calculate the transverse densities of the octet baryons at peripheral distances $b = O(M_\pi^{-1})$ in an approach that combines chiral effective field theory (χEFT) and dispersion analysis. The densities are represented as dispersive integrals of the imaginary parts of the baryon electromagnetic form factors in the timelike region (spectral functions). The spectral functions on the two-pion cut at $t > 4M_\pi^2$ are computed using relativistic χEFT with octet and decuplet baryons in the extended on-mass-shell renormalization scheme. The calculations are extended into the ρ-meson mass region using a dispersive method that incorporates the timelike pion form-factor data. The approach allows us to construct densities at distances $b > 1$ fm with controlled uncertainties. Our results provide insight into the peripheral structure of nucleons and hyperons and can be compared with empirical densities and lattice-QCD calculations.

1 Introduction

Light-front quantization offers a natural framework for formulating the spatial structure of relativistic systems and exploring its connection to the underlying dynamics; see [1] for a review. In this framework, hadronic current matrix elements (vector, axial) are represented in terms of transverse densities, which are two-dimensional Fourier transforms of the invariant form factors and describe the transverse spatial distribution of charge and current in the hadron at fixed light-front time $x^+ \equiv x^0 + x^3$ [2–5]. The transverse densities are frame independent (they are invariant under longitudinal boosts and transform kinematically under transverse boosts) and thus provide an objective spatial representation of the hadron as a relativistic system. In composite models of hadron structure they correspond to proper densities of the light-front wave functions of the system. In the context of QCD the transverse charge and current densities in hadrons can be related to the generalized parton distributions (GPDs) describing the distribution of quarks and antiquarks in longitudinal momentum and transverse position [3,6]. The charge and magnetization densities in the nucleon have been extracted from the

This article belongs to the Topical Collection "Light Cone 2016".

J. M. Alarcón · C. Weiss
Theory Center, Jefferson Lab, Newport News, VA 23606, USA
E-mail: alarcon@jlab.org

C. Weiss
E-mail: weiss@jlab.org

A. N. H. Blin (✉)
Departamento de Física Teórica and IFIC, Centro Mixto Universidad de Valencia-CSIC, Institutos de Investigación de Paterna, 46071 Valencia, Spain
E-mail: astrid.hiller@uv.es

available electromagnetic form-factor data [7,8] and provide interesting insight into the nucleon structure; see [5] for a review. It is worthwhile to explore how the transverse densities could be calculated using theoretical methods, and how the studies could be extended to other baryons.

The transverse densities at a given distance can be connected with the "exchange mechanisms" acting in the hadron form factors—virtual processes in which the current couples to the hadron through the exchange of a hadronic system in the t-channel. This connection can be made rigorous in a dispersive representation of the transverse densities [9,10]. At distances $b = O(M_\pi^{-1})$, the densities are governed by soft-pion exchange between the current and the hadron and can be calculated model-independently using chiral effective field theory (χEFT). Detailed studies of the peripheral densities in the nucleon were performed in [10–12] using χEFT with SU(2) flavor group, and a simple quantum-mechanical interpretation of the results was obtained (chiral light-front wave functions, orbital motion of a peripheral pion). At distances $b \sim 1$ fm, the transverse densities are dominated by vector-meson exchange (isovector ρ; isoscalar ω, ϕ) and offer interesting insight into the duality between quark structure and meson exchange [13].

In these proceedings we report about a study of the peripheral densities of the SU(3) flavor-octet baryons combining methods of χEFT and dispersion analysis. The densities are represented as dispersive integrals over the imaginary parts of the baryon electromagnetic form factors in the timelike region (spectral functions). The isovector spectral functions on the two-pion cut at $t > 4M_\pi^2$ are computed using relativistic χEFT with octet and decuplet baryons in the extended on-mass-shell (EOMS) renormalization scheme. The calculations are extended into the ρ-meson mass region using a dispersive method that incorporates the timelike pion form-factor data. The methods allow us to construct the densities down to distances $b \sim 1$ fm with controlled uncertainties. Details will be reported in a forthcoming publication [14].

2 Transverse Densities and Dispersive Representation

The matrix element of the electromagnetic current between states of a flavor-octet baryon ($B = N, \Sigma, \Xi$) with 4-momenta p and p' is described by two invariant form factors, $F_1^B(t)$ and $F_2^B(t)$ (the Dirac and Pauli form factors; we follow the conventions of [10]). They are functions of the invariant momentum transfer $t = \Delta^2 = (p' - p)^2$, and can be measured and interpreted without specifying the form of relativistic dynamics or the reference frame. In the context of light-front quantization, one considers the current matrix element in a frame where the 4-momentum transfer has only transverse components $\Delta^+ = \Delta^- = 0$, $\boldsymbol{\Delta}_T \neq 0$, and represents the form factors as Fourier transforms of certain two-dimensional densities (here $b \equiv |\mathbf{b}|$),

$$F_i^B(t = -|\boldsymbol{\Delta}_T|^2) = \int d^2b \, e^{i\boldsymbol{\Delta}_T \cdot \mathbf{b}} \, \rho_i^B(b) \qquad (i = 1, 2). \tag{1}$$

The interpretation of $\rho_{1,2}(b)$ as spatial densities is discussed in [3,5] and summarized in [10]. Their spatial integral reproduces the total charge and anomalous magnetic moment of the baryon. In a state where the baryon is localized in transverse space at the origin, $\rho_1^B(b)$ describes the spin-independent J^+ current at light-front time $x^+ = 0$ and transverse position $\mathbf{b}$, while $\widetilde{\rho}_2^B(b) \equiv \partial/\partial b[\rho_2^B(b)/(2M_B)]$ describes the spin-dependent part of the current in a transversely polarized nucleon.

The form factors are analytic functions of t and satisfy unsubtracted dispersion relations

$$F_i^B(t) = \int_{4M_\pi^2}^{\infty} \frac{dt'}{t' - t - i0} \frac{\mathrm{Im}\, F_i^B(t')}{\pi} \qquad (i = 1, 2). \tag{2}$$

The spectral functions are given by the imaginary parts of the form factors on the principal cut starting at $t = 4M_\pi^2$. They describe processes in which the current at timelike $t > 0$ converts to a hadronic state that couples to the $B\bar{B}$ system. Most of the relevant processes are in the unphysical region below the two-baryon threshold at $t = 4M_B^2$, where the spectral function can only be computed theoretically. From Eqs. (1) and (2), one obtains a dispersive representation of the transverse densities [9]

$$\rho_1^B(b) = \int_{4M_\pi^2}^{\infty} dt \, \frac{K_0(\sqrt{t}b)}{2\pi} \frac{\mathrm{Im} F_1^B(t)}{\pi}, \tag{3}$$

$$\widetilde{\rho}_2^B(b) = -\int_{4M_\pi^2}^{\infty} dt \, \frac{\sqrt{t} K_1(\sqrt{t}b)}{4\pi M_B} \frac{\mathrm{Im} F_2^B(t)}{\pi}, \tag{4}$$

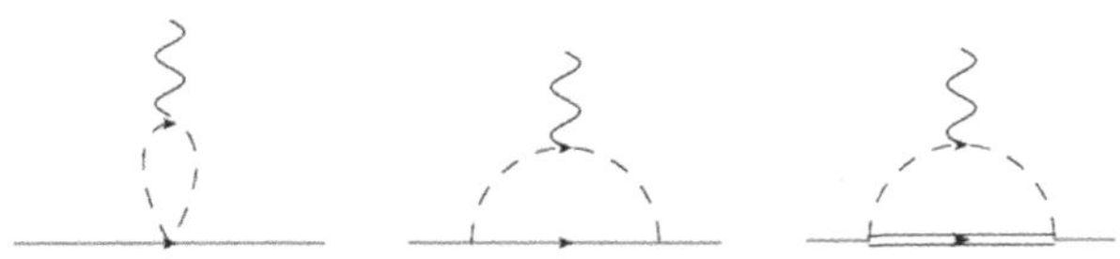

Fig. 1 Chiral processes contributing to the spectral functions of the baryon form factors $\mathrm{Im}F^B_{1,2}(t)$ at $\mathcal{O}(p^3)$. *Dashed lines* pions/kaons. *Solid lines* octet baryons. *Double lines* decuplet baryons

where $K_n (n = 0, 1)$ are the modified Bessel functions. The integrals in Eqs. (3) and (4) converge exponentially at large t, $K_n(\sqrt{t}b) \sim \exp(-\sqrt{t}b)$ at $\sqrt{t}b \gg 1$, strongly suppressing contributions from high-mass hadronic states. Depending on the distance b, the integrals sample the spectral functions in different regions of t [13]. At $b > 2\,\mathrm{fm}$, the integrals extend over the near-threshold region $t = 4M_\pi^2 + \text{few } M_\pi^2$, where the spectral functions arise from soft two-pion exchange between the baryon and the current. At $b \sim 1\,\mathrm{fm}$, the dominant contributions come from the vector-meson region of the spectral functions $t \sim M_V^2$ ($V = \rho^0, \omega, \phi$). At even shorter distances, the integrals extend over the high-mass region $t > 1\,\mathrm{GeV}^2$, where the spectral functions involve multi-hadron states and are poorly known at present [15]. The dispersive representation Eqs. (3) and (4) thus establishes a quantitative connection between the transverse densities and the exchange mechanisms in the form factor.

3 Spectral Functions from Chiral EFT and Dispersion Theory

We compute the isovector spectral functions of the SU(3) octet-baryon form factors above the two-pion threshold using relativistic χEFT. Decuplet baryons are included as dynamical degrees of freedom within the small-scale expansion (SSE) [16,17] (the first form factor calculations in this approach were performed in [18]). The fields and Lagrangian are described in [19–21]. The renormalization of the divergent pieces is performed within the EOMS scheme, which permits consistent power counting [22]. The spectral functions arise from diagrams with a two-pion cut; the $O(p^3)$ diagrams are shown in Fig. 1 (the imaginary parts are actually finite at this order and do not require renormalization). The low-energy constants at this order are given by the nucleon's axial charge (or πNN coupling) and the $\pi N\Delta$ coupling, so that the spectral functions are χEFT predictions free of unknown parameters. While not shown here, we have calculated the entire form factor from the full set of $O(p^3)$ diagrams in order to verify gauge invariance, and reproduce the SU(2) results of [21].

The χEFT expressions by themselves describe the baryon spectral functions only in the near-threshold region $t = 4M_\pi^2 + \text{few } M_\pi^2$; see [23] and references therein for detailed studies of the nucleon case. In order to extend the description to higher t, we use a dispersive technique following [24–26]; see also [27]. On general grounds, the isovector spectral function on the two-pion cut (neglecting the contributions from 4π states) can be expressed as

$$\mathrm{Im}F_i^B(t) = \frac{k_{\mathrm{cm}}^3}{\sqrt{t}}\, \Gamma_i^B(t)\, F_\pi^*(t) \qquad (i = 1, 2), \tag{5}$$

where $k_{\mathrm{cm}} = \sqrt{t/4 - M_\pi^2}$ is the t-channel center-of-mass momentum, $\Gamma_i^B(t)$ is the $I = J = 1$ $\pi\pi \to \bar{B}B$ partial-wave amplitude, and $F_\pi(t)$ is the pion form factor. The expression on the right-hand side of Eq. (5) is real because the complex functions $\Gamma_i^B(t)$ and $F_\pi(t)$ have the same phase on the two-pion cut (Watson theorem). It is convenient to rewrite Eq. (5) as

$$\mathrm{Im}F_i^B(t) = \frac{k_{\mathrm{cm}}^3}{\sqrt{t}}\, \frac{\Gamma_i^B(t)}{F_\pi(t)}\, |F_\pi(t)|^2 \qquad (i = 1, 2). \tag{6}$$

This representation has two major advantages: (1) The function $\Gamma_i^B(t)/F_\pi(t)$ has no two-pion cut, because it is real at $t > 4M_\pi^2$; (2) the squared modulus $|F_\pi(t)|^2$ can be extracted directly from the $e^+e^- \to \pi^+\pi^-$ exclusive annihilation cross section, without determining the phase of the complex function. We now use χEFT to calculate the real function $\Gamma_i^B(t)/F_\pi(t)$ at $t > 4M_\pi^2$, and multiply with the empirical $|F_\pi(t)|^2$ containing the ρ meson resonance. At $O(p^3)$ one has $F_\pi(t) \equiv 1$, so that the χEFT result for the ratio $\Gamma_i^B(t)/F_\pi(t)$ is

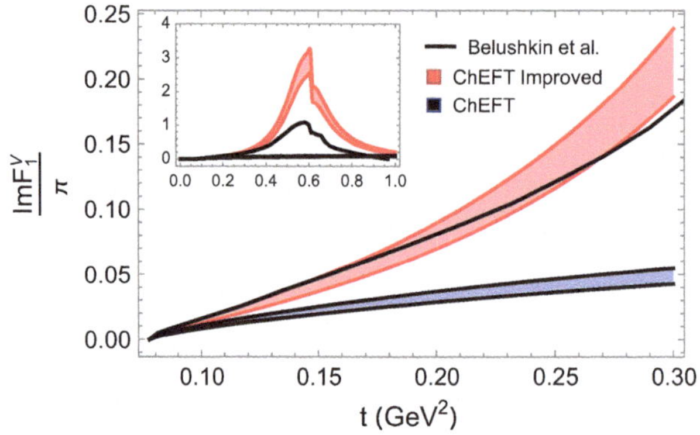

Fig. 2 Dispersive improvement of the χEFT spectral function Im $F_1(t)$, Eq. (7). The main figure shows the spectral functions up to $t = 0.3$ GeV^2, the *inset* figure shows them up to 1 GeV^2. *Blue band* χEFT result at $O(p^3)$, including N and Δ intermediate states. *Red band* Improved spectral function Eq. (7). *Black line* Dispersion-theoretical spectral function from [29]

the same as that for the amplitude $\Gamma_i^B(t)$ itself, and the prescription simply amounts to multiplying the $O(p^3)$ result for the spectral function by $|F_\pi(t)|^2$,

$$\text{Im}F_i^B(t)\,[\text{improved}] = \text{Im}F_i^B(t)\,[O(p^3)] \times |F_\pi(t)|^2 \qquad (i = 1, 2). \tag{7}$$

The prescription Eq. (7) results in a marked improvement of the χEFT predictions for the spectral functions. The improved χEFT results for the nucleon ($B = N$) reproduce the dispersion-theoretical spectral functions (obtained by analytic continuation of the πN phase shifts [28–30]) up to $t \sim 16M_\pi^2$ within errors, and have qualitatively the correct behavior even in the ρ-meson mass region (see Fig. 2). We use this method to calculate the other octet-baryon spectral functions on the two-pion cut. A detailed discussion of the procedure and its applications will be presented in [14].

4 Peripheral Transverse Densities

Using the improved χEFT results for the spectral functions, we calculate the peripheral isovector densities in the octet baryons. The restriction to distances $b > 1$ fm ensures that the dispersion integrals Eqs. (3) and (4) extend only over the region of t where our approximation to the spectral functions is justified. The uncertainties of the isovector densities are estimated by propagating the theoretical uncertainty of the spectral functions [14]. In order to estimate also the isoscalar densities, we parametrize the isoscalar spectral functions of the octet baryons in the mass region $t < 1$ GeV^2 by vector-meson poles ($V = \omega, \phi$). The VBB couplings are obtained from SU(3) symmetry, with certain assumptions regarding the F/D ratio and the empirical VNN couplings (we do not aim for a precise description of the isoscalar sector here, as the peripheral densities are dominated by the isovector component).

Results for the transverse charge densities are shown in Fig. 3. The densities decay exponentially at large b, as dictated by the analytic properties of Eqs. (3) and (4). The comparison of the various components reveals several interesting features. At large distances $b > 3$ fm, the densities are dominated by the isovector component, resulting from two-pion exchange near threshold. The isoscalar component generally becomes comparable to the isovector at distances below $b \sim 2$ fm, due to the similar strength of the spectral functions in the vector-meson region (ρ vs. ω, ϕ). In the neutron, the isovector component dominates for $b > 1.5$ fm, and causes the peripheral charge density to be negative. Studies of empirical neutron densities have shown that at $b \sim 1$ fm the isoscalar takes over, and the neutron charge density becomes positive [4,13]; at such distances the present theoretical calculation has large uncertainties and cannot predict the sign.

In the Λ and Σ^0 charge densities, the isovector component is absent because of the isospin selection rules for the t-channel process current $\to B\bar{B}$: the transitions $|I = 1, I^3 = 0\rangle \to |I = 0, I^3 = 0\rangle|I = 0, I^3 = 0\rangle$ and $|I = 1, I^3 = 0\rangle \to |I = 1, I^3 = 0\rangle|I = 1, I^3 = 0\rangle$ are both forbidden. In contrast, in the $\Lambda\Sigma^0$ transition density, the isoscalar component is absent and the density is pure isovector. The peripheral densities of the Λ and Σ^0 thus provide a means to isolate the low-mass isovector and isoscalar exchanges in the form factors. The Ξ^0 and Ξ^- densities are similar to the proton and neutron in that both isovector and isoscalar exchanges are present. The isovector component is relatively smaller in the Ξ states, due to the suppression

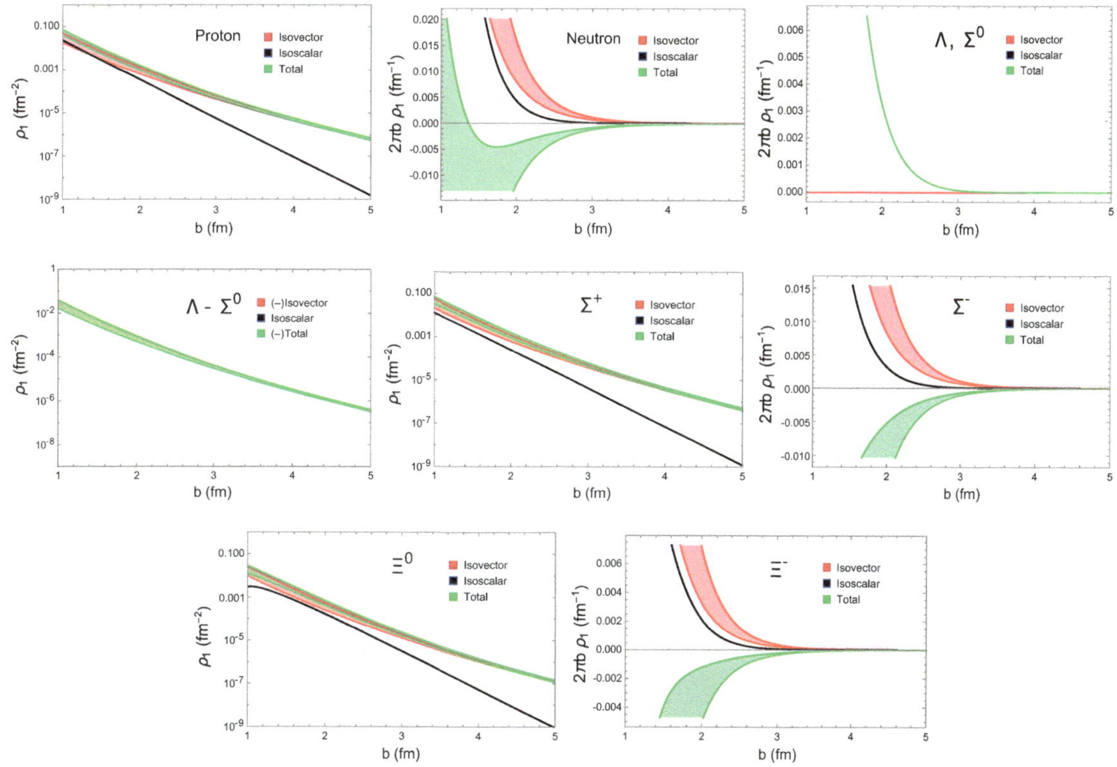

Fig. 3 Transverse charge densities of the octet baryons. *Red* Isovector component calculated using χEFT and dispersive improvement. *Blue* Isoscalar component estimated from vector-meson poles. *Green* Total density (sum or difference of isoscalar and isovector components). For the densities with fixed sign (positive or negative) we plot $\rho_1(b)$ on a logarithmic scale; for those with changing sign we plot the radial densities $2\pi b\rho_1(b)$ on a linear scale

of octet intermediate states in the pion loops for the Ξ form factor (they are of the same order as the decuplet intermediate states).

Similar features are exhibited by the magnetization densities of the octet baryons. Using the same techniques we can also calculate the flavor decomposition of the peripheral densities and study the contribution of individual quark flavors to the charge and magnetization [14].

5 Summary and Outlook

We have presented a new approach to the calculation of the peripheral transverse densities using a combination of relativistic χEFT and dispersion analysis. The dispersive improvement extends the χEFT calculations of the isovector spectral functions into the vector-meson mass region and allows us to compute densities down to distances $\sim$ 1fm with controlled accuracy. The approach can be extended to the octet-baryon form factors of other operators (energy-momentum tensor, moments of GPDs). It can also be used to calculate the transverse densities of the decuplet baryons (especially the Δ isobar), which are studied in Lattice QCD, as well as to the densities of the octet–decuplet transition form factors (especially $N \to \Delta$), which are measured in resonance electroproduction [31].

Our approach consistently includes the contributions from decuplet intermediate states in the π/K loop diagrams. These contributions are numerically important in the transverse densities at distances $b \sim$1–2 fm. They are also essential for ensuring the proper scaling behavior of the densities in the large-N_c limit of QCD. This was demonstrated for the SU(2) χEFT in [10,32], and can be shown for the present SU(3) calculation as well.

Acknowledgements This material is based upon work supported by the U.S. Department of Energy, Office of Science, Office of Nuclear Physics under contract DE-AC05-06OR23177. This work was supported by the Spanish Ministerio de Economía y

Competitividad (MINECO) and the European fund for regional development (FEDER) under contracts FIS2014-51948-C2-2-P and SEV-2014-0398.

References

1. S.J. Brodsky, H.C. Pauli, S.S. Pinsky, Quantum chromodynamics and other field theories on the light cone. Phys. Rep. **301**, 299–486 (1998). arXiv:hep-ph/9705477
2. D.E. Soper, The Parton model and the Bethe-Salpeter wave function. Phys. Rev. D **15**, 1141 (1977)
3. M. Burkardt, Impact parameter dependent parton distributions and off forward parton distributions for $\zeta \rightarrow 0$. Phys. Rev. D **62**, 071,503 (2000), [Erratum: Phys. Rev. D **66**, 119903 (2002)]. arXiv: hep-ph/0005108
4. G.A. Miller, Charge density of the neutron. Phys. Rev. Lett. **99**, 112001 (2007). arXiv:0705.2409 [nucl-th]
5. G.A. Miller, Transverse charge densities. Ann. Rev. Nucl. Part. Sci. **60**, 1 (2010). arXiv:1002.0355 [nucl-th]
6. M. Burkardt, Impact parameter space interpretation for generalized parton distributions. Int. J. Mod. Phys. A **18**, 173 (2003). arXiv:hep-ph/0207047
7. C.F. Perdrisat, V. Punjabi, M. Vanderhaeghen, Nucleon electromagnetic form factors. Prog. Part. Nucl. Phys. **59**, 694–764 (2007). arXiv:hep-ph/0612014
8. V. Punjabi, C.F. Perdrisat, M.K. Jones, E.J. Brash, C.E. Carlson, The structure of the nucleon: elastic electromagnetic form factors. Eur. Phys. J. A **51**, 79 (2015). arXiv:1503.01452
9. M. Strikman, C. Weiss, Quantifying the nucleon's pion cloud with transverse charge densities. Phys. Rev. C **82**(042), 201 (2010). arXiv:1004.3535
10. C. Granados, C. Weiss, Chiral dynamics and peripheral transverse densities. JHEP **01**, 092 (2014). arXiv:1308.1634
11. C. Granados, C. Weiss, Light-front representation of chiral dynamics in peripheral transverse densities. JHEP **1507**, 170 (2015). arXiv:1503.04839 [hep-ph]
12. C. Granados, C. Weiss, Quantum-mechanical picture of peripheral chiral dynamics. Phys. Rev. C **92**, 025206 (2015). arXiv:1503.02055 [hep-ph]
13. G.A. Miller, M. Strikman, C. Weiss, Realizing vector meson dominance with transverse charge densities. Phys. Rev. C **84**(045), 205 (2011). arXiv:1105.6364
14. J.M. Alarcón, A.N. Hiller Blin, M.J. Vicente Vacas, C. Weiss, Peripheral transverse densities of the baryon octet from chiral effective field theory and dispersion analysis (2017). arXiv:1703.04534 [hep-ph]
15. M.A. Belushkin, H.W. Hammer, U.G. Meißner, Dispersion analysis of the nucleon form-factors including meson continua. Phys. Rev. C **75**(035), 202 (2007). arXiv:hep-ph/0608337
16. T.R. Hemmert, B.R. Holstein, J. Kambor, Systematic 1/M expansion for spin 3/2 particles in baryon chiral perturbation theory. Phys. Lett. B **395**, 89–95 (1997). arXiv:hep-ph/9606456
17. T.R. Hemmert, B.R. Holstein, J. Kambor, Chiral Lagrangians and delta (1232) interactions: formalism. J. Phys. **G24**, 1831–1859 (1998). arXiv:hep-ph/9712496
18. V. Bernard, H.W. Fearing, T.R. Hemmert, U.G. Meissner, The form-factors of the nucleon at small momentum transfer. Nucl. Phys. A **635**, 121–145 (1998). arXiv:hep-ph/9801297
19. L.S. Geng, J. Martin Camalich, M.J. Vicente Vacas, Leading-order decuplet contributions to the baryon magnetic moments in chiral perturbation theory. Phys. Lett. B **676**, 63–68 (2009). arXiv:0903.0779
20. L.S. Geng, J. Martin Camalich, M.J. Vicente Vacas, Electromagnetic structure of the lowest-lying decuplet resonances in covariant chiral perturbation theory. Phy. Rev. D **80**(034), 027 (2009). arXiv:0907.0631
21. T. Ledwig, J.M. Camalich, L.S. Geng, M.J.V. Vacas, Octet-baryon axial-vector charges and SU(3)-breaking effects in the semileptonic hyperon decays. Phys. Rev. D **90**(5), 054502 (2014). arXiv:1405.5456
22. T. Fuchs, J. Gegelia, G. Japaridze, S. Scherer, Renormalization of relativistic baryon chiral perturbation theory and power counting. Phys. Rev. D **68**(056), 005 (2003). arXiv:hep-ph/0302117
23. N. Kaiser, Spectral functions of isoscalar scalar and isovector electromagnetic form-factors of the nucleon at two loop order. Phys. Rev. C **68**, 025202 (2003). arXiv:nucl-th/0302072
24. W.R. Frazer, J.R. Fulco, Partial-wave dispersion relations for the process $\pi\pi \rightarrow N\bar{N}$. Phys. Rev. **117**, 1603 (1960)
25. W.R. Frazer, J.R. Fulco, Effect of a pion-pion scattering resonance on nucleon structure II. Phys. Rev. **117**, 1609 (1960)
26. G. Hohler, E. Pietarinen, The ρNN vertex in vector dominance models. Nucl. Phys. B **95**, 210 (1975)
27. H.W. Hammer, D. Drechsel, U.G. Meißner, On the pion cloud of the nucleon. Phys. Lett. B **586**, 291 (2003). arXiv:hep-ph/0310240
28. G. Hohler, E. Pietarinen, I. Sabba Stefanescu, F. Borkowski, G.G. Simon, V.H. Walther, R.D. Wendling, Analysis of electromagnetic nucleon form-factors. Nucl. Phys. **B114**, 505–534 (1976)
29. M.A. Belushkin, H.W. Hammer, U.G. Meißner, Novel evaluation of the two-pion contribution to the nucleon isovector form-factors. Phys. Lett. B **633**, 507 (2006). arXiv:hep-ph/0510382
30. M. Hoferichter, B. Kubis, J.R. de Elvira, H.W. Hammer, U.G. Meißner, On the $\pi\pi$ continuum in the nucleon form factors and the proton radius puzzle. Eur. Phys. J. A **52**, 331 (2016). arXiv:1609.06722 [hep-ph]
31. C.E. Carlson, M. Vanderhaeghen, Empirical transverse charge densities in the nucleon and the nucleon-to-Delta transition. Phys. Rev. Lett. **100**, 032004 (2008). arXiv:0710.0835 [hep-ph]
32. C. Granados, C. Weiss, Light-front representation of chiral dynamics with isobar and large-N_c relations. JHEP **1606**, 075 (2016). arXiv:1603.08881 [hep-ph]

Few-Body Syst (2017) 58:144
DOI 10.1007/s00601-017-1304-4

Nabil Chouika · Cédric Mezrag · Hervé Moutarde · José Rodríguez-Quintero

An Algebraic Model for the Pion's Valence-Quark GPD: A Probe for a Consistent Extension Beyond DGLAP Region Via Radon Transform Inversion

Received: 2 December 2016 / Accepted: 11 May 2017 / Published online: 16 June 2017

Abstract We briefly report on a recent computation, with the help of a fruitful algebraic model, sketching the pion valence dressed-quark generalized parton distribution. Then, preliminary, we introduce on a sensible procedure to get reliable results in both Dokshitzer–Gribov–Lipatov–Altarelli–Parisi and Efremov–Radyushkin–Brodsky–Lepage kinematical regions, grounded on the GPD overlap representation and its parametrization of a Radon transform of the so-called double distribution.

1 Introduction

The Generalized Parton Distributions (GPDs), introduced independently by Müller et al. [1], Ji [2] and Radyushkin [3], are shown to be related to hadron form factors by sum rules, and to contain the usual Parton Distribution Functions (PDFs) as a limiting case. They not only generalize the classical objects describing the static or dynamical content of hadrons; additionally, they also provide unique information about the structure of hadrons, including 3D imaging of its partonic components and access to the quark orbital angular momentum. GPDs have been the object of an intense theoretical and experimental activity ever since (see e.g. [4–9], for the pion case, or the more general reviews [10–15] and references therein).

Most of the constraints that apply to GPDs are fulfilled when the GPD is written in a double distribution representation [1,16,17], which is plainly equivalent to expressing the GPD as a Radon transform [18]. In order to obtain insights into the nature of hadron GPDs, it has been common to model the Radon amplitudes, F, G, following Refs. [19]. This approach has achieved some phenomenological success (see e.g. [14,20]); but more flexible parametrisations enable a better fit to data [21]. Such fits play a valuable role in establishing the GPD framework; However there is no known parameterization of GPDs relying on first principles only. Computing GPDs in a symmetry-preserving framework is a key ingredient for the a priori fulfillment of all GPD theoretical constraints. This observation is highlighted by experience drawn from the simpler case of the pion's valence-quark PDF [22]. In [23,24], steps are made by following a different approach for the computation of hadron GPDs based on the example provided by the pion's valence-quark PDF. We prove therein that, within a

This article belongs to the Topical Collection "Light Cone 2016".

J. Rodríguez-Quintero (✉)
Dpt. of Física Aplicada, Fac. CCEE, Unversidad de Huelva, Huelva 21071, Spain
E-mail: jose.rodriguez@dfaie.uhu.es
Tel.: +34-959-219787

N. Chouika · H. Moutarde
Centre de Saclay, IRFU/Service de Physique Nucléaire, 91191 Gif-sur-Yvette, France

C. Mezrag
Physics Division, Argonne National Laboratory, Argonne, IL 60439, USA

covariant computational approach based on solutions of the Dyson–Schwinger and Bethe–Salpeter equations, observing the Ward–Takahashi identities, that properly account for the chiral symmetry and its dynamical breaking, the pion GPD can be obtained and expressed in terms of Radon transform amplitudes. However, as sketched in [23] and elaborated futher in [25], such a procedure only leaves with a reliable result near the so-called forward limit ($\xi = 0$), within the DGLAP region [26–29].

We will here briefly introduce the above-mentioned algebraic model for the pion's valence-quark GPD [23], obtain the overlap representation for the GPD for the same model and, preliminary, discuss a procedure, based on the Radon-transform technology for its parametrization, allowing for a reliable extension to the ERBL kinematical region [30,31].

2 The Algebraic Model for the Pion's Valence-Quark GPD

The main features of the pion's valence-quark GPD has been recently sketched [23] (although only within the DGLAP kinematical region) with the help of a simple algebraic model, anyhow able of a veracious pion's description, properly grounded in a faithful expression of their symmetries and their breaking patterns [32–34], via the Dyson–Schwinger equations (DSEs). Such is the model we will exploit in the following, aiming at the comparison of its results with the corresponding overlap approach's, in DGLAP region, and at exploring the extension beyond via the Radon transform approach.

Despite its complexity, the pion bound-state is still a $J = 0$ system and hence, in the twist-two approximation, there is only one chirally-even GPD associated with a quark q in the pion ($\pi^{\pm}$, π^0), which is defined by the matrix element

$$H_\pi^q(x,\xi,t) = \int \frac{d^4 z}{4\pi}\, e^{ixP\cdot z}\, \delta(n\cdot z)\, \delta^2(z_\perp)\, \langle \pi(P_+)|\bar{q}\,(-z/2)\, n\cdot\gamma\; \mathcal{W}(-z/2, z/2) q\,(z/2)\,|\pi(P_-)\rangle, \tag{1}$$

where k, n are light-like four-vectors, satisfying $k^2 = 0 = n^2$, $k \cdot n = 1$; $z_\perp$ represents that two-component part of z annihilated by both k, n; and $P_\pm = P \pm \Delta/2$ and $\mathcal{W}(-z/2, z/2)$ represents a Wilson line laid along a light-like path that joins the two listed vectors. In Eq. (1), $\xi = -n \cdot \Delta/[2n \cdot P]$ is the "skewness", $t = -\Delta^2$ is the momentum transfer, and $P^2 = t/4 - m_\pi^2$, $P \cdot \Delta = 0$. The GPD also depends on the resolving scale, ζ. Within the domain upon which perturbation theory is valid, evolution to another scale ζ' is described by the ERBL equations [30,31] for $|x| < \xi$ and the DGLAP equations [26–29] for $|x| > \xi$, where $\xi \geq 0$.

As discussed at length in [23], the valence-quark piece of the GPD expressed by (1) can be first approximated by

$$H_\pi^{\mathrm{v}}(x,\xi,t) = \frac{1}{2}\, N_c \mathrm{tr} \int_{d\ell} \delta_n^{xP}(\ell)\, i\Gamma_\pi\left(\ell_+^{\mathrm{R}}; -P_+\right)\, S(\ell_+)\, in\cdot\Gamma\,(\ell_+,\ell_-)\; S(\ell_-) i\Gamma_\pi\left(\ell_-^{\mathrm{R}}; P_-\right), \tag{2}$$

derived from the incomplete impulse-approximation for the so-called *handbag* diagram [22], and can be next corrected by the additional contribution

$$\begin{aligned} H_\pi^{\mathrm{C}}\left(x,0,-\Delta_\perp^2\right) = \frac{1}{2} N_c \mathrm{tr} \int_{d\ell} \delta_n^{xP}(\ell) \Big[& n\cdot\partial_{\ell_+^{\mathrm{R}}} \Gamma_\pi\left(\ell_+^{\mathrm{R}}; -P_+\right) S(\ell_P) \Gamma_\pi\left(\ell_-^{\mathrm{R}}; P_-\right)\, S(\ell_-) \\ & + \Gamma_\pi\left(\ell_+^{\mathrm{R}}; -P_+\right) S(\ell_P) n\cdot\partial_{\ell_-^{\mathrm{R}}} \Gamma_\pi\left(\ell_-^{\mathrm{R}}; P_-\right) S(\ell_-) \Big] , \end{aligned} \tag{3}$$

within the non-skewed kinematical region, for the transverse momentum $\Delta_\perp^2$. It is very worthwhile to emphasize here that, although (2) paves the way for a fully general covariant computation of the pion's valence-quark GPD, the corrective term given by (3) can be hardly identified beyond DGLAP region and, even for the non-skewed case, some judicious modelling has been required for an appropriate description of the large-momentum transfer domain [23]. In Eqs. (2, 3), $\int_{d\ell} := \int \frac{d^4\ell}{(2\pi)^4}$ is a translationally invariant regularisation of the integral; $\delta_n^{xP}(\ell) := \delta(n\cdot\ell - xn\cdot P)$; the trace is over spinor indices; $\eta \in [0,1]$, $\bar{\eta} = 1-\eta$; $\ell_+^{\mathrm{R}} = \bar{\eta}\ell_+ + \eta\ell_P$, $\ell_-^{\mathrm{R}} = \eta\ell_- + \bar{\eta}\ell_P$, $\ell_\pm = \ell \pm \Delta/2$, $\ell_P = \ell - P$ (N.B. Owing to Poincaré covariance, no observable can legitimately depend on η; i.e., the definition of the relative momentum). In order to gain novel insights into pion structure, in Ref. [23], we used the algebraic model of [35],

Few-Body Syst (2017) 58:144
DOI 10.1007/s00601-017-1304-4

Nabil Chouika · Cédric Mezrag · Hervé Moutarde · José Rodríguez-Quintero

An Algebraic Model for the Pion's Valence-Quark GPD: A Probe for a Consistent Extension Beyond DGLAP Region Via Radon Transform Inversion

Received: 2 December 2016 / Accepted: 11 May 2017 / Published online: 16 June 2017

Abstract We briefly report on a recent computation, with the help of a fruitful algebraic model, sketching the pion valence dressed-quark generalized parton distribution. Then, preliminary, we introduce on a sensible procedure to get reliable results in both Dokshitzer–Gribov–Lipatov–Altarelli–Parisi and Efremov–Radyushkin–Brodsky–Lepage kinematical regions, grounded on the GPD overlap representation and its parametrization of a Radon transform of the so-called double distribution.

1 Introduction

The Generalized Parton Distributions (GPDs), introduced independently by Müller et al. [1], Ji [2] and Radyushkin [3], are shown to be related to hadron form factors by sum rules, and to contain the usual Parton Distribution Functions (PDFs) as a limiting case. They not only generalize the classical objects describing the static or dynamical content of hadrons; additionally, they also provide unique information about the structure of hadrons, including 3D imaging of its partonic components and access to the quark orbital angular momentum. GPDs have been the object of an intense theoretical and experimental activity ever since (see e.g. [4–9], for the pion case, or the more general reviews [10–15] and references therein).

Most of the constraints that apply to GPDs are fulfilled when the GPD is written in a double distribution representation [1,16,17], which is plainly equivalent to expressing the GPD as a Radon transform [18]. In order to obtain insights into the nature of hadron GPDs, it has been common to model the Radon amplitudes, F, G, following Refs. [19]. This approach has achieved some phenomenological success (see e.g. [14,20]); but more flexible parametrisations enable a better fit to data [21]. Such fits play a valuable role in establishing the GPD framework; However there is no known parameterization of GPDs relying on first principles only. Computing GPDs in a symmetry-preserving framework is a key ingredient for the a priori fulfillment of all GPD theoretical constraints. This observation is highlighted by experience drawn from the simpler case of the pion's valence-quark PDF [22]. In [23,24], steps are made by following a different approach for the computation of hadron GPDs based on the example provided by the pion's valence-quark PDF. We prove therein that, within a

This article belongs to the Topical Collection "Light Cone 2016".

J. Rodríguez-Quintero (✉)
Dpt. of Física Aplicada, Fac. CCEE, Unversidad de Huelva, Huelva 21071, Spain
E-mail: jose.rodriguez@dfaie.uhu.es
Tel.: +34-959-219787

N. Chouika · H. Moutarde
Centre de Saclay, IRFU/Service de Physique Nucléaire, 91191 Gif-sur-Yvette, France

C. Mezrag
Physics Division, Argonne National Laboratory, Argonne, IL 60439, USA

covariant computational approach based on solutions of the Dyson–Schwinger and Bethe–Salpeter equations, observing the Ward–Takahashi identities, that properly account for the chiral symmetry and its dynamical breaking, the pion GPD can be obtained and expressed in terms of Radon transform amplitudes. However, as sketched in [23] and elaborated futher in [25], such a procedure only leaves with a reliable result near the so-called forward limit ($\xi = 0$), within the DGLAP region [26–29].

We will here briefly introduce the above-mentioned algebraic model for the pion's valence-quark GPD [23], obtain the overlap representation for the GPD for the same model and, preliminary, discuss a procedure, based on the Radon-transform technology for its parametrization, allowing for a reliable extension to the ERBL kinematical region [30, 31].

2 The Algebraic Model for the Pion's Valence-Quark GPD

The main features of the pion's valence-quark GPD has been recently sketched [23] (although only within the DGLAP kinematical region) with the help of a simple algebraic model, anyhow able of a veracious pion's description, properly grounded in a faithful expression of their symmetries and their breaking patterns [32–34], via the Dyson–Schwinger equations (DSEs). Such is the model we will exploit in the following, aiming at the comparison of its results with the corresponding overlap approach's, in DGLAP region, and at exploring the extension beyond via the Radon transform approach.

Despite its complexity, the pion bound-state is still a $J = 0$ system and hence, in the twist-two approximation, there is only one chirally-even GPD associated with a quark q in the pion ($\pi^{\pm}$, π^0), which is defined by the matrix element

$$H_\pi^q(x, \xi, t) = \int \frac{d^4 z}{4\pi}\, e^{ixP\cdot z}\, \delta(n \cdot z)\, \delta^2(z_\perp)\, \langle \pi(P_+)|\bar{q}\,(-z/2)\, n \cdot \gamma\, \mathcal{W}(-z/2, z/2) q\,(z/2)\,|\pi(P_-)\rangle, \tag{1}$$

where k, n are light-like four-vectors, satisfying $k^2 = 0 = n^2$, $k \cdot n = 1$; $z_\perp$ represents that two-component part of z annihilated by both k, n; and $P_\pm = P \pm \Delta/2$ and $\mathcal{W}(-z/2, z/2)$ represents a Wilson line laid along a light-like path that joins the two listed vectors. In Eq. (1), $\xi = -n \cdot \Delta/[2n \cdot P]$ is the "skewness", $t = -\Delta^2$ is the momentum transfer, and $P^2 = t/4 - m_\pi^2$, $P \cdot \Delta = 0$. The GPD also depends on the resolving scale, ζ. Within the domain upon which perturbation theory is valid, evolution to another scale ζ' is described by the ERBL equations [30, 31] for $|x| < \xi$ and the DGLAP equations [26–29] for $|x| > \xi$, where $\xi \geq 0$.

As discussed at length in [23], the valence-quark piece of the GPD expressed by (1) can be first approximated by

$$H_\pi^{\rm v}(x, \xi, t) = \frac{1}{2}\, N_c {\rm tr} \int_{d\ell} \delta_n^{xP}(\ell)\, i\Gamma_\pi\left(\ell_+^{\rm R}; -P_+\right)\, S(\ell_+)\, i n \cdot \Gamma\,(\ell_+, \ell_-)\, S(\ell_-) i\Gamma_\pi\left(\ell_-^{\rm R}; P_-\right), \tag{2}$$

derived from the incomplete impulse-approximation for the so-called *handbag* diagram [22], and can be next corrected by the additional contribution

$$\begin{aligned} H_\pi^{\rm C}\left(x, 0, -\Delta_\perp^2\right) = \frac{1}{2} N_c {\rm tr} \int_{d\ell} \delta_n^{xP}(\ell) \Big[& n \cdot \partial_{\ell_+^{\rm R}} \Gamma_\pi\left(\ell_+^{\rm R}; -P_+\right) S(\ell_P) \Gamma_\pi\left(\ell_-^{\rm R}; P_-\right) S(\ell_-) \\ & + \Gamma_\pi\left(\ell_+^{\rm R}; -P_+\right) S(\ell_P) n \cdot \partial_{\ell_-^{\rm R}} \Gamma_\pi\left(\ell_-^{\rm R}; P_-\right) S(\ell_-) \Big], \end{aligned} \tag{3}$$

within the non-skewed kinematical region, for the transverse momentum $\Delta_\perp^2$. It is very worthwhile to emphasize here that, although (2) paves the way for a fully general covariant computation of the pion's valence-quark GPD, the corrective term given by (3) can be hardly identified beyond DGLAP region and, even for the non-skewed case, some judicious modelling has been required for an appropriate description of the large-momentum transfer domain [23]. In Eqs. (2, 3), $\int_{d\ell} := \int \frac{d^4\ell}{(2\pi)^4}$ is a translationally invariant regularisation of the integral; $\delta_n^{xP}(\ell) := \delta(n \cdot \ell - x n \cdot P)$; the trace is over spinor indices; $\eta \in [0, 1]$, $\bar{\eta} = 1 - \eta$; $\ell_+^{\rm R} = \bar{\eta}\ell_+ + \eta\ell_P$, $\ell_-^{\rm R} = \eta\ell_- + \bar{\eta}\ell_P$, $\ell_\pm = \ell \pm \Delta/2$, $\ell_P = \ell - P$ (N.B. Owing to Poincaré covariance, no observable can legitimately depend on η; i.e., the definition of the relative momentum). In order to gain novel insights into pion structure, in Ref. [23], we used the algebraic model of [35],

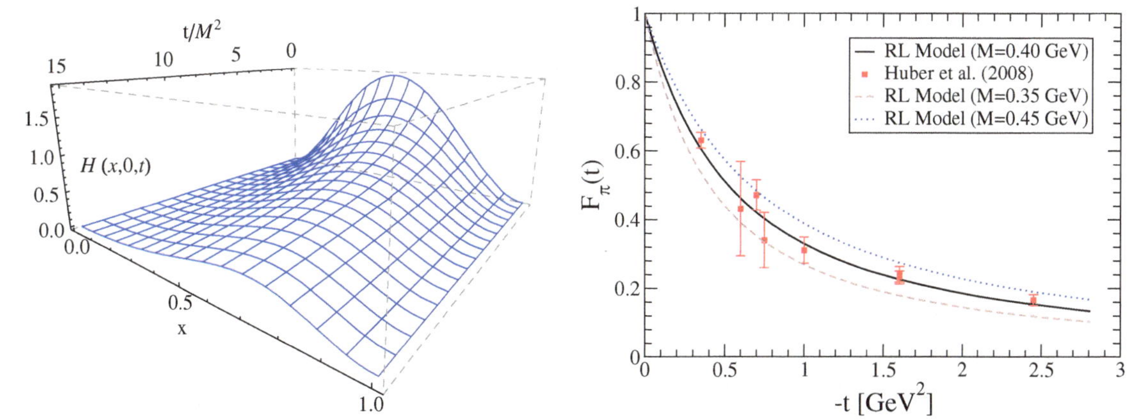

Fig. 1 *Left panel* Pion valence dressed-quark GPD, $H_\pi^{\rm v}(x, 0, -\Delta_\perp^2)$, defined by the addition of Eqs. (2, 3), obtained as explained in [23] and plotted as a function of $t/M^2 = \Delta_\perp^2/M^2$, where M is a mass scale for the dressed-quark in the algebraic model of [35] (result obtained at the model scale, $\zeta_H = 0.51$ GeV). *Right panel* Pion electromagnetic form factor obtained from the integtation over x (sum rule) of $H_\pi^{\rm v}(x, 0, -\Delta_\perp^2)$, resulting from Eqs. (2), (4) and associated definitions, in Ref. [23]. The data are described in Ref. [39]. The most favourable comparison is obtained with $M = 0.40$ GeV in Eq. (4) and the band shows results with $M = 0.40 \pm 0.05$ GeV

$$S(\ell) = [-i\gamma \cdot \ell + M]\Delta_M(\ell^2) \tag{4a}$$

$$\rho_\nu(z) = \frac{1}{\sqrt{\pi}} \frac{\Gamma(\nu + 3/2)}{\Gamma(\nu + 1)} (1 - z^2)^\nu \tag{4b}$$

$$n_\pi \Gamma_\pi(\ell_\pm^{\rm R}; \pm P) = i\gamma_5 \int_{-1}^{1} dz\, \rho_\nu(z)\, \hat{\Delta}_M^\nu(\ell_{z\pm}^2) \tag{4c}$$

for the dressed-quark and pion elements in Eqs. (2, 3); where $\Delta_M(\ell^2) = 1/(\ell^2 + M^2)$, M is a dressed-quark mass-scale; $\hat{\Delta}_M(\ell^2) = M^2\Delta_M(\ell^2)$; $\ell_{z\pm} = \ell_\pm^{\rm R} + (z \pm 1)P/2$ and we work in the chiral limit ($P^2 = 0 = \hat{m}$, where $\hat{m}$ is the current-quark mass); and n_π is the Bethe–Salpeter amplitude's canonical normalisation constant. We then applied the algebra and approximations described in [23,24] to be finally left with the results displayed in Fig. 1 (left panel). Basically, we write the Mellin moments of the GPD, make the analytical integration over momenta with the help of the Feynman decompositon and formally express them as integrals over Feynman parameters, defined in a compact domain [0, 1]. We then rearrange the integrals with the appropriate change of variables for these compact parameters and identify two of them, defined in the appropriate support domain, with the arguments of the Radon amplitudes. Thus, we can determine the Radon amplitudes that are consistent with those Mellin moments. Notably, the so-computed non-skewed GPD shows all the properties expected on the basis of the GPD overlap representation [12,36–38], discussed in [23]. Additionaly, integrated over x, it produces an estimate for the pion electromagnetic form factor which compares very well with experimental data (see right panel of Fig. 1).

Let us, however, focus on the forward limit for the non-skewed GPD obtained with Eqs. (2, 3), as described in [23]. It naturally reproduces the pion valence dressed-quark distribution function (PDF) found in [22], observing, in particular, the symmetry under the exchange $x \to 1-x$ that owes to the two-body-problem nature for the pion's distribution at zero-momentum transfer when isospin-breaking contributions are not considered. Furthermore, as will be seen in the next section, the so-computed PDF can be compared with that obtained, for the same algebraic model, within the approach resulting from the GPD overlap representation [12,36–38].

3 Overlap Representation, Radon Transform and Skewed GPD

As it has been previously discussed, in the aim of obtaining results beyond $\xi = 0$, both in DGLAP and ERBL kinematical regions, a sensible extension of (3) correcting (2) is far from being obvious. A different approach based on the representation of the pion GPD as overlap of light-cone wave functions (LCWF),

$$H^q_\pi(x,\xi,t)_{\xi\le x\le 1} = C^q \int d^2\mathbf{k}^2_\perp \Psi^*\left(\frac{x-\xi}{1-\xi}, \mathbf{k}_\perp + \frac{1-x}{1-\xi}\frac{\Delta_\perp}{2}; P_-\right)\Psi\left(\frac{x+\xi}{1+\xi}, \mathbf{k}_\perp - \frac{1-x}{1+\xi}\frac{\Delta_\perp}{2}; P_+\right), \tag{5}$$

can be followed, where C^q is a normalization constant and the LCWF can be computed by integrating over k^- the pion Bethe–Salpeter wave function $\chi_\pi(k, P)$ projected onto $\gamma^+\gamma_5$

$$\Psi\left(k^+, \mathbf{k}_\perp; P\right) = -\frac{1}{2\sqrt{3}}\int \frac{dk^-}{2\pi}\mathrm{Tr}\left[\gamma^+\gamma_5\,\chi_\pi(k,P)\right]. \tag{6}$$

The latter is only true within the DGLAP kinematical region, as only there the GPD results from the overlap of two N-body LCWF's (truncated to two-body in our pion's GPD case). Had the GPD overlap been lying within the ERBL region, it would have connected N- and $N+2$-body wave functions instead. One interesting remark in order here. The GPD can be represented by the Radon transform of DD's owing to fundamental properties it must fulfill, as Lorentz invariance. In addition, information from both DGLAP and ERBL regions is required in order to fully determine the DD representing the GPD. Consequently, by means of the Radon-transform representation, non-trivial connection between N- and $N+2$-body wave functions have to emerge from the fundamental symmetries of the physical world. Indeed, one can take advantage of the usually called one-component DD (1CDD) or, specifically, BMKS scheme [40] (see also discussion in [20] and references therein), and the GPD can be expressed as the Radon transform, $\mathcal{R}f$, of the distribution $f(\beta,\alpha,t)$,

$$\frac{\sqrt{1+\xi^2}}{x}H(x,\xi,t) = \sqrt{1+\xi^2}\int_\Omega d\beta\, d\alpha\, f(\beta,\alpha,t)\,\delta(x-\beta-\alpha\xi) \tag{7}$$

$$= \mathcal{R}f(s,\varphi,t) = \int_\Omega d\beta\, d\alpha\, f(\beta,\alpha,t)\,\delta(s-\beta\cos\varphi-\alpha\sin\varphi), \tag{8}$$

where the two DD variables (β,α) appear supported on a rotated square $\Omega \equiv \{|\alpha|+|\beta|\le 1\}$ and (8) makes explicit the geometrical interpretation of the Radon transform by the use of the polar coordinates which relate to standard GPD ones as $s = x\cos\varphi$ and $\xi = \tan\varphi$. Owing to its connection with the computerized tomography problem, mathematicians have paid enough attention to the Radon transform and provided us with a valuable tool for the characterization of solutions and their uniqueness under some smoothness and consistency conditions [41]. Therefore, a natural way in order to make the extension for the GPD beyond DGLAP to ERBL region comes out from inverting (7) in order to get the corresponding DD from the DGLAP-constrained GPD.

We will preliminary discuss how this can be acomplished in the next section, in particular by applying this program to the results obtained with the previously described algebraic model. The first step is to use Eq. (6) to get a Bethe–Salpeter wave function from (4), which will be subsequently plugged into Eq. (5) and one will be finally left with the corresponding DGLAP overlap GPD. Thus, we obtain [25]

$$H^u_{\pi^+}(x,\xi,0)\big|_{\mathrm{DGLAP}} = \mathcal{N}_\nu \frac{(1-x)^{2\nu}(x^2-\xi^2)^\nu}{(1-\xi^2)^{2\nu}}. \tag{9}$$

If one takes now the forward limit ($t=0$, $\xi=0$) in (9), the result reads

$$q_\pi(x) = H^u_{\pi^+}(x,0,0) = \mathcal{N}_\nu\, x^{2\nu}(1-x)^{2\nu}, \tag{10}$$

where, in the case $\nu = 1$, the normalization factor is $k_1 = 30$ and the result for the pion dressed-quark distribution function (PDF) has been proved to be numerically consistent with that obtained from Eqs. (2, 3), also in the forward limit. This can be seen in Fig. 2, where both results appear depicted and, to the eye, can be barely distinguishable from each other. In Ref. [22], the same result of (10) with $\nu = 1$ was introduced as an excellent and efficacious approximation to the pion's valence dressed-quark PDF, while also resulted, in [23], from a heuristic LCWF implemented in (5). Here, it appears as the natural result from the algebraic model of Eqs. (4) [35] and the overlap representation of pion's GPD.

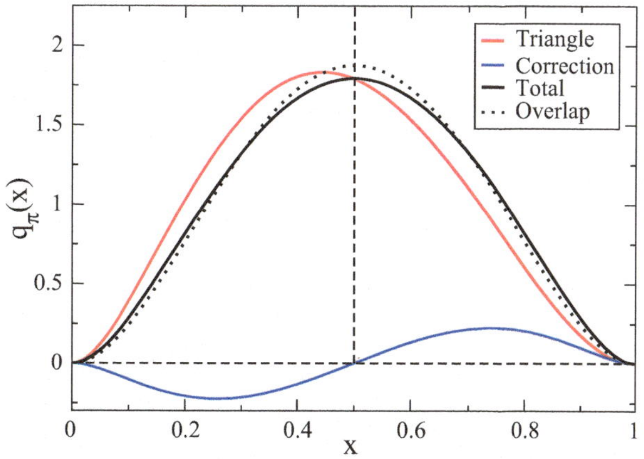

Fig. 2 Pion dressed-quark distribution function (PDF) obtained from the GPD in the forward limit ($t = 0, \xi = 0$); *solid red* and *blue lines* stand for the results from (2) and (3), respectively, and *black solid one* is for their addition; the *dotted line* corresponds to the overlap result given by (10), obtained from Eqs. (5, 6), for $\nu = 1$

4 Beyond DGLAP

The final step of the computational program is then to make the inversion of the Radon transform in (7) with the DGLAP GPD given by Eq. (9). One will so obtain a DD representation for the GPD which determines it fully in both DGLAP and ERBL regions. It was anyhow discovered [18,42] that many parametrizations for DDs yield the same GPDs. These physically equivalent parametrizations can be mapped into each other by means of the so-called *gauge* transformations. In particular, there is a transformation that brings the DD to a particular representation, $f_P(\beta, \alpha, t)$, due to Pobylitsa [43] that, owing to a specific choice of the D-term, allows for the GPD to be recast as

$$H(x, \xi, t) = (1 - x) \int_\Omega d\beta \, d\alpha \; f_P(\beta, \alpha, t) \, \delta(x - \beta - \alpha\xi), \tag{11}$$

which will be particularly useful below. We will now specialize for the asymptotic value of $\nu = 1$ in (9) and will only consider here the case $t = 0$. Then, in the following and for the sake of economy in the notation, t will be dropped from the GPDs and DDs arguments.

We have afforded the inversion of the Radon transform by properly discretizing the support domain Ω and so be left with a linear inversion problem in a large-dimension matrix space, $Ax = y$, the dimension determined by our discretization choice, where the linear operator, A, expresses the action of the Radon transform over the discretized DD space. This is anyhow an *ill-posed* problem, at least in a mild sens, fom the mathematical point of view and much care needs to be taken in facing it numerically. In our case, the DD solution has been constrained by implementing two physically-grounded properties as (i) the parity in α, $f(\beta, -\alpha) = f(\beta, \alpha)$; and, provided that we deal with a valence-quark GPD, the kinematics requires that $f(\beta, \alpha) = 0$ for any $\beta < 0$. Both requirements work for any choice of the gauge and have been externally imposed. In particular, parity in α reduces the dimension of the matrix and requires a symmetrization of the discrete contributions coming from both sides of the $\alpha = 0$ line. More details of the computation, in particular concerning the numerics for the Radon transform inversion, and a full kinematical analysis including the impact-parameter-dependent GPD will be found in a forthcoming work [44].

Apart from toy examples, as a constant DD over the support domain where $\beta > 0$, we have also first succesfully checked our numerics with a simplified version of the so-called Radyushkin Double Distribution Ansatz (RDDA) [19],

$$\begin{matrix} \text{Pobylitsa} & (1-\beta) \\ \text{BMKS} & \beta \end{matrix} \left. \vphantom{\begin{matrix} a \\ b \end{matrix}} \right\} \times f_{RDDA}(\beta, \alpha) = \frac{\Gamma(N + 3/2)}{\sqrt{\pi}\,\Gamma(N+1)} \frac{\left[(1 - |\beta|)^2 - \alpha^2\right]^N}{(1 - |\beta|)^{2N+1}} \, q(\beta) \,, \tag{12}$$

where we specialized for $N = 1$ and for a simple PDF resulting from $q(\beta) = \beta^2(1 - \beta)^2$. The GPD associated to this RDDA can be analytically computed within the DGLAP region, and then numerically inverted and found to agree pretty well with the original DD given by (12), as can be seen in Fig. 3.

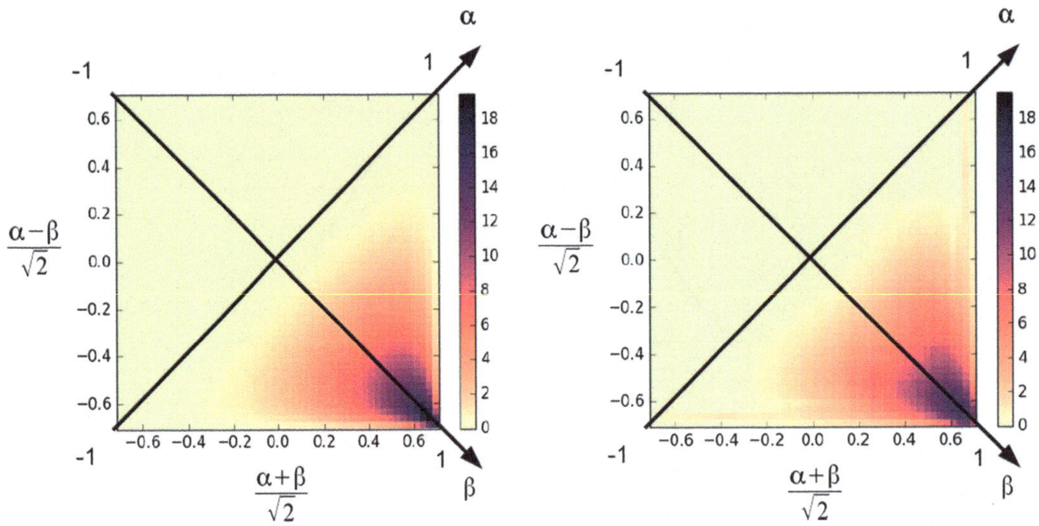

Fig. 3 Pobylitsa-gauge RDDA for $t = 0$ obtained from the analytical expression, Eq. (12), in the *left panel*; wich compares pretty well to the result of the inversed Radon-transform of its associated GPD in the DGLAP region, numerically obtained from (11) as explained in the text, displayed in the *right panel*. The support for the parameters (α,β) corresponds to a rotated square that we have rotated back by the choice of the two combinations, $(\alpha+\beta)/\sqrt{2}$ and $(\alpha-\beta)/\sqrt{2}$, for the horizontal and vertical axes, respectively

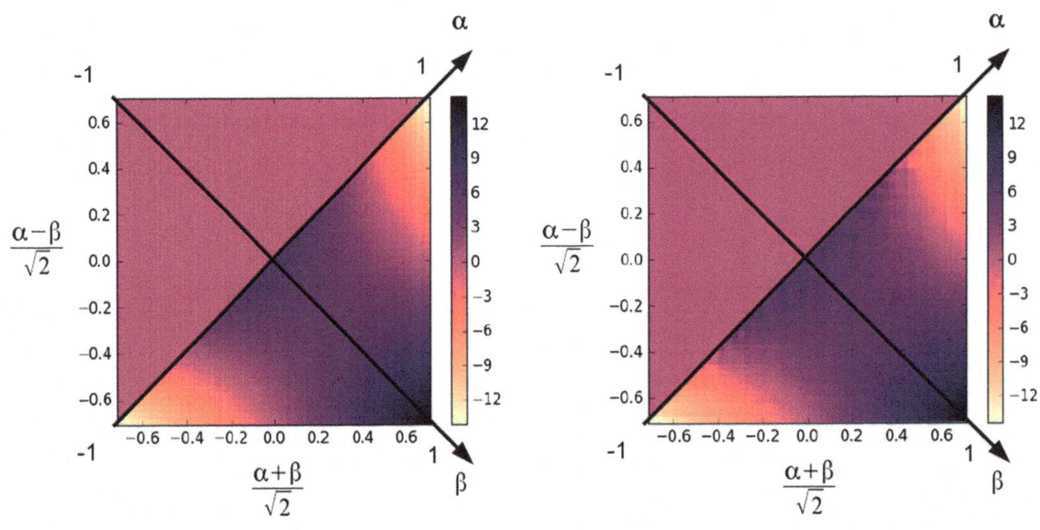

Fig. 4 Pion's valence dressed-quark overlap GPD at $t = 0$ given by Eq. (13) in the Pobylitsa representation, in the *left panel*; which also compares pretty well to the result of the inversed Radon-transform of its associated GPD in the DGLAP region, numerically obtained from (11) as explained in the text, displayed in the *right panel*

Then, after succeeding in inverting the DGLAP GPD derived from the RDDA, we have a solid basis to rely on the following results obtained from applying the same numerical procedure to the overlap GPD given by (9). In that case, we found a much smoother solution by inverting the GPD expressed as a Radon transform representation in the Pobylitsa gauge, Eq. (11). Indeed, the numerical inversion yielded, in this case, a solution that can be easily accomodated in a polynomial in both, α and β, which can be suggested as an ansatz for the direct Radon transform and so get a solution to be compared with (9). Thus, we can conclude that

$$f_P(\beta,\alpha) = \frac{30}{4}\left(1 - 3\alpha^2 - 2\beta + 3\beta^2\right) \tag{13}$$

is the analytical Pobylitsa DD for pion's valence-quark overlap GPD of the algebraic model given by (4) for the $t = 0$ case (see Fig. 4). The mapping from Pobylitsa into BMKS gauge is then possible [45] and we find

$$f(\beta,\alpha) = \frac{30}{4}\left(\alpha^2\left(3 - \frac{3}{(|\alpha|+\beta)^3}\right) + (3\beta^2 - 4\beta)\left(\frac{1}{(|\alpha|+\beta)^4} - 1\right) + \frac{2}{(|\alpha|+\beta)^2} - 2 \right.$$
$$\left. + (1-\alpha^2)\delta(\beta)\right), \tag{14}$$

which shows a singular behaviour in the $\beta = 0$ line (this is precisely why the Pobylitsa gauge appears to offer a more friendly DD representation for the GPD concerning the inversion problem).

Finally, once the Radon-transform inversion of (9) has furnished us with any of the two DDs in Eqs. (13, 14) for Pobylitsa and BMKS gauges (analytical expressions in this case, for $t = 0$, but numerical results in general), we can apply either (11) or (7) to obtain the Radon transform everywhere.

5 Conclusions

We have briefly reported on the first steps recently made towards the computation of the pion's valence dressed-quark GPD, within the symmetry-preserving framework provided by DSEs. In addition, we have also preliminarly introduced a reliable procedure, on the basis of the GPD representation as the overlap of LCWFs and Radon-transform technology, that can be fruitful exploited in order to extend previous results; in particular, those for the pion's GPD in DGLAP kinematical region and near the forward limit to non-zero skewness and to the ERBL domain. In addition, this procedure happens to be an open avenue for generic GPD modelling and, in the future, for the ambitious goal of computing GPDs from very first principles.

Acknowledgements This work has been partially supported by Spanish ministry projects FPA-2014-53631-C2-2-P, French GDR 3034 PH-QCD "Cromodynamique Quantique et Physique des Hadrons" and ANR-12-MONU-0008-01 "PARTONS" and US Department of Energy, Office of Science, Office of Nuclear Physics, under contract no. DE-AC02-06CH11357. We thank for their very valuable remarks to L. Chang, C.D. Roberts, F. Sabatié and S.M. Schmidt.

References

1. D. Mueller, D. Robaschik, B. Geyer, F. Dittes, J. Hořejsi, Wave functions, evolution equations and evolution kernels from light ray operators of QCD. Fortsch. Phys. **42**, 101–141 (1994). doi:10.1002/prop.2190420202. arxiv:hep-ph/9812448
2. X.D. Ji, Deeply virtual compton scattering. Phys. Rev. **D55**, 7114–7125 (1997). doi:10.1103/PhysRevD.55.7114. arxiv:hep-ph/9609381
3. A. Radyushkin, Nonforward parton distributions. Phys. Rev. **D56**, 5524–5557 (1997). doi:10.1103/PhysRevD.56.5524. arxiv:hep-ph/9704207
4. W. Broniowski, E. Ruiz Arriola, Impact parameter dependence of the generalized parton distribution of the pion in chiral quark models. Phys. Lett. **B574**, 57–64 (2003). doi:10.1016/j.physletb.2003.09.009. arxiv:hep-ph/0307198
5. W. Broniowski, E. Ruiz Arriola, K. Golec-Biernat, Generalized parton distributions of the pion in chiral quark models and their QCD evolution. Phys. Rev. **D77**(034), 023 (2008). doi:10.1103/PhysRevD.77.034023. arxiv:0712.1012
6. T. Frederico, E. Pace, B. Pasquini, G. Salme, Pion generalized parton distributions with covariant and light-front constituent quark models. Phys. Rev. **D80**(054), 021 (2009). doi:10.1103/PhysRevD.80.054021. arxiv:0907.5566
7. C.R. Ji, Y. Mishchenko, A. Radyushkin, Higher Fock state contributions to the generalized parton distribution of pion. Phys. Rev. **D73**(114), 013 (2006). doi:10.1103/PhysRevD.73.114013. arxiv:hep-ph/0603198
8. L. Theussl, S. Noguera, V. Vento, Generalized parton distributions of the pion in a Bethe-Salpeter approach. Eur. Phys. J. **A20**, 483–498 (2004). doi:10.1140/epja/i2003-10174-3. arxiv:nucl-th/0211036
9. B. Tiburzi, G. Miller, Generalized parton distributions and double distributions for q anti-q pions. Phys. Rev. **D67**(113), 004 (2003). doi:10.1103/PhysRevD.67.113004. arxiv:hep-ph/0212238
10. A. Belitsky, A. Radyushkin, Unraveling hadron structure with generalized parton distributions. Phys. Rep. **418**, 1–387 (2005). doi:10.1016/j.physrep.2005.06.002. arxiv:hep-ph/0504030
11. S. Boffi, B. Pasquini, Generalized parton distributions and the structure of the nucleon. Riv. Nuovo Cim. **30**, 387 (2007). doi:10.1393/ncr/i2007-10025-7. arxiv:0711.2625
12. M. Diehl, Generalized parton distributions. Phys. Rep. **388**, 41–277 (2003). doi:10.1016/j.physrep.2003.08.002. arxiv:hep-ph/0307382
13. K. Goeke, M.V. Polyakov, M. Vanderhaeghen, Hard exclusive reactions and the structure of hadrons. Prog. Part. Nucl. Phys. **47**, 401–515 (2001). doi:10.1016/S0146-6410(01)00158-2. arxiv:hep-ph/0106012
14. M. Guidal, H. Moutarde, M. Vanderhaeghen, Generalized parton distributions in the valence region from deeply virtual compton scattering. Rep. Prog. Phys. **76**(066), 202 (2013). doi:10.1088/0034-4885/76/6/066202. arxiv:1303.6600
15. X.D. Ji, Off forward parton distributions. J. Phys. **G24**, 1181–1205 (1998). doi:10.1088/0954-3899/24/7/002. arxiv:hep-ph/9807358
16. A. Radyushkin, Double distributions and evolution equations. Phys. Rev. **D59**(014), 030 (1999). doi:10.1103/PhysRevD.59.014030. arxiv:hep-ph/9805342

17. A. Radyushkin, Symmetries and structure of skewed and double distributions. Phys. Lett. **B449**, 81–88 (1999). doi:10.1016/S0370-2693(98)01584-6. arxiv:hep-ph/9810466
18. O. Teryaev, Crossing and radon tomography for generalized parton distributions. Phys. Lett. **B510**, 125–132 (2001). doi:10.1016/S0370-2693(01)00564-0. arxiv:hep-ph/0102303
19. I. Musatov, A. Radyushkin, Evolution and models for skewed parton distributions. Phys. Rev. **D61**(074), 027 (2000). doi:10.1103/PhysRevD.61.074027. arxiv:hep-ph/9905376
20. C. Mezrag, H. Moutarde, F. Sabatié, Test of two new parameterizations of the generalized parton distribution H. Phys. Rev. **D88**(014), 001 (2013). doi:10.1103/PhysRevD.88.014001. arxiv:1304.7645
21. K. Kumericki, D. Mueller, K. Passek-Kumericki, Sum rules and dualities for generalized parton distributions: is there a holographic principle? Eur. Phys. J. **C58**, 193–215 (2008). doi:10.1140/epjc/s10052-008-0741-0. arxiv:0805.0152
22. L. Chang, C. Mezrag, H. Moutarde, C.D. Roberts, J. Rodríguez-Quintero, P.C. Tandy, Basic features of the pion valence-quark distribution function. Phys. Lett. B **737**, 23–29 (2014). doi:10.1016/j.physletb.2014.08.009. arxiv:1406.5450
23. C. Mezrag, L. Chang, H. Moutarde, C.D. Roberts, J. Rodríguez-Quintero, F. Sabatié, S.M. Schmidt, Sketching the pion's valence-quark generalised parton distribution. Phys. Lett. B **741**, 190–196 (2014). doi:10.1016/j.physletb.2014.12.027. arxiv:1411.6634
24. C. Mezrag, H. Moutarde, J. Rodríguez-Quintero, F. Sabatié, Towards a pion generalized parton distribution model from Dyson–Schwinger Equations (2014). arxiv:1406.7425
25. C. Mezrag, H. Moutarde, J. Rodriguez-Quintero, From Bethe–Salpeter wave functions to generalised parton distributions. Few Body Syst. **57**(9), 729–772 (2016). doi:10.1007/s00601-016-1119-8. arxiv:1602.07722
26. G. Altarelli, G. Parisi, Asymptotic freedom in parton language. Nucl. Phys. B **126**, 298 (1977). doi:10.1016/0550-3213(77)90384-4
27. Y.L. Dokshitzer, Calculation of the structure functions for deep inelastic scattering and e+ e− annihilation by perturbation theory in quantum chromodynamics. Sov. Phys. JETP **46**, 641–653 (1977). [Zh. Eksp. Teor. Fiz.73,1216(1977)]
28. V.N. Gribov, L.N. Lipatov, Deep inelastic e p scattering in perturbation theory. Sov. J. Nucl. Phys. **15**, 438–450 (1972). [Yad. Fiz.15,781(1972)]
29. L.N. Lipatov, The parton model and perturbation theory. Sov. J. Nucl. Phys. **20**, 94–102 (1975). [Yad. Fiz.20,181(1974)]
30. A.V. Efremov, A.V. Radyushkin, Factorization and asymptotical behavior of pion form-factor in QCD. Phys. Lett. B **94**, 245–250 (1980). doi:10.1016/0370-2693(80)90869-2
31. G.P. Lepage, S.J. Brodsky, Exclusive processes in perturbative quantum chromodynamics. Phys. Rev. D **22**, 2157 (1980). doi:10.1103/PhysRevD.22.2157
32. A. Bashir, L. Chang, I.C. Cloet, B. El-Bennich, Y.X. Liu et al., Collective perspective on advances in Dyson–Schwinger Equation QCD. Commun. Theor. Phys. **58**, 79–134 (2012). doi:10.1088/0253-6102/58/1/16. arxiv:1201.3366
33. L. Chang, C.D. Roberts, P.C. Tandy, Selected highlights from the study of mesons. Chin. J. Phys. **49**, 955–1004 (2011). arxiv:1107.4003
34. S.X. Qin, C.D. Roberts, S.M. Schmidt, Ward–Green–Takahashi identities and the axial-vector vertex. Phys. Lett. B **733**, 202–208 (2014). doi:10.1016/j.physletb.2014.04.041. arxiv:1402.1176
35. L. Chang, I. Cloet, J. Cobos-Martinez, C. Roberts, S. Schmidt et al., Imaging dynamical chiral symmetry breaking: pion wave function on the light front. Phys. Rev. Lett. **110**(132), 001 (2013). doi:10.1103/PhysRevLett.110.132001. arxiv:1301.0324
36. M. Burkardt, Impact parameter dependent parton distributions and off forward parton distributions for zeta > 0. Phys. Rev. **D62**(071), 503 (2000). doi:10.1103/PhysRevD.62.071503. arxiv:hep-ph/0005108
37. M. Burkardt, Impact parameter space interpretation for generalized parton distributions. Int. J. Mod. Phys. A **18**, 173–208 (2003). doi:10.1142/S0217751X03012370. arxiv:hep-ph/0207047
38. M. Diehl, T. Feldmann, R. Jakob, P. Kroll, The overlap representation of skewed quark and gluon distributions. Nucl. Phys. **B596**, 33–65 (2001). doi:10.1016/S0550-3213(00)00684-2. [Erratum: Nucl. Phys. B605,647(2001)], arxiv:hep-ph/0009255
39. G. Huber et al., Charged pion form-factor between Q**2 = 0.60-GeV**2 and 2.45-GeV**2. II. Determination of, and results for, the pion form-factor. Phys. Rev. **C78**, 045203 (2008). doi:10.1103/PhysRevC.78.045203. arxiv:0809.3052
40. A.V. Belitsky, D. Mueller, A. Kirchner, A. Schafer, Twist three analysis of photon electroproduction off pion. Phys. Rev. **D64**(116), 002 (2001). doi:10.1103/PhysRevD.64.116002. arxiv:hep-ph/0011314
41. H. Moutarde, Nucleon reverse engineering. Structuring hadrons with colored degrees of freedom. Habilitation thesis disertation (to be published) (2017)
42. B. Tiburzi, Double distributions: loose ends. Phys. Rev. **D70**(057), 504 (2004). doi:10.1103/PhysRevD.70.057504. arxiv:hep-ph/0405211
43. P.V. Pobylitsa, Virtual compton scattering in the generalized Bjorken region and positivity bounds on generalized parton distributions. Phys. Rev. D **70**(034), 004 (2004). doi:10.1103/PhysRevD.70.034004. arxiv:hep-ph/0211160
44. N. Chouika et al., Covariant Extension of the GPD overlap representation at low Fock state. (2017)
45. D. Mueller, Generalized parton distributions: visions, basics, and realities. Few Body Syst. **55**, 317–337 (2014). doi:10.1007/s00601-014-0894-3. arxiv:1405.2817

Few-Body Syst (2017) 58:31
DOI 10.1007/s00601-016-1208-8

Ho-Meoyng Choi · Chueng-Ryong Ji

Twist-3 Distribution Amplitudes of Pion in the Light-Front Quark Model

Received: 10 November 2016 / Accepted: 28 December 2016 / Published online: 11 January 2017

Abstract We analyzed two twist-3 distribution amplitudes of pion, i.e. pseudoscalar $\phi^P_{3;\pi}(x)$ and pseudotensor $\phi^\sigma_{3;\pi}(x)$, within the LFQM. Our LFQM descriptions both for twist-3 $\phi^P_{3;\pi}$ and $\phi^\sigma_{3;\pi}$ obtained from the Gaussian radial wave function not only satisfy the fundamental constraint required from the isospin symmetry, but also reproduce exactly the asymptotic forms anticipated from QCD's conformal limit.

1 Introduction

Hadronic light-cone distribution amplitudes (DAs) have been known to play an essential role in the QCD description of hard exclusive processes via the factorization theorem [1]. Taking the transverse separation to zero, the factorization takes the form of a convolution of a perturbatively calculable hard-scattering amplitude and the process-independent nonperturbative DAs. These nonperturbative DAs motivated many theoretical studies to calculate meson DAs using nonperturbative methods such as the QCD sum rule [2–4], the chiral-quark model from the instanton vacuum [5,6], the Dyson-Schwinger equations (DSE) approach [7,8], and the light-front quark model (LFQM) [9–12]. Among them, the LFQM appears to be one of the most efficient and effective tools in studying hadron physics as it takes advantage of the distinguished features of the light-front dynamics (LFD) [13]. Specifically, the rational energy-momentum dispersion relation of LFD, namely $p^- = (\mathbf{p}^2_\perp + m^2)/p^+$, yields the sign correlation between the LF energy $p^-(=p^0 - p^3)$ and the LF longitudinal momentum $p^+(=p^0 + p^3)$ and leads to the suppression of vacuum fluctuations in LFD. This simplification is a remarkable advantage in LFD and facilitates the partonic interpretation of the amplitudes. Based on the advantage of LFD, the LFQM has been quite successful in describing various static and non-static properties of hadrons such as meson mass spectra [14], the decay constants [15], electromagnetic and weak transition form factors [16] and generalized parton distribution (GPDs) [17].

In Ref. [9], we have analyzed twist-2 DAs of pseudoscalar ($\phi^A_{2;M}(x)$) and vector ($\phi^{||}_{2;V}(x)$) mesons using the LFQM [14]. In more recent works [10,11], we have extended our LFQM to analyze twist-3 pseudoscalar ($\phi^P_{3;M}(x)$) DAs of pseudoscalar mesons [11] and chirality-even twist-3 ($\phi^\perp_{3;V}(x)$) DAs of vector mesons [10] and

This article belongs to the Topical Collection "Light Cone 2016".

This work was supported by the Korean Research Foundation Grant funded by the Korean Government (No. NRF-2014R1A1A2057457).

H.-M. Choi (✉)
Department of Physics, Teachers College, Kyungpook National University, Daegu 41566, Korea
E-mail: homyoung@knu.ac.kr

C.-R. Ji
Department of Physics, North Carolina State University, Ralegh 27695-8202, New Caledonia
E-mail: crji@ncsu.edu

discussed the link between the chiral symmetry of QCD and the numerical results of the LFQM. In particular, we have discussed a wave function dependence of the LF zero-mode [18–20] contributions to $\phi^P_{3;M}(x)$ and $\phi^{\perp}_{3;V}(x)$ not only for the exactly solvable manifestly covariant Bethe-Salpeter (BS) model but also for the more phenomenologically accessible realistic LFQM [9,14] using the standard LF (SLF) approach. We also linked the covariant BS model to the standard LFQM providing the correspondence relation [10,11] between the two models. Effectively, we prescribed a consistent substitution for the LF vertex function in the covariant BS model with the more phenomenologically accessible Gaussian wave function provided by the LFQM analysis of meson mass [14]. The remarkable finding is that the zero-mode contribution as well as the instantaneous contribution revealed in the covariant BS model become absent in the LFQM. Without involving the zero-mode and instantaneous contributions, our LFQM result of twist-3 DAs $\phi^P_{3;M}(x)$ and $\phi^{\perp}_{3;V}(x)$ provides the consistency with the chiral symmetry anticipated from QCD's conformal limit [2,21].

The purpose of this work is to extend our previous work [11] to analyze the twist-3 pseudotensor DA $\phi^{\sigma}_{3;\pi}(x)$ of a pion within the LFQM. The paper is organized as follows: In Sect. 2, we compute the twist-3 pseudotensor DA $\phi^{\sigma}_{3;\pi}(x)$ in an exactly solvable model based on the covariant BS model of (3+1)-dimensional fermion field theory. We then linked the covariant BS model to the standard LFQM following the correspondence relation [11] between the covariant BS and LFQM models and present the form of $\phi^{\sigma}_{3;\pi}(x)$ as well as $\phi^P_{3;\pi}(x)$ in our LFQM. In Sect. 3, we present our numerical results. Summary and discussion follow in Sect. 4.

2 Model Description

The $\phi^P_{2;M}$ and $\phi^{\sigma}_{3;M}$ are defined in terms of the following matrix elements of gauge invariant nonlocal operators in the light-cone gauge [2,3]:

$$\phi^P_{3;M}(x) = \frac{2(P\cdot\eta)}{f_M\mu_M}\int_{-\infty}^{\infty}\frac{d\tau}{2\pi}e^{-i\zeta\tau(P\cdot\eta)}\langle 0|\bar{q}(\tau\eta)i\gamma_5 q(-\tau\eta)|M(P)\rangle, \tag{1}$$

$$\phi^{\sigma}_{3;M}(x) = -\frac{12}{f_M\mu_M}\int_{-\infty}^{\infty}\frac{d\tau}{2\pi}\int_0^x dx' e^{-i\zeta'\tau(P\cdot\eta)}\langle 0|\bar{q}(\tau\eta)i(\not{P}\not{\eta} - P\cdot\eta)\gamma_5 q(-\tau\eta)|M(P)\rangle, \tag{2}$$

respectively, where $\eta = (1,0,0,-1)$ and P is the four-momentum of the meson ($P^2 = m^2_M$) and x corresponds to the longitudinal momentum fraction carried by the quark and $\zeta = 2x-1$. The normalization parameter $\mu_M = m^2_M/(m_q + m_{\bar{q}})$ results from quark condensate. For the pion, $\mu_\pi = -2\langle\bar{q}q\rangle/f^2_\pi$ from the Gell-Mann-Oakes-Renner relation [22]. The nonlocal matrix elements $\mathcal{M}_\alpha \equiv \langle 0|\bar{q}(\tau\eta)i\Gamma_\alpha q(-\tau\eta)|M(P)\rangle$ for pseudoscalar ($\Gamma_\alpha = \gamma_5$) and pseudotensor ($\Gamma_\alpha = (\not{P}\not{\eta} - P\cdot\eta)\gamma_5$) channels are given by the following momentum integral in two-point function of the manifestly covariant BS model

$$\mathcal{M}_\alpha = N_c\int\frac{d^4k}{(2\pi)^4}e^{-i\tau k\cdot\eta}e^{-i\tau(k-P)\cdot\eta}\frac{\mathrm{Tr}\left[i\Gamma_\alpha\left(\not{p}+m_q\right)\gamma_5\left(-\not{k}+m_{\bar{q}}\right)\right]}{(p^2-m^2_q+i\varepsilon)(k^2-m^2_{\bar{q}}+i\varepsilon)}H_0, \tag{3}$$

where N_c denotes the number of colors. The quark propagators of mass m_q and $m_{\bar{q}}$ carry the internal four-momenta $p = P-k$ and k, respectively. In order to regularize the covariant loop, we use the usual multipole ansatz [10,11] for the $q\bar{q}$ bound-state vertex function $H_0 = H_0(p^2,k^2)$ of a meson: $H_0(p^2,k^2) = g/(p^2 - \Lambda^2 + i\varepsilon)^2$, where g and Λ are constant parameters. After a little manipulation, we obtain for the pion ($m_q = m_{\bar{q}}$)

$$\phi^P_{3;\pi}(x) = \frac{N_c}{f_\pi\mu_\pi}\int\frac{d^2\mathbf{k}_\perp}{8\pi^3}\frac{\chi(x,\mathbf{k}_\perp)}{(1-x)}M^2_0, \tag{4}$$

$$\phi^{\sigma}_{3;\pi}(x) = -\frac{6N_c}{f_\pi\mu_\pi}\int\frac{d^2\mathbf{k}_\perp}{8\pi^3}\int_0^x dx'\frac{(2x'-1)}{x'(1-x')}\frac{\chi(x',\mathbf{k}_\perp)}{(1-x')}(\mathbf{k}^2_\perp + m^2_q), \tag{5}$$

where $\chi(x,\mathbf{k}_\perp) = \frac{g}{[x(m^2_M - M^2_0)][x(m^2_M - M^2_\Lambda)]^2}$ and $M^2_{0(\Lambda)} = \frac{\mathbf{k}^2_\perp + m^2_q(\Lambda^2)}{x} + \frac{\mathbf{k}^2_\perp + m^2_{\bar{q}}}{(1-x)}$.

In the standard LFQM, the wave function of a ground state pseudoscalar meson as a $q\bar{q}$ bound state is given by $\Psi_{\lambda\bar{\lambda}}(x,\mathbf{k}_\perp) = \Phi_R(x,\mathbf{k}_\perp)\mathcal{R}_{\lambda\bar{\lambda}}(x,\mathbf{k}_\perp)$, where Φ_R is the radial wave function and the spin-orbit wave

function $\mathcal{R}_{\lambda\bar{\lambda}}$ with the helicity $\lambda(\bar{\lambda})$ of a quark(antiquark) that is obtained by the interaction-independent Melosh transformation [23,24] from the ordinary spin-orbit wave function assigned by the quantum numbers J^{PC}. For the radial wave function Φ_R, we use both the gaussian or harmonic oscillator (HO) wave function $\Phi_{\rm HO}$ and the power-law (PL) type wave function $\Phi_{\rm PL}$ [25] as follows

$$\Phi_{\rm HO}(x,\mathbf{k}_\perp) = \frac{4\pi^{3/4}}{\beta^{3/2}}\sqrt{\frac{\partial k_z}{\partial x}}e^{-\mathbf{k}^2/2\beta^2},\ \Phi_{\rm PL}(x,\mathbf{k}_\perp) = \sqrt{\frac{128\pi}{\beta^3}}\sqrt{\frac{\partial k_z}{\partial x}}\frac{1}{(1+\mathbf{k}^2/\beta^2)^2}, \tag{6}$$

where $\mathbf{k}^2 = \mathbf{k}_\perp^2 + k_z^2$ and $k_z = (x-1/2)M_0 + (m_{\bar{q}}^2 - m_q^2)/2M_0$ and β is the variational parameter fixed by the analysis of meson mass spectra [14]. The normalization of Φ_R is given by $\int \frac{dxd^2\mathbf{k}_\perp}{16\pi^3}|\Phi_R(x,\mathbf{k}_\perp)|^2 = 1$.

In our previous analyses of twist-2 and pseudoscalar twist-3 DAs of a pion [11], we have shown that the SLF results in the standard LFQM is obtained by the the replacement of the LF vertex function χ in BS model with our LFQM wave function Φ_R as follows (see Eq. (35) in [11])

$$\sqrt{2N_c}\frac{\chi(x,\mathbf{k}_\perp)}{1-x} \to \frac{\Phi_R(x,\mathbf{k}_\perp)}{\sqrt{\mathbf{k}_\perp^2+m_q^2}}. \tag{7}$$

The correspondence in Eq. (7) is valid again in this analysis of a pseudotensor twist-3 DA $\phi_{3;\pi}^{\sigma}(x)$. We now apply this correspondence to both pseudoscalar DA $\phi_{3;\pi}^{P}(x)$ in Eq. (4) and pseudotensor DA $\phi_{3;\pi}^{\sigma}(x)$ in Eq. (5) to obtain them in our LFQM as follows:

$$\phi_{3;\pi}^{P}(x) = \frac{\sqrt{2N_c}}{f_\pi\mu_\pi}\int\frac{d^2\mathbf{k}_\perp}{16\pi^3}\frac{\Phi_R(x,\mathbf{k}_\perp)}{\sqrt{\mathbf{k}_\perp^2+m_q^2}}M_0^2, \tag{8}$$

$$\phi_{3;\pi}^{\sigma}(x) = \frac{6\sqrt{2N_c}}{f_\pi\mu_\pi}\int\frac{d^2\mathbf{k}_\perp}{16\pi^3}\int_0^x dx'\frac{(2x'-1)}{x'(1-x')}\frac{\Phi_R(x',\mathbf{k}_\perp)}{\sqrt{\mathbf{k}_\perp^2+m_q^2}}(\mathbf{k}_\perp^2+m_q^2), \tag{9}$$

respectively.

3 Numerical Results

In the numerical computations, we use $m_q = (0.22, 0.25)$ GeV and $\beta_{q\bar{q}} = (0.3659, 0.3194)$ GeV for the linear (HO) confining potential model parameters for the Gaussian wave function, which were obtained from the calculation of meson mass spectra using the variational principle in our LFQM [9,14]. For the PL wave function, we use $m_q = 0.25$ GeV and $\beta_{q\bar{q}} = 0.335$ GeV adopted from [25]. We first compute the second transverse moments $\langle\mathbf{k}_\perp^2\rangle_\pi^{P(\sigma)}$ for both pseudoscalar (P) and pseudotensor (σ) channels, and the results obtained from the linear [HO] parameters are given by $\langle\mathbf{k}_\perp^2\rangle_\pi^P = (553\ \text{MeV})^2[(480\ \text{MeV})^2]$ and $\langle\mathbf{k}_\perp^2\rangle_\pi^\sigma = (481\ \text{MeV})^2[(394\ \text{MeV})^2]$, respectively.

Figure 1 shows the two-particle twist-3 pion $\phi_{3;\pi}^{P}(x)$ (left panel) and $\phi_{3;\pi}^{\sigma}(x)$ (right panel) obtained from the nonzero constituent quark masses using Gaussian wave functions with HO (solid lines) model parameters and PL wave functions (dashed lines). We also plot our results in the chiral symmetry ($m_{u(d)} \to 0$) limit for both Gaussian (dotted lines) and PL (dot-dashed lines) wave functions and compare them with the chiral-limit prediction of DSE approach employing the dynamical chiral symmetry breaking-improved kernels [8] (double-dot-dashed line) as well as the asymptotic result $6x(1-x)$ (circled data) for the case $\phi_{3;\pi}^{\sigma}(x)$. Our results for both $\phi_{3;\pi}^{P}(x)$ and $\phi_{3;\pi}^{\sigma}(x)$ are normalized without the momentum cutoff (i.e. $|\mathbf{k}_\perp| \to \infty$).

For $\phi_{3;\pi}^{P}(x)$ case, our results with nonzero constituent quark masses show rather convex shapes for both Gaussian and PL wave functions but they show quite different end point behaviors, i.e. the end points are more enhanced for the PL wave function than the Gaussian wave function. The difference between the two wave functions are more drastic in the chiral symmetry limit, where the result of Gaussian wave function reproduces the result $\phi_{3;\pi}^{P}(x) \to 1$ anticipated from the QCD's conformal limit [2] but the result of PL wave function shows the concave shape similar to the result of DSE approach [8], in which the following asymptotic form

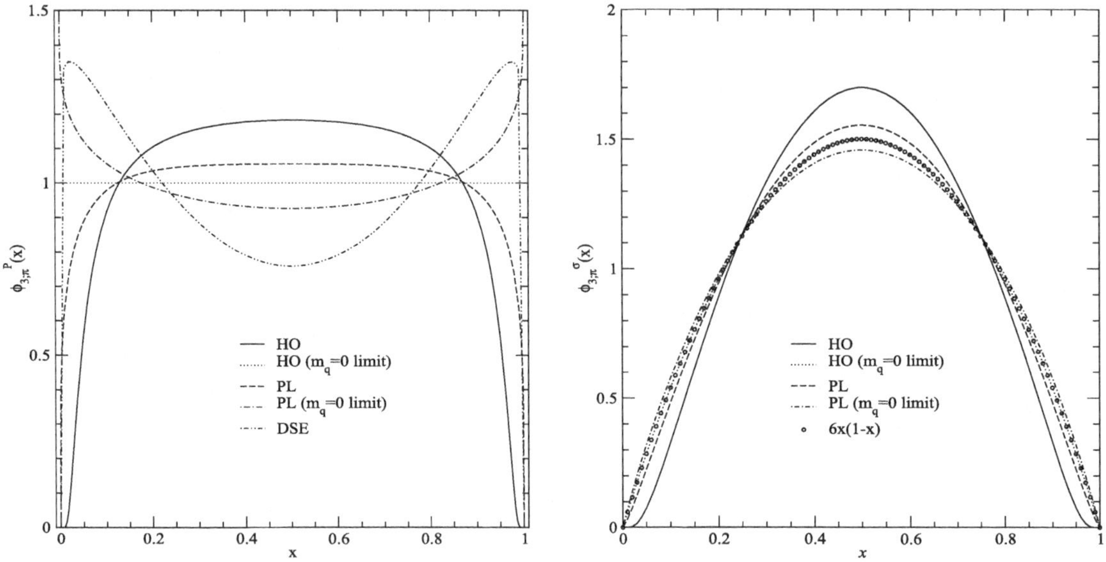

Fig. 1 The twist-3 DAs $\phi^P_{3;\pi}(x)$ (*left panel*) and $\phi^\sigma_{3;\pi}(x)$ (*right panel*) of pion

$\phi^P_{3;\pi}(x) \to 1 + (1/2)C_2^{(1/2)}(2x-1)$ was obtained. While the authors in [8] explained that the difference, i.e. $(1/2)C_2^{(1/2)}(2x-1)$ term in chiral symmetry limit may come from the mixing effect between the two- and three-particle twist-3 amplitudes, we observe that this difference may come from the cutoff scale of the transverse momentum scale associated with the different choice of LF wave functions. Especially, as on can see from Fig. 1, while the end-point suppressed form of $\phi^P_{3;\pi}(x)$ obtained from HO wave function shows constant shape (dotted line) in the chiral symmetry limit, the end-point enhanced form of $\phi^P_{3;\pi}(x)$ obtained from PL wave function shows concave shape (dot-dashed line) in the chiral symmetry limit. The cutoff dependent behaviors of $\phi^P_{3;\pi}(x)$ obtained from both Gaussian and PL wave functions are also presented in [12], where the concave shape for the Gaussian wave function can also be seen with the cutoff scale $\mu = 1$ GeV or less being taken but the cutoff dependence was shown to be more sensitive for the PL wave function than the Gaussian one.

For $\phi^\sigma_{3;\pi}(x)$ case, our results with nonzero constituent quark masses for both Gaussian (solid line) and PL (dashed line) show again different end point behaviors, i.e. the end points are more enhanced for the PL wave function than the Gaussian wave function. However, in the chiral symmetry limit, Gaussian (dotted line) and PL (dot-dashed line) wave functions show very similar shapes each other. Furthermore, the result from Gaussian wave function reproduces exactly the asymptotic form $6x(1-x)$. The same chiral-limit behavior was also obtained from the DSE approach [8]. As one can see from Fig. 1, the twist-3 pseudoscalar $\phi^P_{3;\pi}(x)$ is more sensitive to the shape of the model wave functions (Gaussian vs. PL) than the twist-3 pseudotensor $\phi^\sigma_{3;\pi}(x)$. It is quite interesting to note in the chiral symmetry limit that while $\phi^P_{3;\pi}(x)$ is sensitive to the shapes of model wave functions, $\phi^\sigma_{3;\pi}(x)$ is insensitive to them.

The twist-3 pseudoscalar DA $\phi^P_{3;M}(x)$ and pseudotensor DA $\phi^\sigma_{3;M}(x)$ are usually expanded in terms of the Gegenbauer polynomials $C_n^{1/2}$ and $C_n^{3/2}$, respectively, as follows [6]: $\phi^P_{3;M} = \sum_{n=0}^{\infty} a^P_{n,M} C_n^{1/2}(2x-1)$, and $\phi^\sigma_{3;M} = 6x(1-x)\sum_{n=0}^{\infty} a^\sigma_{n,M} C_n^{3/2}(2x-1)$. The coefficients $a^{P(\sigma)}_{n,M}$ are called the Gegenbauer moments, which describe how much the DAs deviate from the asymptotic one. In addition to the Gegenbauer moments, one can also define the expectation value of the longitudinal momentum, so-called ξ-moments, as follows: $\langle \xi^n \rangle^{P(\sigma)}_M = \int_0^1 dx \xi^n \phi^{P(\sigma)}_{3;M}(x)$. The Gegenbauer- and ξ- moments of the pseudoscalar twist-3 DAs $\phi^P_{3;\pi}(x)$ and twist-2 DA $\phi^A_{2;\pi}(x)$ can be found in our previous works [9,11].

In Table 1, we list the calculated Gegenbauer- and ξ-moments of pseudotensor twist-3 pion DA $\phi^\sigma_{3;\pi}(x)$ obtained from the Gaussian wave functions with linear and HO potential models and PL wave function. We also compare our results with other model predictions, e.g. QCD sum rules (SR) [3], DSE approach [8] and the chiral quark model (χQM) [6]. As expected from the isospin symmetry, all odd Gegenbauer and ξ moments are all zero. It is interesting to note that the sign of $a^\sigma_{2,\pi}$ is negative from our LFQM and χQM predictions but

Table 1 The Gegenbauer moments and ξ moments of twist-3 pion DAs obtained from the linear and HO potential models compared other model estimates

Models	$a^{\sigma}_{2,\pi}$	$a^{\sigma}_{4,\pi}$	$a^{\sigma}_{6,\pi}$	$\langle\xi^2\rangle^{\sigma}_{\pi}$	$\langle\xi^4\rangle^{\sigma}_{\pi}$	$\langle\xi^6\rangle^{\sigma}_{\pi}$
HO	−0.1155	−0.0268	−0.0046	0.1604	0.0565	0.0263
Linear	−0.0803	−0.0256	−0.0082	0.1725	0.0647	0.0318
PL	−0.0375	−0.0092	−0.0031	0.1871	0.0762	0.0406
SR [3]	0.0979	−0.0016	−0.0011	0.2325	0.1075	0.0624
DSE [8]	···	···	···	0.20	0.085	0.047
χQM [6]	−0.0984	−0.0192	−0.0037	0.1663	0.0612	−0.0015
$6x(1-x)$	···	···	···	0.20	0.086	0.048

Table 2 Asymptotic forms of various DAs of π and ρ

Model in $m_q \to 0$	$\phi^{P}_{3;\pi}$	$\phi^{\sigma}_{3;\pi}$	$\phi^{\parallel}_{2;\rho}(x)$	$\phi^{\perp}_{3;\rho}(x)$
LFQM (HO model) [10,11]	1	$6x(1-x)$	$6x(1-x)$	$\frac{3}{4}[1+(2x-1)^2]$
DSE approach [8]	$1+\frac{1}{2}C^{1/2}2(2x-1)$	$6x(1-x)$	–	–
QCD's conformal limit [2,21]	1	$6x(1-x)$	$6x(1-x)$	$\frac{3}{4}[1+(2x-1)^2]$

is positive for QCDSR prediction. Larger positive value of $a^{\sigma}_{2,\pi}$ leads to more flat shape of DA but the larger negative value leads to more narrower shape of DA. Knowing our LFQM results from the HO model are exact to the asymptotic result in the chiral-symmetry limit, i.e. $[\langle\xi^2\rangle^{\sigma}_{\pi}]_{\rm HO} = [\langle\xi^2\rangle^{\sigma}_{\pi}]_{\rm asy}$ in $m_q \to 0$ limit, one can see that the ξ-moments are reduced when the chiral symmetry is broken. We also should note for the same reason that our LFQM results are in good agreement with DSE results in the chiral-symmetry limit of $\phi^{\sigma}_{3;\pi}(x)$.

4 Summary

We analyzed the two twist-3 DAs of pion, i.e. pseudoscalar $\phi^{P}_{3;\pi}(x)$ and pseudotensor $\phi^{\sigma}_{3;\pi}(x)$, within the LFQM. We also investigated the discrepancy of the asymptotic forms of $\phi^{P}_{3;\pi}(x)$ between DSE approach [8] and QCD's conformal limit expression [2] from the perspective of dependence of DA on the form of LF wave functions, e.g. Gaussian vs. PL wave functions. In order to compute the twist-3 pseudotensor DA $\phi^{\sigma}_{3;\pi}(x)$, we utilized the same manifestly covariant BS model used in [10–12] and then substituted the LF vertex function in the covariant BS model with the more phenomenologically accessible Gaussian and PL wave functions. Linking the covariant BS model to the standard LFQM, we used the same correspondence relation Eq. (7) between the two models as the one found in [10,11]. The remarkable finding in linking the covariant BS model to the standard LFQM is that the treacherous points such as the zero-mode contributions and the instantaneous ones existed in the covariant BS model become absent in the LFQM with the Gaussian or PL wave function.

Our LFQM descriptions for both twist-3 $\phi^{P}_{3;\pi}$ and $\phi^{\sigma}_{3;\pi}$ satisfy the fundamental constraint(i.e. symmetric form with respect to x) anticipated from the isospin symmetry. For the $\phi^{P}_{3;\pi}(x)$ case, our results with nonzero constituent quark masses show rather convex shapes for both Gaussian and PL wave functions but they show quite different end point behaviors, i.e. the end points are more enhanced for the PL wave function than the Gaussian wave function. The difference between the two wave functions are more drastic in the chiral symmetry limit, where the result of Gaussian wave function reproduces the result $\phi^{P}_{3;\pi}(x) \to 1$ anticipated from the QCD's conformal limit [2] but the result of PL wave function shows the concave shape similar to the result of DSE approach [8]. For the $\phi^{\sigma}_{3;\pi}(x)$ case, our results from both Gaussian and PL wave functions in the chiral symmetry limit, show very similar shapes each other. Especially, the result from Gaussian wave function reproduces exactly the asymptotic form $6x(1-x)$ anticipated from QCD's conformal limit. The same chiral-limit behavior was also obtained from the DSE approach [8]. The remarkable thing is that our predictions for the two twist-3 DAs of π and chirality-even twist-2 DA $\phi^{\parallel}_{2;\rho}(x)$ and twist-3 DA $\phi^{\perp}_{3;\rho}(x)$ of ρ [10] obtained from the Gaussian wave function in the chiral limit exactly reproduce the forms anticipated from QCD's conformal limit. We summarize in Table 2 the asymptotic forms for the two twist-3 DAs of π and chirality-even twist-2 and twist-3 DAs of rho meson compared with DSE approach [8] and QCD's conformal limit expression [2,21].

The idea of our LFQM is to provide the nonperturbative wave functions at the momentum scale consistent with the use of constituent quark mass. The DAs determined from this nonperturbative wave functions can be fed into the QCD evolution equation to provide the shorter distance information of the corresponding hadrons. While our results for both $\phi^P_{3;\pi}$ and $\phi^\sigma_{3;\pi}$ in Fig. 1 were obtained without the momentum cutoffs (i.e. $|\mathbf{k}_\perp| \to \infty$), the scales of both DAs obtained from the Gaussian wave functions and that of $\phi^\sigma_{3;\pi}$ obtained from PL wave function are estimated to be $|\mathbf{k}_\perp| \simeq 1$ GeV. However, the scale of $\phi^P_{3;\pi}$ obtained from PL wave function are estimated to $|\mathbf{k}_\perp| \simeq 2$ GeV due to the high momentum tail compared to other wave functions. The DAs obtained without the cutoff should not be regarded as the fully evolved DAs but still be nonperturbative as they just mean that the cutoff dependence becomes marginal beyond a certain nonperturbative cutoff scale.

References

1. G.P. Lepage, S.J. Brodsky, Exclusive processes in perturbative quantum chromodynamics. Phys. Rev. D. **22**, 2157 (1980)
2. V.M. Braun, I.E. Filyanov, Conformal invariance and pion wave functions of nonleading twist. Z. Phys. C **48**, 239 (1990)
3. P. Ball, V.M. Braun, A. Lenz, Higher-twist distribution amplitudes of the K meson in QCD. J. High Energy Phys. **05**, 004 (2006)
4. S.V. Mikhailov, A.V. Pimikov, N.G. Stefanis, Endpoint behavior of the pion distribution amplitude in QCD sum rules with nonlocal condensates. Phys. Rev. D. **82**, 054020 (2010)
5. V.Y. Petrov, M.V. Polyakov, R. Ruskov, C. Weiss, K. Goeke, Pion and photon light-cone wave functions from the instanton vacuum. Phys. Rev. D. **59**, 114018 (2006)
6. S.I. Nam, H-Ch. Kim, Twist-3 pion and kaon distribution amplitudes from the instanton vacuum with flavor SU(3) symmetry breaking. Phys. Rev. D. **74**, 096007 (2006)
7. L. Chang, C.D. Roberts, S.M. Schmidt, Light front distribution of the chiral condensate. Phys. Lett. B. **727**, 255 (2013)
8. C. Shi, C. Chen, L. Chang, C.D. Roberts, S.M. Schmidt, H.-S. Zong, Kaon and pion parton distribution amplitudes to twist three. Phys. Rev. D. **92**, 014035 (2015)
9. H.-M. Choi, C.-R. Ji, Distribution amplitudes and decay constants for (π, K, ρ, K^*) mesons in the light-front quark model. Phys. Rev. D. **75**, 034019 (2007)
10. H.-M. Choi, C.-R. Ji, Self-consistent covariant description of vector meson decay constants and chirality-even quark-antiquark distribution amplitudes up to twist 3 in the light-front quark model. Phys. Rev. D. **89**, 033011 (2014)
11. H.-M. Choi, C.-R. Ji, Consistency of the light-front quark model with chiral symmetry in the pseudoscalar meson analysis. Phys. Rev. D. **91**, 014018 (2015)
12. H.-M. Choi, C.-R. Ji, Light-front quark model phenomenology consistent with chiral symmetry. EPJ Web of Conference **113**, 05009 (2016)
13. S.J. Brodsky, H.-C. Pauli, S. Pinsky, Quantum chromodynamics and other field theories on the light cone. Phys. Rep. **301**, 299 (1998)
14. H.-M. Choi, C.-R. Ji, Mixing angles and electromagnetic properties of ground state pseudoscalar and vector meson nonets in the light-cone quark model. Phys. Rev. D. **59**, 074015 (1999)
15. H.-M. Choi, Decay constants and radiative decays of heavy mesons in light-front quark model. Phys. Rev. D. **75**, 073016 (2007)
16. H.-M. Choi, C.-R. Ji, Semileptonic and radiative decays of the B(c) meson in light-front quark model. Phys. Rev. D. **80**, 054016 (2009)
17. H.-M. Choi, C.-R. Ji, L.S. Kisslinger, Skewed quark distribution of the pion in the light front quark model. Phys. Rev. D. **64**, 093006 (2001)
18. S.J. Brodsky, D.S. Hwang, Exact light cone wave function representation of matrix elements of electroweak currents. Nucl. Phys. B. **543**, 239 (1998)
19. J.P.B.C. de Melo, J.H.O. Sales, T. Frederico, P.U. Sauer, Pairs in the light front and covariance. Nucl. Phys. A. **631**, 574c (1998)
20. H.-M. Choi, C.-R. Ji, Nonvanishing zero modes in the light front current. Phys. Rev. D. **58**, 071901(R) (1998)
21. P. Ball, V.M. Braun, Y. Koike, K. Tanaka, Higher twist distribution amplitudes of vector mesons in QCD: formalism and twist–three distributions. Nucl. Phys. B. **529**, 323 (1998)
22. M. Gell-Mann, R. Oakes, B. Renner, Behavior of current divergences under SU(3) $\times$ SU(3). Phys. Rev. **175**, 2195 (1968)
23. H.J. Melosh, Quarks: currents and constituents. Phys. Rev. D. **9**, 1095 (1974)
24. P.L. Chung, F. Coester, B.D. Keister, W.N. Polyzou, Hamiltonian light front dynamics of elastic electron deuteron scattering. Phys. Rev. D. **37**, 2000 (1988)
25. F. Schlumpf, Charge form factors of pseudoscalar mesons. Phys. Rev. D. **50**, 6895 (1994)

Few-Body Syst _#####################_
DOI 10.1007/s00601-017-1298-y

Cristiano Fanelli

The GLUEX Experiment: First Results

Received: 22 January 2017 / Accepted: 10 April 2017

Abstract GLUEX is a nuclear physics experiment located at the Thomas Jefferson National Accelerator Facility designed to study and understand the nature of confinement in QCD by mapping the spectrum of exotic mesons. The experiment will be able to probe new areas by using photoproduction, i.e. the scattering on nucleon of ~9 GeV linearly polarized photons derived from the recently upgraded CEBAF with a 12 GeV electron beam. Spring 2016 has been characterized by a continued detector commissioning and initial running at the full design energy. The current status of the GLUEX detector performance and data collection will be discussed, with a brief overview of first physics results, future run plans, and long term upgrades.

1 Introduction

The primary goal of the GLUEX experiment is to understand the nature of confinement in Quantum Chromodynamics (QCD) by studying the light meson spectrum and the properties of hybrid mesons for which the gluonic field would directly contribute to the J^{PC} quantum numbers [1].

Lattice QCD calculations suggest that several of the nonets of these hybrid mesons have exotic quantum numbers, corresponding to J^{PC} values forbidden for a simple fermion-antifermion system [2]. The expected masses for the lightest hybrids are suited to the energy and kinematics accessible to the GLUEX experiment, which is located in Hall D at the end of a new beamline from the Continuous Electron Beam Accelerator Facility (CEBAF) at Jefferson Lab (JLab), using 12 GeV electrons to deliver linearly-polarized photons. Here the photon beam impinges on a hydrogen target contained within a detector characterized by a near-complete charged and neutral particle coverage. In spring 2016, GLUEX has completed its commissioning and taken its first substantial data in the design configuration.

In this document we present some of the first results of the GLUEX experiment with 2016 data as well as an overview of the future physics program which will benefit from the next upgrade of the particle identification (PID) system with the addition of the DIRC detector.

2 The Beamline

The Hall D takes advantage of the 12 GeV electron beam from the upgraded CEBAF. The electrons pass through a bremsstrahlung radiator and then into the dipole magnet for tagging the energy of the scattered electrons (see

This article belongs to the Topical Collection "Light Cone 2016".

On behalf of the GLUEX Collaboration.

C. Fanelli (✉)
Laboratory for Nuclear Science, Massachusetts Institute of Technology, 77 Massachusetts Ave, Cambridge, MA 02139, USA
E-mail: cfanelli@mit.edu

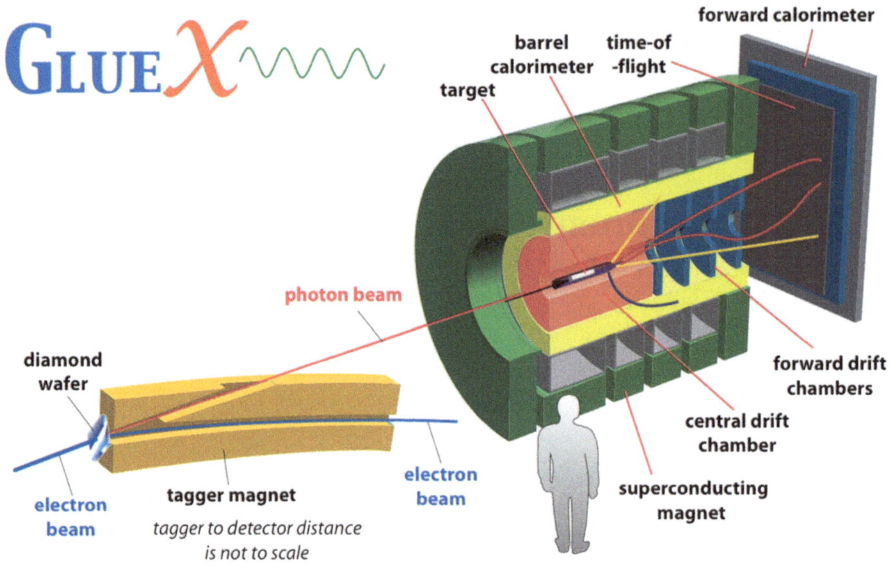

Fig. 1 A sketch of the photon tagger and the GLUEX detector at JLab

Fig. 1 on the left). For the electrons producing a bremsstrahlung photon in the region between 25 and 98% of the primary electron beam energy, the scattered electrons are detected in a pair of tagger hodoscopes, hence tagging the energy of the photon. For 12 GeV/c electrons, the fine-grained Tagger Microscope (TAGM) has been placed to tag photons in the range between 8.2 and 9.2 GeV, with each microscope element spanning 10 MeV in energy. For both higher and lower energy photons, the Tagger Hodoscope (TAGH) tags photons using elements about 30 MeV wide.

A 20 μm diamond crystal is utilized as a radiator to produce a linearly polarized coherent bremsstrahlung photon beam. The photons in the coherent peak range from 8.4 to 9 GeV and are linearly polarized relative to the crystal axes in the diamond with an expected polarization of ~40%. The diamond can be rotated to produce two perpendicular directions of the polarization. The photons travel through an 80 m long vacuum beamline before entering Hall D. Since the coherent bremsstrahlung photons are produced in the direction of the incident electrons, while the incoherent bremsstrahlung photons are produced within a cone around this direction, the off-axis photons are removed by passing the beam through a 3.4 mm diameter collimator. The photons pass then through a *Triplet Polarimeter* which is used to monitor the linear polarization, an information needed in the measurements of the beam asymmetry as explained in the following. Successively they enter a *Pair Spectrometer* system [3] that is used to monitor both the energy and intensity of the beam. Finally the photons reach the GLUEX spectrometer to impinge on the liquid hydrogen target to produce the physics events that will be detected. The initial operation of GLUEX was with beam intensity of 10^7 γ/s in the coherent peak. The ultimate rate is expected to be an order of magnitude higher.

3 The GLUEX Detector and PID Upgrade

The GLUEX detector is represented schematically on the right part of Fig. 1. It is characterized by a near uniform acceptance for both charged and neutral particles in the polar angle range from 1° to 120°. The 30 cm long liquid hydrogen target is located in the barrel part of the detector. The solenoid magnet provides a magnetic field of about 2 T along the direction of the beam. Particles produced in the interaction are first detected by a scintillator-based *Start Counter* surrounding the target and the beamline in the forward direction. The Central Drift Chamber (CDC) [4] based on straw-tube technology immediately surrounds the start counter and provides the position measurement along the charged tracks with 150 μm accuracy in the $r - \phi$ plane and at the mm level along the beam axis. It also provides a measurement of dE/dx for charged tracks which allows to separate π/p up to a momentum of 1 GeV/c. The Forward Drift Chamber (FDC) [5] complements the tracking system in the forward region and performs PID by measuring dE/dx. The Barrel Calorimeter (BCAL) [6,7] surrounds the tracking devices inside the solenoid and covers the polar angle

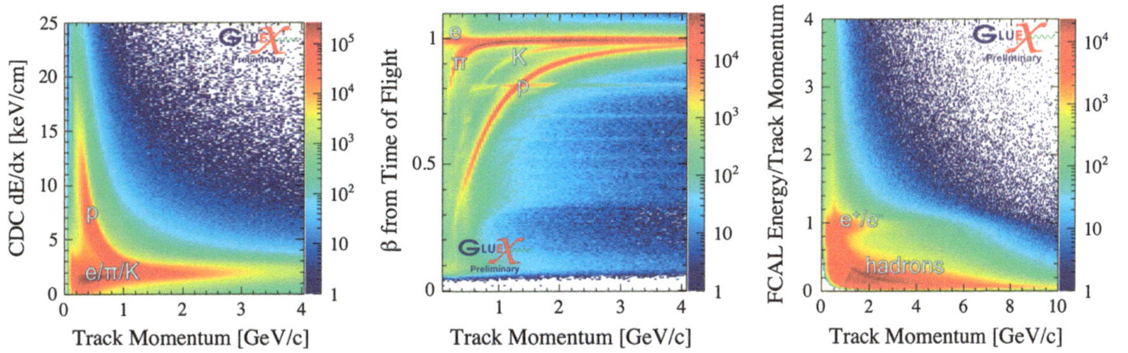

Fig. 2 (*left*) Energy loss dE/dx in the CDC (*middle*) β as determined from TOF and (*right*) the energy measured by the FCAL divided by the momentum of the associated track. All these quantities are shown as a function of the reconstructed momentum of the track

acceptance between 11° and 120°. The energy resolution is $\sigma_E/E \approx 6\%/\sqrt{E} \oplus 2\%$. The timing information is used to provide time-of-flight for particles interacting in the BCAL contributing to the global PID for both charged particles and photons. The Forward Calorimeter (FCAL) [8] is located downstream of the solenoid magnet to cover a range in polar angles between 1° and 11°. FCAL is sensitive to photons with energy above 100 MeV and has a similar energy resolution to BCAL. The Time-Of-Flight (TOF) wall located upstream of FCAL comprises two layers of scintillator bars which are read out on both sides using photomultiplier tubes. The TOF contributes to the PID of forward going particles by providing the timing and energy-loss information, with a timing resolution of 100 ps. All of the detector systems are now performing near the design specifications.

An example of the PID capabilities are shown in Fig. 2 for CDC, TOF, and FCAL respectively, as a function of the particle momentum. The energy loss in CDC (Fig. 2 (left)) allows to separate the proton from the other positively charged particles up to 1 GeV/c. The TOF β (Fig. 2 (middle)) is characterized by clear bands for for π^+, K^+, e^+, and p, with a distinguishable proton up to 3 GeV/c and effective π/K separation up to about 2 GeV/c. Figure 2 (right) shows the ratio of the energy deposited in FCAL normalized to the associated track momentum as a function of the track momentum. This quantity allows to separate electrons/positrons ($E/p \sim 1$) from hadrons in the FCAL.

Different types of measurements are planned to take advantage of the potential of the GLUEX detector. The physics program began in 2016, the first data being already sufficient to measure the beam asymmetry for several photo-produced reactions (e.g. $\gamma p \rightarrow pX, X = \pi, \eta, \eta', \rho, \omega, \phi$), as reported briefly in the following section. The physics data from 2017 and 2018 (planned at low intensity with 10 MHz in the peak) will allow to measure cross sections, spin-density matrix elements, as well as the start of the exotic hybrids search and the partial wave analysis program. The GLUEX strangeness program complements the mapping of the hybrid spectrum by identifying final states with kaons. In particular, the quark flavor composition of a meson can be experimentally inferred by comparing the branching ratios for strange and non-strange decays. A systematic study of the kaon final states is planned for 2018 after the upgrade of the PID system.

A new Detection of Internally Reflected Cherenkov (DIRC) light detector [9] will be installed in the forward region of the spectrometer. After the installation is completed, GLUEX will run in a high intensity mode (50 MHz in the peak) focusing on the identification of states with hidden strangeness and hyperon resonances. The four existing BaBar DIRC [10] bar boxes filled with long and narrow radiators ($1.75 \times 3.5 \times 490\,\text{cm}^3$) will be reused in the GLUEX DIRC without any modifications. Each pair of boxes will be attached to a new compact photon camera, inspired by [11]. Further details on the DIRC design can be found in [12,13]. The GLUEX DIRC is located about 4 m away from the target and forms a thin wall of fused silica radiators attached to photon cameras read-out by an array of multi-anode photomultipliers (MaPMT).

The DIRC will cover the polar angle range from 2° to 11° and requires four unaltered BaBar bar boxes each made of 12 radiators. The bar boxes will be oriented horizontally and installed upstream of the TOF detector, with two bar boxes above and two below the beam and each pair attached to a photon camera (one of which located to the left, the other to the right of the beam respectively). The construction has been started, and the calibration system is being studied. The GLUEX DIRC will provide a clean separation between π/K of at least 3σ for momenta less than 4 GeV/c.

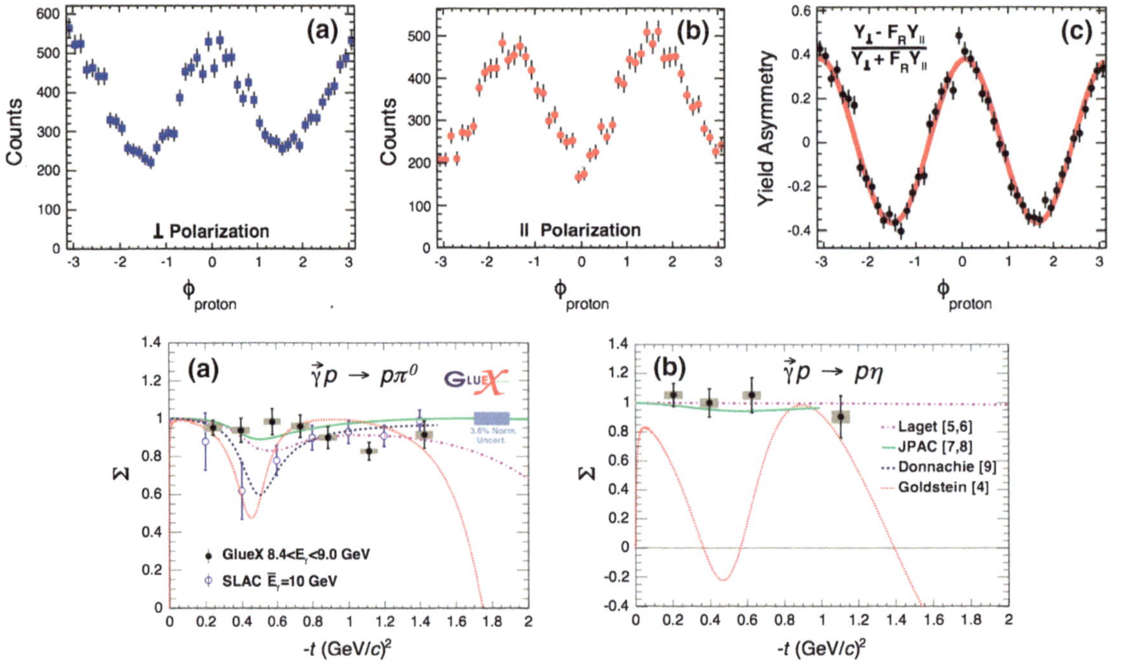

Fig. 3 *(top)* $\gamma p \to \pi^0 p$ yield versus ϕ_p integrated over $-t$ for **a** PERP and **b** PARA. **c** The yield asymmetry to extract the beam asymmetry Σ. *(bottom)* **a** Σ for $\gamma p \to \pi^0 p$ and **b** $\gamma p \to \eta p$. See for more details [15]

4 Preliminary Results

The primary goal of the GLUEX experiment is to search for and ultimately map out the spectrum of light quark hybrid mesons using partial wave analysis. In the next years the beam intensity will be increased to cover all parts of the GLUEX exotic hybrid program. An extensive research program complements and enhances these physics goals. This ranges from studies of photo-produced resonances to physics beyond the Standard Model. The first data collected in Spring 2016 allowed for several physics opportunities (most of which shown at DNP, Vancouver 2016). Part of these topics is briefly reported in what follows.

4.1 Early Spectroscopy and Asymmetries

The beam asymmetry Σ provides insight into the dominant production mechanism [14], whose understanding is crucial to disentangle the J^{PC} of observed states in the exotic hybrid search. GLUEX has measured the photon beam asymmetry for the reactions $\gamma p \to \pi^0 p$ and $\gamma p \to \eta p$ [15] using the 9 GeV linearly-polarized, tagged photon beam incident on a liquid hydrogen target. It's worth mentioning that the asymmetries, measured as a function of the proton momentum transfer, possess greater precision than previous π^0 measurements, and are the first η measurements in this energy regime. The photon beam linear polarization is measured using the Triplet Polarimeter, with the azimuthal angle of the beam photon's linear polarization plane in two configurations, PARA ($\phi_\gamma{}^{lin} = 0°$) and PERP ($\phi_\gamma{}^{lin} = 90°$). Σ is extracted as the only free parameter of the fit of the yield asymmetry defined by combining the two perpendicular configurations as a function of the azimuthal angle of the production plane defined by the final state proton, ϕ_p, as shown in Fig. 3. The results on the asymmetry are compared to existing theoretical predictions based on t-channel, quasi-particle exchange [16–18], and are thus expected to contribute to our comprehension of the production mechanism in photo-production at high energy.

4.2 Transition Form Factors

Radiative decays of mesons are powerful probes of the hadron structure. The GLUEX experiment will produce and record huge samples of light mesons, opening the possibility to study their radiative decays and to measure the transition form factors of the η and η' mesons [19].

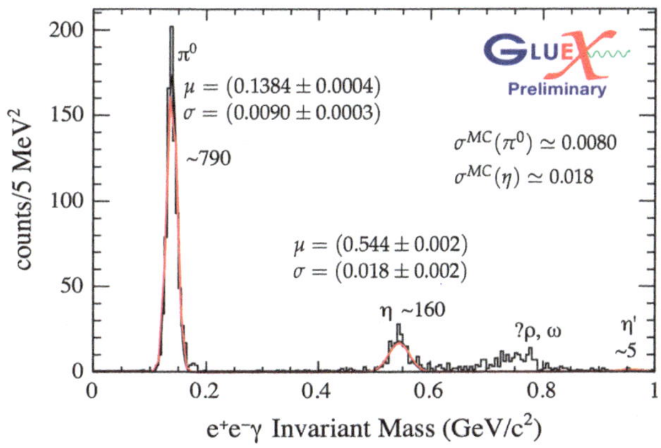

Fig. 4 M($e^+e^-\gamma$) invariant mass from the process γp$\rightarrow$pe$^+e^-\gamma$ with part of the data collected in 2016 at GLUEX, corresponding approximately to 1% of the total integrated luminosity that will be collected in the next years (cf. 'Phase IV')

The TFF's are important in studies of the muon anomalous magnetic moment, $a_\mu = (g_\mu - 2)/2$, since the hadronic light-by-light (HLbL) scattering contribution to a_μ includes two meson-photon-photon vertices that can be related to the form factors in $P \rightarrow \gamma\gamma^* \rightarrow \gamma e^+e^-$ decays [20]. Models describing these transitions should be tested as precisely as possible to reduce the uncertainty in the SM prediction for a_μ. The low Q^2 slope of the TFF provides a unique method to measure the pseudo-scalar meson radius, whose physical interpretation has still not completely been explored by the theory. A recent study [21] including first results with real data suggests that GLUEX will be competitive in the next years with the current experimental results of the $\eta(')$ TFF's measurements [22,23]. A preliminary plot of the invariant mass distribution in the $e^+e^-\gamma$ channel is shown in Fig. 4.

4.3 Physics Beyond the Standard Model

Dark Matter (DM) dominates the matter density in the universe, but very little is known about it. Its existence and stability provides a strong hint that there may be a dark sector, which consists of a rich symmetry structure with new forces and new particles not interacting by the known forces (with the exception of gravity). Consequently, the discovery of these particles would have a profound impact in our understanding of the universe. There are only a few well-motivated interactions that provide a "portal" from the SM sector into the dark sector [24], among these the dark vector portal has inspired a global effort at high intensity frontier centers at intermediate energies like Jefferson Lab. A compelling dark-force scenario consist of promoting the baryon number to a gauge symmetry $U(1)_B$, in this way extending the SM to produce new bosons with direct coupling to quarks and not to leptons (viz. "leptophobic B bosons").

If $U(1)_B$ is broken, B is a massive vector boson with quark couplings. This B boson has $I^G(J^{PC}) = (0^-)1^{--}$, similarly to ω and ϕ. Recently the direct photoproduction of leptophobic bosons that couple to quarks has been calculated [25] and it has been estimated that $\gamma p \rightarrow pB$ will provide the best sensitivity for B masses above 0.5 GeV, as shown in Fig. 5. GLUEX will discover or tightly constrain the existence of a directly photo-produced B boson in this mass range. Preliminary results based on 2016 data and scaled to the higher integrated luminosity of GLUEX Phase IV are consistent with these predictions [27].

5 Summary

The GLUEX experiment has been successfully commissioned, and the first analyses based on the available data are in progress. This paper in Sect. 4 reports on some of the first physics results obtained: (i) early spectroscopy and the beam asymmetry measurements [15], (ii) transition form factors with radiative decay studies showing encouraging results and (iii) beyond Standard Model searches with scaled preliminary results consistent with predictions.

The first stage of GLUEX running at low intensity (10 MHz in the peak) is dedicated mainly to measurements of cross-sections, asymmetries, and spin-density matrix elements, along with the identification of known

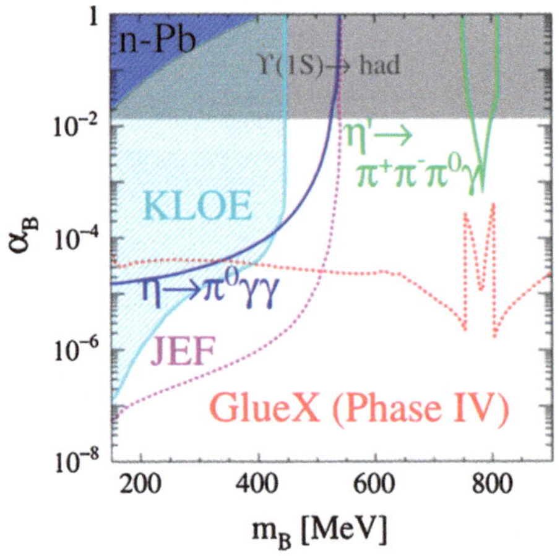

Fig. 5 Existing limits on B bosons compared to the expected sensitivity obtained from η decays at JEF [26] and photoproduction [25]

resonances and hybrids using partial wave analysis. After 2018 it is planned to increase the beam intensity by a factor of 5 to cover all parts of the GLUEX exotic hybrid program and upgrade the PID system by installing the DIRC detector.

Acknowledgements Funding was provided by U.S. Department of Energy (Grant No. DE-SC0010497).

References

1. C.A. Meyer, E.S. Swanson, Prog. Part Nucl. Phys. **82**, 21–58 (2015)
2. J.J. Dudek et al., Phys. Rev. D **88**(9), 094505 (2013). arXiv:1309.2608 [hep-lat]
3. F. Barbosa et al., Nucl. Instrum. Methods A **795**, 376 (2015)
4. Y. Van Haarlem et al., Nucl. Instrum. Methods A **622**, 142 (2010). arXiv:1004.3796 [nucl-ex]
5. V.V. Berdnikov et al., Instrum. Exp. Tech. **58**, 25 (2015)
6. Z. Papandreou et al., Nucl. Instrum. Methods A **596**, 338 (2008)
7. A.E. Baulin et al., Nucl. Instrum. Methods A **715**, 48 (2013)
8. K. Moriya et al., Nucl. Instrum. Methods A **726**, 60 (2013)
9. J. Stevens et al., JINST **11**(07), C07010 (2016). arXiv:1606.05645 [physics.ins-det]
10. I. Adam et al., Nucl. Instrum. Methods A **538**, 281 (2005)
11. B. Dey et al., Nucl. Instrum. Methods A **775**, 112 (2015). arXiv:1410.0075 [physics.ins-det]
12. GlueX Collaboration, GlueX DIRC Technical Design Report (2015). https://halldweb.jlab.org/DocDB/0028/002809/003/dirc_tdr.pdf
13. J. Hardin, M. Williams, JINST **11**(10), P10007 (2016)
14. I.S. Barker, A. Donnachie, J.K. Storrow, Nucl. Phys. B **95**, 347 (1975)
15. H. Al Ghoul et al., (GlueX Collaboration), arXiv:1701.08123 (2017)
16. J.M. Laget, Phys. Lett. B **695**, 199 (2011)
17. V. Mathieu, G. Fox, A.P. Szczepaniak, Phys. Rev. D **92**, 074013 (2015)
18. A. Donnachie, YuS Kalashnikova, Phys. Rev. C **93**, 025203 (2016)
19. A.M. Bernstein, PoS CD15 (2016) 046 SISSA (2016-02-26)
20. T. Blum et al., The Muon (g-2) Theory value: present and future (2013). arXiv:1311.2198
21. C. Fanelli, M. Williams, (GlueX Collaboration), Study of the $\eta(') \to e^+e^-\gamma$ decay at GlueX and transition form factors. APS Division of Nuclear Physics Vancouver Meeting (2016)
22. P. Aguar-Bartolome et al., Phys. Rev. C **89**, 044608 (2014)
23. M. Ablikim et al., (BESIII Collaboration), Phys. Rev. D **92**, 012001 (2015)
24. R. Essig et al., Dark sectors and new, light, weakly-coupled particles (2013). arXiv:1311.0029
25. C. Fanelli, M. Williams, J. Phys. G **44**(1), 014002 (2016)
26. The GlueX Collaboration and Other Participants, JLab proposal PR12-14-004 (2014). https://www.jlab.org/exp_prog/proposals/14/PR12-14-004.pdf
27. J. Hardin, C. Fanelli, M. Williams, (GlueX Collaboration), Leptophobic Boson Searches. APS Division of Nuclear Physics Vancouver Meeting (2016)

Few-Body Syst (2017) 58:122
DOI 10.1007/s00601-017-1278-2

Christopher E. Thomas

Meson Spectroscopy from Lattice QCD

Received: 8 December 2016 / Accepted: 7 March 2017 / Published online: 24 March 2017

Abstract Some recent progress in using lattice QCD to perform first-principles calculations of the spectra of mesons is discussed. In particular, I highlight some new results on resonances, near-threshold states and related scattering phenomena—this is a theoretically and experimentally interesting area where we have made significant advances in the last few years.

1 Introduction

There has been a resurgence of interest in hadron spectroscopy in the last decade or so driven by the emergence of a wealth of high-quality experimental data. In particular, a number of unexpected states have been observed, e.g. various charmonium and bottomonium-like structures ("X, Y, Z's") and the charm-strange $D_{s0}(2317)$ [49,53], and there are also longer-standing puzzles such as the makeup of the light scalar mesons [55,57]. There has been a lot of theoretical work and speculation about their nature [6,50]—possibilities include tetraquarks (containing two quarks and two antiquarks), molecular states of hadrons, hadro-quarkonia and hybrid mesons (where the gluonic field is excited). States with exotic quantum numbers, i.e. those which cannot arise from solely a quark-antiquark pair, are particularly interesting because they are a smoking gun for physics beyond a model of a quark and an antiquark moving in a potential. For example, exotic spin (J), parity (P), charge-conjugation (C) combinations (e.g. $J^{PC} = 0^{--}, 0^{+-}, 1^{-+}, 2^{+-}$) or exotic flavour states (such as charged charmonium and bottomonium-like states).

Lattice Quantum Chromodynamics (Lattice QCD) is a method which enables first-principles computations of the spectra and properties of hadrons: four-dimensional space-time is discretised on a finite four-dimensional hypercubic lattice and the calculation of quantities in the path integral formulation then becomes an ordinary (but very large) integration problem. If a Euclidean (imaginary-time) space-time metric is used, the integrals can be evaluated effectively using importance-sampling Monte Carlo methods. The masses and other properties of hadrons then follow from analyzing correlation functions involving interpolating operators built from quark and gluon fields. Precise calculations of the spectrum of low-lying hadrons have long been benchmarks of lattice QCD (e.g. see Refs. [17,23]) but only relatively recently has there been significant progress in using lattice QCD to study excited hadrons.

Most hadrons are unstable with respect to the strong interaction and appear as resonances in the scattering of two or more lighter hadrons. In particular, many of the recently-observed puzzling states appear close to or above the threshold for strong decay—this must be taken into account in theoretical approaches and when analyzing experimental data. In recent years there have been significant advances in using lattice QCD to study resonances, near-threshold states and related scattering phenomena and I will discuss some highlights of this

This article belongs to the Topical Collection "Light Cone 2016".

C. E. Thomas (✉)
DAMTP, University of Cambridge, Wilberforce Road, Cambridge CB3 0WA, UK
E-mail: c.e.thomas@damtp.cam.ac.uk

work in these proceedings. These calculations often use unphysically-heavy light quarks, corresponding to a pion mass, m_π, larger than the experiment pion mass and do not usually have a precise control over all the systematic uncertainties—details will not be given here and I refer to the references for further information.

After a brief description of some of the lattice QCD methodology in Sect. 2, in Sect. 3 I discuss calculations of the ρ resonance in $\pi\pi$ scattering. Some other channels relevant for light and strange mesons are discussed in Sect. 4 followed by heavy-light and charmonium-like resonances in Sect. 5. I conclude in Sect. 6 with an outlook.[1]

2 Excited Spectroscopy and Scattering in Lattice QCD

In lattice QCD the discrete spectrum of states in a finite volume follows from analyzing the dependence on the time separation, t, of two-point correlation functions,

$$C_{ij}(t) = \langle 0|\mathcal{O}_i(t)\mathcal{O}_j^\dagger(0)|0\rangle \, .$$

Here the interpolating operator $\mathcal{O}_j^\dagger(0)$ creates the states of interest at time 0 and $\mathcal{O}_i(t)$ annihilates them at time t—all states with the quantum numbers of the operators contribute to the correlation function. The variational method [5,43,44] enables excited states to be extracted reliably: a matrix of correlation functions involving a basis of N operators with the required quantum numbers is computed, $C_{ij}(t)$, where $i, j = 1, 2, \ldots, N$. A generalised eigenvalue problem,

$$C_{ij}(t)v_j^{(n)} = \lambda^{(n)}(t)C_{ij}(t_0)v_j^{(n)} \, ,$$

is then solved for an appropriate reference time t_0. The time dependence of an eigenvalue, $\lambda^{(n)}(t)$, is related to the energy of the n'th state. The eigenvectors, $v_i^{(n)}$, are related to the operator-state overlaps, $Z_i^{(n)} \equiv \langle 0\,|\mathcal{O}_i|\,n\rangle$, which can be used to probe the structure of states and also give the optimal (in a variational sense) linear combination of operators to create the n'th state. To accurately determine the spectrum, the basis of operators must have a sufficiently wide range of structures to disentangle the various states that appear.

Although it is not possible to compute scattering properties directly in the Euclidean formulation of lattice QCD, the Lüscher method [40–42] and its extensions [7,8,16,25,27,32,38,39,62] allow indirect access to *infinite-volume* scattering amplitudes from *finite-volume* spectra, at least in the case of any number of two-hadron coupled channels. For elastic scattering, there is a one-to-one correspondence[2] between an energy level with centre-of-mass energy E_{cm} and the scattering phase shift δ at E_{cm} or equivalently the scattering t-matrix, $t(E_{\mathrm{cm}})$. For coupled-channel two-hadron scattering, extracting an energy level E_{cm} constrains the scattering t-matrix at E_{cm} but this is in general an under-constrained problem. For example, with two coupled channels there are three energy-dependent parameters but only one constraint at each E_{cm}. One solution is to parameterize the E_{cm}-dependence of the t-matrix with a relatively small number of parameters—the spectrum resulting from this t-matrix is fit to the computed spectrum by varying these parameters. With the t-matrix in hand, its singularity structure can be investigated to determine the bound state and resonant content. The former corresponds to a pole below threshold on the real axis of the physical Riemann sheet, whereas the latter corresponds to poles away from the real axis on unphysical sheets.

3 The ρ Resonance in $\pi\pi$ Scattering

One of the simplest resonances, and the one that has attracted the most initial attention from lattice QCD [1, 2,24,33,56], is the $\rho(770)$ which appears in $\pi\pi$ scattering in $L = 1$ (P-wave) with $J^{PC} = 1^{--}$ and isospin $I = 1$; empirically the ρ decays almost entirely to $\pi\pi$ [53]. The results of a recent lattice QCD calculation [65] of the elastic $I = 1$ $\pi\pi$ scattering phase shift by the Hadron Spectrum Collaboration (HSC) are shown in Fig. 1. The computations have dynamical strange and degenerate up and down quarks, "$N_f = 2 + 1$", corresponding to $m_\pi \approx 240$ MeV, and used 469 gauge field configurations; the spatial (temporal) lattice extent is $L \sim 4$ fm ($T \sim 9$ fm). The plot also shows results from an earlier calculation with $m_\pi \approx 390$ MeV. A combination of

[1] This talk was based on a similar talk I gave at the Confinement XII conference and this write-up is similar to Ref. [63].

[2] Ignoring complications arising from the mixing of partial waves due to the reduced symmetry of a finite box which we will not discuss here.

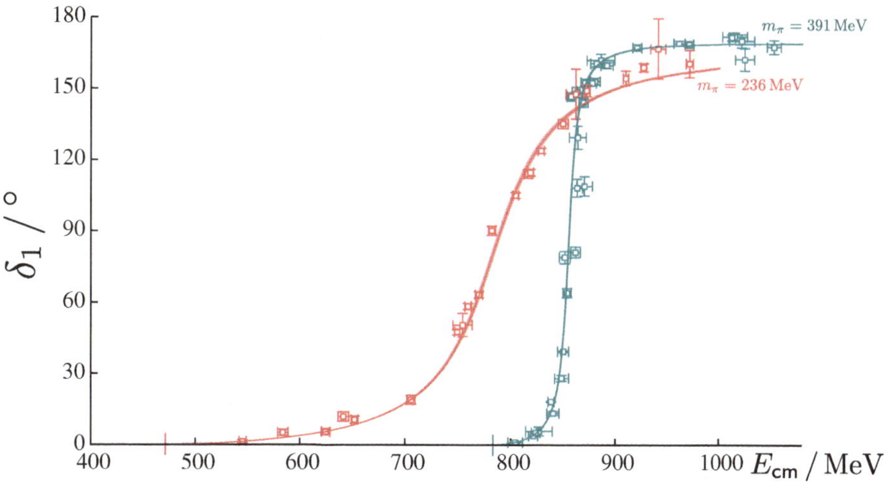

Fig. 1 From Ref. [65]. The P-wave $I = 1$ $\pi\pi$ elastic scattering phase shift plotted from $\pi\pi$ threshold to the inelastic $K\bar{K}$ threshold, from dynamical lattice QCD calculations with $m_\pi \approx 240$ MeV [65] and $m_\pi \approx 390$ MeV [20]. Points are from an energy-level by energy-level analysis and the curves are from a relativistic Breit Wigner parameterization as described in [65]

techniques were used to calculate extensive finite-volume spectra with high statistical precision: the distillation approach [54] was used to compute correlation matrices involving large bases of carefully-constructed fermion-bilinear [18,64] and multi-hadron interpolating operators [19,20] (between 10 and 37 operators in each of 12 different irreducible representations of the relevant symmetry group). The correlation matrices were analysed using the variational method and the resulting spectra, with 22 energy levels in the elastic region, enabled the energy dependence of the scattering phase shift to be mapped out in detail via the the Lüscher method.

The figure shows convincingly the rapid rise in the phase shift from 0° through 90° to 180° that is expected in the presence of a single isolated resonance. The resonance mass ($M_R = 790 \pm 2$ MeV) is not too different from experiment (775.49 ± 0.3 MeV) but the width is significantly smaller ($\Gamma = 87 \pm 2$ MeV compared to 149.1 ± 0.8 MeV experimentally). This is simply because the pion mass in this study is larger than the physical value and so the phase space for decay is reduced. The coupling, g, with the phase space factor divided out, defined by $\Gamma = \frac{g^2}{6\pi}\frac{p_{\rm cm}^3}{M_R}$, where $p_{\rm cm}$ is the cm-frame scattering momentum, shows little dependence on m_π and is in reasonable agreement with experiment.

Ref. [65] also extends the analysis to the coupled-channel $\pi\pi$, $K\bar{K}$ energy region using 34 energy levels. Negligible coupling between the channels is found in the energy range explored.

The RQCD Collaboration have recently studied the ρ resonance with $m_\pi \approx 150$ MeV, close to the physical value [3]. However, in these "$N_f = 2$" computations only the light up and down quarks were dynamical whereas the strange quarks were quenched (not present in the sea). As can be seen from the comparison of different lattice QCD results in Ref. [3], the resonance mass is sensitive to whether or not the strange quarks are quenched (unlike g which appears to be insensitive to this and to m_π). The same effect was observed in recent $N_f = 2$ computations by Guo et. al. [26] with $m_\pi = 226$ and 315 MeV, as is discussed in Ref. [31].

Going beyond the resonance mass and width, recently the form factor of an unstable hadron has been calculated for the first time in lattice QCD [10,12]. From computations of the $\pi\pi \to \pi\gamma^*$ transition amplitude, the resonant $\rho \to \pi\gamma^*$ transition form factor and the radiative decay width were extracted.

4 Other Scattering Channels Relevant for Light and Strange Mesons

Moving to other scattering channels relevant for light and strange mesons, using a similar setup to that described above with $m_\pi \approx 390$ MeV and three lattice volumes, the HSC investigated coupled-channel $K\pi$, $K\eta$ scattering with $I = 1/2$ [21,66], the first study of coupled-channel scattering on the lattice. In total 73 energy levels were extracted and these enabled the energy dependence of the scattering matrix to be mapped out. A number of interesting features emerged including a broad $J^P = 0^+$ scalar resonance in S-wave [c.f. the $K_0^*(1430)$], a P-wave 1^- vector bound state [c.f. the $K^*(892)$], a narrow 2^+ tensor resonance in D-wave [c.f. the $K_2^*(1430)$] and a suggestion that the κ, a scalar resonance with physical-mass light quarks, corresponds to a virtual bound

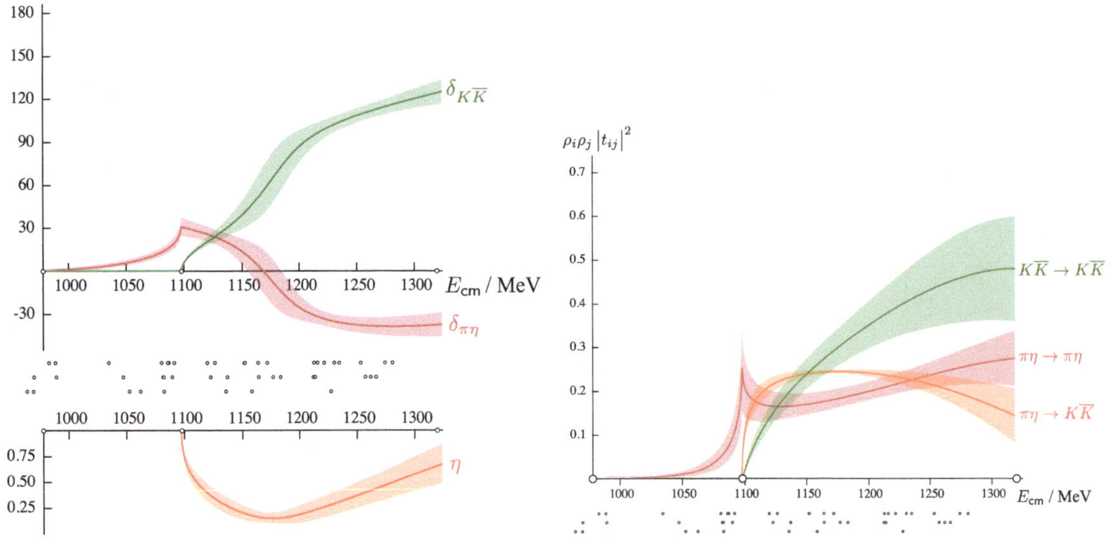

Fig. 2 From Ref. [22]. The S-wave $I = 1$ $\pi\eta$, $K\bar{K}$ coupled-channel scattering amplitudes ($J^{PC} = 0^{++}$) from a lattice calculation with $m_\pi \approx 390$ MeV. *Left panel* shows phase shifts, δ_i, and inelasticity, η, and the *right panel* shows quantities proportional to cross sections. *Open circles* on the *horizontal axis* indicate $\pi\eta$, $K\bar{K}$ and $\pi\eta'$ thresholds. The *dots* below the figures show energy levels used to constrain the amplitudes

state[3] when $m_\pi \approx 390$ MeV. The latter observation is consistent with how the κ behaves in unitarised chiral perturbation theory as the π mass is varied [48]. The exotic-flavour $I = 3/2$ $K\pi$ channel was also studied.

The RQCD Collaboration have studied P-wave elastic $K\pi$ scattering in a $N_f = 2$ calculation with light quarks corresponding to $m_\pi \approx 150$ MeV [3]. They find a vector resonance, consistent with the expectation that the bound state found by the HSC with $m_\pi \approx 390$ MeV becomes a resonance as the light-quark masses are reduced.

The HSC have also studied coupled $\pi\eta$, $K\bar{K}$, $\pi\eta'$ scattering in $I = 1$ with $m_\pi \approx 390$ MeV on three lattice volumes: $L \approx 1.9$, 2.4 and 2.9 fm using respectively 479, 603 and 553 configurations [22]. The S-wave coupled-channel $\pi\eta$, $K\bar{K}$ scattering amplitudes constrained by 47 energy levels are shown in Fig. 2. Here the phase shifts, δ_i, and inelasticity, η, are defined in terms of elements of the t-matrix,[4]

$$t_{ij} = \begin{cases} \frac{\eta\, e^{2i\delta_i} - 1}{2i\, \rho_i} & (i = j) \\ \frac{\sqrt{1-\eta^2}\, e^{i(\delta_i + \delta_j)}}{2\sqrt{\rho_i\, \rho_j}} & (i \neq j) \end{cases} \tag{1}$$

where i, j label scattering channels and $\rho_i = 2p_{\mathrm{cm},i}/E_{\mathrm{cm}}$ is the phase space factor for channel i. The amplitudes show a cusp-like enhancement in $\pi\eta \to \pi\eta$ near $K\bar{K}$ threshold and a rapid turn on of amplitudes to $K\bar{K}$. This behaviour originates from an $a_0(980)$-like resonance which is strongly coupled to both $\pi\eta$ and $K\bar{K}$—the first strongly-coupled meson-meson scattering system extracted in a lattice QCD calculation. The corresponding pole appears on a single unphysical Riemann sheet, unlike a canonical two-channel resonance where poles would be expected on two unphysical sheets, and this may be a sign that the state binds through the long-range interaction between a pair of mesons. Ref. [22] also presents results with the $\pi\eta'$ channel included and considers D-wave scattering where a narrow tensor resonance is found.

Using a similar setup again, the HSC have recently determined the elastic $\pi\pi$ S-wave $I = 0$ scattering phase shift for the first time in a lattice QCD computation [11]. Two different light quark masses were used corresponding to $m_\pi \approx 240$ MeV and 390 MeV. For the lighter mass the scattering amplitude has a pole on the unphysical Riemann sheet with a large imaginary part, corresponding to a broad resonance, in qualitative agreement with experiment. At the heavier mass a pole is found below threshold on the real axis of the physical sheet, corresponding to a bound state. Work is ongoing to extend this calculation into the coupled-channel region above $K\bar{K}$ threshold where a $f_0(980)$-like resonance is expected, the final component of a study of the scalar meson nonet using lattice QCD.

[3] A pole below threshold on the real axis of an unphysical Riemann sheet.

[4] Note that $\eta = 1$ would indicate no coupling between channels.

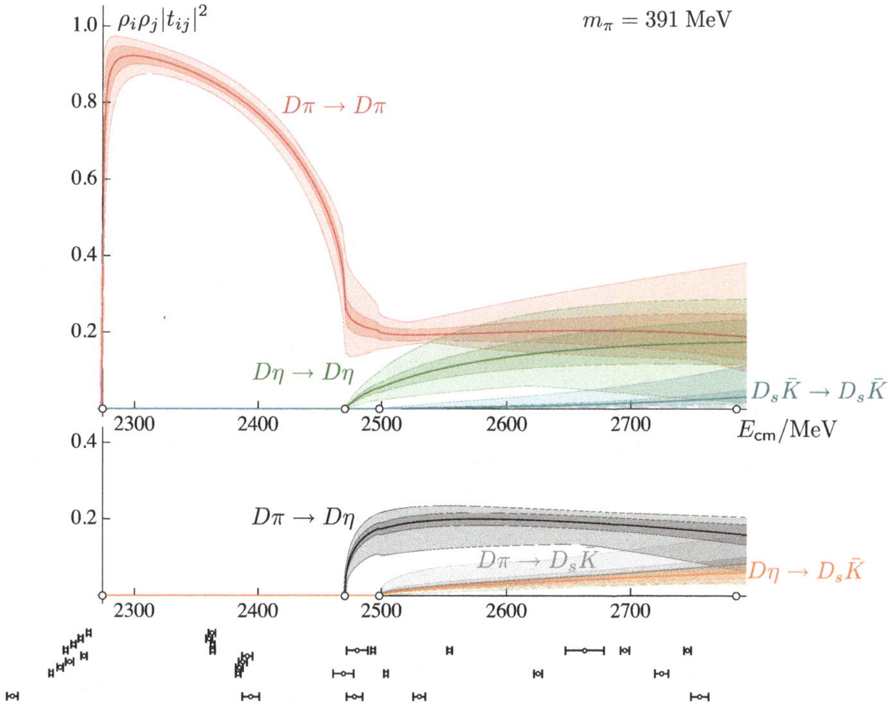

Fig. 3 From Ref. [47]. The S-wave $I = 1/2$ $D\pi$, $D\eta$, $D_s\bar{K}$ coupled-channel scattering amplitudes from a lattice calculation with $m_\pi \approx 390$ MeV. Quantities proportional to the diagonal (off-diagonal) cross sections are plotted in the *top* (*bottom*) panel. *Open circles* on the *horizontal axes* indicate $D\pi$, $D\eta$, $D_s\bar{K}$ and $D^*\pi\pi$ thresholds. The *black points* below the plot show energy levels used to constrain the amplitudes

5 Heavy-Light and Charmonium-Like Resonances

Turning to channels relevant for heavy-light and quarkonium mesons, we begin with the charm-light (D) and charm-strange (D_s) sectors. The $D^*_{s0}(2317)$ and $D_{s1}(2460)$ present a long standing-puzzle: they were expected to be broad resonances, like the analogous $D^*_0(2400)$ and $D_1(2430)$ mesons, but were observed to be narrow states below the respective DK and D^*K thresholds [53]. Some early lattice calculations of elastic $D^*\pi$ and D^*K scattering can be found in Refs. [34,45,46].

The HSC have recently investigated $I = 1/2$ coupled-channel $D\pi$, $D\eta$, $D_s\bar{K}$ scattering with light-quark masses corresponding to $m_\pi \approx 390$ MeV on three lattice volumes: $L \approx 1.9$, 2.4 and 2.9 fm using respectively 479, 603 and 553 configurations [47]. Figure 3 shows the S-wave scattering amplitudes constrained by 47 energy levels—the broad feature in the $D\pi \to D\pi$ amplitude is associated with a $J^P = 0^+$ bound state very close to $D\pi$ threshold. Although, by contrast, the associated experimental state, $D^*_0(2400)$, is a resonance, they are similar in that they influence a broad energy range and couple predominantly to $D\pi$. The pole position lies between what is observed experimentally for the $D^*_{s0}(2317)$ and the $D^*_0(2400)$. In P-wave ($J^P = 1^-$), a deeply bound state was found and can be compared to the experimentally observed narrow resonance, $D^*(2007)$. A narrow 2^+ resonance was found in D-wave. The $I = 3/2$ elastic $D\pi$ channel was also investigated and a weakly repulsive interaction found in S-wave.

In the charmonium sector there have been a number of lattice QCD investigations [13–15,35,51,52,59–61] and there have been a couple in the bottom sector [36,37]. However, the calculations are somewhat less advanced than in the other sectors and studies have generally been exploratory, inconclusive or have not determined the phenomena in detail or robustly, and I will not discuss these here. More robust and detailed computations are required—one challenge is the number of channels to which a resonance can potentially couple and all these channels need to be considered in the calculations.

6 Conclusions

I have highlighted some recent lattice QCD studies of resonances and related scattering phenomena, an area which has seen significant advances in the last few years and where prospects for further progress are very good. Although it may be seen as a disadvantage that many of the computations used unphysically-heavy light quarks, corresponding to m_π larger than the experimental value, and so cannot be compared quantitatively to experiment, it turns out that studying how states evolve as m_π varies is a useful tool in discerning their nature. As we go further up in the spectrum and undertake calculations with m_π closer to the physical value, in general calculations become more difficult because additional channels become relevant and, in particular, channels involving more than two hadrons open up. Work on the formalism for such situations is ongoing [4, 9, 28–30, 58].

To stringently probe the structure of states we must go beyond solely computing masses and consider other properties of hadrons such as form factors and related quantities. There are also good prospects here with the recent appearance of the first lattice QCD calculation of a transition form factor involving a resonance [10, 12].

Acknowledgements I thank the organisers for the invitation to give a talk at this meeting and my colleagues in the Hadron Spectrum Collaboration. I acknowledge support from the UK Science and Technology Facilities Council (STFC) [Grant ST/L000385/1].

References

1. S. Aoki et al., Lattice QCD calculation of the ρ meson decay width. Phys. Rev. D **76**(094), 506 (2007). doi:10.1103/PhysRevD.76.094506. arXiv:0708.3705
2. S. Aoki et al., ρ meson decay in 2 + 1 flavor lattice QCD. Phys. Rev. **D84**(094), 505 (2011). doi:10.1103/PhysRevD.84.094505. arXiv:1106.5365
3. G.S. Bali, S. Collins, A. Cox, G. Donald, M. Gckeler, C.B. Lang, A. Schfer, ρ and K^* resonances on the lattice at nearly physical quark masses and $N_f = 2$. Phys. Rev. D **93**(5), 054509 (2016). doi:10.1103/PhysRevD.93.054509. arXiv:1512.08678
4. S.R. Beane, W. Detmold, M.J. Savage, n-boson energies at finite volume and three-boson interactions. Phys. Rev. D **76**(074), 507 (2007). doi:10.1103/PhysRevD.76.074507. arXiv:0707.1670
5. B. Blossier, M. Della Morte, G. von Hippel, T. Mendes, R. Sommer, On the generalized eigenvalue method for energies and matrix elements in lattice field theory. JHEP **04**, 094 (2009). doi:10.1088/1126-6708/2009/04/094. arXiv:0902.1265
6. N. Brambilla et al., QCD and strongly coupled gauge theories: challenges and perspectives. Eur. Phys. J. **C74**(10), 2981 (2014). doi:10.1140/epjc/s10052-014-2981-5. arXiv:1404.3723
7. R.A. Briceno, Two-particle multichannel systems in a finite volume with arbitrary spin. Phys. Rev. D **89**(7), 074507 (2014). doi:10.1103/PhysRevD.89.074507. arXiv:1401.3312
8. R.A. Briceno, Z. Davoudi, Moving multichannel systems in a finite volume with application to proton–proton fusion. Phys. Rev. D **88**(9), 094507 (2013a). doi:10.1103/PhysRevD.88.094507. arXiv:1204.1110
9. R.A. Briceno, Z. Davoudi, Three-particle scattering amplitudes from a finite volume formalism. Phys. Rev. D **87**(9), 094507 (2013b). doi:10.1103/PhysRevD.87.094507. arXiv:1212.3398
10. R.A. Briceno, J.J. Dudek, R.G. Edwards, C.J. Shultz, C.E. Thomas, D.J. Wilson, The resonant $\pi^+\gamma \to \pi^+\pi^0$ amplitude from quantum chromodynamics. Phys. Rev. Lett. **115**(242), 001 (2015). doi:10.1103/PhysRevLett.115.242001. arXiv:1507.06622
11. R.A. Briceno, J.J. Dudek, R.G. Edwards, D.J. Wilson, Isoscalar $\pi\pi$ scattering and the σ meson resonance from QCD. Phys. Rev. Lett. **118**(2), 022002 (2017). doi:10.1103/PhysRevLett.118.022002
12. R.A. Briceo, J.J. Dudek, R.G. Edwards, C.J. Shultz, C.E. Thomas, D.J. Wilson, The $\pi\pi \to \pi\gamma^\star$ amplitude and the resonant $\rho \to \pi\gamma^\star$ transition from lattice QCD. Phys. Rev. D **93**(11), 114508 (2016). doi:10.1103/PhysRevD.93.114508. arXiv:1604.03530
13. T. Chen et al., A lattice study of $(\bar{D}_1 D^*)^\pm$ near-threshold scattering. Phys. Rev. D **93**(11), 114501 (2016). doi:10.1103/PhysRevD.93.114501. arXiv:1602.00200
14. Y. Chen et al., Low-energy scattering of the $(D\bar{D}^*)^\pm$ system and the resonance-like structure $Z_c(3900)$. Phys. Rev. D **89**(9), 094506 (2014). doi:10.1103/PhysRevD.89.094506. arXiv:1403.1318
15. Y. Chen et al., Low-energy scattering of $(D^*\bar{D}^*)^\pm$ system and the resonance-like structure $Z_c(4025)$. Phys. Rev. D **92**(5), 054507 (2015). doi:10.1103/PhysRevD.92.054507. arXiv:1503.02371
16. N.H. Christ, C. Kim, T. Yamazaki, Finite volume corrections to the two-particle decay of states with non-zero momentum. Phys. Rev. **D72**(114), 506 (2005). doi:10.1103/PhysRevD.72.114506. arXiv:hep-lat/0507009
17. R. Dowdall, C. Davies, T. Hammant, R. Horgan, Precise heavy-light meson masses and hyperfine splittings from lattice QCD including charm quarks in the sea. Phys. Rev. **D86**(094), 510 (2012). doi:10.1103/PhysRevD.86.094510. arXiv:1207.5149
18. J.J. Dudek, R.G. Edwards, M.J. Peardon, D.G. Richards, C.E. Thomas, Toward the excited meson spectrum of dynamical QCD. Phys. Rev. D **82**, 034508 (2010). doi:10.1103/PhysRevD.82.034508. arXiv:1004.4930
19. J.J. Dudek, R.G. Edwards, C.E. Thomas, S and D-wave phase shifts in isospin-2 pi pi scattering from lattice QCD. Phys. Rev. D **86**(034), 031 (2012). doi:10.1103/PhysRevD.86.034031. arXiv:1203.6041
20. J.J. Dudek, R.G. Edwards, C.E. Thomas, Energy dependence of the ρ resonance in $\pi\pi$ elastic scattering from lattice QCD. Phys. Rev. D **87**(3), 034505, (2013). doi:10.1103/PhysRevD.87.034505, doi:10.1103/PhysRevD.90.099902, [Erratum: Phys. Rev.D90,no.9,099902(2014)], arXiv:1212.0830
21. J.J. Dudek, R.G. Edwards, C.E. Thomas, D.J. Wilson, Resonances in coupled $\pi K - \eta K$ scattering from quantum chromodynamics. Phys. Rev. Lett. **113**(18), 182001 (2014). doi:10.1103/PhysRevLett.113.182001. arXiv:1406.4158

22. J.J. Dudek, R.G. Edwards, D.J. Wilson, An a_0 resonance in strongly coupled $\pi\eta$, $K\overline{K}$ scattering from lattice QCD. Phys. Rev. D **93**(9), 094506 (2016). doi:10.1103/PhysRevD.93.094506. arXiv:1602.05122
23. S. Durr et al., Ab-initio determination of light hadron masses. Science **322**, 1224–1227 (2008). doi:10.1126/science.1163233. arXiv:0906.3599
24. X. Feng, K. Jansen, D.B. Renner, Resonance parameters of the rho-meson from lattice QCD. Phys. Rev. D **83**, 094505 (2011). doi:10.1103/PhysRevD.83.094505. arXiv:1011.5288
25. Z. Fu, Rummukainen–Gottlieb's formula on two-particle system with different mass. Phys. Rev. D **85**(014), 506 (2012). doi:10.1103/PhysRevD.85.014506. arXiv:1110.0319
26. D. Guo, A. Alexandru, R. Molina, M. Dring, Rho resonance parameters from lattice QCD. Phys. Rev. D **94**(3), 034501 (2016). doi:10.1103/PhysRevD.94.034501. arXiv:1605.03993
27. M.T. Hansen, S.R. Sharpe, Multiple-channel generalization of Lellouch–Luscher formula. Phys. Rev. D **86**(016), 007 (2012). doi:10.1103/PhysRevD.86.016007. arXiv:1204.0826
28. M.T. Hansen, S.R. Sharpe, Relativistic, model-independent, three-particle quantization condition. Phys. Rev. D **90**(11), 116003 (2014). doi:10.1103/PhysRevD.90.116003. arXiv:1408.5933
29. M.T. Hansen, S.R. Sharpe, Expressing the three-particle finite-volume spectrum in terms of the three-to-three scattering amplitude. Phys. Rev. D **92**(11), 114509 (2015). doi:10.1103/PhysRevD.92.114509. arXiv:1504.04248
30. M.T. Hansen, S.R. Sharpe, Perturbative results for two and three particle threshold energies in finite volume. Phys. Rev. D **93**(014), 506 (2016). doi:10.1103/PhysRevD.93.014506. arXiv:1509.07929
31. B. Hu, R. Molina, M. Dring, A. Alexandru, Two-flavor simulations of the $\rho(770)$ and the role of the $K\bar{K}$ channel. Phys. Rev. Lett. **117**(12), 122001 (2016). doi:10.1103/PhysRevLett.117.122001. arXiv:1605.04823
32. C. Kim, C. Sachrajda, S.R. Sharpe, Finite-volume effects for two-hadron states in moving frames. Nucl. Phys. **B727**, 218–243 (2005). doi:10.1016/j.nuclphysb.2005.08.029. arXiv:hep-lat/0507006
33. C. Lang, D. Mohler, S. Prelovsek, M. Vidmar, Coupled channel analysis of the rho meson decay in lattice QCD. Phys. Rev. **D84**(054), 503 (2011). doi:10.1103/PhysRevD.84.054503. arXiv:1105.5636
34. C.B. Lang, L. Leskovec, D. Mohler, S. Prelovsek, R.M. Woloshyn, Ds mesons with DK and D*K scattering near threshold. Phys. Rev. D **90**(3), 034510 (2014). doi:10.1103/PhysRevD.90.034510. arXiv:1403.8103
35. C.B. Lang, L. Leskovec, D. Mohler, S. Prelovsek, Vector and scalar charmonium resonances with lattice QCD. JHEP **09**, 089 (2015). doi:10.1007/JHEP09(2015)089. arXiv:1503.05363
36. C.B. Lang, D. Mohler, S. Prelovsek, R.M. Woloshyn, Predicting positive parity B_s mesons from lattice QCD. Phys. Lett. B **750**, 17–21 (2015). doi:10.1016/j.physletb.2015.08.038. arXiv:1501.01646
37. C.B. Lang, D. Mohler, S. Prelovsek, $B_s\pi^+$ scattering and search for X(5568) with lattice QCD. Phys. Rev. D **94**(074), 509 (2016). doi:10.1103/PhysRevD.94.074509. arXiv:1607.03185
38. L. Leskovec, S. Prelovsek, Scattering phase shifts for two particles of different mass and non-zero total momentum in lattice QCD. Phys. Rev. **D85**(114), 507 (2012). doi:10.1103/PhysRevD.85.114507. arXiv:1202.2145
39. C. Liu, X. Feng, S. He, Two particle states in a box and the S-matrix in multi-channel scattering. Int. J. Mod. Phys. A **21**, 847–850 (2006). doi:10.1142/S0217751X06032150. arXiv:hep-lat/0508022
40. M. Luscher, Volume dependence of the energy spectrum in massive quantum field theories. 2. Scattering States. Commun. Math. Phys. **105**, 153–188 (1986). doi:10.1007/BF01211097
41. M. Luscher, Signatures of unstable particles in finite volume. Nucl. Phys. **B364**, 237–254 (1991a). doi:10.1016/0550-3213(91)90584-K
42. M. Luscher, Two particle states on a torus and their relation to the scattering matrix. Nucl. Phys. **B354**, 531–578 (1991b). doi:10.1016/0550-3213(91)90366-6
43. M. Luscher, U. Wolff, How to calculate the elastic scattering matrix in two- dimensional quantum field theories by numerical simulation. Nucl. Phys. **B339**, 222–252 (1990). doi:10.1016/0550-3213(90)90540-T
44. C. Michael, Adjoint sources in lattice gauge theory. Nucl. Phys. B **259**, 58 (1985). doi:10.1016/0550-3213(85)90297-4
45. D. Mohler, C.B. Lang, L. Leskovec, S. Prelovsek, R.M. Woloshyn, $D^*_{s0}(2317)$ meson and D-meson–Kaon scattering from lattice QCD. Phys. Rev. Lett. **111**(22), 222001 (2013a). doi:10.1103/PhysRevLett.111.222001. arXiv:1308.3175
46. D. Mohler, S. Prelovsek, R.M. Woloshyn, $D\pi$ scattering and D meson resonances from lattice QCD. Phys. Rev. D **87**(3), 034501 (2013b). doi:10.1103/PhysRevD.87.034501. arXiv:1208.4059
47. G. Moir, M. Peardon, S.M. Ryan, C.E. Thomas, D.J. Wilson, Coupled-channel $D\pi$, $D\eta$ and $D_s\bar{K}$ scattering from lattice QCD. JHEP **10**, 011 (2016). doi:10.1007/JHEP10(2016)011. arXiv:1607.07093
48. J. Nebreda, Strange and non-strange quark mass dependence of elastic light resonances from SU(3) unitarized chiral perturbation theory to one loop. Phys. Rev. D **81**, 054035 (2010). doi:10.1103/PhysRevD.81.054035. arXiv:1001.5237
49. S. Neubert, Review on exotic hadrons. In: 12th Conference on Quark Confinement and the Hadron Spectrum (Confinement XII) Thessaloniki, Greece, 29 Aug–3 Sept (2016)
50. M. Nielsen, R.M. Albuquerque, J.M. Dias, K.P. Khemchandani, A. Martinez Torres, F.S. Navarra, C.M. Zanetti , X, Y and Z States. In: *12th Conference on Quark Confinement and the Hadron Spectrum (Confinement XII) Thessaloniki, Greece, 29 August–3 September*, 2016, http://inspirehep.net/record/1496993/files/arXiv:1611.03300, arXiv:1611.03300 (2016)
51. S. Ozaki, S. Sasaki, Lúscher's finite size method with twisted boundary conditions: an application to J/ψ-ϕ system to search for narrow resonance. Phys. Rev. **D87**(014), 506 (2013). doi:10.1103/PhysRevD.87.014506. arXiv:1211.5512
52. M. Padmanath, C.B. Lang, S. Prelovsek, X(3872) and Y(4140) using diquark-antidiquark operators with lattice QCD. Phys. Rev. D **92**(3), 034501 (2015). doi:10.1103/PhysRevD.92.034501. arXiv:1503.03257
53. C. Patrignani et al., Review of particle physics. Chin. Phys. **C40**(10), 100001 (2016). doi:10.1088/1674-1137/40/10/100001
54. M. Peardon et al., A novel quark-field creation operator construction for hadronic physics in lattice QCD. Phys. Rev. D **80**(054), 506 (2009). doi:10.1103/PhysRevD.80.054506. arXiv:0905.2160
55. J.R. Pelaez, From controversy to precision on the sigma meson: a review on the status of the non-ordinary $f_0(500)$ resonance. Phys. Rep. **658**, 1 (2016). doi:10.1016/j.physrep.2016.09.001. arXiv:1510.00653
56. C. Pelissier, A. Alexandru, Resonance parameters of the rho-meson from asymmetrical lattices. Phys. Rev. **D87**(014), 503 (2013). doi:10.1103/PhysRevD.87.014503. arXiv:1211.0092

57. M.R. Pennington, Impressions of the meson spectrum: hybrids and exotics, present and future. EPJ Web Conf. **113**(01), 014 (2016). doi:10.1051/epjconf/201611301014. arXiv:1509.02555
58. K. Polejaeva, A. Rusetsky, Three particles in a finite volume. Eur. Phys. J. A **48**, 67 (2012). doi:10.1140/epja/i2012-12067-8. arXiv:1203.1241
59. S. Prelovsek, L. Leskovec, Evidence for X(3872) from DD* scattering on the lattice. Phys. Rev. Lett. **111**(192), 001 (2013a). doi:10.1103/PhysRevLett.111.192001. arXiv:1307.5172
60. S. Prelovsek, L. Leskovec, Search for $Z_c^+(3900)$ in the 1^{+-} channel on the lattice. Phys. Lett. B **727**, 172–176 (2013). doi:10.1016/j.physletb.2013.10.009. arXiv:1308.2097
61. S. Prelovsek, C.B. Lang, L. Leskovec, D. Mohler, Study of the Z_c^+ channel using lattice QCD. Phys. Rev. D **91**(1), 014504 (2015). doi:10.1103/PhysRevD.91.014504. arXiv:1405.7623
62. K. Rummukainen, S.A. Gottlieb, Resonance scattering phase shifts on a nonrest frame lattice. Nucl. Phys. **B450**, 397–436 (1995). doi:10.1016/0550-3213(95)00313-H. arXiv:hep-lat/9503028
63. C.E. Thomas. Meson spectroscopy, resonances and scattering on the lattice. In: 12th Conference on Quark Confinement and the Hadron Spectrum (Confinement XII) Thessaloniki, Greece, 29 Aug–3 Sept (2016)
64. C.E. Thomas, R.G. Edwards, J.J. Dudek, Helicity operators for mesons in flight on the lattice. Phys. Rev. D **85**(014), 507 (2012). arXiv:1107.1930
65. D.J. Wilson, R.A. Briceno, J.J. Dudek, R.G. Edwards, C.E. Thomas, Coupled $\pi\pi$, $K\bar{K}$ scattering in P-wave and the ρ resonance from lattice QCD. Phys. Rev. D **92**(9), 094502 (2015a). doi:10.1103/PhysRevD.92.094502. arXiv:1507.02599
66. D.J. Wilson, J.J. Dudek, R.G. Edwards, C.E. Thomas, Resonances in coupled πK, ηK scattering from lattice QCD. Phys. Rev. D **91**(5), 054008 (2015b). doi:10.1103/PhysRevD.91.054008. arXiv:1411.2004

Few-Body Syst (2017) 58:104
DOI 10.1007/s00601-017-1265-7

CrossMark

Giannis Koutsou

Nucleon Structure and the Neutron Electric Dipole Moment from Twisted Mass Lattice QCD

Received: 26 November 2016 / Accepted: 21 February 2017 / Published online: 4 March 2017

Abstract Recent results on nucleon structure using twisted mass lattice QCD simulated with a quark mass yielding the physical pion mass are presented. Namely, the nucleon axial charges and moments of parton distribution functions are discussed as well as the implication of these results on the decomposition of the nucleon spin. Results on the electromagnetic structure are also presented, including a recent study of the neutron Electric Dipole Moment using simulations with pion mass of 370 MeV.

1 Introduction

With the successful reproduction of the low-lying hadron spectrum, lattice QCD has entered an exciting new era in which precision hadron structure calculations are in reach. Nucleon structure in particular is seeing tremendous progress with the recent availability of simulations at physical values of the pion mass [24,25]. The availability of precision nucleon structure experiments offers a number of benchmark quantities for lattice QCD, such as the nucleon axial charge and the electromagnetic form factors, while lesser well known quantities such as the tensor charge, the nucleon σ-terms and the neutron Electric Dipole Moment (nEDM) allow for lattice QCD to make predictions and contact new physics searches.

The twisted mass formulation is particularly suited for nucleon structure calculations, since after tuning the bare quark mass to its critical value (maximal twist), all discretization errors appear to second order in the lattice spacing ($\mathcal{O}(a)$) improvement) without the need for further tuning of lattice operators. In what follows we review nucleon structure results using twisted mass fermions, including a simulation with quarks with masses tuned to yield a physical pion mass.

2 Nucleon Matrix Elements on the Lattice

In lattice QCD, one extracts nucleon matrix elements by calculating a three-point correlation function:

$$G_{\mu,\mu_1,\dots}(\Gamma;\mathbf{p}',\mathbf{p};t_s,t_{\rm ins}) = \sum_{\mathbf{x}_s\mathbf{x}_{\rm ins}} e^{-i\mathbf{x}_s\mathbf{p}'}e^{i\mathbf{x}_{\rm ins}(\mathbf{p}'-\mathbf{p})}\Gamma^{\alpha\beta}\left\langle\bar{\chi}^{\beta}_N(\mathbf{x}_s;t_s)|\mathcal{O}^{\mu,\mu_1,\dots}(\mathbf{x}_{\rm ins};t_{\rm ins})|\chi^{\alpha}_N(\mathbf{0};0)\right\rangle \xrightarrow[t_{\rm ins}\gg]{t_{\rm s}-t_{\rm ins}\gg}$$
$$\sum_{ss'}\Gamma^{\alpha\beta}\left\langle\chi^{\beta}_N|N(p',s')\right\rangle\left\langle N(p,s)|\bar{\chi}^{\alpha}_N\right\rangle\left\langle N(p',s')|\mathcal{O}_{\mu,\mu_1,\dots}(q)|N(p,s)\right\rangle\times e^{-E_N(\mathbf{p})t_{\rm ins}}e^{-E_N(\mathbf{p}')(t_{\rm s}-t_{\rm ins})}, \quad (1)$$

This article belongs to the Topical Collection "Light Cone 2016".

G. Koutsou (✉)
Computation-Based Science and Technology Research Center, The Cyprus Institute, 20 Konstantinou Kavafi Str., 2121 Nicosia, Cyprus
E-mail: g.koutsou@cyi.ac.cy
Tel.: +357 22 208633

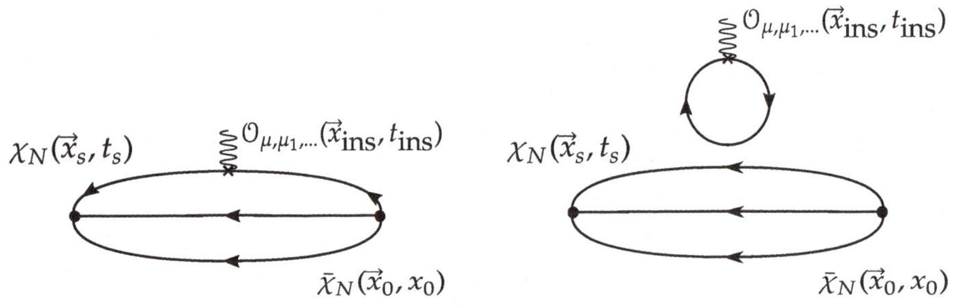

Fig. 1 Connected (*left*) and disconnected (*right*) contributions to the nucleon three-point function

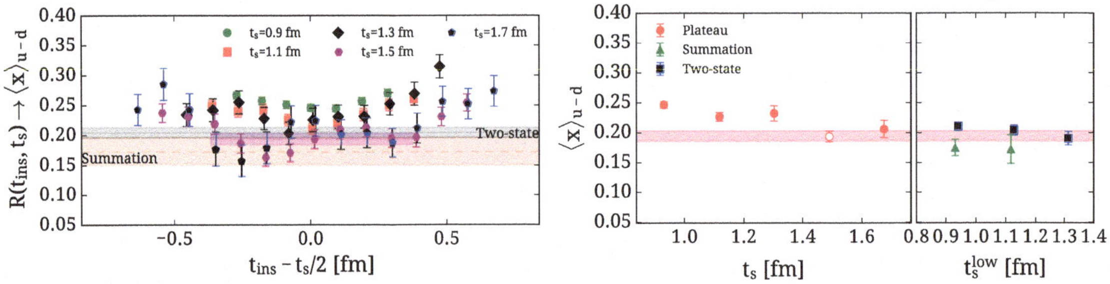

Fig. 2 Extraction of $\langle x \rangle_{u-d}$ using five sink-source separations. *Left*: the ratio is plotted for t_s =0.9 fm (*green circles*), 1.1 fm (*red squares*), 1.3 fm (*black diamonds*), 1.5 fm (*magenta hexagons*) and 1.7 fm (*blue pentagons*). For the three smallest separations, statistics of 9280 (580 gauge-field configurations times 16 source positions per configuration) were used, for $t_s \simeq 1.5$ fm about 44,000 (500 configurations times 88 source positions) and for $t_s \simeq 1.7$ fm about 66,000 (750 configurations times 88 source positions). For $t_s \simeq 1.5$ fm we show a fit to a constant with fit range indicated by the range of the magenta band. The *dashed line* shows the result of applying the summation method. The *solid line* spanning the entire horizontal range is the extracted matrix element from the two-state fit method. *Right*: Fit results when using the plateau method (*red circles*) for all five t_s, the two-state fit method (*blue squares*) for three choices for the smallest t_s included in the fit ($t_s^{\rm low}$) and the summation method for two choices of $t_s^{\rm low}$. The band and *open symbol* show the plateau method result for the smallest t_s for which agreement is found with all other methods

where χ_N are quark interpolating operators with the quantum numbers of the nucleon and $\mathcal{O}$ is a quark operator. The three-point function in general takes both quark-connected and -disconnected contributions, as shown in Fig. 1, the latter requiring an order of magnitude more computational resources to compute to similar accuracy as the connected. In the case of isovector operators, such as for example the isovector axial operator $\mathcal{O}_\mu = \bar{u}\gamma_5\gamma_\mu u - \bar{d}\gamma_5\gamma_\mu d$, which yields the axial charge, and given degenerate up- and down-quarks, the disconnected contributions cancel.

After Fourier transform of the three-point function over *sink* ($\mathbf{x}_s$) and *insertion* ($\mathbf{x}_{\rm ins}$) coordinates, taking large time separations isolates the ground-state matrix element of the nucleon, shown in the second line of Eq. (1), with $q = p' - p$ the momentum transfer from initial momentum p to final momentum p'. The unknown overlaps $\langle \chi_N | N(p,s)\rangle$ can be canceled using nucleon two-point correlation functions: $G(\Gamma; \mathbf{p}; t) = \sum_{\mathbf{x}} e^{-i\mathbf{x}\mathbf{p}} \Gamma^{\alpha\beta} \langle \bar{\chi}_N^\beta(\mathbf{x}; t) | \chi_N^\alpha(\mathbf{0}; 0)\rangle$ with appropriately selected momentum, time separation and projection matrix Γ.

In Fig. 2 we show as an example the extraction of the first moment of the isovector unpolarized PDF $\langle x \rangle_{u-d}$, obtained using the one-derivative operator $\mathcal{O}_V^{\mu\mu_1} = \bar{\psi}\gamma_{\{\mu} i D_{\mu_1\}} \frac{\tau^3}{2} \psi$ where the doublet, $\bar{\psi} = (\bar{u}, \bar{d})$, τ_3 is the third Pauli matrix acting on flavor space and the brackets $\{\mu, \mu_1\}$ denote symmetrization over indices and subtraction of trace. The statistical error grows with the time separation, meaning that separations cannot be taken arbitrarily large. Three methods are used to extract the matrix element using the smallest possible time separations while ensuring ground-state dominance: i) in the standard *plateau* method a ratio $R^{\mu\mu_1}(t_{\rm ins}, t_s)$ of three- to two-point functions is formed to cancel unknown overlaps and excited states, and fitted to a constant with respect to $t_{\rm ins}$ for multiple t_s. ii) In the *summation* method, the sum of the ratio over $t_{\rm ins}$ is taken: $\sum_{t_{\rm ins}=1}^{t_s-1} R^{\mu\mu_1}(t_{\rm ins}, t_s) \xrightarrow{t_s\to\infty} C + t_s\mathcal{M}$ and fitted to a linear form to extract the desired matrix element $\mathcal{M}$. iii) in the *two-state* fit method we fit three- and two-point functions to a form which includes the first excited state. Agreement among the three methods is an indication of suppression of excited states.

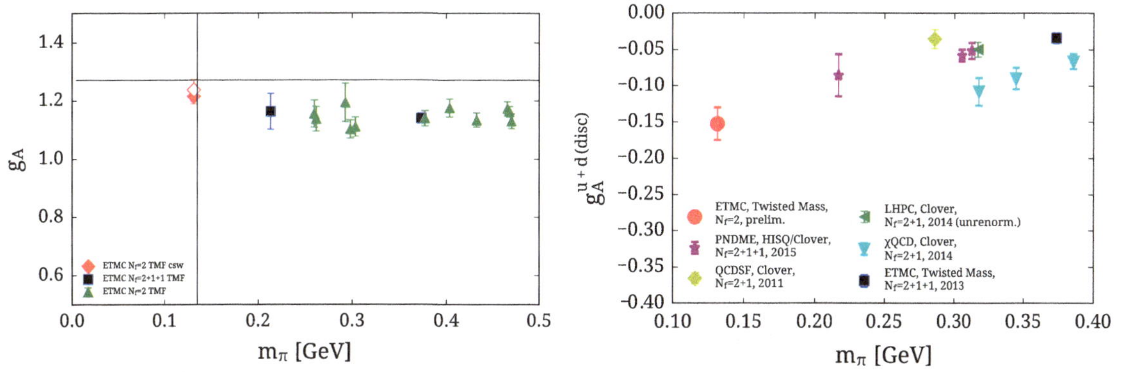

Fig. 3 *Left*: isovector axial charge using $N_f = 2$ twisted mass fermions (*green triangles*) [5], $N_f = 2+1+1$ twisted mass fermions (*blue squares*) [8] and the physical point ensemble (*red diamonds*) [3]. The *open diamond* is the result of the two-state fit. The *vertical line* shows the physical pion mass and the horizontal line the experimental value from Ref. [33]. *Right*: disconnected contribution to the isoscalar axial charge from $N_f = 2 + 1 + 1$ twisted mass fermions (*blue squares*) [1] and the physical point ensemble (*red circle*) [15], compared to Clover on HISQ (*magenta asterisk*) from Ref. [22], Clover fermions from QCDSF (*yellow diamond*) [20] and LHPC (*green, left-pointing triangles*) [29] and overlap on domain wall (*cyan downwards-pointing triangles*) from Ref. [39]

The results in Fig. 2 were obtained on an ensemble of twisted mass fermions with a clover term simulated with two degenerate light quarks ($N_f = 2$) with masses tuned to yield a pion mass of about 131 MeV [4], which we will refer to as the *physical point* lattice in what follows. The lattice volume is $48^3 \times 96$ and the lattice spacing was determined to be $a = 0.093(1)$ fm using the nucleon mass [3].

3 Nucleon Axial Charge, Moments of PDFs and Spin

The nucleon axial charge g_A is considered a benchmark quantity in lattice QCD since it is well known experimentally and being an isovector quantity does not depend on disconnected contributions. Results from twisted mass fermions from pion mass of 470 MeV down to the physical point are shown in the left panel of Fig. 3, with mild pion mass dependence observed. For obtaining the isoscalar axial charge, disconnected contributions, which exhibit large statistical fluctuations, need to be evaluated. Recent methods for their evaluation [1,18] and the increase in computing resources brought by the use of GPUs [7,9] have allowed their reliable evaluation. Lattice results are shown in the right panel of Fig. 3. For the physical point ensemble, statistics of over 2×10^5 were used to obtain this result [2].

In the left panel of Fig. 4 we show the strange quark contribution to the axial charge which is entirely disconnected and shows milder dependence on the pion mass than observed for the case of the light quarks. Having the isovector and isoscalar light axial charges and the strange axial charge we can extract the individual up-, down-, and strange-quark intrinsic spin contributions to the nucleon spin: $\Delta\Sigma^q = g_A^q$ with $q = u, d, s$. This is shown in the right panel of Fig. 4 using twisted mass fermions. At the physical point, one observes the small effect of including the disconnected contributions, which brings the lattice results closer to the experimental values.

On the lattice, moments of parton distribution functions (PDF) are accessible via matrix elements of derivative operators. The first moment of the unpolarized PDF, the momentum fraction $\langle x\rangle_q$, is shown in Fig. 5. In the left panel we show results from various collaborations, with pion masses from 500 MeV down to the physical point, with a pion mass dependence observed towards the experimental value. In the right panel of Fig. 5 we show plateau fits for the disconnected isoscalar and strange contributions to the momentum fraction, where we observe noisy but non-negligible contributions. About 1500 configurations were used with 100 source positions per configuration, averaged over forwards and backwards propagating proton and neutron for a total of about 600,000 statistics, an order of magnitude of what was used for the connected contributions in the left panel of Fig. 5.

Having the individual up-, down- and strange-quark contributions for the one derivative operator allows for further study of the spin decomposition of the nucleon. According to the spin sum by Ji [27] we separate between quark and gluon contributions: $\frac{1}{2} = J^q + J^g$, where the quark contribution is given with respect to the generalized form factors A_{20} and B_{20}: $J^q = \frac{1}{2}\Delta\Sigma^q + L^q = \frac{1}{2}[A_{20}(0) + B_{20}(0)]$. $A_{20}(Q^2)$ and $B_{20}(Q^2)$ are obtained by the vector derivative matrix element:

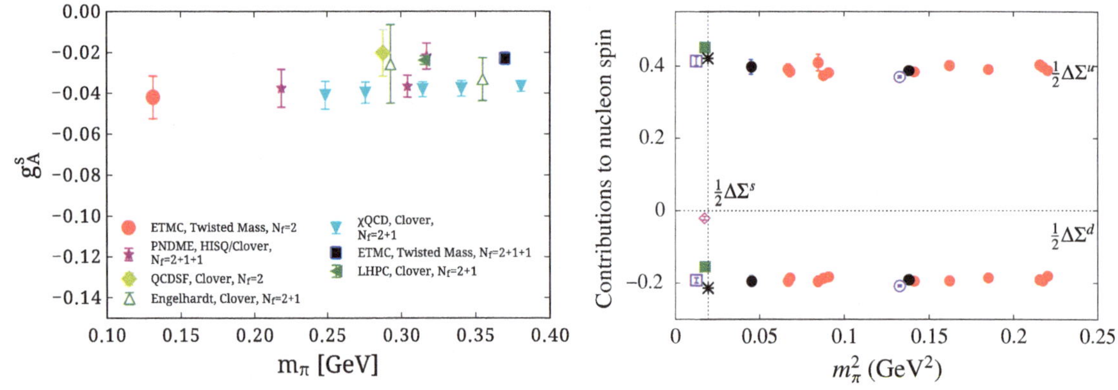

Fig. 4 *Left*: g_A^s, with *open triangles* from Ref. [26]. The rest of the notation is as in Fig. 3. *Right*: $\Delta\Sigma^u$ and $\Delta\Sigma^d$ from $N_f = 2$ (*red circles*) and $N_f = 2+1+1$ (*blue circles*) twisted mass fermions and from the physical point twisted mass ensemble (*squares*). *Open symbols* include disconnected contributions. The *diamond* shows the strange-quark contribution from the latter ensemble. *Asterisks* show experimental determinations

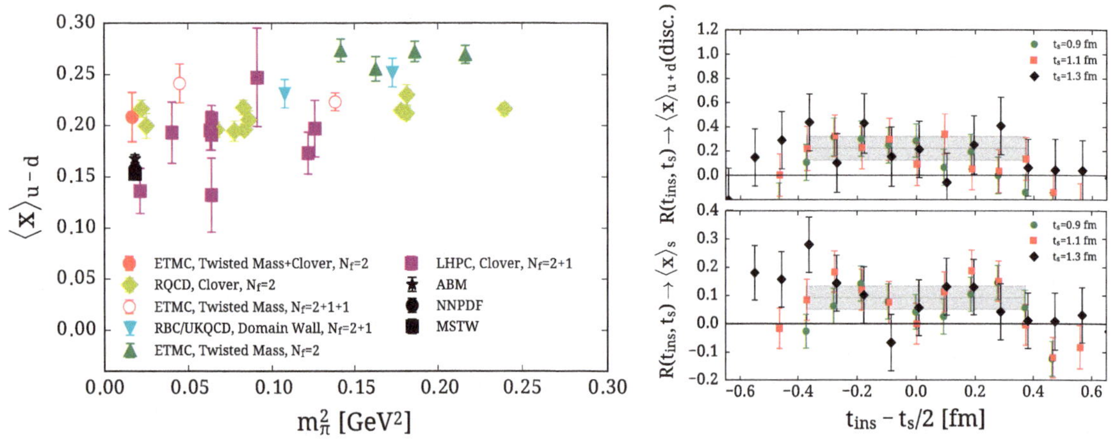

Fig. 5 *Left*: $\langle x\rangle_{u-d}$ from our physical point ensemble (*red filled circles*) [4], from $N_f = 2+1+1$ twisted mass (*open red circles*) [8] and using $N_f = 2$ twisted mass (*green triangles*) [6], compared to results from RQCD (*yellow diamonds*) [19], RBC/UKQCD (*cyan, downwards-pointing triangles*) [16] and LHPC (*magenta squares*) [30]. *Right*: Plateaus to the disconnected isoscalar (*top*) and strange (*bottom*) contributions to the nucleon quark momentum fraction for three sink-source separations and corresponding fit (*black band*)

$$\langle N(p',s')|\mathcal{O}_V^{\mu\nu}|N(p,s)\rangle = \bar{u}_N(p',s')\left[A_{20}(q^2)\gamma^{\{\mu}P^{\nu\}} + B_{20}(q^2)\frac{i\sigma^{\{\mu\alpha}q_\alpha P^{\nu\}}}{2m_N} + C_{20}(q^2)\frac{1}{m_N}q^{\{\mu}q^{\nu\}}\right]u_N(p,s) \quad (2)$$

with $P = p' + p$ and u_N are nucleon spinors. $A_{20}^q(0) = \langle x\rangle_q$ is obtained directly at zero momentum on the lattice while $B_{20}(0)$ requires extrapolating $B_{20}(q^2)$ from finite momenta [8]. Using the physical point ensemble, we obtain $J^{u+d+s} = 0.374(51)$. Combining with a direct calculation of the gluon momentum fraction on the same ensemble in Ref. [12], we obtain: $J^{u+d+s+g} = 0.510(50)$, where the mixing between the quark and gluon singlet operators has been taken into account to 1-loop in lattice perturbation theory.

4 Nucleon Electromagnetic form Factors and Neutron Electric Dipole Moment

The nucleon electromagnetic form factors are obtained from the local vector matrix element $\mathcal{O}_V^\mu$:

$$\langle N(p',s')|\mathcal{O}_V^{\mu}|N(p,s)\rangle = \bar{u}_N(p',s')\left[\gamma_\mu F_1(q^2) + \frac{i\sigma_{\mu\nu}q^\nu}{2m_N}F_2(q^2)\right]u_N(p,s), \quad (3)$$

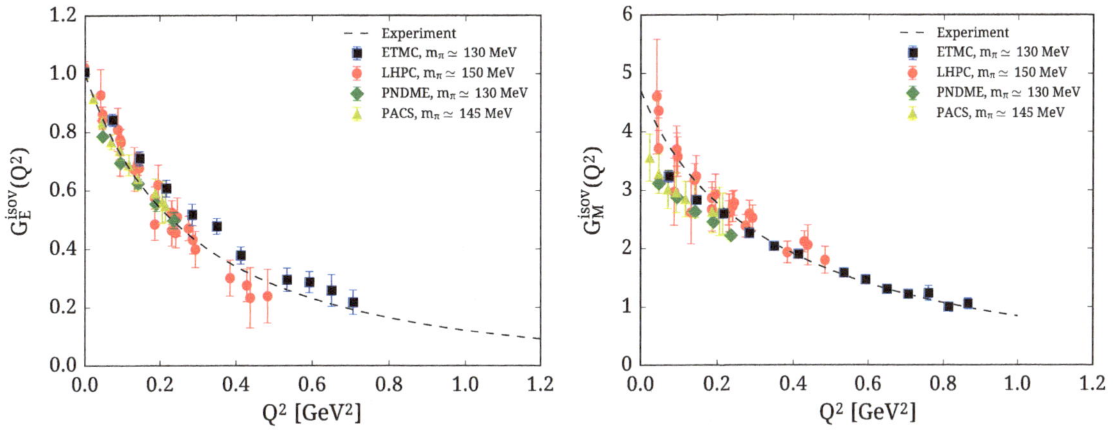

Fig. 6 Isovector electric (*left*) and magnetic (*right*) Sachs form factors using the physical point ensemble (*blue squares*) [14] compared to results by LHPC (*red circles*) [28], PNDME (*green diamonds*) [21] and PACS (*yellow triangles*) [38]. The *dashed line* is a parametrization to experimental results

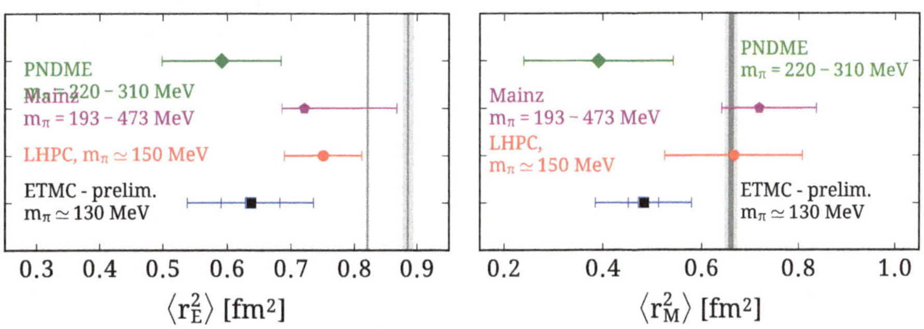

Fig. 7 Nucleon isovector electric (*left*) and magnetic (*right*) radii. The magenta pentagons are from Ref. [23]. The rest of the notation is the same as in Fig. 6

where F_1 and F_2 are the Dirac and Pauli form factors. Alternatively, these can be given in terms of the electric G_E and magnetic G_M Sachs form factors via $G_E(q^2) = F_1(q^2) + \frac{q^2}{(2m_N)^2} F_2(q^2)$ and $G_M(q^2) = F_1(q^2) + F_2(q^2)$. Recent results on the isovector Sachs form factors with pion mass near or at the physical point are shown in Fig. 6. As can be observed, lattice results agree with each other and there is general agreement with experiment for G_E. For G_M, lattice results reproduce the momentum dependence at momentum transfer larger than about 0.3 GeV^2, however there appears to be an underestimation of $G_M(0)$ for most lattice results shown. Calculations are under way at larger volumes, for accessing finer momenta, which can shed light on the low-Q^2 dependence of $G_M(0)$.

The electric and magnetic radii $\langle r_E^2 \rangle$ and $\langle r_M^2 \rangle$ are obtained by the slope of the form factors at zero momentum: $\frac{\partial}{\partial Q^2} G_i(Q^2)|_{Q^2=0} = -\frac{1}{6} G_i(0) \langle r_i^2 \rangle$. This requires modeling the momentum dependence of the lattice data which introduces a systematic error, however methods have been proposed for model independent extraction [13] with preliminary results in Refs [14,32]. Comparison of isovector electric and magnetic radii from recent lattice QCD calculations and with experiment is shown in Fig. 7. The underestimation of experimental results observed may be due to residual excited state contamination. Further investigation at larger time separations and with more statistics is required.

In Eq. 3, the decomposition of the nucleon matrix element assumes a CP-symmetric vacuum. Alternatively, in the presence of CP-violation, the electromagnetic matrix element is decomposed into an additional two CP-odd form factors, namely the dipole F_3 and anapole F_A form factors. The dipole form factor at zero momentum yields the electric dipole moment (nEDM) of the nucleon $|\mathbf{d}_N| = \theta \lim_{Q^2 \to 0} \frac{|F_3(Q^2)|}{2m_N}$ to leading order in the parameter θ, which is a bare parameter of the action multiplying the CP-violating term. Experimentally, a non-zero $|\mathbf{d}_N|$ is sought for as a signal of CP-violation, with current experimental bounds at $|\mathbf{d}_N| < 2.9 \cdot 10^{-13}e\cdot$fm (90% CL) [17].

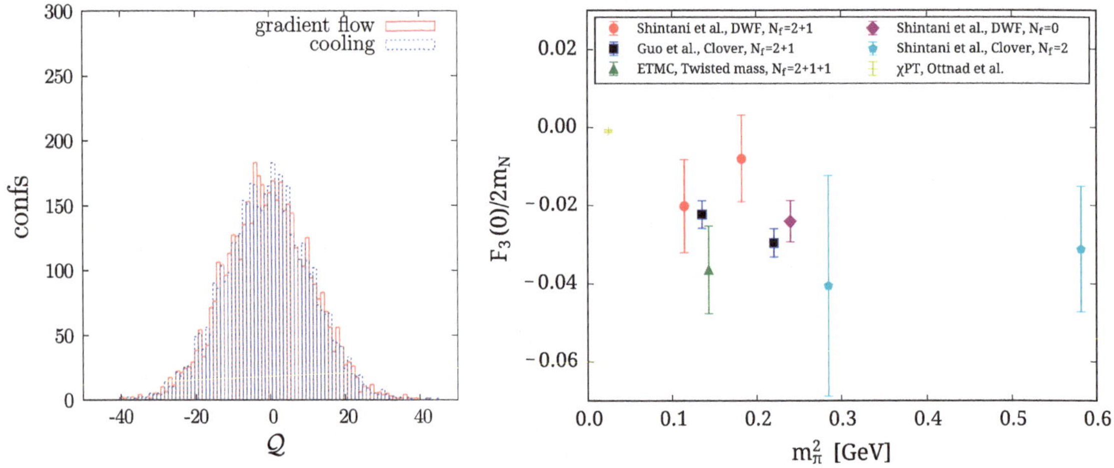

Fig. 8 *Left*: Histogram of the topological charge from B55.32 using cooling (*blue*) and gradient flow (*red*). *Right*: Results for the nEDM using B55.32 (*green triangles*) from direct evaluation of $F_3(0)$ to $O(\theta)$ [11] compared to an $N_\mathrm{f} = 2 + 1$ and a quenched domain wall fermion calculation using the same method [35,37] (*red circles* and *magenta diamonds*), a calculation of $F_3(0)$ after Wick rotation of θ in the action [31] (*blue squares*) and using $N_\mathrm{f} = 2$ clover fermions by calculating the energy shift in the nucleon mass under an external electric field [36] (*cyan pentagons*). A result from chiral perturbation theory is also included [34] (*yellow cross*)

In lattice QCD, studying the nEDM requires inducing parity violation in the QCD action. A sign problem arises since the CP-violating Chern-Simons term is imaginary: $\mathcal{L}_{\mathrm{CS}}(x) = -i\theta q(x) = -\frac{i\theta}{32\pi^2}\epsilon_{\mu\nu\sigma\rho}\mathrm{Tr}[G_{\mu\nu}(x)G_{\rho\sigma}(x)]$. Approaches to this problem include a Wick rotation to imaginary θ [31] or assuming θ is small and expanding the action to leading order in θ [11,35,37]. Furthermore, alternative to the already mentioned calculation of F_3, one can study the nEDM by calculating the shift in the nucleon mass under an external electric field [36]. A calculation of the nEDM from lattice QCD is shown in Fig. 8, using $N_\mathrm{f} = 2 + 1 + 1$ fermions with a pion mass of 373 MeV and lattice spacing of about 0.082 fm (B55.32 ensemble). In the left panel, the distribution of the topological charge: $\mathcal{Q} = \int d^4x q(x)$, needed for weighting to order θ is shown over about 4600 configurations from B55.32. Two link smoothing techniques are shown, giving equivalent results [10], and for each we use the Wilson, the Symanzik tree-level improved and the Iwasaki definition of the lattice gluon action which carry different discretization errors. We observe consistent values within statistical errors from all combinations indicating that lattice artifacts are small. We note that the signal for F_3 is inherently noisier than for the parity symmetric form factors since it is weighted with the topological charge, a Gaussian distribution around zero. This is reflected in the error of our result using B55.32 [11] shown in the right panel of Fig. 8, for which we use statistics of about 5000. Our result is compared to other lattice results, as explained in the caption of Fig. 8, showing qualitative agreement.

5 Summary and Outlook

Lattice QCD simulations using physical values of the quark masses have removed the need for chiral extrapolations thus eliminating an important systematic error in nucleon structure calculations. We have seen how the next significant source of systematic uncertainty contributions from excited states and the challenges in accumulating O(10^5-10^6) statistics for obtaining accurate results, especially for the case of disconnected diagrams and the nEDM. Future lattice QCD calculations will need to address further the issue of excited state effects, including new methods for dealing with unstable particles, as well as extrapolations to the continuum and infinite volume.

Acknowledgements I would like to thank all members of the ETMC for an enjoyable collaboration. I am especially indebted to my collaborators: C. Alexandrou, A. Athenodorou, M. Constantinou, K. Hadjiyiannakou, C. Kallidonis, K. Jansen, K. Ottnad, M. Petschlies, F. Steffens, A. Vaquero and C. Wiese. Results were obtained using Jureca via NIC allocation ECY00, HazelHen at HLRS and SuperMUC at LRZ via Gauss allocations 44066 and 10862 and Piz Daint at CSCS via projects s540 and s625. I thank the staff of these centers for access to the computational resources and for their support.

References

1. A. Abdel-Rehim, C. Alexandrou, M. Constantinou, V. Drach, K. Hadjiyiannakou et al., Disconnected quark loop contributions to nucleon observables in lattice QCD. Phys. Rev. **D89**(3), 034,501 (2014). doi:10.1103/PhysRevD.89.034501. arxiv:1310.6339
2. A. Abdel-Rehim, C. Alexandrou, M. Constantinou, K. Hadjiyiannakou, K. Jansen, C. Kallidonis, G. Koutsou, A. Vaquero Aviles-Casco, Direct evaluation of the quark content of nucleons from lattice QCD at the physical point. Phys. Rev. Lett. **116**(25), 252,001 (2016). doi:10.1103/PhysRevLett.116.252001. arxiv:1601.01624
3. A. Abdel-Rehim, et al., Nucleon and pion structure with lattice QCD simulations at physical value of the pion mass. Phys. Rev. D (in press) (2015a). arxiv:1507.04936
4. A. Abdel-Rehim, et al., Simulating QCD at the physical point with $N_f = 2$ Wilson twisted mass fermions at maximal twist (2015b). arxiv:1507.05068
5. C. Alexandrou, M. Brinet, J. Carbonell, M. Constantinou, P.A. Harraud, P. Guichon, K. Jansen, T. Korzec, M. Papinutto, Axial Nucleon form factors from lattice QCD. Phys. Rev. **D83**, 045,010 (2011a). doi:10.1103/PhysRevD.83.045010. arxiv:1012.0857
6. C. Alexandrou, J. Carbonell, M. Constantinou, P.A. Harraud, P. Guichon, K. Jansen, C. Kallidonis, T. Korzec, M. Papinutto, Moments of nucleon generalized parton distributions from lattice QCD. Phys. Rev. **83**, 114,513 (2011b). doi:10.1103/PhysRevD.83.114513. arxiv:1104.1600
7. C. Alexandrou, K. Hadjiyiannakou, G. Koutsou, A. O'Cais, A. Strelchenko, Evaluation of fermion loops applied to the calculation of the η' mass and the nucleon scalar and electromagnetic form factors. Comput. Phys. Commun. **183**, 1215–1224 (2012). doi:10.1016/j.cpc.2012.01.023. arxiv:1108.2473
8. C. Alexandrou, M. Constantinou, S. Dinter, V. Drach, K. Jansen, C. Kallidonis, G. Koutsou, Nucleon form factors and moments of generalized parton distributions using $N_f = 2 + 1 + 1$ twisted mass fermions. Phys. Rev. **D88**(1), 014,509 (2013). doi:10.1103/PhysRevD.88.014509. arxiv:1303.5979
9. C. Alexandrou, M. Constantinou, V. Drach, K. Hadjiyiannakou, K. Jansen et al., Evaluation of disconnected quark loops for hadron structure using GPUs. Comput. Phys. Commun. **185**, 1370–1382 (2014). doi:10.1016/j.cpc.2014.01.009. arxiv:1309.2256
10. C. Alexandrou, A. Athenodorou, K. Jansen, Topological charge using cooling and the gradient flow. Phys. Rev. D **92**(12), 125,014 (2015). doi:10.1103/PhysRevD.92.125014. arxiv:1509.04259
11. C. Alexandrou, A. Athenodorou, M. Constantinou, K. Hadjiyiannakou, K. Jansen, G. Koutsou, K. Ottnad, M. Petschlies, Neutron electric dipole moment using $N_f = 2+1+1$ twisted mass fermions. Phys. Rev. D **93**(7), 074,503 (2016a). doi:10.1103/PhysRevD.93.074503. arxiv:1510.05823
12. C. Alexandrou, M. Constantinou, K. Hadjiyiannakou, C. Kallidonis, G. Koutsou, K. Jansen, H. Panagopoulos, F. Steffens, A. Vaquero, C. Wiese, Recent results for the proton spin decomposition from lattice QCD [PoSDIS2016,240(2016)], (2016b). arxiv:1609.00253
13. C. Alexandrou, M. Constantinou, G. Koutsou, K. Ottnad, M. Petschlies, A position space method for the nucleon magnetic moment in lattice QCD (2016c). arxiv:1605.07327
14. C. Alexandrou, M. Constantinou, K. Hadjiyiannakou, K. Jansen, C. Kallidonis, G. Koutsou, K. Ottnad, A. Vaquero, Nucleon electromagnetic and axial form factors with $N_f = 2$ twisted mass fermions at the physical point. PoS **LATTICE2016**, 154 (2016d). arXiv:1702.00984
15. C. Alexandrou, M. Constantinou, K. Hadjiyiannakou, K. Jansen, C. Kallidonis, G. Koutsou, A. Vaquero Avilés-Casco, C. Wiese, Nucleon spin and quark content at the physical point. PoS **LATTICE2016**, 153 (2016e). arXiv:1611.09163
16. Y. Aoki, T. Blum, H.W. Lin, S. Ohta, S. Sasaki, R. Tweedie, J. Zanotti, T. Yamazaki, Nucleon isovector structure functions in (2 + 1)-flavor QCD with domain wall fermions. Phys. Rev. D **82**, 014,501 (2010). doi:10.1103/PhysRevD.82.014501. arxiv:1003.3387
17. C.A. Baker et al., An Improved experimental limit on the electric dipole moment of the neutron. Phys. Rev. Lett. **97**, 131,801 (2006). doi:10.1103/PhysRevLett.97.131801. arxiv:hep-ex/0602020
18. G.S. Bali, S. Collins, A. Schafer, Effective noise reduction techniques for disconnected loops in Lattice QCD. Comput. Phys. Commun. **181**, 1570–1583 (2010). doi:10.1016/j.cpc.2010.05.008. arxiv:0910.3970
19. G.S. Bali, S. Collins, B. Glle, M. Gckeler, J. Najjar, R.H. Rdl, A. Schfer, R.W. Schiel, A. Sternbeck, W. Sldner, The moment $\langle x\rangle_{u-d}$ of the nucleon from $N_f = 2$ lattice QCD down to nearly physical quark masses. Phys. Rev. D **90**(7), 074,510 (2014). doi:10.1103/PhysRevD.90.074510. arxiv:1408.6850
20. G.S. Bali et al., Strangeness contribution to the proton spin from lattice QCD. Phys. Rev. Lett. **108**, 222,001 (2012). doi:10.1103/PhysRevLett.108.222001. arxiv:1112.3354
21. T. Bhattacharya, S.D. Cohen, R. Gupta, A. Joseph, H.W. Lin, B. Yoon, Nucleon charges and electromagnetic form factors from 2+1+1-flavor lattice QCD. Phys. Rev. D **89**(9), 094,502 (2014). doi:10.1103/PhysRevD.89.094502. arxiv:1306.5435
22. T. Bhattacharya, V. Cirigliano, S. Cohen, R. Gupta, H.W. Lin, B. Yoon, Axial, scalar and tensor charges of the nucleon from 2 + 1 + 1-flavor lattice QCD. Phys. Rev. D **94**(5), 054,508 (2016). doi:10.1103/PhysRevD.94.054508. arxiv:1606.07049
23. S. Capitani, M. Della Morte, D. Djukanovic, G. von Hippel, J. Hua, B. Jger, B. Knippschild, H.B. Meyer, T.D. Rae, H. Wittig, Nucleon electromagnetic form factors in two-flavor QCD. Phys. Rev. D **92**(5), 054,511 (2015). doi:10.1103/PhysRevD.92.054511. arxiv:1504.04628
24. S. Collins, Hadron Structure, *34th International Symposium on Lattice Field Theory (Lattice 2016)*, Southampton, UK, July 24–30 (2016)
25. M. Constantinou, Hadron Structure. PoS LATTICE2014:001 (2015). arXiv:1411.0078
26. M. Engelhardt, Strange quark contributions to nucleon mass and spin from lattice QCD. Phys. Rev. D **86**, 114,510 (2012). doi:10.1103/PhysRevD.86.114510. arxiv:1210.0025
27. B.W. Filippone, X.D. Ji, The Spin structure of the nucleon. Adv. Nucl. Phys. 26:1 (2001). doi:10.1007/0-306-47915-X-1, arxiv:hep-ph/0101224

28. J. Green, J. Negele, A. Pochinsky, S. Syritsyn, M. Engelhardt et al., Nucleon electromagnetic form factors from lattice QCD using a nearly physical pion mass. Phys. Rev. **D90**(7), 074,507 (2014a). doi:10.1103/PhysRevD.90.074507. arxiv:1404.4029
29. J. Green, N. Hasan, S. Meinel, M. Engelhardt, S. Krieg, J. Laeuchli, J. Negele, K. Orginos, A. Pochinsky, S. Syritsyn, Light and strange axial form factors of the nucleon at pion mass 317 MeV. 34th International Symposium on Lattice Field Theory (Lattice 2016), Southampton, UK, July 24–30 (2016)
30. J.R. Green, M. Engelhardt, S. Krieg, J.W. Negele, A.V. Pochinsky, S.N. Syritsyn, Nucleon structure from lattice QCD using a nearly physical pion mass. Phys. Lett. B **734**, 290–295 (2014). doi:10.1016/j.physletb.2014.05.075. arxiv:1209.1687
31. F.K. Guo, R. Horsley, U.G. Meissner, Y. Nakamura, H. Perlt, P.E.L. Rakow, G. Schierholz, A. Schiller, J.M. Zanotti, The electric dipole moment of the neutron from $2 + 1$ flavor lattice QCD. Phys. Rev. Lett. **115**(6), 062,001 (2015). doi:10.1103/PhysRevLett.115.062001. arxiv:1502.02295
32. N. Hasan, M. Engelhardt , J. Green , S. Krieg , S. Meinel, J. Negele, A. Pochinsky, S. Syritsyn, Computing the nucleon Dirac radius directly at $Q^2 = 0$. In: Proceedings, 34th International Symposium on Lattice Field Theory (Lattice 2016): Southampton, UK, July 24–30, (2016). https://inspirehep.net/record/1495845/files/arXiv:1611.01383.pdf, arxiv:1611.01383
33. M.P. Mendenhall et al., Precision measurement of the neutron β-decay asymmetry. Phys. Rev. C **87**(3), 032,501 (2013). doi:10.1103/PhysRevC.87.032501. arxiv:1210.7048
34. K. Ottnad, B. Kubis, U.G. Meissner, F.K. Guo, New insights into the neutron electric dipole moment. Phys. Lett. B **687**, 42–47 (2010). doi:10.1016/j.physletb.2010.03.005. arxiv:0911.3981
35. E. Shintani, S. Aoki, N. Ishizuka, K. Kanaya, Y. Kikukawa, Y. Kuramashi, M. Okawa, Y. Tanigchi, A. Ukawa, T. Yoshie, Neutron electric dipole moment from lattice QCD. Phys. Rev. D **72**, 014,504 (2005). doi:10.1103/PhysRevD.72.014504. arxiv:hep-lat/0505022
36. E. Shintani, S. Aoki, Y. Kuramashi, Full QCD calculation of neutron electric dipole moment with the external electric field method. Phys. Rev. D **78**, 014,503 (2008). doi:10.1103/PhysRevD.78.014503. arxiv:0803.0797
37. E. Shintani, T. Blum, A. Soni, T. Izubuchi, neutron and proton edm with $n_f = 2+1$ domain-wall fermion. PoS Lattice **2013**, 298 (2014)
38. T. Yamazaki, Light Nuclei and Nucleon Form Factors in $N_f = 2 + 1$ Lattice QCD. PoS LAT-TICE2015:081 (2016). arxiv:1511.09179
39. Y.B. Yang, M. Gong, K.F. Liu, M. Sun, Quark Spin in Proton from Anomalous Ward Indentity. PoSLATTICE2014:138, (2014). arxiv:1504.04052

Few-Body Syst (2017) 58:99
DOI 10.1007/s00601-017-1269-3

O. Oliveira · A. G. Duarte · D. Dudal · P. J. Silva

Gluon and Ghost Dynamics from Lattice QCD

Received: 11 November 2016 / Accepted: 20 February 2017 / Published online: 4 March 2017

Abstract The two point gluon and ghost correlation functions and the three gluon vertex are investigated, in the Landau gauge, using lattice simulations. For the two point functions, we discuss the approach to the continuum limit looking at the dependence on the lattice spacing and volume. The analytical structure of the propagators is also investigated by computing the corresponding spectral functions using an implementation of the Tikhonov regularisation to solve the integral equation. For the three point function we report results when the momentum of one of the gluon lines is set to zero and discuss its implications.

1 Introduction and Motivation

In this proceeding we report on ongoing work concerning the computation of the Landau gauge two and three point functions in pure Yang–Mills SU(3) theory using lattice QCD methods.

The gluon propagator started to be investigated within the lattice QCD approach in the late eighties of the last century [1]. At the time, the focus was on the nature of the propagator and in particular if the gluon behaves as a massive particle. In the nineties, there was a renewed interest due to various proposals for the non-perturbative quantisation of the Yang–Mills theory, some of which made precise predictions for the behaviour of gluon and ghost propagators at zero momentum.

In what concerns the evaluation of the ghost propagator using lattice QCD methods, a first calculation was reported only during the 1990s [2].

On the other hand, in the late nineties of the last and by the beginning of this century, the evolution of the continuum approach to QCD based on the Dyson–Schwinger equations provided another non-perturbative

This article belongs to the Topical Collection "Light Cone 2016".

O. Oliveira (✉)· A. G. Duarte · P. J. Silva
CFisUC, Department of Physics, University of Coimbra, 3004-516 Coimbra, Portugal
E-mail: orlando@fis.uc.pt

A. G. Duarte
E-mail: agd077@hotmail.com

P. J. Silva
E-mail: psilva@uc.pt

D. Dudal
Department of Physics, KU Leuven Campus Kortrijk - KULAK, Etienne Sabbelaan 51 bus 7800, 8500 Kortrijk, Belgium
E-mail: david.dudal@kuleuven.be

D. Dudal
Department of Physics and Astronomy, Ghent University, Krijgslaan 281-S9, 9000 Gent, Belgium

approach which, together with lattice QCD and the increase in computer power, enabled a better comprehension of the fundamental QCD Green functions. Furthermore, new developments on the theoretical side and new numerical simulations gave an important contribution towards our current understanding of the propagators and, partially, on higher order Green functions of QCD.

For the gluon propagator it is now clear that it is finite over the whole range of momenta and it does not vanish at $p = 0$. The non-linear dynamics of QCD is translated into a gluon running mass whose value is about $\sim 2\ \Lambda_{QCD}$ at zero momentum. For the ghost propagator, lattice QCD methods and continuum methods provided essentially the same results and suggest that the propagator follows closely the behavior of a free massless (ghost) particle. Useful reference works on the analytical study of QCD Green's functions are [3–7].

The ghost propagator being essentially the propagator of massless particle has profound implications for the Green functions with higher number of gluon legs. Indeed, if one requires that the renormalised Dyson–Schwinger equations are finite, then the full three gluon vertex should change sign for momenta $\sim$200 MeV [8]. On the other hand, the vanishing of the three gluon vertex also implies that the gluon propagator should have a local maximum at about the same momenta. According to the predictions of the Dyson–Schwinger equations this maximum is very mild and, therefore, it will be very hard to observe in lattice simulations. Anyway, if our current understanding of the QCD dynamics is not that far from the "true exact dynamics" one expects to observe, at least, a change in sign of the three gluon vertex.

Our motivation to revisit the calculation of the gluon and ghost propagators is twofold. Previous simulations which helped to improve the understanding of these two point functions were done on large lattices but using the quite large lattice spacings of the order $\sim$0.2 fm [9, 10]. Also the approach to the continuum limit relied on relative large lattice spacings. Therefore, it is important to check the results of the previous simulations and also to understand how the results of the lattice simulations approach the continuum. Having access to the two propagators, we can also check how the strong coupling changes with the lattice spacing and the lattice volume. We call the reader's attention to the observation that what we call here the strong coupling constant is a particular combination of the gluon and ghost propagators which is a renormalisation group invariant quantity but other definitions for $\alpha_s(p^2)$ are possible. Further, from the gluon and ghost propagators one can also investigate the so-called spectral representation via a Källén–Lehmann representation. The computation of the spectral representation requires inverting an integral equation and this can be achieved via the so-called (linear) Tikhonov regularisation. The spectral representation for the gluon and the ghost can provide us information about the nature of these fields and, in particular, if the Hilbert space of QCD $\mathcal{H}$ includes single gluon and/or ghost particle states. If such states are not within $\mathcal{H}$, then gluons and ghosts can only belong to the physical states as constituents of other particles and, in this sense, are confined.

The three gluon vertex is an important element of the continuum approach to QCD and it provides an important test to the full tower of Green functions and, in particular, a test of the nature of the ghost propagator. Our final aim is to provide a complete as possible description of the three gluon vertex but herein we just report results for a single kinematical configuration.

The current proceeding summarises the results reported before in [11–13] and the interested reader should consult these references for further details. Here we also report on our investigation of the gluon and ghost spectral density computed as described in [14].

2 The Gluon Propagator

In Euclidean space and in the Landau gauge the gluon and ghost propagators are given by

$$D^{ab}_{\mu\nu}(p) = \delta^{ab}\left(\delta_{\mu\nu} - \frac{p_\mu p_\nu}{p^2}\right) D(p^2) \quad \text{and} \quad G^{ab} = -\delta^{ab} G(p^2), \tag{1}$$

where the latin indices mean color degrees of freedom and the greek indices Lorentz degrees of freedom.

The general features of the gluon and ghost propagators can be seen on Fig. 1. The gluon propagator in the infrared region can be described by a Yukawa like propagator [15] and, therefore, its infrared behaviour can be approximated by a quasi-particle massive type of propagator. In the ultraviolet region the lattice data is well described by one-loop renormalisation group improved perturbation theory. Moreover, as a function of p^2, the function $D(p^2)$ shows two different curvatures which can be linked to different signs for its spectral density. The different curvatures can be interpreted as a sign of positivity violation and, therefore, $\mathcal{H}$ should not contain one gluon states. On the other side, as the figure shows the ghost propagator retains essentially its perturbative form and, in this sense, behaves as a massless particle. Indeed, we have tried to fit $G(p^2)$

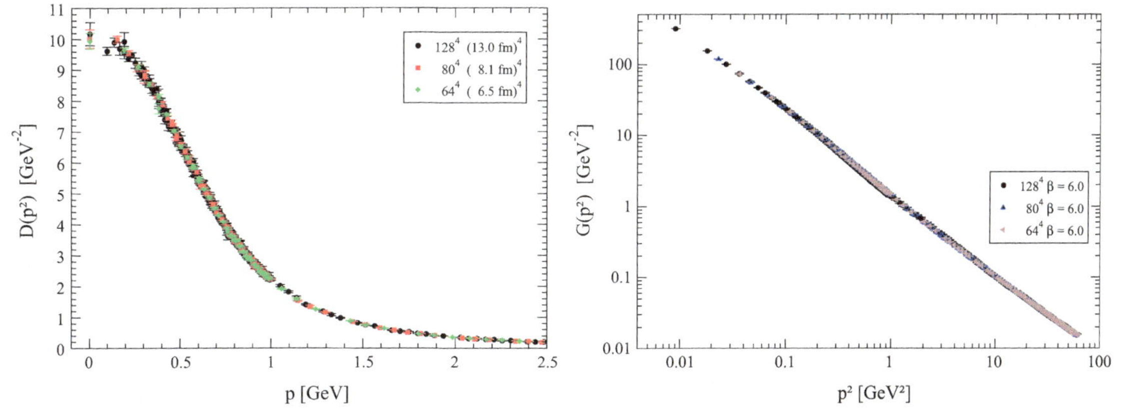

Fig. 1 Gluon (*left*) and ghost (*right*) propagator compute for $\beta = 6.0$ and various lattice volumes and renormalized at $\mu = 4\,\text{GeV}$. The lattice spacing associated with these simulations being $a = 0.1016\,\text{fm}$

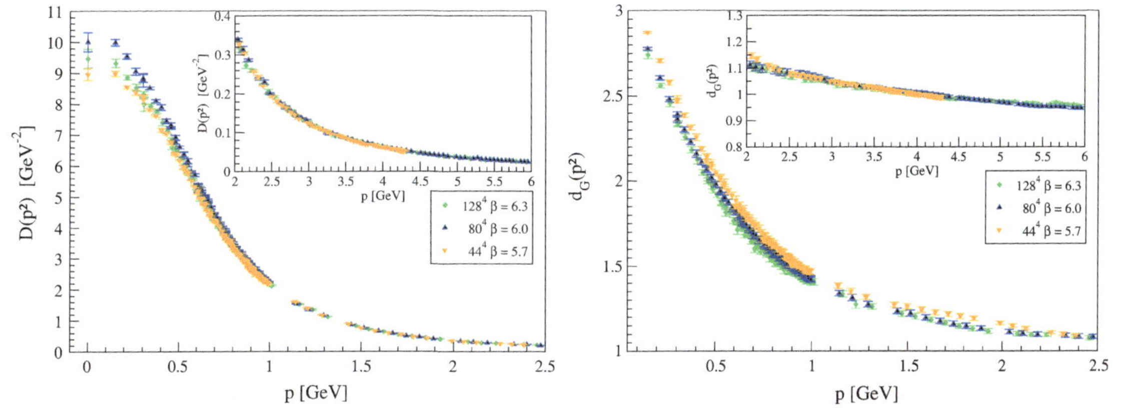

Fig. 2 Gluon propagator (*left*) and ghost dressing function $d_G(p^2) = p^2 G(p^2)$ (*right*) compute using different lattice spacings and the same physical volume $V \sim (8\,\text{fm})^4$. All data is renormalized at $\mu = 4\,\text{GeV}$. The lattice spacing associated with these simulations being $a = 0.1838\,\text{fm}$ for $\beta = 5.7$, $a = 0.1016\,\text{fm}$ for $\beta = 6.0$ and $a = 0.0627$ for $\beta = 6.3$

to the one-loop renormalisation group improved perturbative expression and verified that the lattice data and perturbation theory are already compatible for momenta above $\sim$800 MeV.

Figure 1 also shows that the volume dependence of the lattice results for lattices with physical volumes greater than $(6.5\,\text{fm})^4$ is very mild.

The dependence of the lattice propagators with the lattice spacing, for the same physical volume, is not trivial and can be seen in Fig. 2. For the gluon propagator the differences show up for momenta smaller than $\sim$500 MeV and, as the figure shows, the dependence is not monotonous. The simulation using the larger value for the lattice spacing lies always below all the remaining data. On the other hand, the data for the ghost propagator points towards a decrease of the lattice propagator as one approaches the continuum limit. For the ghost, the differences between the various sets of data appear at all momenta. Our simulations suggest that the results for $\beta = 5.7$ provide lower bounds for the gluon propagator and upper bounds for the ghost propagator.

In the Landau gauge one can define a renormalisation group invariant strong coupling constant from the gluon and ghost propagators. The definition is[1]

$$\alpha_s(p^2) = \frac{g^2}{4\pi}\, d_D(p^2)\, d_G^2(p^2), \tag{2}$$

where $d_D(p^2) = p^2 D(p^2)$ and $d_G(p^2) = p^2 G(p^2)$ are the gluon and ghost dressing functions, respectively, and the lattice data can be seen in Fig. 3. As the figure shows, $\alpha_s(p^2)$ has a mild or almost no dependence on

[1] This is one of the many possibilities to define a strong coupling constant over the whole momentum range, see the discussions in e.g. [16–18], which all share the same property of having the desired UV perturbative behaviour. The upshot of using the specific definition of (2) is that this can be accessed from lattice data.

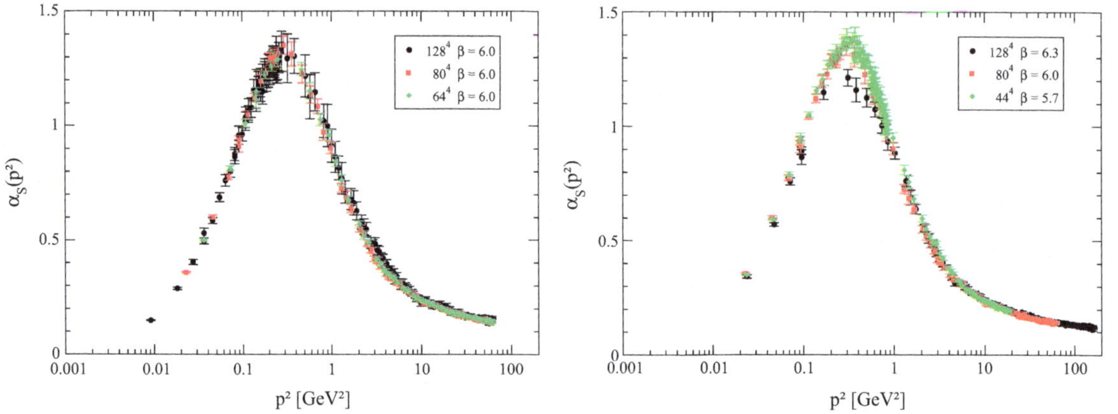

Fig. 3 Volume dependence (*left*) and lattice spacing dependence (*right*) of the strong coupling constant as defined by Eq. (2). For the values of the lattice spacing see caption of Fig. 2

the physical volume used in the simulation. However, the strong coupling constant seems to have a smaller maximum, although its position occurs at the same momentum, when the lattice spacing is decreased.

We would like to call the reader's attention that, although we see nontrivial dependences of the lattice results with the lattice spacing, the functional forms of the propagators and the strong coupling constant seem to be same. Indeed, the lattice propagators and, therefore, the strong coupling are well described by the same function forms.

3 Gluon and Ghost Spectral Representation

The Källén–Lehmann representation associated to the gluon propagator is given by

$$D(p^2) = \int_0^{+\infty} d\mu \, \frac{\rho(\mu)}{p^2 + \mu}, \tag{3}$$

where the spectral function $\rho(\mu)$ would be a positive definite function if the Hilbert space of states of QCD includes asymptotical one gluon states. If $\mathcal{H}$ does not include such gluon states, then it is not clear if the spectral integral (3) even holds to begin with. For the remainder of this proceeding, we will nevertheless assume that the gluon and ghost propagators have an associated spectral function defined in the usual way and we shall try to get some information on the function $\rho(\mu)$ from the lattice data. A non-positive spectral function will be interpreted as a sign that gluons are confined, i.e. can only be part of composite states of $\mathcal{H}$.

According to renormalization group improved perturbation theory (leading log resummation), for sufficiently high momenta, the gluon propagator is given by

$$D(p^2) \propto \frac{1}{p^2} \left[\frac{\ln \frac{p^2}{\Lambda_R^2}}{\ln \frac{\mu_R^2}{\Lambda_R^2}} \right]^{-\gamma}, \tag{4}$$

where μ_R is the renormalisation scale, and Λ_R the renormalization group invariant QCD scale in a given scheme R, here we have $\gamma = 13/22$, up to the sign, the gluon anomalous dimension rescaled by the leading order β-function coefficient. Given that $D(p^2)$ decreases faster than $1/p^2$ at high momenta, then the spectral density satisfies the following sum rule

$$\int d\mu \, \rho(\mu) = 0; \tag{5}$$

see [19] for a review on spectral representations for non-Abelian gauge theories and for some further details.

In order to invert the integral equation (3) one needs to regularise it. Our calculation of $\rho(\mu)$ uses the Tikhonov regularisation, it follows [14] and it corresponds to minimising the following functional

$$\mathcal{J}_\lambda = \sum_{i=1}^{N} \left[\int_{\mu_0}^{+\infty} d\mu \, \frac{\rho(\mu)}{p_i^2 + \mu} - D(p_i^2) \right]^2 + \lambda(\mu_0) \int_{\mu_0}^{+\infty} d\mu \, \rho^2(\mu) \tag{6}$$

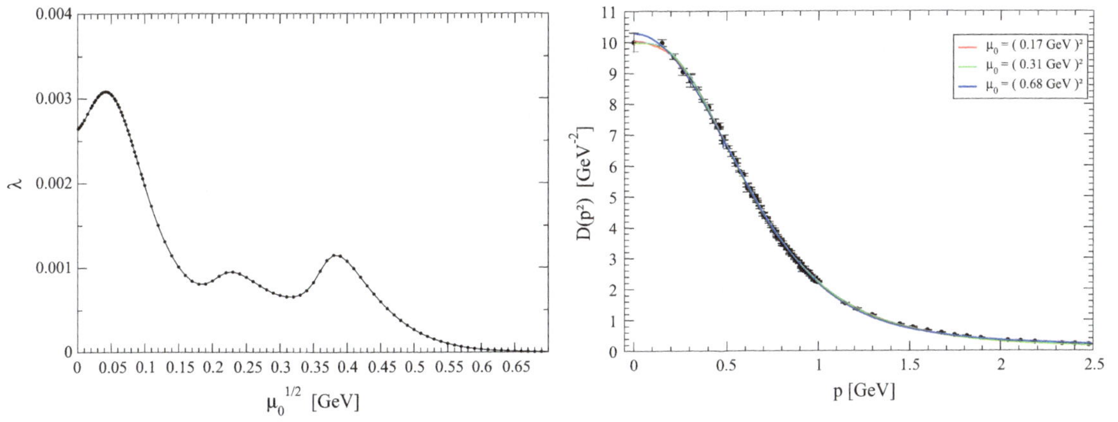

Fig. 4 The curve $\lambda(\mu_0)$ (*left*) together with the lattice gluon data (80^4 and $\beta = 6.0$ simulation) and the rebuilt propagator for several optimal λ parameters

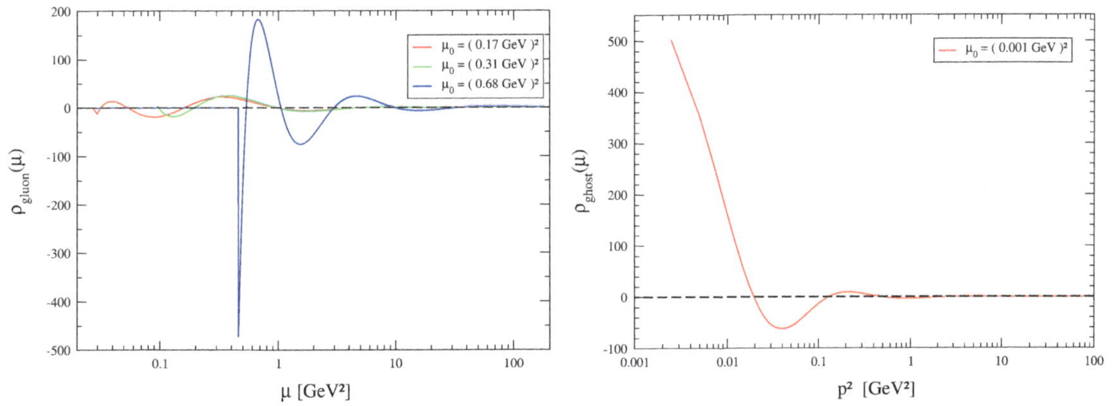

Fig. 5 The gluon (*left*) and ghost (*right*) spectral functions as foreseen by the Tikhonov regularisation and using the Morozov criterium

where N is the number of lattice momenta p_i^2 where the lattice propagator is defined, μ_0 an infrared cutoff and λ the regularisation parameter to be determined by the Morozov discrepancy principle, which is a way to select the optimal λ while keeping into account the quality of the data ($\sim$ total error margin). We refer to the literature for additional details. Afterwards, we choose the optimal λ's at the minimum of the curve $\lambda(\mu_0)$. Note that the minimisation procedure favours spectral functions with small $L_2(\mu_0, +\infty)$ norms. This is open to improvement for which we refer to future work. The minimisation procedure reduces the problem to the solution of a linear system of equations. From its solution one can compute both the spectral function and rebuild the propagator using (3).

From the simulation using the 80^4 lattice and $\beta = 6.0$, the function $\lambda(\mu_0)$ together with the propagator data and the rebuilt propagator for the various minima of the curve $\lambda(\mu_0)$ for the gluon propagator can be seen in Fig. 4. The corresponding spectral function can be seen in Fig. 5 and it clearly show regions of momenta where the sign of ρ is either positive or negative. According to our reasoning, this means that the Hilbert space of QCD does not contain single gluon states and, therefore, gluons have to be confined in composite states belonging to $\mathcal{H}$.

In Fig. 4 we also report, for the first time, the preliminary result for the spectral function associated to the ghost propagator. As can be seen, the lattice data favours a small or vanishing infrared cutoff and, furthermore, $\rho(\mu)$ shows a steepest increase as one approaches the zero momentum suggesting that indeed $\rho(\mu) = \delta(\mu) + \cdots$, as one would expect from an essentially massless propagator. In this sense, it seems that our inversion procedure is able to capture the main features of the spectral function. We will report on this in a forthcoming larger paper.

In what concerns the sum rule (5), our preliminary analysis suggests that indeed the spectral function $\rho(\mu)$ computed using the Tikhonov regularisation is satisfied within the statistical precision of the simulation. A more detailed study, including error analysis and continuum extrapolation is underway.

As a final comment we would like to point out that our estimate for $\rho(\mu)$ does not agree very well with that predicted by solving the Dyson–Schwinger equations in the complex p^2 plane for the gluon and ghost propagators [20]. Furthermore, the gluon propagator data is compatible with the tree level expression as predicted by the refined Gribov–Zwanziger expression [4,21,22],

$$D(p^2) = z \frac{p^2 + m_1^2}{p^4 + m_2^2\, p^2 + m_3^4}, \tag{7}$$

where m_i are parameters with dimensions of mass, and, according to this expression, the gluon propagator has a pair of complex conjugate poles at $p^2 \sim -0.3 \pm i\, 0.5\,\mathrm{GeV}^2$ for the simulations reported here. This result implies that the spectral decomposition (3) with a real $\rho(\mu)$ does not hold for the gluon. A more elaborate discussion of these issues will be discussed elsewhere, including possible generalizations of the Källén–Lehmaan spectral integral.

For other recent works related to gluon, ghost and quark spectral functions at zero temperature, let us refer to [23,24].

4 The Three Gluon Vertex

The three gluon one particle irreducible (1PI) Green function $\Gamma^{a_1 a_2 a_3}_{\mu_1 \mu_2 \mu_3}(p_1, p_2, p_3)$, our convention being all momenta are incoming which means that $p_1 + p_2 + p_3 = 0$, can be decomposed in terms of six Lorentz scalar form factors [25]. On the lattice one can not measure directly 1PI functions but, instead, one can access the complete Green function given by

$$G^{(n)}\,{}^{a_1 a_2 a_3}_{\mu_1 \mu_2 \mu_3}(p_1, p_2, p_3) = \left\langle A^{a_1}_{\mu_1}(p_1) A^{a_2}_{\mu_2}(p_2) A^{a_3}_{\mu_3}(p_3) \right\rangle, \tag{8}$$

which can be written in terms of propagators and the three gluon irreducible Green function and, therefore, one can only measure combinations of the form factors appearing in Γ. Of the possible contractions and momenta configurations, we will report results for

$$G^{(n)}\,{}_{\alpha\mu\alpha}(p, 0, -p)\, p_\mu = \mathrm{Tr}\langle A_\alpha(p)\, A_\mu(0) A_\alpha(-p)\rangle\, p_\mu = \frac{N_c(N_c^2-1)}{4} \left[D(p^2)\right]^2\, D(0)\, \Gamma(p^2)\, p^2 \tag{9}$$

where Tr stands for trace over color indices. This function was also investigated in the first lattice calculation of the three gluon vertex [26]. According to the analysis of the Dyson–Schwinger equations, one expects the form factor $\Gamma(p^2)$ to change sign for momenta ~200 MeV. We recall that such a change of sign was already observed in the simulations of 3D SU(2) pure gauge theories [27] and recently also for 4D SU(3) pure Yang–Mills [13,28].

For the computation of $\Gamma(p^2)$ we considered two sets of pure gauge ensembles both generated with the Wilson action and at $\beta = 6.0$: (i) a smaller lattice 64^4, which corresponds to a physical volume of $(6.5\,\mathrm{fm})^4$, but with better statistics (2000 configurations); (ii) a larger lattice of 80^4, which has a physical volume of $(8.1\,\mathrm{fm})^4$, but using less (279) configurations. As discussed in [13], if at the level of the gluon propagator we see no difference for different types of momenta, for the three point function, due to the breaking of rotational symmetry, the results depend on the type of momenta considered. In order to reduce possible effects associated to the breaking of the group $O(4)$, we report results only for momenta of types $(n_x, n_y, 0, 0)$, $(n_x, n_y, n_z, 0)$ and (n_x, n_y, n_z, n_t).

The form factors associated to $G^{(n)}$ can be seen in Fig. 6. On the left hand side, $\Gamma(p^2)$ is plotted for momenta below 2 GeV and the data suggest that this form factors changes sign for $p \sim 200\,\mathrm{MeV}$ as expected from the analysis made in [8,29]. A more careful analysis of the lattice data suggests that the zero crossing for 4D SU(3) Yang–Mills theory should take place in the interval 216–264 MeV in good agreement also with the results for 3D SU(2) Yang–Mills simulations [27].

On the right side plot of Fig. 6, we display the results of our simulation for the 80^4 over the full range of momenta, together with the predictions of perturbation theory. As the figure shows, at sufficiently high momenta the results of the lattice simulation are in good agreement with renormalisation group improved perturbation theory.

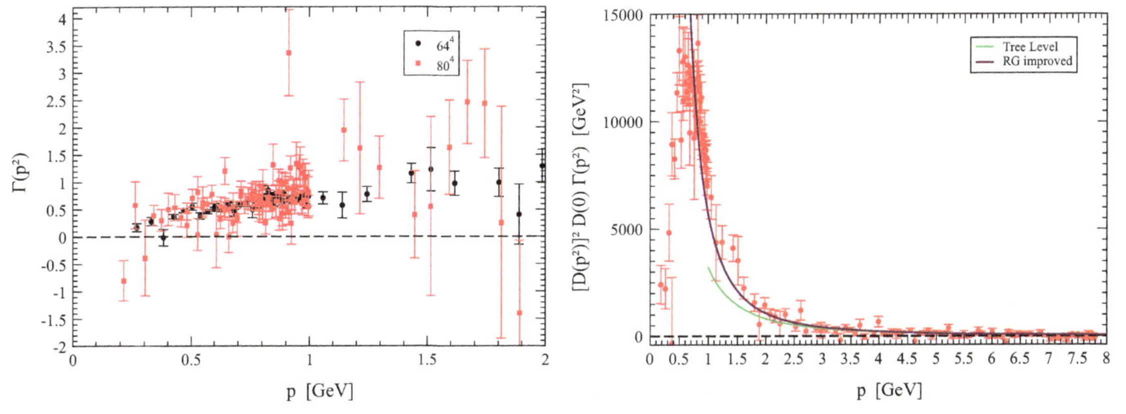

Fig. 6 The bare form factor associated to the complete Green function (9). The *left plot* refers to $\Gamma(p^2)$ up to momenta 2 GeV. The *right plot* includes the propagator form factors together with the predictions of perturbation theory using $\Lambda_{QCD} = 200$ MeV

5 Summary and Conclusions

In this work we report the results of simulating pure Yang–Mills theory on a finite lattice. Our simulation confirms that the gluon propagator is finite and non-vanishing at all momenta and that it is the propagator of a massive-like boson. On the other hand, the ghost reproduces the propagator of an essentially massless degree of freedom. Further, the simulations show that for the propagators the finite volume effects are under control for lattice volumes of $(6.5\,\text{fm})^4$ and above and that the most important lattice artifacts are related to the value of the lattice spacing. While the gluon propagator seems to approach its continuum value from below, the ghost propagator seems to approach its continuum limit from above.

We have also investigated the gluon and ghost spectral representations. The two spectral functions have positive and negative values for different ranges of momenta, meaning that the spectrum of QCD does not account for single gluon (and evidently) ghost states. Gluon and ghost only show up in the eigenstates of $\mathcal{H}$ as constituents of other particles and, in this sense, they are confined particles. For what concerns the spectral functions, there are still a number of problems that need to be solved, in particular in relation to the ultraviolet tails which are no well reproduced by the method used in the current work. This can be understood, at least partially, due to the large number of data in the low momenta region and relatively scarce number of data points for momenta above two or three GeV. Ways to overcome these problems are currently under investigation.

The good agreement between lattice simulations and theoretical analyses (refined Gribov–Zwanziger, Curci–Ferrari models, Dyson–Schwinger equations, functional renormalization group equations, etc.) for the propagators allows us to conclude that we have now a good theoretical picture for the glue and ghost dynamics.

For what concerns the investigation of the three gluon vertex, our simulations support that a change of sign for the form factor Γ takes place at momenta $\sim$200 MeV. The analysis of the simulation shows that the artifacts due to the breaking of rotational symmetry are larger for the three point function than for the two point functions. The simulations also show that in the high momenta regime they agree well with the results from perturbation theory. We aim to proceed our investigation of the three point function in order to describe the form factors that are associated to the three gluon vertex and we hope that our results will also open the road to study gluon correlation functions with a higher number of external legs.

Acknowledgements O. Oliveira, A. G. Duarte and P. J. Silva acknowledge financial support from FCT Portugal under contract with reference UID/FIS/04564/2016. P. J. Silva acknowledges supported by FCT under Contracts No. SFRH/BPD/40998/2007 and SFRH/BPD/109971/2015. The computing time was provided by the Laboratory for Advanced Computing at the University of Coimbra and by PRACE Projects COIMBRALATT (DECI-9) and COIMBRALATT2 (DECI-12).

References

1. J.E. Mandula, M. Ogilvie, The gluon is massive: a lattice calculation of the gluon propagator in the Landau gauge. Phys. Lett. B **185**, 127 (1987)
2. H. Suman, K. Schilling, First lattice study of ghost propagators in SU(2) and SU(3) Gauge theories. Phys. Lett. B **373**, 314 (1996)

3. R. Alkofer, L. von Smekal, The infrared behavior of QCD Green's functions: confinement dynamical symmetry breaking, and hadrons as relativistic bound states. Phys. Rept. **353**, 281 (2001)
4. D. Dudal, J.A. Gracey, S.P. Sorella, N. Vandersickel, H. Verschelde, A refinement of the Gribov–Zwanziger approach in the Landau gauge: infrared propagators in harmony with the lattice results. Phys. Rev. D **78**, 065047 (2008)
5. D. Binosi, J. Papavassiliou, Pinch technique: theory and applications. Phys. Rept. **479**, 1 (2009)
6. P. Boucaud, J.P. Leroy, A.L. Yaouanc, J. Micheli, O. Pene, J. Rodriguez-Quintero, The infrared behaviour of the pure Yang–Mills Green functions. Few Body Syst. **53**, 387 (2012)
7. A. Bashir, L. Chang, I.C. Cloet, B. El-Bennich, Y.X. Liu, C.D. Roberts, P.C. Tandy, Collective perspective on advances in Dyson–Schwinger equation QCD. Commun. Theor. Phys. **58**, 79 (2012)
8. D. Binosi, D. Ibañez, J. Papavassiliou, QCD effective charge from the three-gluon vertex of the background-field method. Phys. Rev. D **87**, 125026 (2013)
9. A. Cucchieri, T. Mendes, What's up with IR gluon and ghost propagators in Landau gauge? A puzzling answer from huge lattices. PoS (LAT2007), 297 (2007). arXiv:0710.0412
10. I.L. Bogolubsky, E.M. Ilgenfritz, M. Muller-Preussker, A. Sternbeck, Lattice gluodynamics computation of Landau gauge Green's functions in the deep infrared. Phys. Lett. B **676**, 69 (2009)
11. O. Oliveira, P.J. Silva, The lattice Landau gauge gluon propagator: lattice spacing and volume dependence. Phys. Rev. D **86**, 114513 (2012)
12. A.G. Duarte, O. Oliveira, P.J. Silva, Lattice gluon and ghost propagators, and the strong coupling in pure SU(3) Yang–Mills theory: finite lattice spacing and volume effects. Phys. Rev. D **94**, 014502 (2016)
13. A.G. Duarte, O. Oliveira, P.J. Silva, Further evidence for zero crossing on the three gluon vertex. Phys. Rev. D **94**, 074502 (2016)
14. D. Dudal, O. Oliveira, P.J. Silva, Källén–Lehmann spectroscopy for (un)physical degrees of freedom. Phys. Rev. D **89**, 014010 (2014)
15. O. Oliveira, P. Bicudo, Running gluon mass from Landau gauge lattice QCD propagator. J. Phys. G **38**, 045003 (2011)
16. C. Lerche, L. von Smekal, On the infrared exponent for gluon and ghost propagation in Landau gauge QCD. Phys. Rev. D **65**, 125006 (2002)
17. A.C. Aguilar, D. Binosi, J. Papavassiliou, J. Rodriguez-Quintero, Non-perturbative comparison of QCD effective charges. Phys. Rev. D **80**, 085018 (2009)
18. D. Binosi, C. Mezrag, J. Papavassiliou, C.D. Roberts, J. Rodriguez-Quintero, Process-independent strong running coupling (2016). arXiv:1612.04835 [nucl-th]
19. J.M. Cornwall, Positivity violations in QCD. Mod. Phys. Lett A **28**, 133005 (2013)
20. S. Strauss, C.S. Fischer, C. Kellermann, Analytic structure of the Landau gauge gluon propagator. Phys. Rev. Lett. **109**, 252001 (2012)
21. D. Dudal, O. Oliveira, N. Vandersickel, Indirect lattice evidence for the Refined Gribov–Zwanziger formalism and the gluon condensate $\langle A^2 \rangle$ in the Landau gauge. Phys. Rev. D **81**, 074505 (2010)
22. A. Cucchieri, D. Dudal, T. Mendes, N. Vandersickel, Modeling the gluon propagator in Landau gauge: lattice estimates of pole masses and dimension-two condensates. Phys. Rev. D **85**, 094513 (2012)
23. F. Siringo, Analytic structure of QCD propagators in Minkowski space (2016). arXiv:1605.07357 [hep-ph]
24. A. Rothkopf, Bayesian inference of non-positive spectral functions in quantum field theory (2016). arXiv:1611.00482 [hep-ph]
25. J.S. Ball, T.-W. Chiu, Analytic properties of the vertex function in gauge theories. 2. Phys. Rev. D **22**, 2550 (1980). Erratum, Phys. Rev. D **23**, 3085 (1981)
26. B. Allés, D.S. Henty, H. Panagopoulos, C. Parrinello, C. Pittori, D.G. Richards, α_s from the nonperturbatively renormalised three gluon vertex. Nucl. Phys. B **502**, 325 (1997)
27. A. Cucchieri, A. Maas, T. Mendes, Three-point vertices in Landau-gauge Yang–Mills theory. Phys. Rev. D **77**, 094510 (2008)
28. A. Athenodorou, D. Binosi, Ph Boucaud, F. De Soto, J. Papavassiliou, J. Rodríguez-Quintero, S. Zafeiropoulos, On the zero crossing of the three-gluon vertex. Phys. Lett. B **761**, 444 (2016)
29. G. Eichmann, R. Williams, R. Alkofer, M. Vujinovic, Three-gluon vertex in Landau gauge. Phys. Rev. D **89**, 105014 (2014)

Few-Body Syst (2017) 58:140
DOI 10.1007/s00601-017-1301-7

Marina Krstić Marinković

Lattice Calculations and the Muon Anomalous Magnetic Moment

Received: 28 March 2017 / Accepted: 17 April 2017 / Published online: 26 May 2017

Abstract Anomalous magnetic moment of the muon, $a_\mu = (g_\mu - 2)/2$, is one of the most precisely measured quantities in particle physics and it provides a stringent test of the Standard Model. The planned improvements of the experimental precision at Fermilab and at J-PARC propel further reduction of the theoretical uncertainty of a_μ. The hope is that the efforts on both sides will help resolve the current discrepancy between the experimental measurement of a_μ and its theoretical prediction, and potentially gain insight into new physics. The dominant sources of the uncertainty in the theoretical prediction of a_μ are the errors of the hadronic contributions. I will discuss recent progress on determination of hadronic contributions to a_μ from lattice calculations.

1 Introduction

The contributions of new physics (NP) beyond the Standard Model (SM) to the anomalous magnetic moment of a lepton, $a_\ell = \frac{g_l - 2}{2}$, are expected to scale as $a_\ell - a_\ell^{\mathrm{SM}} \propto (m_\ell^2/\Lambda_{\mathrm{NP}}^2)$ for leptons $\ell = e, \mu, \tau$ and some new physics scale Λ_{NP}. The muon magnetic moment, a_μ, is therefore m_μ^2/m_e^2 times more sensitive to NP than the one of the electron, while the magnetic moment of τ is not yet experimentally accessible. Additionally, a_μ can be obtained with high precision both from the experiment [6], as well as from the SM prediction [26,42] and as such serves as one of the most sensitive tests of the SM.

The experimental measurement of a_μ currently shows a 3σ to 4σ deviation from the SM estimate [26,42]. The summary of various SM contributions to a_μ given in Table 1 shows that the dominating uncertainty comes from the hadronic contributions. The value of the hadronic vacuum polarization (HVP) entering the theoretical average in Table 1 is obtained via a dispersion integral of the total hadronic e^+e^- annihilation and τ-decays cross sections (for comprehensive reviews see [42,51]), while the value of hadronic Light by Light contribution (HLbL) quoted in Table 1 is purely based on low energy effective QCD hadronic models. Due to the plans of Fermilab [57] and J-PARC [55] experiments to reduce the experimental uncertainty by a factor of four, performing an independent check of the phenomenological estimate of the hadronic contributions is of crucial importance in order to confirm or resolve current discrepancy between theory and experiment, and potentially gain insight into new physics beyond the SM. Lattice discretization of gauge theories provides a way to compute the contributions to a_μ with prevailing uncertainties, HVP and HLbL, from first principles. We

This article belongs to the Topical Collection "Light Cone 2016".

M. K. Marinković
Trinity College Dublin, School of Mathematics, Dublin 2, Ireland
E-mail: mmarina@maths.tcd.ie

M. K. Marinković (✉)
Theoretical Physics Department, CERN, Geneva, Switzerland
E-mail: marina.marinkovic@cern.ch

Table 1 Comparison between a SM prediction and the experimental result for a_μ. The last column contains example Feynman diagrams for the individual contributions to the total theoretical estimate

Contribution	Value	Uncertainty	Sample diagrams
QED $O(\alpha)$	116,140,973.21	0.03	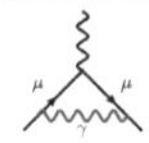
QED $O(\alpha^2)$	413,217.63	0.01	
QED $O(\alpha^3)$	30,141.90	0.00	
QED $O(\alpha^4)$	381.01	0.02	
QED $O(\alpha^5)$	5.09	0.01	
QED, total [2]	116,584,718.95	0.04	
Electroweak, total [34]	153.6	1.0	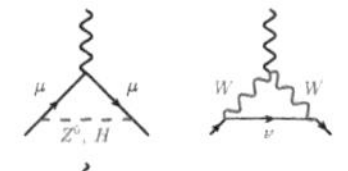
HVP (LO) [39]	6949.9	43.0	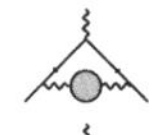
HVP (NLO) [39]	−98.0	1.0	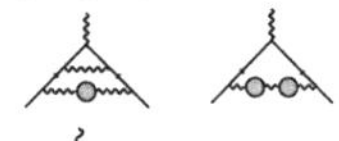
HLbL [43]	116.0	40.0	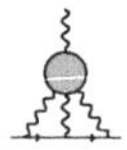
HVP(NNLO) [46]	12.4	0.1	
HLbL(NLO) [22]	3.0	2.0	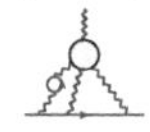
Theory, total	116,591,855.0	59.0	
Experiment [50]	116,592,089.0	63.0	

will present in the following sections the ongoing progress of these calculations, aiming to achieve sub-percent precision for the HVP and ≈10% precision for HLbL.

2 Hadronic Vacuum polarization in Euclidean Space–Time

Ever since the first attempt to perform a lattice QCD computation of the HVP is made by T. Blum [7], a precise computation of this quantity has become one of the long-standing goals of the lattice comunity. The leading order HVP correction to a_μ at Euclidean space–time is obtained as [7,47]:

$$a_\mu^{\rm HVP} = \left(\frac{\alpha}{\pi}\right)^2 \int_0^\infty dQ^2 f(Q^2) \times \hat{\Pi}(Q^2), \qquad (1)$$

where $\hat{\Pi}(Q^2) = \Pi(Q^2) - \Pi(0)$ is the renormalized scalar vacuum polarization, α is the fine structure constant, m_μ is the muon mass and $f(Q^2)$ is analytically known[1] function. Gauge and Lorentz invariance relate scalar hadronic vacuum polarization function and vacuum polarization tensor $\Pi_{\mu\nu}(Q) = (Q^2\delta_{\mu\nu} - Q_\mu Q_\nu)\Pi(Q^2)$, which can be computed using a lattice discretization of the vector current J_μ^{em}:

$$\Pi_{\mu\nu}(Q) = a^4 \sum_x e^{iQx} \langle J_\mu^{em}(x) J_\nu^{em}(0)\rangle. \qquad (2)$$

The vector two-point correlator in Eq. 2 consists of the quark-connected and quark-disconnected contributions illustrated in the leftmost panel of Fig. 1, whose continuum and infinite volume limits are usually computed independently.

[1] $f(Q^2) = m_\mu^2 Q^2 Z^3(Q^2) \frac{1-Q^2 Z(Q^2)}{1+m_\mu^2 Q^2 Z^2(Q^2)}$, where $Z(Q^2) = \frac{\sqrt{(Q^2)^2+4m_\mu^2 Q^2}-Q^2}{2m_\mu^2 Q^2}$.

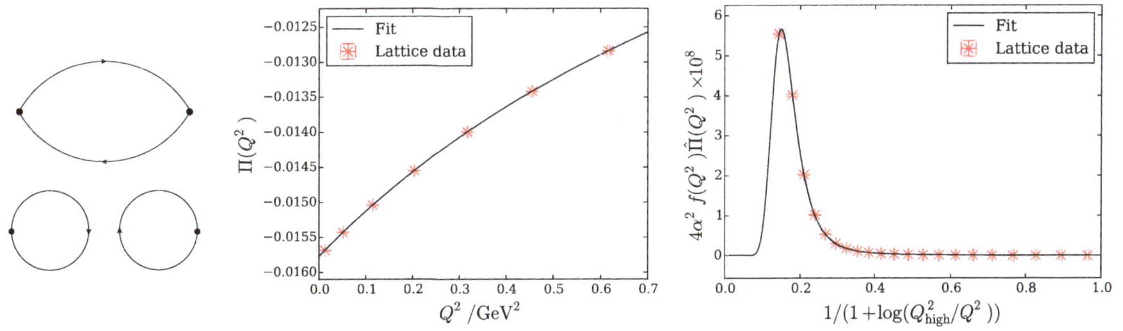

Fig. 1 *Left panel* Connected HVP (up) and disconnected HVP contribution (down). The *dots* represent the insertion of the conserved and/or local vector currents. *Central panel* Strange HVP function ($\Pi_{\mu\nu}(Q) = a^4 \sum_x e^{iQx} \langle J_\mu^{em,s}(x) J_\nu^{em,s}(0)\rangle$) reproduced from Ref. [8], measured on $N_f = 2+1$ configurations with Domain Wall fermions. *Right panel* The strange quark contribution to the integrand of Eq. 1, reproduced from the strange HVP study of Ref. [8]

The integrand of Eq. 1 is strongly peaked at $Q^2 \approx \frac{m_\mu^2}{4}$ and the integral corresponding to the a_μ^{HVP} is therefore dominated by the low momentum region (see Fig. 1). Lattice computation is in practice performed in a finite box, with time and spatial extents L_μ and periodic boundary conditions. The lattice momenta are thus quantized: $Q_\mu = \frac{2\pi}{L_\mu} n_\mu$, $n_\mu = 0, \ldots, L_\mu - 1$, $\mu = 0 \ldots 3$, and this condition leaves us with few lattice points in the most important region for the evaluation of the HVP integral: around and bellow the peak of the integrand. The latter calls for the usage of models or parametrizations that would estimate the value of $\Pi(Q^2)$ in the low momentum region. Moreover, the signal to noise ratio worsens at small Q^2. For all these reasons, a number of tricks for noise reduction and control of the systematic uncertainties needed to be developed in order to bring the accuracy of the lattice computation of a_μ^{HVP} closer to the precision of the estimates coming from the phenomenology and experiment.

2.1 Strategies for the Evaluation of the HVP on the Lattice

Initially, computing a_μ^{HVP} in lattice QCD required performing a fit of the lattice data in order to obtain the renormalized amplitude $\hat{\Pi}(Q^2)$ [4,7,15]. The fits were introducing model dependence that made systematics difficult to control. We discuss several recent proposals for the non-perturbative determinations of the HVP, whose development was motivated by overcoming the aforementioned difficulty of model-dependent extraction of the additive renormalization $\Pi(0)$ and a slope of the HVP near $Q^2 = 0$.

Padé Approximants In order to avoid uncontrolled systematics originating from performing model dependent fits of the HVP (usually based on Vector Meson Dominance), a model independent approach for constructing the functional form of $\Pi(Q^2)$ has been proposed by the authors of Ref. [4]. The suggested method is based only on known mathematical properties of the vacuum polarization, and consists of fitting a sequence of Padé approximants to the lattice data. The Padés are such that the full sequence converges to the actual polarization for any compact region in the complex plane excluding the cut along the negative real axis. Increasing the order of the Padé approximation allows for testing the stability of the fits and guarantees a more reliable estimate of the systematics that come mainly from the decreasing precision of lattice data at low Q^2.

Time Moments Recently proposed approach by the HPQCD collaboration [19] obtains the HVP by reconstructing the Adler function from its derivatives. From the space-averaged current-current correlator $G(x_0) = -\frac{a^3}{3}\sum_{\mathbf{x}} \langle J_\mu^{em}(x) J_\nu^{em}(0)\rangle$, one computes the time-moments of the renormalized HVP by taking the Q^2-derivatives at zero spatial momenta: $G_{2n} = (-1)^n \frac{\partial^{2n}}{\partial Q^{2n}} \{Q^2 \hat{\Pi}(Q^2)\}|_{Q^2=0} = a \sum_{x_0} x_0^{2n} G(x_0)$. The relation of the HVP form factor with the constructed time-moments allows us to determine the coefficients of a given HVP function parametrization (Padé approximants, conformal polynomials etc.). Further developments of this idea, which replace continuous time-moments with discrete- and/or spatial-moments, have been applied in more recent works by RBC/UKQCD, Mainz group and BMW collaboration [8,14,41].

Time-Momentum Representation An approach revived in many recent HVP computations [10,30,33] proposes an alternative to the computation of the HVP in momentum space captured in Eq. 2. One starts from the spatially summed vector correlator $G(x_0)$, and writes a renormalized HVP as $\hat{\Pi}(Q^2) = \frac{1}{Q^2}\int_0^{\infty} dx_0 G(x_0)[Q^2x_0^2 - 4\sin^2(\frac{1}{2}Qx_0)]$. The challenge is here transferred to understanding the large time behavior of $G(x_0)$. It can be shown that the TMR representation is equivalent to the previously discussed Time Moments approach [58].

Analytic Continuation The method discussed in Ref. [30] computes the $a_\mu^{\rm HVP}$ at small space-like momenta, simultaneously with the time-like momenta. This is achieved by keeping the spatial Fourier transform of Eq. 2 and only integrating in the time direction with a factor e^{wt}, such that continuous values of Q^2, Q=(**q**,-iw), are accessed: $\bar{\Pi}(Q^2)(\delta_{\mu\nu}Q^2 - Q_\mu Q_\nu) = \int dt\, e^{wt} \int d^3\mathbf{x}\, e^{i\mathbf{qx}} \langle J_\mu^{em}(\mathbf{x},t) J_\nu^{em}(\mathbf{0},0)\rangle$. w is a continuous parameter, and $\bar{\Pi}(Q^2)$ is a modified HVP form factor, which has been proven to lead to an evaluation of the physical $a_\mu^{\rm HVP}$. The integration range is divided into two parts: the interval $[0, Q_{max}]$, where the analytic continuation is applied and the integration range $[Q_{max}, \infty)$ that is treated in a standard way. An advantage of this approach is that the model dependence comes only from the region $[Q_{max}, \infty)$. Unfortunately, the method does not seem to lead to an increased precision in the overall calculation of $a_\mu^{\rm HVP}$. The described approach is utilized in the computation of $a_\mu^{\rm HVP}$ for $N_f = 2+1+1$ QCD with twisted mass fermions by ETMC [17].

Hybrid Method The approach proposed in Ref. [35] allows for better control of the systematics arising from the decreasing precision of lattice data in the range of low-Q^2. The integral that gives $a_\mu^{\rm HVP}$ is this time divided into three non-overlapping regions: $[0, Q_{low}^2]$,$[Q_{low}^2, Q_{high}^2]$ and $[Q_{high}^2, \infty)$ (cf. Fig. 2). One then applies different fits/moment methods in the range $[0, Q_{low}^2]$, and performs a simple numerical integration in the region $[Q_{low}^2, Q_{high}^2]$. Taking into account that roughly 80% of the contribution to the HVP integral comes from $Q^2 < 0.2\mathrm{GeV}^2$ domain [35], this approach helps us give a better estimate of the systematics in the computation of $a_\mu^{\rm HVP}$ by probing different integration boundaries Q_{low}^2 and Q_{high}^2. In the application of this method by RBC/UKQCD [8], the parametrization in the low-Q^2 region is varied, alongside the method used to match the parametrization in this region and the numerical methods used to integrate the mid-Q^2 region. This allowed obtaining a reliable estimate of the overall systematic error, which for the strange quark, interestingly, came out to be smaller than the statistical error.

2.2 Summary of Current Lattice Evaluations of the Connected HVP

Figure 2 shows the summary of the determinations of different contributions to the HVP by several lattice collaborations previously presented at Lattice 2016 [58]. In the left panel the strange contributions by Mainz group on CLS ensembles [41], RBC/UKQCD [8], ETM [17] and HPQCD [19] collaborations are compared and the results are mutually consistent. The right panel contrasts the PDG estimate of the HVP with the results by various lattice collaborations, which are treating two, three or all four (u,d,s,c) quark flavors dynamically. Lattice estimates agree with the PDG value, although the accuracy is still several times lower than the one from non-lattice methods that are the sole contributors to the PDG result.

Recent lattice works on the connected HVP front that are not covered in this review include the computation of the HVP by introducing a background magnetic field [5], a suggestion to perform a subtraction of $\Pi(0)$

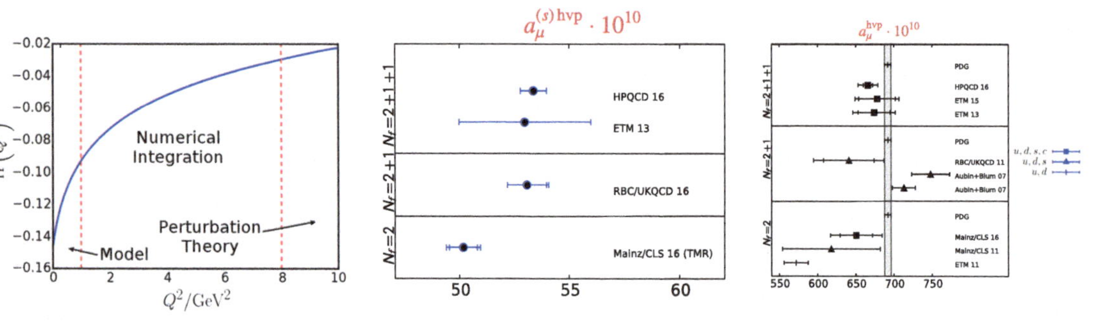

Fig. 2 *Left panel* Illustration of the hybrid method [35] reproduced from Ref. [8]. *Vertical dashed lines* represent the integration boundaries: Q_{low}, a variable frontier between the parametrization of the low-Q^2 region and the intermediate numerical integration region; and the second boundary Q_{high}, where the contact with perturbation theory is made. *Center* and *right panels* show the summary of strange quark and light quark HVP results by several collaborations using methods described in Sect. 2.1. The two figures are reproduced from the plenary talk by H. Wittig at Lattice 2016 [58]

using the derivative of the twisting angle [29], computing HVP directly at an arbitrary momentum and in a finite volume (FV) [54], as well as a systematic study of the FV effects on the BMW ensembles [49]. Additionally, Ref. [3] gives an estimate of FV effects for the connected HVP from chiral perturbation theory (ChPT) and an interesting proposal to reduce FV effects by twist averaging has been described in Ref. [48]. The most recent addition to the rapidly growing literature on lattice determinations of the HVP is the systematic study of the slope and curvature of the HVP by the BMW collaboration [14], where for the first time continuum limit of these quantities is performed using six different lattice spacings.

2.3 Disconnected HVP Contribution

Quark-disconnected contributions to vector correlation functions (see Fig. 1) are known to have a bad signal to noise ratio, and represented one of the main sources of unquantified uncertainties in the earlier computations of the HVP from the lattice [15,27]. An estimate of the disconnected HVP from one-loop FV lattice ChPT [28] and an earlier ETMC work [31] that included the disconnected contribution for almost half of the ensembles used in the study, have been followed by several contemporary computations of this quantity [5,20,38,56]. The latter have rendered this contribution relevant in achieving the aimed sub-percent precision for the total HVP from the lattice. Recent work by RBC/UKQCD [9] is the first to statistically resolve the signal of the disconnected HVP with physical light quarks. The success of the latter is due to pinpointing the mechanisms related solely with the nature of the disconnected contribution, in addition to the application of the noise reduction techniques such as all-mode-averaging strategy [13], all-to-all quark propagators [32], and sparse random sources. Moreover, light-strange quark difference is computed directly, as previously proposed in Ref. [38], and a result for the disconnected HVP amounting to $\approx$1.5% of the total HVP at 3σ level is obtained [9].

2.4 Isospin Breaking Correction to the HVP

Similarly to the disconnected piece, the isospin breaking corrections to the HVP are expected to be at a few percent level. However, the goal precision for the HVP from the lattice cannot be achieved before we control the effect of performing the lattice computation in the isosymmetric limit: with up ad down quark masses degenerate and no electromagnetic effects taken into account. Up to now, only a limited number of efforts have been invested in lattice community, in order to determine the isospin breaking effects to the hadronic observables in high precision lattice simulations (for a recent review see [52]). The existing estimates of the isospin breaking corrections to the HVP based on phenomenological input [21,44], speak in favor of the importance of this correction for reaching a sub-percent precision of the HVP from the lattice. In particular, the error budget of [21] quotes precisely the unaccounted isospin breaking effects as the largest contribution to the total uncertainty. The definite answer regarding the significance of these effects can be obtained only by carrying out a lattice calculation and several such studies are on the way [16,45].

2.5 HVP from New Experiments Combined with Lattice QCD

Note that an alternative way to write Eq. 1 features the running of the fine structure constant: $a_\mu^{\rm HVP} = (\frac{\alpha}{\pi})\int_0^1 dx(1-x)\Delta\alpha_{had}(-Q^2)$ [47], where $Q^2 = \frac{x^2 m_\mu^2}{1-x}$. Recently, a couple of novel approaches to perform a direct space-like measurement of the hadronic contributions to the running of the fine structure constant, $\Delta\alpha_{had}(-Q^2)$, have been proposed [1,18]. A precise value of $\Delta\alpha_{had}(-Q^2)$ could be obtained either in Bhabha scattering with a small modification of the existing experiments [18], or in the recently proposed measurement of the scattering of the high-energy muons on the fixed electron target [1]. The statistical precision of the value of $a_\mu^{\rm HVP}$ obtained from the latter of the two proposed experiments is estimated to be O(0.3%) after two years of data taking, while the systematics still need to be accounted.

The original proposals assume using time-like dispersive data and/or perturbation theory beyond the maximal space-like momenta that can be achieved in space-like scattering experiments, Q^2_{exp}. Similarly to the Hybrid Method described in Sect. 2.1, one could instead combine the input from the newly proposed experiments in some low momentum integration range $[0, Q^2_{exp}]$ with lattice QCD calculation of the contribution to the $a_\mu^{\rm HVP}$ from the region $[Q^2_{exp}, Q^2_{high}]$, and finally make contact with perturbation theory at some value

Q^2_{high}. Due to the complementarity of the integration ranges where the proposed space-like experimental measurements and lattice QCD give precise results, the value of $a_\mu^{\rm HVP}$ obtained by combining the two methods is expected to be comparable to the projected accuracy of the $g-2$ experiments at Fermilab and J-PARC.

3 Hadronic Light by Light contribution

The value for the Hadronic Light by Light contribution entering the phenomenological estimate of a_μ [43] (cf. Table 1) is based on model calculations and lattice methods allow for the model independent evaluation of these estimates. In the previously described HVP calculations, one would first separate QED and QCD contributions, such that only the QCD piece is computed on the lattice. Following a similar approach for HLbL would require to resolve the four point correlation function $a_\mu^{HLbL} \propto \int dQ^2 f'(Q^2) \sum_{x_1,x_2,x_3,x_4} e^{iQx} \langle J_\mu(x_1) J_\nu(x_2) J_\rho(x_3) J_\sigma(x_4)\rangle$ from the lattice, which is computationally very demanding. A feasible methodology to compute the HLbL contribution to a_μ by factorizing the QED part has been first proposed by T.Blum, T. Izubuchi and M. Hayakawa [40]. An original approach of combining lattice QCD and stochastically generated photon propagators [10,40] has recently been superseded with a more promising approach of replacing stochastic QED with the analytic photon propagators[11,12]. On the other hand, recent progress in the dispersive-based analysis inspired by a data-driven HVP calculations [23–25,53] has motivated lattice practitioners to revisit the mentioned approach of calculating HLbL by computing the four-point correlator on the lattice. Its recent implementation by the Mainz group assumes the computation of the multidimensional integral over a position-space QED kernel in the continuum [36]. This proposal has recently been tested by evaluating the π_0-pole contribution to the HLbL scattering amplitude [37] and a more complete calculation is underway.

In the lattice calculation of the HLbL contribution, again both the connected and the disconnected pieces have to be taken into account. Despite the tremendous advances the computations of connected HLbL have seen in the previous years, the disconnected contribution still remains a challenge, particularly due to the fact that unlike for the HVP, for the HLbL there is no reason to expect its suppression with respect to the connected part. This concern has been confirmed in a recent calculation by RBC [11], which estimates the leading disconnected piece for HLbL to be as as much as $\approx$50% of the connected one.

4 Summary and Outlook

Due to its high sensitivity to the potential physics beyond the Standard Model, muon magnetic moment is an optimal channel to unravel new physics. The dominating uncertainties in the theoretical estimate of a_μ come from the hadronic contributions: the leading order HVP and HLbL. The computation of the connected contribution to the HVP has become a mainstream calculation for lattice QCD by now, and it is particularly encouraging to see that more and more lattice studies are focused on computing the disconnected and isospin breaking corrections to the HVP, as well as the HLbL contributions. It is expected that once the effect of the isospin breaking to the HVP is known with a precision of O(10%), the sub-percent accuracy of the HVP contribution from the lattice will already be achieved. The experimental precision of the anomalous magnetic moment will be improved four times in the next couple of years at the Fermilab new experiment E989 [57] and at J-PARC, E34 [55], making the verification and improvement of the current theoretical prediction of a_μ a timely task. For all these reasons, it is crucial to invest in further development of the lattice methods described in this review and to devise new methods, in order to take full advantage of the first principles calculations of the hadronic contributions to a_μ.

Acknowledgements I would like to thank the members of the RBC/UKQCD for an enjoyable collaboration on $a_\mu^{\rm HVP}$ related topics, and to Nazario Tantalo and Fred Jegerlehner for enlightening discussions on the muon g-2. Many thanks to the organizers of Light Cone 2016 for the friendly and stimulating atmosphere of the meeting. Furthermore, I am especially indebted to Massimo Passera and Marija Vranic for valuable comments on the manuscript and to Hartmut Wittig, who generously allowed reproduction of the unpublished figures.

References

1. G. Abbiendi et al., Measuring the leading hadronic contribution to the muon g-2 via μe scattering. Eur. Phys. J. C **77**(3), 139 (2017). doi:10.1140/epjc/s10052-017-4633-z
2. T. Aoyama, M. Hayakawa, T. Kinoshita, M. Nio, Tenth-order electron anomalous magnetic moment—contribution of diagrams without closed lepton loops. Phys. Rev. **D91**(3), 033,006 (2015). doi:10.1103/PhysRevD.91.033006
3. C. Aubin, T. Blum, P. Chau, M. Golterman, S. Peris, C. Tu, Finite-volume effects in the muon anomalous magnetic moment on the lattice. Phys. Rev. **D93**(5), 054,508 (2016). doi:10.1103/PhysRevD.93.054508
4. C. Aubin, T. Blum, M. Golterman, S. Peris, Model-independent parametrization of the hadronic vacuum polarization and g-2 for the muon on the lattice. Phys. Rev. **D86**, 054,509 (2012). doi:10.1103/PhysRevD.86.054509
5. G. Bali, G. Endrodi, Hadronic vacuum polarization and muon g-2 from magnetic susceptibilities on the lattice. Phys. Rev. **D92**(5), 054,506 (2015). doi:10.1103/PhysRevD.92.054506
6. G.W. Bennett et al., Final report of the muon E821 anomalous magnetic moment measurement at BNL. Phys. Rev. **D73**, 072,003 (2006). doi:10.1103/PhysRevD.73.072003
7. T. Blum, Lattice calculation of the lowest order hadronic contribution to the muon anomalous magnetic moment. Phys. Rev. Lett. **91**, 052,001 (2003). doi:10.1103/PhysRevLett.91.052001
8. T. Blum, P. Boyle, L. Del Debbio, R. Hudspith, T. Izubuchi, A. Juettner, C. Lehner, R. Lewis, K. Maltman, M. Krstic Marinkovic, A. Portelli, M. Spraggs, Lattice calculation of the leading strange quark-connected contribution to the muon g -2. JHEP **04**, 063 (2016). doi:10.1007/JHEP04(2016)063
9. T. Blum, P.A. Boyle, T. Izubuchi, L. Jin, A. Jttner, C. Lehner, K. Maltman, M. Marinkovic, A. Portelli, M. Spraggs, Calculation of the hadronic vacuum polarization disconnected contribution to the muon anomalous magnetic moment. Phys. Rev. Lett. **116**(23), 232,002 (2016). doi:10.1103/PhysRevLett.116.232002
10. T. Blum, S. Chowdhury, S. Hayakawa, T. Izubuchi, Hadronic light-by-light scattering contribution to the muon anomalous magnetic moment from lattice QCD. Phys. Rev. Lett. **114**(1), 012,001 (2015). doi:10.1103/PhysRevLett.114.012001
11. T. Blum, N. Christ, M. Hayakawa, T. Izubuchi, L. Jin, C. Jung, C. Lehner, Connected and leading disconnected hadronic light-by-light contribution to the muon anomalous magnetic moment with a physical pion mass. Phys. Rev. Lett. **118**(2), 022,005 (2017). doi:10.1103/PhysRevLett.118.022005
12. T. Blum, N. Christ, M. Hayakawa, T. Izubuchi, L. Jin, C. Lehner, Lattice calculation of hadronic light-by-light contribution to the muon anomalous magnetic moment. Phys. Rev. **D93**(1), 014,503 (2016). doi:10.1103/PhysRevD.93.014503
13. T. Blum, T. Izubuchi, E. Shintani, New class of variance-reduction techniques using lattice symmetries. Phys. Rev. **D88**(9), 094,503 (2013). doi:10.1103/PhysRevD.88.094503
14. S. Borsanyi, Z. Fodor, T. Kawanai , S. Krieg, L. Lellouch, R. Malak, K. Miura, K.K. Szabo, C. Torrero, B. Toth, Slope and curvature of the hadron vacuum polarization at vanishing virtuality from lattice. QCD (2016). doi:10.1103/PhysRevLett.117.082001. ArXiv:1612.02364
15. P. Boyle, L. Del Debbio, E. Kerrane, J. Zanotti, Lattice determination of the hadronic contribution to the muon $g-2$ using dynamical domain wall fermions. Phys. Rev. **D85**, 074,504 (2012). doi:10.1103/PhysRevD.85.074504
16. P. Boyle, V. Gulpers, J. Harrison, A. Juttner, A. Portelli, C. Sachrajda, Electromagnetic corrections to meson masses and the HVP. PoS **LATTICE2016**, 172 (2016)
17. F. Burger, X. Feng, G. Hotzel, K. Jansen, M. Petschlies, D.B. Renner, Four-flavour leading-order hadronic contribution to the muon anomalous magnetic moment. JHEP **02**, 099 (2014). doi:10.1007/JHEP02(2014)099
18. C.M. Carloni Calame, M. Passera, L. Trentadue, G. Venanzoni, A new approach to evaluate the leading hadronic corrections to the muon g-2. Phys. Lett. B **746**, 325–329 (2015). doi:10.1016/j.physletb.2015.05.020
19. B. Chakraborty, C.T.H. Davies, G.C. Donald, R.J. Dowdall, J. Koponen, G.P. Lepage, T. Teubner, Strange and charm quark contributions to the anomalous magnetic moment of the muon. Phys. Rev. **D89**(11), 114,501 (2014). doi:10.1103/PhysRevD.89.114501
20. B. Chakraborty, C.T.H. Davies, J. Koponen, G.P. Lepage, M.J. Peardon, S.M. Ryan, Estimate of the hadronic vacuum polarization disconnected contribution to the anomalous magnetic moment of the muon from lattice QCD. Phys. Rev. **D93**(7), 074,509 (2016). doi:10.1103/PhysRevD.93.074509
21. B. Chakraborty, C.T.H. Davies, P.G. de Oliviera, J. Koponen, G.P. Lepage, The hadronic vacuum polarization contribution to a_μ from full lattice QCD (2016)
22. G. Colangelo, M. Hoferichter, A. Nyffeler, M. Passera, P. Stoffer, Remarks on higher-order hadronic corrections to the muon g-2. Phys. Lett. B **735**, 90–91 (2014). doi:10.1016/j.physletb.2014.06.012
23. G. Colangelo, M. Hoferichter, M. Procura, P. Stoffer, Dispersive approach to hadronic light-by-light scattering. JHEP **09**, 091 (2014). doi:10.1007/JHEP09(2014)091
24. G. Colangelo, M. Hoferichter, M. Procura, P. Stoffer, Dispersion relation for hadronic light-by-light scattering: theoretical foundations. JHEP **09**, 074 (2015). doi:10.1007/JHEP09(2015)074
25. G. Colangelo, M. Hoferichter, M. Procura, P. Stoffer, Dispersion relation for hadronic light-by-light scattering: two-pion contributions (2017)
26. M. Davier, A. Hoecker, B. Malaescu, Z. Zhang, Hadron contribution to vacuum polarisation. Adv. Ser. Direct. High Energy Phys. **26**, 129–144 (2016)
27. M. Della Morte, B. Jager, A. Juttner, H. Wittig, Towards a precise lattice determination of the leading hadronic contribution to $(g-2)_\mu$. JHEP **03**, 055 (2012). doi:10.1007/JHEP03(2012)055
28. M. Della Morte, A. Juttner, Quark disconnected diagrams in chiral perturbation theory. JHEP **1011**, 154 (2010). doi:10.1007/JHEP11(2010)154
29. G. de Divitiis, R. Petronzio, N. Tantalo, On the extraction of zero momentum form factors on the lattice. Phys. Lett. B **718**, 589–596 (2012). doi:10.1016/j.physletb.2012.10.035
30. X. Feng, S. Hashimoto, G. Hotzel, K. Jansen, M. Petschlies, D.B. Renner, Computing the hadronic vacuum polarization function by analytic continuation. Phys. Rev. **D88**, 034,505 (2013). doi:10.1103/PhysRevD.88.034505

31. X. Feng, K. Jansen, M. Petschlies, D.B. Renner, Two-flavor QCD correction to lepton magnetic moments at leading-order in the electromagnetic coupling. Phys. Rev. Lett. **107**, 081,802 (2011)
32. J. Foley, K. Jimmy Juge, A. O'Cais, M. Peardon, S.M. Ryan, J.I. Skullerud, Practical all-to-all propagators for lattice QCD. Comput. Phys. Commun. **172**, 145–162 (2005). doi:10.1016/j.cpc.2005.06.008
33. A. Francis, V. Gülpers, B. Jäger, H. Meyer, G. von Hippel et al., The leading disconnected contribution to the anomalous magnetic moment of the muon. PoS **LATTICE2014**, 128 (2014)
34. C. Gnendiger, D. Stckinger, H. Stckinger-Kim, The electroweak contributions to $(g-2)_\mu$ after the Higgs boson mass measurement. Phys. Rev. **D88**, 053,005 (2013). doi:10.1103/PhysRevD.88.053005
35. M. Golterman, K. Maltman, S. Peris, New strategy for the lattice evaluation of the leading order hadronic contribution to $(g-2)_\mu$. Phys. Rev. **D90**(7), 074,508 (2014). doi:10.1103/PhysRevD.90.074508
36. J. Green, N. Asmussen, O. Gryniuk, G. von Hippel, H.B. Meyer, A. Nyffeler, V. Pascalutsa, Direct calculation of hadronic light-by-light scattering. PoS **LATTICE2015**, 109 (2016)
37. J. Green, N. Asmussen, H.B. Meyer, A. Nyffeler, Position-space approach to hadronic light-by-light scattering in the muon $g-2$ on the lattice. PoS **LATTICE2016**, 164 (2017)
38. V. Gulpers, A. Francis, B. Jager, H. Meyer, G. von Hippel, H. Wittig, The leading disconnected contribution to the anomalous magnetic moment of the muon. PoS **LATTICE2014**, 128 (2014)
39. K. Hagiwara, R. Liao, A.D. Martin, D. Nomura, T. Teubner, $(g-2)\_mu$ and alpha(M_Z$\hat{2}$) re-evaluated using new precise data. J. Phys. **G38**, 085,003 (2011). doi:10.1088/0954-3899/38/8/085003
40. M. Hayakawa, T. Blum, T. Izubuchi, N. Yamada, Hadronic light-by-light scattering contribution to the muon g-2 from lattice QCD: Methodology. PoS **LAT2005**, 353 (2006)
41. H. Horch, The leading order hadronic contribution of the anomalous magnetic moment of the muon with O(a)-improved Wilson fermions with Pade approximants from fits and time moments. PoS **LATTICE2016**, 165 (2017)
42. F. Jegerlehner, Leading-order hadronic contribution to the electron and muon g-2. EPJ Web Conf. **118**, 01,016 (2016). doi:10.1051/epjconf/201611801016
43. F. Jegerlehner, A. Nyffeler, The muon g-2. Phys. Rep. **477**, 1–110 (2009). doi:10.1016/j.physrep.2009.04.003
44. F. Jegerlehner, R. Szafron, $\rho^0-\gamma$ mixing in the neutral channel pion form factor F_π^e and its role in comparing e^+e^- with τ spectral functions. Eur. Phys. J. C **71**, 1632 (2011). doi:10.1140/epjc/s10052-011-1632-3
45. M. Krstic Marinkovic, Leading isospin breaking correction to the hadronic vacuum polarisation. Talk presented at Lattice 2015, Kobe (2015)
46. A. Kurz, T. Liu, P. Marquard, M. Steinhauser, Hadronic contribution to the muon anomalous magnetic moment to next-to-next-to-leading order. Phys. Lett. B **734**, 144–147 (2014). doi:10.1016/j.physletb.2014.05.043
47. B.E. Lautrup, E. De Rafael, Calculation of the sixth-order contribution from the fourth-order vacuum polarization to the difference of the anomalous magnetic moments of muon and electron. Phys. Rev. **174**, 1835–1842 (1968). doi:10.1103/PhysRev.174.1835
48. C. Lehner, T. Izubuchi, Towards the large volume limit—a method for lattice QCD + QED simulations. PoS **LATTICE2014**, 164 (2015)
49. R. Malak, Z. Fodor, C. Hoelbling, L. Lellouch, A. Sastre et al., Finite-volume corrections to the leading-order hadronic contribution to $g_\mu-2$ **LATTICE2014**, 161 (2015)
50. P.J. Mohr, B.N. Taylor, D.B. Newell, CODATA recommended values of the fundamental physical constants: 2014. Rev. Mod. Phys. **88**, (2016). doi:10.1103/RevModPhys.84.1527
51. M. Passera, The Standard model prediction of the muon anomalous magnetic moment. J. Phys. **G31**, R75–R94 (2005). doi:10.1088/0954-3899/31/5/R01
52. A. Patella, QED corrections to hadronic observables. PoS **LATTICE2016**, 020 (2017)
53. V. Pauk, M. Vanderhaeghen, Anomalous magnetic moment of the muon in a dispersive approach. Phys. Rev. **D90**(11), 113,012 (2014). doi:10.1103/PhysRevD.90.113012
54. A. Portelli, Finite volume hadronic vacuum polarisation at arbitrary momenta. Talk presented at Lattice 2015, Kobe (2015)
55. N. Saito, A novel precision measurement of muon g-2 and EDM at J-PARC. AIP Conf. Proc. **1467**, 45–56 (2012). doi:10.1063/1.4742078
56. B. Toth, Disconnected contribution to hadron correlation functions. Talk presented at Lattice 2015, Kobe (2015)
57. G. Venanzoni, The new muon g2 experiment at Fermilab. Nucl. Part. Phys. Proc. **273–275**, 584–588 (2016). doi:10.1016/j.nuclphysbps.2015.09.087
58. H. Wittig, Hadronic contributions to the muon (g-2) from Lattice QCD. Plenary talk presented at Lattice 2016, Southampton (2016)

Few-Body Syst
DOI 10.1007/s00601-017-1276-4

Nuno Cardoso · Pedro Bicudo · Marco Cardoso

Confinement at Finite Temperature

Received: 14 November 2016 / Accepted: 7 March 2017

Abstract We show the flux tubes produced by static quark–antiquark, quark–quark and quark–gluon charges at finite temperature. The sources are placed on the lattice with fundamental and adjoint Polyakov loops. We compute the squared strengths of the chromomagnetic and chromoelectric fields above and below the critical temperature. Our results are for pure gauge SU(3) gauge theory, they are invariant and all computations are done with GPUs using CUDA.

1 Introduction

We study the color field distributions inside the flux tubes formed by Polyakov loops in the static $Q\bar{Q}$, QQ and QG systems. We address how the flux tube evolves with the distance between quarks and when the temperature increase beyond the deconfinement temperature. In Sect. 2, we describe the lattice formulation. We briefly review the Polyakov loop for these systems and show how to compute the color fields as well as the Lagrangian distribution. In Sect. 3, the numerical results are shown. Finally, we conclude in Sect. 4.

2 Computation of the Chromo-Fields in the Flux Tube

The central observables that govern the physics in the flux tube can be extracted from the correlation of a plaquette, $\square_{\mu\nu}$, with the Polyakov loops, L. To reduce the fluctuations of the observables $\mathcal{O}\,\square_{\mu\nu}(x)$, we measure the following quantity [6],

$$f_{\mu\nu}(r,x) = \frac{\beta}{a^4}\left[\frac{\langle \mathcal{O}\,\square_{\mu\nu}(x)\rangle - \langle \mathcal{O}\,\square_{\mu\nu}(x_R)\rangle}{\langle \mathcal{O}\rangle}\right], \tag{1}$$

This article belongs to the Topical Collection "Light Cone 2016".

Talk presented by NC at the Light Cone 2016 Conference, Lisbon, Portugal, 5–8 September 2016. NC and MC are supported by FCT under the contracts SFRH/BPD/109443/2015 and SFRH/BPD/73140/2010 respectively. We also acknowledge the use of CPU and GPU servers of PtQCD, supported by NVIDIA, CFTP and FCT Grant UID/FIS/00777/2013.

N. Cardoso (✉) · P. Bicudo · M. Cardoso
CFTP, Departamento de Física, Instituto Superior Técnico, Universidade de Lisboa, Avenida Rovisco Pais 1, 1049-001 Lisbon, Portugal
E-mail: nuno.cardoso@tecnico.ulisboa.pt

P. Bicudo
E-mail: bicudo@tecnico.ulisboa.pt

M. Cardoso
E-mail: mjdcc@cftp.ist.utl.pt

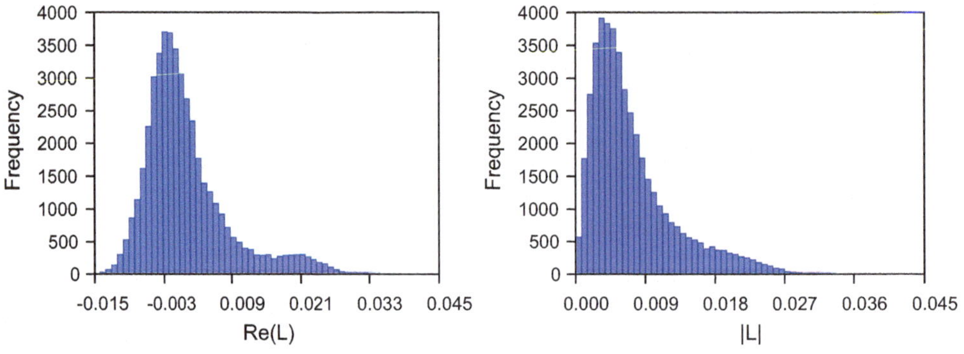

Fig. 1 Histogram of the Polyakov loop history for $\beta = 6.055$, $T = 0.988\, T_c$

Table 1 Lattice simulations for a $48^3 \times 8$ volume

β	T/T_c	$a\sqrt{\sigma}$	# config.
5.96	0.845	0.235023	5990
6.0534	0.986	0.201444	5990/5110*
6.13931	1.127	0.176266	5990
6.29225	1.408	0.141013	5990
6.4249	1.690	0.117513	5990

We denote with an * the number of remaining configurations after we remove the configurations that lie above the phase transition

where x_R is the reference point placed far from the sources, and

$$\begin{aligned} \mathcal{O} &= L(0)\, L^{\dagger}(r) && \text{for the } Q\bar{Q} \text{ system},\\ \mathcal{O} &= L(0)\, L(r) && \text{for the } QQ \text{ system},\\ \mathcal{O} &= \left(L(0)L^{\dagger}(0) - 1\right) L(r) && \text{for the } Qg \text{ system}. \end{aligned} \tag{2}$$

Moreover x denotes the distance of the plaquette from the line connecting the sources, r is the separation between the sources, $L(r) = \frac{1}{3}\mathrm{Tr}\Pi_{t=1}^{N_t} U_4(r,t)$ and N_t is the number of time slices of the lattice. We also use the periodicity in the time direction for the Polyakov loops, $\Box_{\mu\nu}(x) = \frac{1}{N_t}\sum_{t=1}^{N_t} \Box_{\mu\nu}(x,t)$, to average the plaquette over the time direction.

Therefore, using the plaquette orientations $(\mu,\nu) = (2,3), (1,3), (1,2), (1,4), (2,4), (3,4)$, we can relate the six components in Eq. (1) to the components of the chromoelectric and chromomagnetic fields,

$$f_{\mu\nu} \to \frac{1}{2}\left(-\langle B_x^2\rangle, -\langle B_y^2\rangle, -\langle B_z^2\rangle, \langle E_x^2\rangle, \langle E_y^2\rangle, \langle E_z^2\rangle\right), \tag{3}$$

and also calculate the total action (Lagrangian) density, $\langle \mathcal{L}\rangle = \frac{1}{2}\left(\langle E^2\rangle - \langle B^2\rangle\right)$.

In order to improve the signal over noise ratio in the $Q\bar{Q}$ and QQ systems, we use the multihit technique [1,5], replacing each temporal link by it's thermal average,

$$U_4 \to \bar{U}_4 = \frac{\int dU_4 U_4\, e^{\beta \mathrm{Tr}[U_4 F^{\dagger}]}}{\int dU_4\, e^{\beta \mathrm{Tr}[U_4 F^{\dagger}]}} \tag{4}$$

and the extended multihit technique, [3].

The extended multihit consists in replacing each temporal link by it's thermal average with the first N neighbours fixed, rather than just taking the thermal average of a temporal link with the first neighbors. We apply the heat-bath algorithm to all these links, averaging the central link,

$$U_4 \to \bar{U}_4 = \frac{\int [\mathcal{D}U_4]_{\Omega}\, U_4\, e^{\beta \sum_{\mu,s} \mathrm{Tr}[U_\mu(s) F^{\dagger}(s)]}}{\int [\mathcal{D}U_4]_{\Omega}\, e^{\beta \sum_{\mu,s} \mathrm{Tr}[U_\mu(s) F^{\dagger}(s)]}}. \tag{5}$$

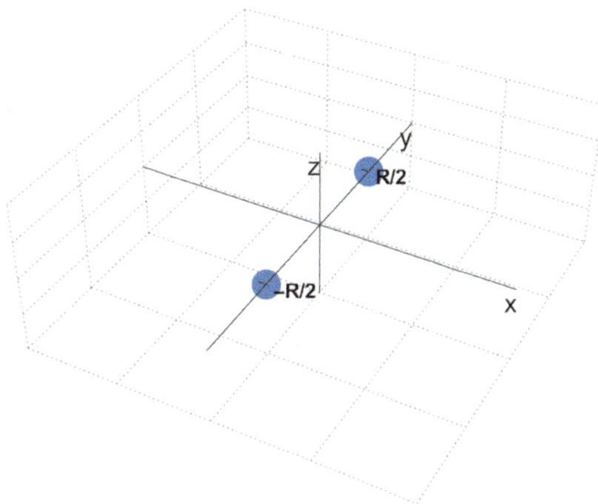

Fig. 2 Scheme of the charges displacement in space

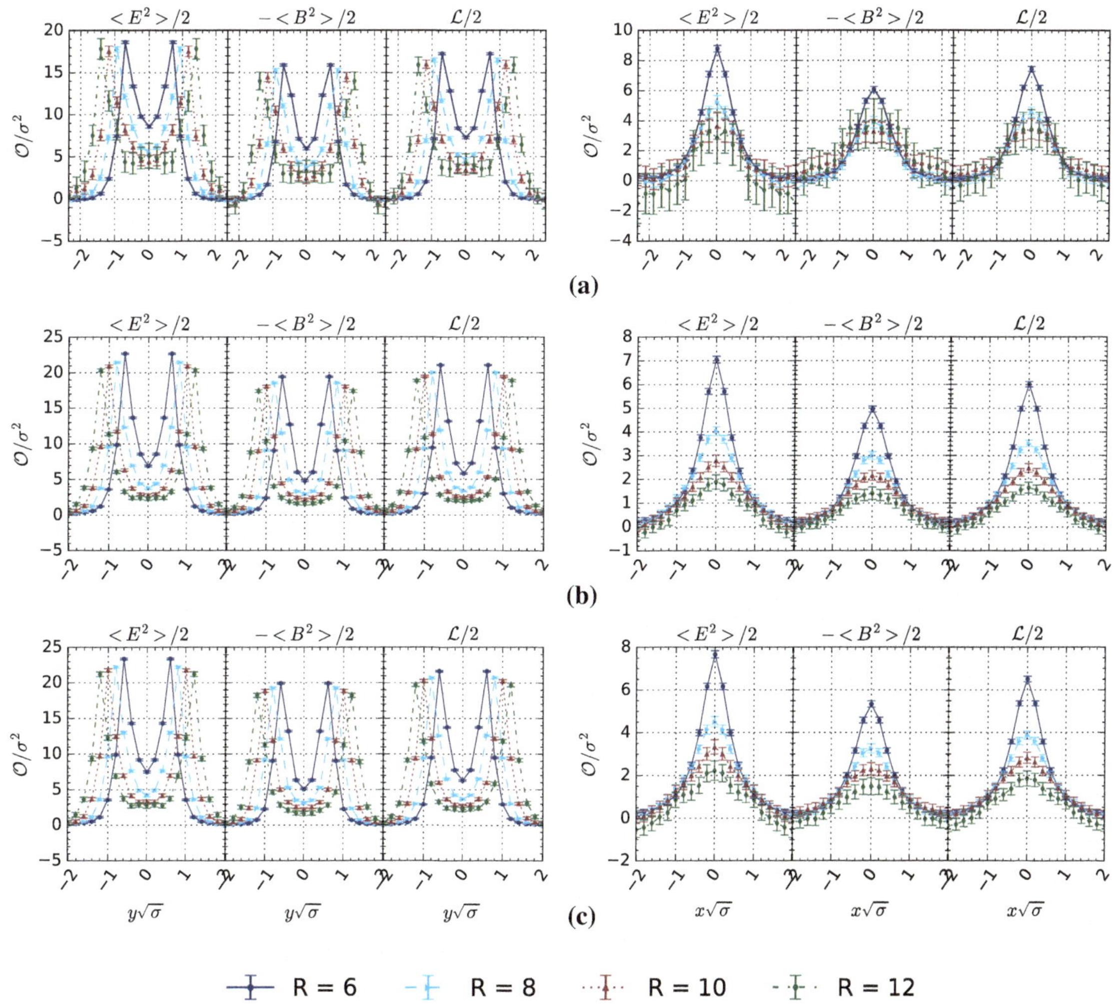

Fig. 3 The results for the $Q\bar{Q}$ system at $T < T_c$. The results in the *left column* correspond to the fields along the sources (plane XY) and the *right column* to the results in the middle of the flux tube (plane XZ). R is the distance between the sources in lattice units. **a** $\beta = 5.96$, $T = 0.845\, T_c$. **b** $\beta = 6.0534$, $T = 0.986\, T_c$, with contaminated configurations. **c** $\beta = 6.0534$, $T = 0.986\, T_c$, without contaminated configurations

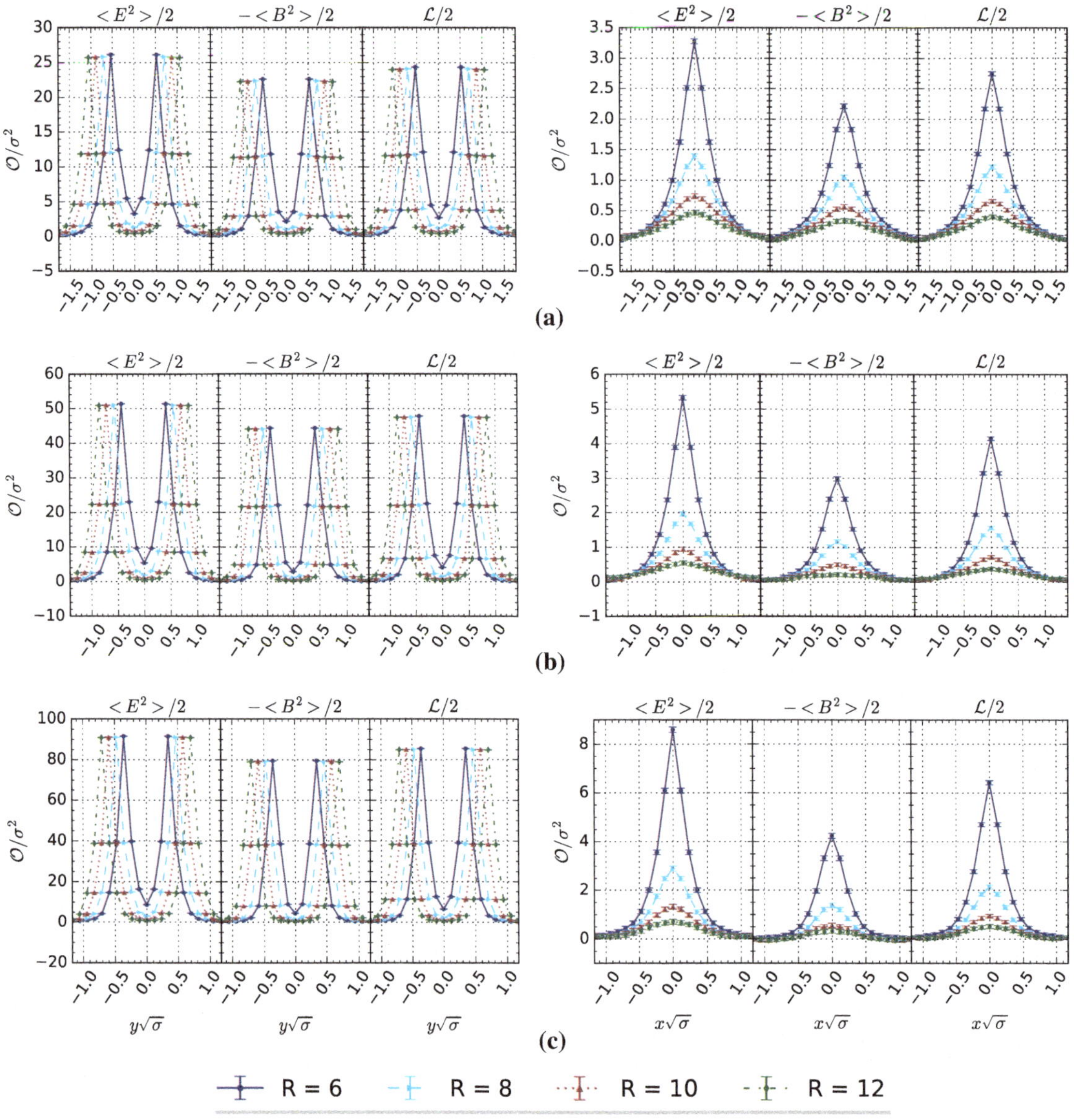

Fig. 4 The results for the $Q\bar{Q}$ system at $T > T_c$. The results in the *left column* correspond to the fields along the sources (plane XY) and the *right column* to the results in the middle of the flux tube (plane XZ). R is the distance between the sources in lattice units. **a** $\beta = 6.13931$, $T = 1.127\,T_c$. **b** $\beta = 6.29225$, $T = 1.408\,T_c$. **c** $\beta = 6.4249$, $T = 1.690\,T_c$

By using $N = 2$ we are able to greatly improve the signal, when compared with the error reduction achieved with the simple multihit. Of course, this technique is more computer intensive than simple multihit, while being simpler to implement than multilevel. The only restriction is $R > 2N$ for this technique to be valid.

Moreover, just below the phase transition, we need to make sure that we don't have contaminated configurations as already mentioned in [2]. By plotting the histogram of the Polyakov loop for $\beta = 6.055$, Fig. 1, we are able to identify a second peak, and thus we remove all the configurations that lie on the second peak. Therefore, in Table 1 the value with asterisk corresponds to the configurations after removing these contaminated configurations.

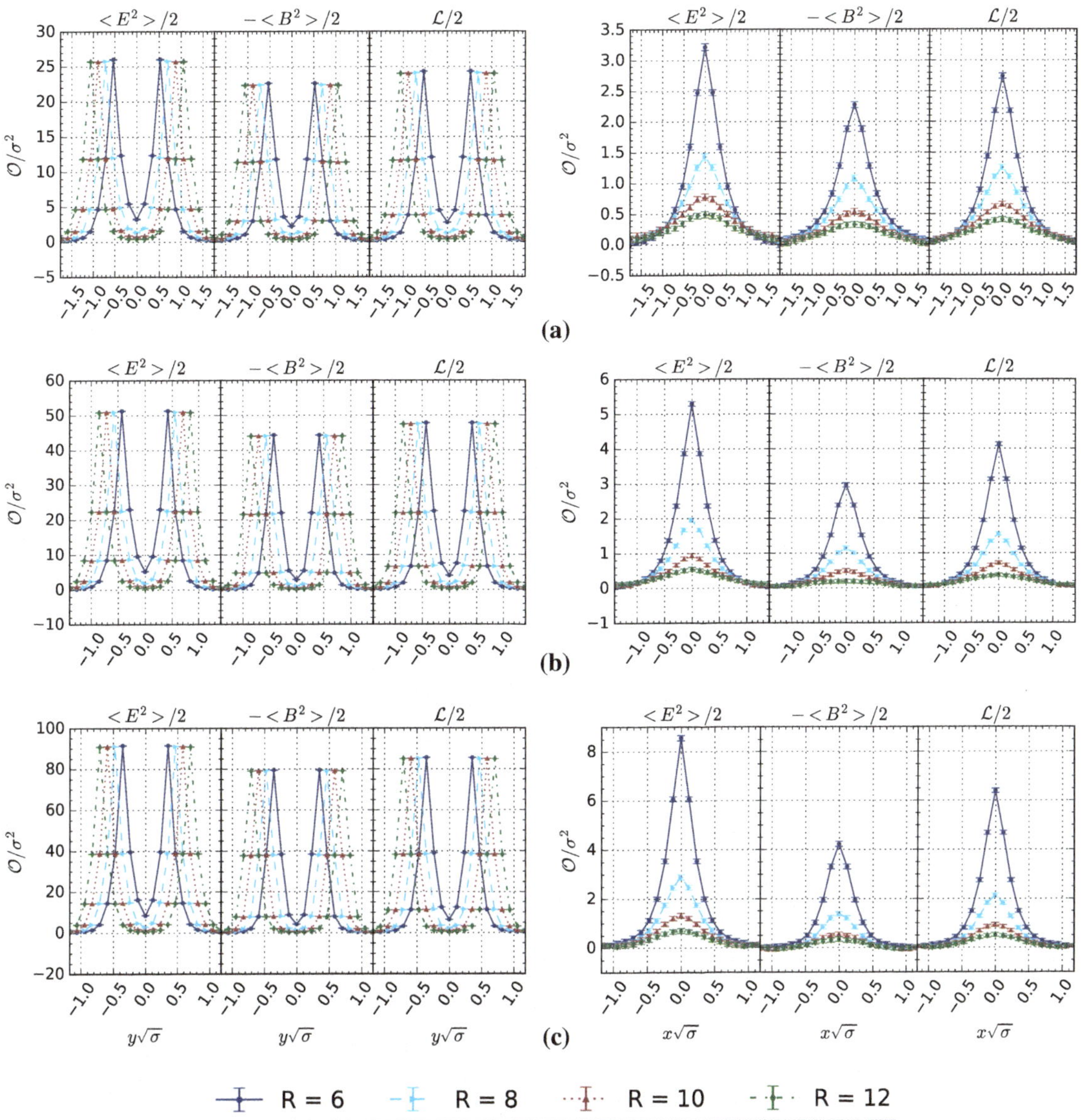

Fig. 5 The results for the QQ system at $T > T_c$. The results in the *left column* correspond to the fields along the sources (plane XY) and the *right column* to the results in the middle of the flux tube (plane XZ). R is the distance between the sources in lattice units. **a** $\beta = 6.13931$, $T = 1.127\,T_c$. **b** $\beta = 6.29225$, $T = 1.408\,T_c$. **c** $\beta = 6.4249$, $T = 1.690\,T_c$

3 Results

In this section, we present the results for a fixed lattice volume of $48^3 \times 8$ with different β, Table 1. The lattice spacing was computed using the parametrization from Ref. [4] in units of the string tension at zero temperature. All our computations are fully performed in NVIDIA GPUs using CUDA.

The two charges, Q, $\bar{Q}$ or G, are located at $(0, -R/2, 0)$ and $(0, R/2, 0)$ for $R = 4, 6, 8, 10$ and 12 in lattice spacing units, see Fig. 2.

In Figs. 3 and 4, we show the results for the $Q\bar{Q}$ system. As expected the strength of the fields decrease with the temperature. Also, in the confined phase the width in the middle of the flux tube increases with the distance between the sources, while above the phase transition the width seems to decrease with the distance.

In Fig. 5, we show the results for the QQ system at $T > T_c$ (below T_c the Polyakov loop of non color-singlet systems vanish). It is remarkable that these results are identical, modulo statistic errors, to the ones of the $Q\bar{Q}$ system. We interpret this as an evidence that the imaginary part of the Polyakov loops is vanishing, since the

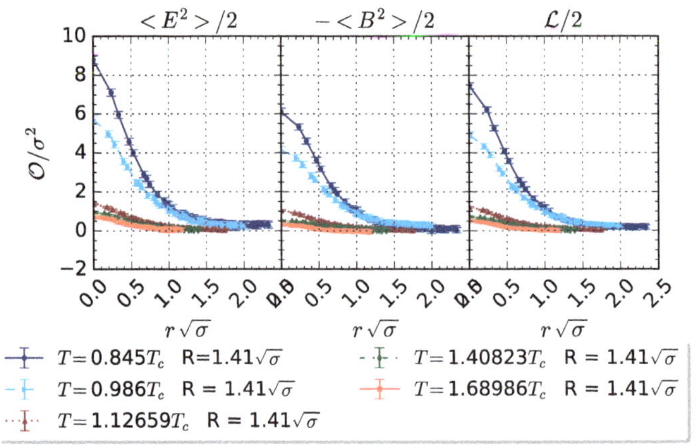

Fig. 6 Results for the fields of the $Q\bar{Q}$ system in the middle of the flux tube in the plane XZ

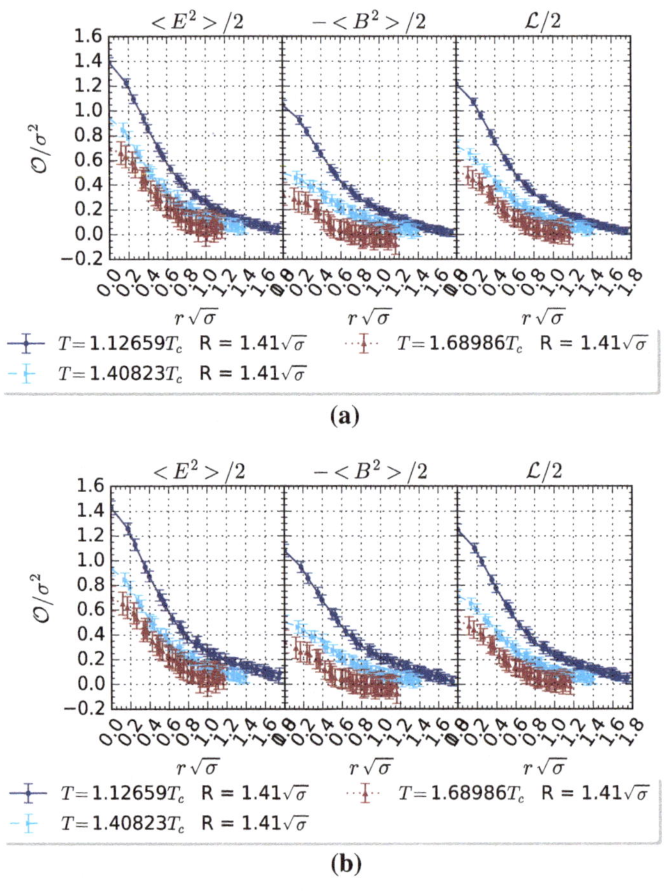

Fig. 7 Results for the fields of the $Q\bar{Q}$ and of the QQ systems at $T > T_c$ in the middle of the flux tube in the plane XZ. **a** $Q\bar{Q}$. **b** QQ

Polyakov loop and anti-loop are hermitian conjugate, as in Eq. (2), and thus only differ in the imaginary part. For more details, in Figs. 6 and 7, we also show the results for the $Q\bar{Q}$ system and the QQ system in the middle of the flux tube in the plane XZ.

Moreover, in Fig. 8 we study the effect of including a static gluon in the system with an adjoint Polyakov loop as in Eq. (2). As in the case with quark sources only, we find evidence for no flux tube at $T > T_c$.

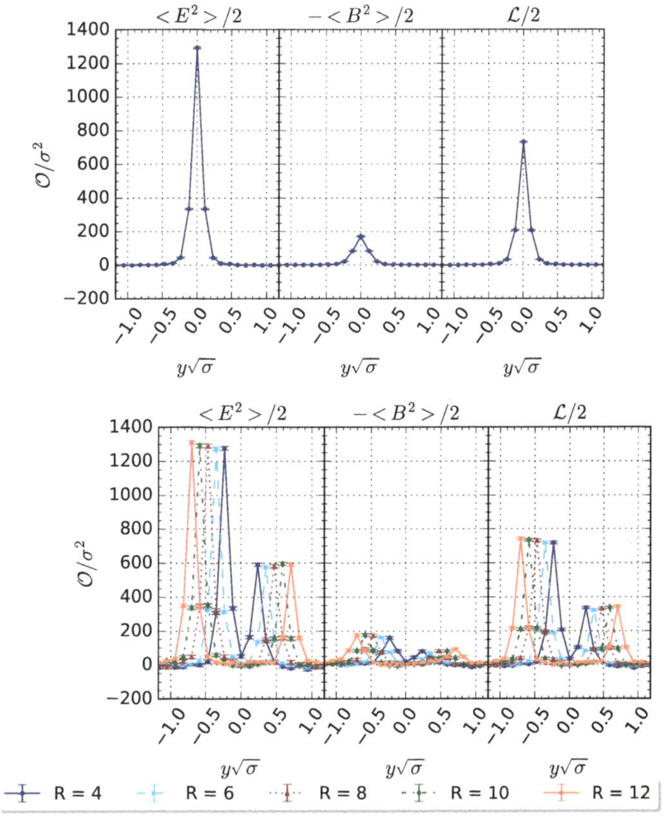

Fig. 8 The results in the *top plot* correspond to the single gluon system, and in *bottom* to the QG system, both for $\beta = 6.4249$, $T = 1.690\,T_c$

4 Conclusions

Using CUDA codes and computations in NVIDIA GPUs, we compute the square densities of the chromomagnetic and chromoelectric fields produced by different Polyakov loop sources, above and below the phase transition.

As the distance increase between the sources, the fields square densities decrease. Below the deconfinement critical temperature, this decrease is moderate and is consistent with the widening of the flux tube as already seen in studies at zero temperature [3], moreover the field strength clearly decreases as the temperature increases as expected from the critical curve for the string tension [2]. Above the deconfinement critical temperature, the fields rapidly decrease to zero as the quarks are pulled apart, qualitatively consistent with screened Coulomb-like fields. While the width of the flux tube below the phase transition temperature increases with the separation between the quark-antiquark, above the phase transition the width seems to decrease with the separation.

As an outlook, we plan to complete the present study with a quantitative study of the widening of the flux tubes as a function of the inter-charge distance, R, and the temperature, T.

References

1. R. Brower, P. Rossi, C.I. Tan, The external field problem for QCD. Nucl. Phys. B **190**, 699 (1981). doi:10.1016/0550-3213(81)90046-8
2. N. Cardoso, P. Bicudo, Lattice QCD computation of the SU(3) string tension critical curve. Phys. Rev. D **85**(077), 501 (2012). doi:10.1103/PhysRevD.85.077501. arXiv:1111.1317
3. N. Cardoso, M. Cardoso, P. Bicudo, Inside the SU(3) quark–antiquark QCD flux tube: screening versus quantum widening. Phys. Rev. D **88**(054), 504 (2013). doi:10.1103/PhysRevD.88.054504. arXiv:1302.3633
4. R.G. Edwards, U.M. Heller, T.R. Klassen, Accurate scale determinations for the Wilson gauge action. Nucl. Phys. **B517**, 377–392 (1998). doi:10.1016/S0550-3213(98)80003-5. arXiv:hep-lat/9711003

5. G. Parisi, R. Petronzio, F. Rapuano, A measurement of the string tension near the continuum limit. Phys. Lett. B **128**, 418 (1983). doi:10.1016/0370-2693(83)90930-9
6. Y. Peng, R.W. Haymaker, Su(2) flux distributions on finite lattices. Phys. Rev. D **47**, 5104–5112 (1993). doi:10.1103/PhysRevD.47.5104. http://link.aps.org/doi/10.1103/PhysRevD.47.5104

Few-Body Syst (2017) 58:127
DOI 10.1007/s00601-017-1281-7

P. J. Silva · O. Oliveira · D. Dudal · P. Bicudo · N. Cardoso

Gluons at Finite Temperature

Received: 15 November 2016 / Accepted: 9 March 2017 / Published online: 3 April 2017

Abstract The gluon propagator is investigated at finite temperature via lattice simulations. In particular, we discuss its interpretation as a massive-type bosonic propagator. Moreover, we compute the corresponding spectral density and study the violation of spectral positivity. Finally, we explore the dependence of the gluon propagator on the phase of the Polyakov loop.

1 Introduction

The study of the dynamics of QCD at finite temperature and density has been the subject of an intensive study, motivated by the heavy-ion experiments running e.g. at CERN [1] and RHIC [2]. From the theoretical point of view, the lattice formulation has been one of the most promising frameworks which allows to investigate the properties of the non-perturbative regime of QCD at non-zero temperature. For pure SU(3) Yang–Mills theory at finite temperature, a first-order transition is found at a critical temperature $T_c \sim 270\,\text{MeV}$—see for example [3], and references therein. For temperatures below T_c, the gluons are confined within color-singlets,

This article belongs to the Topical Collection "Light Cone 2016".

Talk presented by P. J. Silva at the Light Cone 2016 Conference, Lisbon, Portugal, 5–8 September 2016.

P. J. Silva (✉) · O. Oliveira
CFisUC, Department of Physics, University of Coimbra, 3004-516 Coimbra, Portugal
E-mail: psilva@uc.pt

O. Oliveira
E-mail: orlando@fis.uc.pt

D. Dudal
Department of Physics, KU Leuven Campus Kortrijk - KULAK, Etienne Sabbelaan 51 bus 7800, 8500 Kortrijk, Belgium
E-mail: david.dudal@kuleuven.be

D. Dudal
Department of Physics and Astronomy, Ghent University, Krijgslaan 281-S9, 9000 Ghent, Belgium

P. Bicudo · N. Cardoso
CFTP, Departamento de Física, Instituto Superior Técnico, Universidade de Lisboa, Avenida Rovisco Pais 1, 1049-001 Lisbon, Portugal
E-mail: bicudo@tecnico.ulisboa.pt

N. Cardoso
E-mail: nunocardoso@cftp.ist.utl.pt

Table 1 Lattice setup

Temp. (MeV)	β	N_s	N_t	a (fm)	1/a (GeV)
121	6.0000	64	16	0.1016	1.943
162	6.0000	64	12	0.1016	1.943
194	6.0000	64	10	0.1016	1.943
243	6.0000	64	8	0.1016	1.943
260	6.0347	68	8	0.09502	2.0767
265	5.8876	52	6	0.1243	1.5881
275	6.0684	72	8	0.08974	2.1989
285	5.9266	56	6	0.1154	1.7103
290	6.1009	76	8	0.08502	2.3211
305	5.9640	60	6	0.1077	1.8324
305	6.1326	80	8	0.08077	2.4432
324	6.0000	64	6	0.1016	1.943
366	6.0684	72	6	0.08974	2.1989
397	5.8876	52	4	0.1243	1.5881
428	5.9266	56	4	0.1154	1.7103
458	5.9640	60	4	0.1077	1.8324
486	6.0000	64	4	0.1016	1.943

The columns refer to the temperature in MeV, the Wilson gauge action parameter β, the spatial N_s and temporal N_t lattice sizes, the lattice spacing a and its inverse $1/a$ in GeV

whereas for $T > T_c$ the gluons become deconfined and behave as massive quasiparticles. The order parameter for the deconfinement phase transition is the Polyakov loop, defined on the lattice as

$$L(\mathbf{x}) = \mathrm{Tr} \prod_{t=0}^{N_t-1} \mathcal{U}_4(\mathbf{x}, t), \quad L = \langle L(\mathbf{x}) \rangle \propto e^{-F_q/T} \tag{1}$$

where $\mathcal{U}_4$ is the time-oriented link. Its space-averaged value L is a measure of the free energy of a static quark. The behaviour of the Polyakov loop as a function of the temperature is connected with a spontaneous breaking of the center symmetry. The Wilson gauge action and the integration measure are invariant under a center transformation, where the temporal links on a hyperplane $x_4 = const$ are multiplied by an element of the SU(3) center group $z \in Z_3 = \{e^{-i2\pi/3}, 1, e^{i2\pi/3}\}$. Under such a center transformation, the Polyakov loop transforms as $L(\mathbf{x}) \to zL(\mathbf{x})$. For temperatures below T_c, the local phase of the Polyakov loop is equally distributed among the three sectors, and therefore $L = \langle L(\mathbf{x}) \rangle \approx 0$. For $T > T_c$, the Z_3 sectors are not equally populated (a manifestation of a spontaneous breaking of the center symmetry) and thus $L \neq 0$.

In this proceeding, we focus on several aspects of the Landau gauge gluon propagator at finite temperature, computed via lattice QCD simulations. Like other propagators of fundamental fields (e.g. quark and ghost propagators), gluon two-point functions encode information about non-perturbative phenomena, such as confinement and deconfinement.

2 Lattice Setup and Propagators

At finite temperature, the gluon propagator has the following tensor structure

$$D^{ab}_{\mu\nu}(\hat{q}) = \delta^{ab} \left(P^T_{\mu\nu} D_T(q_4, \mathbf{q}) + P^L_{\mu\nu} D_L(q_4, \mathbf{q}) \right) \tag{2}$$

where P^T and P^L are the transverse and longitudinal projectors respectively:

$$P^T_{\mu\nu} = (1 - \delta_{\mu 4})(1 - \delta_{\nu 4}) \left(\delta_{\mu\nu} - \frac{q_\mu q_\nu}{\mathbf{q}^2} \right), \quad P^L_{\mu\nu} = \left(\delta_{\mu\nu} - \frac{q_\mu q_\nu}{q^2} \right) - P^T_{\mu\nu}. \tag{3}$$

In Table 1 we describe our lattice setup used in the next two sections. The temperature is defined by adjusting the lattice temporal size, $T = 1/(N_t a)$. Following our first reports [4], where we saw a measurable dependence of the gluon propagator at finite T on the lattice volume, we have carefully chosen the lattice parameters of the various Monte Carlo simulations in order to keep the physical (spatial) volume at a constant

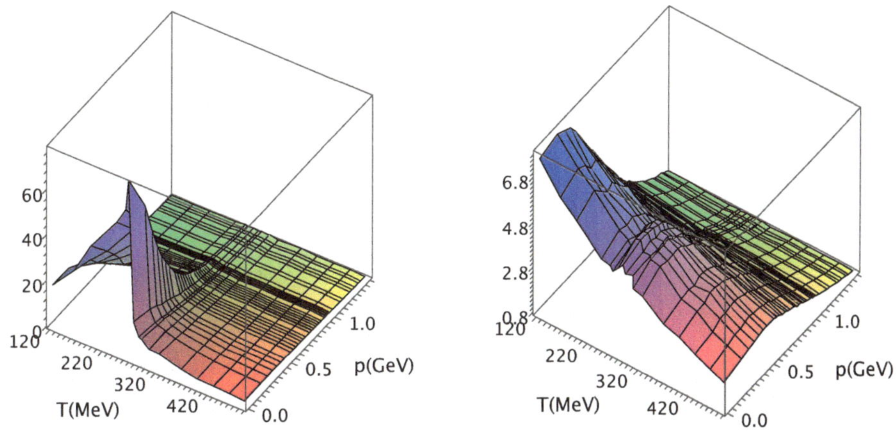

Fig. 1 Longitudinal (*left*) and transverse (*right*) components of the gluon propagator as functions of momentum and temperature

value $\sim$(6.5 fm)^3. Simulations have been made in Coimbra [5] with the help of Chroma [6] and PFFT [7] libraries.

The results shown here are for renormalized longitudinal and transverse propagators, and for a renormalization scale $\mu = 4\,\text{GeV}$. For details about the renormalization procedure see [8].

In Fig. 1 we show how the propagators behave as functions of momentum and temperature. Note the sharp transition for the longitudinal component at $T \sim T_c$, and the turnover of the transverse component in the infrared region, for $T \gg T_c$.

3 Positivity Violation and Spectral Densities

It is well known that a Euclidean momentum-space propagator of a (scalar) physical degree of freedom

$$\mathcal{G}(p^2) \equiv \langle \mathcal{O}(p)\mathcal{O}(-p)\rangle \tag{4}$$

ought to have a Källén–Lehmann spectral representation

$$\mathcal{G}(p^2) = \int_0^\infty \mathrm{d}\mu \frac{\rho(\mu)}{p^2+\mu}, \quad \text{with } \rho(\mu) \geq 0 \text{ for } \mu \geq 0. \tag{5}$$

where the spectral density $\rho(\mu)$ contains information on the masses of physical states described by the operator $\mathcal{O}$. In [9] a method is presented which allows to compute the spectral density of gluons and other (un)physical degrees of freedom, for which the spectral density is not strictly positive. The method relies on Tikhonov regularization combined with the Morozov discrepancy principle. Here we discuss some preliminary results [10] for the spectral density associated to the gluon propagator at finite temperature, together with the Schwinger function

$$C(t) = \int_{-\infty}^{\infty} \frac{dp}{2\pi} D(p^2)\exp(-ipt) = \int_0^\infty d\omega \rho(\omega^2) e^{-\omega t}. \tag{6}$$

Note that $C(t) < 0$ in some range of t implies a negative spectral density, hence positivity violation and gluon confinement. On the other hand, a positive $C(t)$ says nothing about the sign of $\rho(\mu)$.

In Fig. 2 we plot $C(t)$ for the transverse component. We conclude that the positivity is violated for all temperatures. Furthermore, a careful inspection reveals that the time scale for positivity violation decreases with the temperature. In Fig. 3 we plot, for a number of selected temperatures, the spectral density of the longitudinal component. The data in Fig. 3 suggests that the momentum scale where ρ changes sign increases with T. All these results suggest that, for sufficiently high temperatures, the spectral density can be strictly positive, which means that the gluons would behave as quasi-particles.

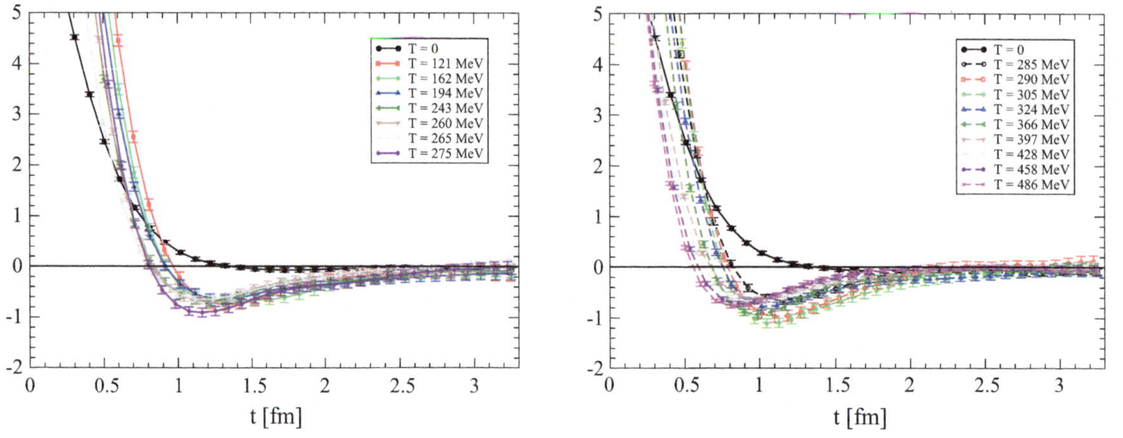

Fig. 2 Schwinger function (Eq. 6) associated to the transverse gluon propagator

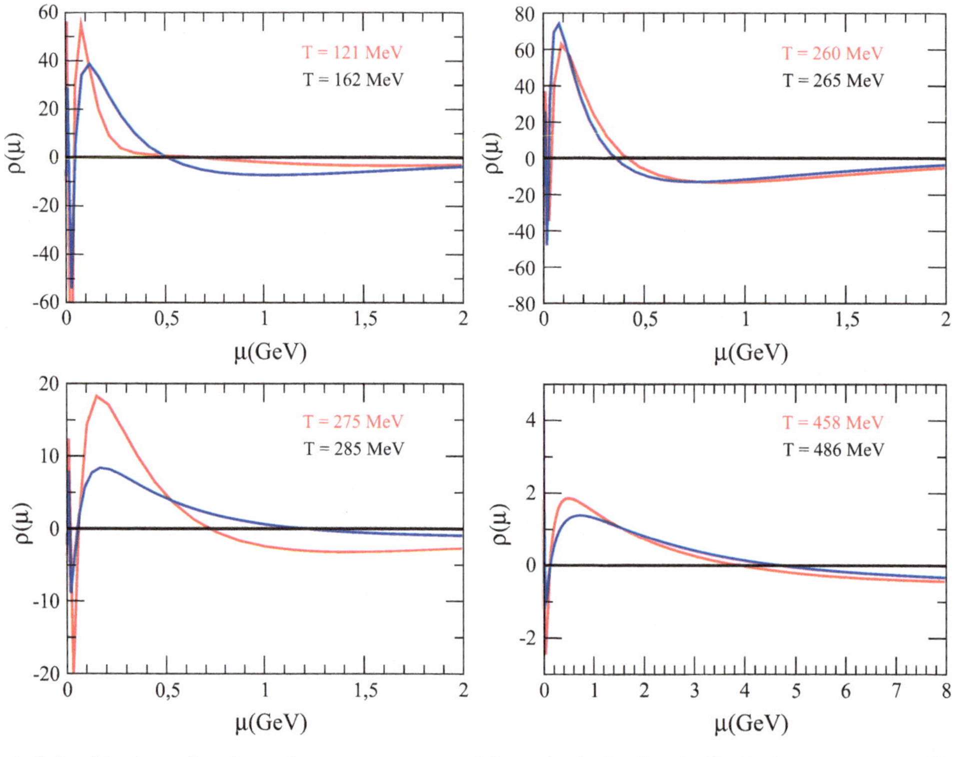

Fig. 3 Spectral densities ρ as functions of momenta computed from the lattice longitudinal gluon propagator. Note that the maximum ρ decreases with the temperature for $T > T_c$

4 Gluon Mass

In this section we investigate whether the gluon propagator at finite temperature behaves as a massive-type bosonic propagator [8]. We consider a Yukawa-type ansatz

$$D(p) = \frac{Z}{p^2 + M^2} \tag{7}$$

where M is the gluon mass and $Z^{\frac{1}{2}}$ the overlap between the gluon state and the quasi-particle massive state.

The simplest definition for a gluon mass scale is given by $M = 1/\sqrt{D(0)}$. Our results for a mass scale considering this definition can be seen in the left plot of Fig. 4. Of course, a more realistic estimate for the gluon mass can be obtained by fitting our lattice data in the infrared region to the ansatz described in Eq. (7).

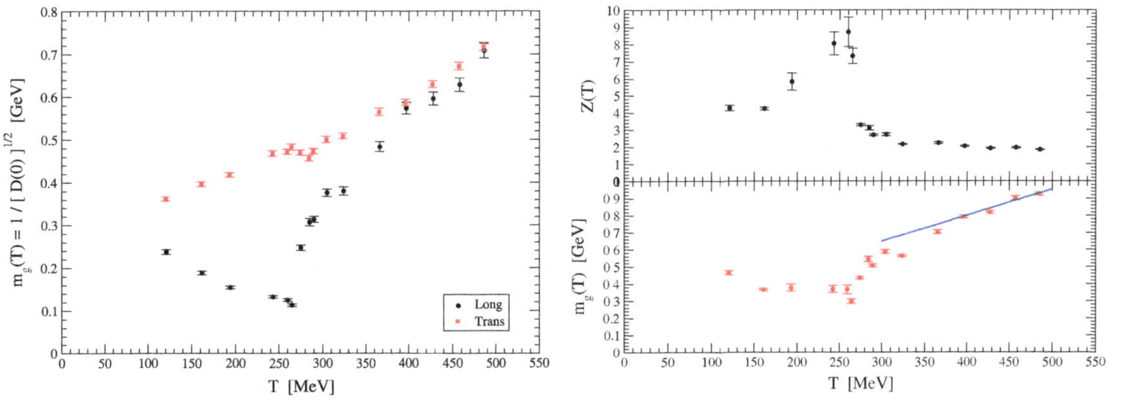

Fig. 4 Gluon mass scales as functions of the temperature. See the text for details

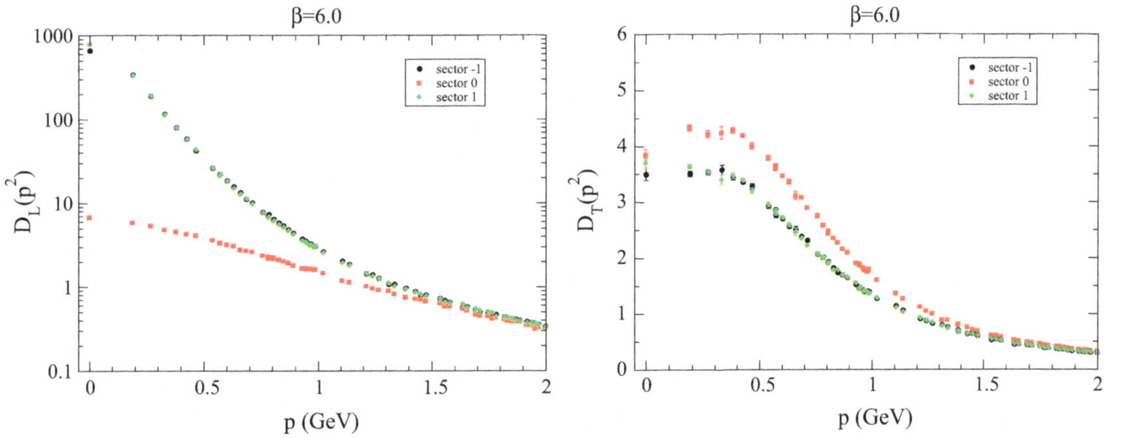

Fig. 5 Longitudinal (*left*) and transverse (*right*) propagators for the different sectors at $T = 324$ MeV

It turned out that the lattice data for the transverse propagator is not compatible with such ansatz. The results for the longitudinal propagator can be seen in the right plot of Fig. 4. Note the linear behaviour of the mass scale for high T, as predicted by perturbation theory.

5 Z_3 Dependence

As seen in Fig. 1, D_L and D_T show quite different behaviours with T. The gluon propagator is usually computed such that $-\pi/3 < \arg(L) \leq \pi/3$ i.e. in the so-called Z_3 sector 0. In this section we investigate the behaviour of the gluon propagator in the other ± 1 Z_3 sectors[1] and compare with the results for the zero sector. To achieve such goal, for each configuration in a given ensemble, we applied a center transformation considering all $z \in Z_3$, thus obtaining 3 different configurations related by Z_3 transformations, with the very same value for the Wilson action. Each configuration is then rotated to the Landau gauge. Finally, each of the three gauge configurations is classified according to the phase of $L = |L|e^{i\theta}$.

In Fig. 5, we can see the typical behaviour of the gluon propagator in the different Z_3 sectors for a temperature well above T_c. For the longitudinal component, the propagator in the ± 1 sectors is strongly enhanced relative to the 0 sector. On the other hand, the tranverse propagator in the ± 1 sectors is suppressed if compared with the 0 sector. However, for temperatures below T_c, the picture changes—see Fig. 6. In this case, the three propagators for the different Z_3 sectors are indistinguishable.

[1] The Z_3 sector -1 corresponds to $-\pi < \arg(L) \leq -\pi/3$, and the Z_3 sector 1 is defined by $\pi/3 < \arg(L) \leq \pi$.

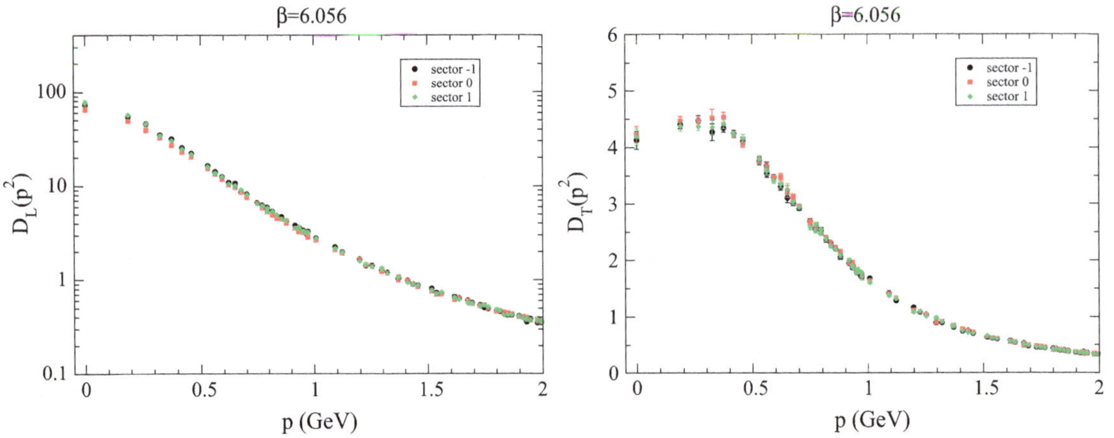

Fig. 6 Longitudinal (*left*) and transverse (*right*) propagators for the different sectors for $T = 269\,\text{MeV}$, slightly below T_c

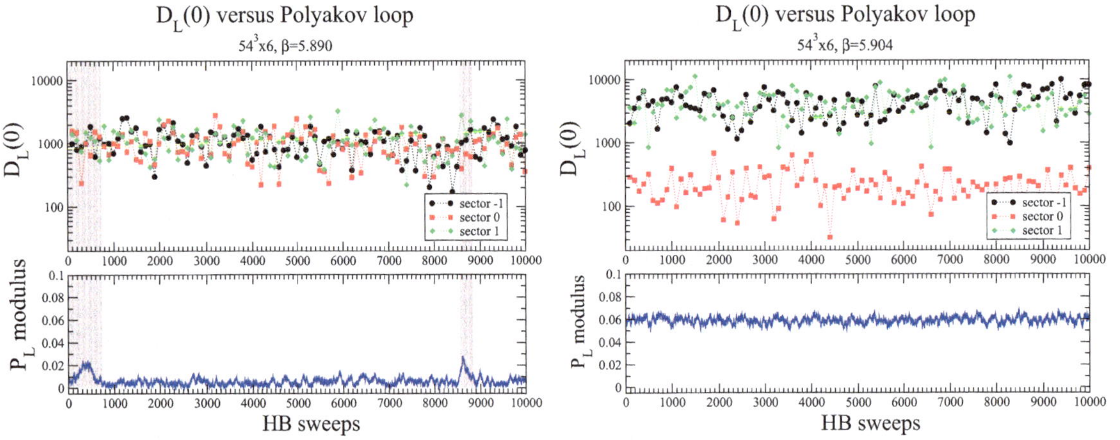

Fig. 7 Monte Carlo history of the bare values of $D_L(0)$ and $|L|$ for ensembles with $T = 266\,\text{MeV} < T_c$ (*left*) and $T = 273\,\text{MeV} > T_c$ (*right*). Note that configurations inside the *shaded bands* are not used to compute the gluon propagator. See [11] for details

A comparison of the Markov chain history for temperatures below and above T_c—see Fig. 7—allows one to conclude that the difference of the longitudinal propagator between the different Z_3 sectors can be used as a criterion to identify whether a given configuration is in the confined or deconfined phase.

More details about this work may be found in [11].

Acknowledgements O. Oliveira, and P. J. Silva acknowledge financial support from FCT Portugal under contract with reference UID/FIS/04564/2016. P. J. Silva acknowledges support by FCT under Contracts SFRH/BPD/40998/2007 and SFRH/BPD/109971/2015. P. Bicudo and N. Cardoso thank CFTP with FCT Grant UID/FIS/00777/2013. N. Cardoso is supported by FCT under Contract SFRH/BPD/109443/2015. The computing time was provided by the Laboratory for Advanced Computing at the University of Coimbra [5].

References

1. C. Roland, Jets in heavy ion collisions at the LHC. Int. J. Mod. Phys. A **30**, 1546010 (2015)
2. T.A. Trainor, A critical review of RHIC experimental results. Int. J. Mod. Phys. E **23**, 1430011 (2014)
3. B. Lucini, M. Teper, U. Wenger, The high temperature phase transition in SU(N) gauge theories. J. High Energy Phys. **01**, 061 (2004)
4. O. Oliveira, P.J. Silva, The lattice Landau gauge gluon propagator at zero and finite temperature. Acta Phys. Polon. Supp. **5**, 1039 (2012)
5. http://www.uc.pt/lca/
6. R.G. Edwards, B. Joó, The Chroma software system for lattice QCD. Nucl. Phys. Proc. Suppl. **140**, 832 (2005)
7. M. Pippig, PFFT: an extension of FFTW to massively parallel architectures. SIAM J. Sci. Comput. **35**, C213 (2013)

8. P.J. Silva, O. Oliveira, P. Bicudo, N. Cardoso, Gluon screening mass at finite temperature from the Landau gauge gluon propagator in lattice QCD. Phys. Rev. D **89**, 074503 (2014)
9. D. Dudal, O. Oliveira, P.J. Silva, Källén–Lehmann spectroscopy for (un)physical degrees of freedom. Phys. Rev. D **89**, 014010 (2014)
10. P.J. Silva, D. Dudal, O. Oliveira, *Spectral Densities from the Lattice*, PoS(LATTICE 2013)366 (2014). https://pos.sissa.it/archive/conferences/187/366/LATTICE%202013_366.pdf
11. P.J. Silva, O. Oliveira, Gluon dynamics, center symmetry, and the deconfinement phase transition in SU(3) pure Yang–Mills theory. Phys. Rev. D **93**, 114509 (2016)

Few-Body Syst (2017) 58:92
DOI 10.1007/s00601-017-1254-x

Sabrina Cotogno

Parametrization of the Transverse Momentum Dependent Light-Front Correlator for Gluons

Received: 19 January 2017 / Accepted: 31 January 2017 / Published online: 17 February 2017

Abstract We study the transverse momentum dependent light-front correlator for gluons. At the operator level this is expressed as a matrix element containing nonlocal field strength operators and gauge links bridging the nonlocality. We parametrize the leading (twist-2) gluon–gluon correlator in terms of transverse momentum dependent distribution functions for unpolarized, vector and tensor polarized targets (the latter being relevant for spin-1 targets). For a tensor polarized target there are eleven functions among which two are time reversal odd. We discuss bounds on some functions which might become useful for future applications.

1 Introduction

The cross section of high energy processes can be written as a convolution of hadronic matrix elements and partonic hard cross section, so long as the validity of factorization theorems is assumed. Nonlocal forward matrix elements containing quarks and gluon fields are studied in terms of Parton distribution functions (PDFs). Extending the non locality to the transverse directions offers possibilities to study transverse momentum depended (TMD) PDFs, or simply TMDs. The hadronic matrix element containing two gluon fields was parametrized in terms of TMD PDFs for the first time in [1] and later on proposed in a different form in [2]. These works describe unpolarized and vector polarized targets, relevant for spin-0 and spin-$\frac{1}{2}$ particles. Recently in [3] the parametrization of the gluon–gluon correlator has been extended to tensor polarization, allowing in fact the inclusion of spin-1 hadrons in the description. The outline of this work is as follows: in Sect. 2 we show the parametrization of the gluon–gluon correlator in terms of gluon TMDs for hadrons with spin up to 1 and in Sect. 3 we present some bounds on the distributions.

2 Parametrization of the TMD Gluon–Gluon Correlation Function for Hadrons of Spin ≤ 1

The correlation function for gluons in hadrons is defined as:

$$\Gamma^{[U,U']\,\mu\nu;\rho\sigma}(k,P,n,\ldots) \equiv \int \frac{d^4\xi}{(2\pi)^4}\, e^{ik\cdot\xi}\, \langle P,\ldots|\, F^{\mu\nu}(0)U_{[0,\xi]}F^{\rho\sigma}(\xi)U'_{[\xi,0]}\,|P,\ldots\rangle\,, \tag{1}$$

This article belongs to the Topical Collection "Light Cone 2016".

S. Cotogno (✉)
Nikhef, Science Park 105, 1098 XG Amsterdam, The Netherlands
E-mail: scotogno@nikhef.nl

S. Cotogno
Department of Physics and Astronomy, VU University Amsterdam, De Boelelaan 1081, 1081 HV Amsterdam, The Netherlands

where a trace in color space (Tr_c) is implicitly assumed. The Wilson lines $U_{[0,\xi]}$ and $U'_{[\xi,0]}$ guarantee color gauge invariance as they bridge the nonlocality of the two gluon field tensors. The dots indicate the spin vectors and/or tensors needed to deal with polarization of hadrons with spin ≤ 1.

The relevant momentum variables are P and k, i.e. the hadron and parton momentum respectively. We use the lightlike vector n, i.e. $n^2 = 0$, and we parametrize P and k such that they satisfy:

$$P^2 = M^2, \qquad P{\cdot}n = 1, \qquad k^\mu = xP^\mu + k_T^\mu + \left(k{\cdot}P - xM^2\right) n^\mu. \tag{2}$$

where M is the mass of the hadron.

The target spin states are characterized by the density matrix. For hadrons of spin 0 and $\frac{1}{2}$, one needs to introduce a spacelike vector S^μ which parametrizes the density matrix [4,5]. By demanding $S^2 = -1$ and $P{\cdot}S = 0$ one can parametrize S in terms of the dimensionful vectors P and n:

$$S^\mu = S_L \frac{P^\mu}{M} + S_T^\mu - M S_L\, n^\mu. \tag{3}$$

To extend the description to spin-1 particles a polarization tensor $T^{\mu\nu}$ is needed, which satisfies $P_\mu T^{\mu\nu} = 0$. The parametrization of the tensorial part reads:

$$\begin{aligned} T^{\mu\nu} = \frac{1}{2}\Bigg[& \frac{2}{3} S_{LL}\, g_T^{\mu\nu} + \frac{4}{3} S_{LL} \frac{P^\mu P^\nu}{M^2} + \frac{S_{LT}^{\{\mu} P^{\nu\}}}{M} + S_{TT}^{\mu\nu} \\ & - \frac{4}{3} S_{LL} P^{\{\mu} n^{\nu\}} - M S_{LT}^{\{\mu} n^{\nu\}} + \frac{4}{3} M^2 S_{LL}\, n^\mu n^\nu \Bigg]. \end{aligned} \tag{4}$$

We have used the metric tensor in transverse space defined as $g_T^{\mu\nu} \equiv g^{\mu\nu} - P^{\{\mu} n^{\nu\}}$ (curly brackets denote symmetrization of the indices), with nonvanishing elements $g_T^{11} = g_T^{22} = -1$. The spin tensor is symmetric and traceless and it has five independent parameters, namely S_{LL}, the two components of the transverse vector S_{LT}, and the two independent components of the symmetric traceless transverse tensor S_{TT}.

Looking at Eq. (1) we note that, even though for unspecified processes the path could run along arbitrary paths, we have included a dependence on the lightlike four-vector n, that enters upon consideration of staple-like gauge links running along the light-front ($\xi{\cdot}n = 0$) via lightlike $\xi{\cdot}P = \pm\infty$. Integrating Eq. (1) over $k{\cdot}P$, one obtains the TMD (light-front) correlator

$$\Gamma^{[U,U']\mu\nu;\rho\sigma}(x, \boldsymbol{k}_T; P, n) \equiv \int \frac{d\xi{\cdot}P\, d^2\xi_T}{(2\pi)^3}\, e^{ik\cdot\xi}\, \langle P |\, F^{\mu\nu}(0) U_{[0,\xi]} F^{\rho\sigma}(\xi) U'_{[\xi,0]}\, | P \rangle \Big|_{\xi\cdot n = 0}. \tag{5}$$

Simple power counting leads to the definition of the leading (usually referred to as leading twist or twist-2) correlator:

$$\Gamma^{ij}(x, \boldsymbol{k}_T) \equiv \Gamma^{[U,U']ni;nj}(x, \boldsymbol{k}_T; P, n). \tag{6}$$

The latter is indeed the operator considered as a starting point in [3], and we repeat the main results on the parametrization therein derived.

In the decomposition we employ symmetric and traceless tensors $k_T^{i_1 \ldots i_n}$ that are built from the partonic momentum k_T. Up to rank $n = 4$, these are given by

$$k_T^{ij} \equiv k_T^i k_T^j + \frac{1}{2} \boldsymbol{k}_T^2 g_T^{ij}, \tag{7}$$

$$k_T^{ijk} \equiv k_T^i k_T^j k_T^k + \frac{1}{4} \boldsymbol{k}_T^2 \left(g_T^{ij} k_T^k + g_T^{ik} k_T^j + g_T^{jk} k_T^i \right), \tag{8}$$

$$\begin{aligned} k_T^{ijkl} \equiv\; & k_T^i k_T^j k_T^k k_T^l + \frac{1}{6} \boldsymbol{k}_T^2 \left(g_T^{ij} k_T^{kl} + g_T^{ik} k_T^{jl} + g_T^{il} k_T^{jk} + g_T^{jk} k_T^{il} + g_T^{jl} k_T^{ik} + g_T^{kl} k_T^{ij} \right) \\ & - \frac{1}{8} \boldsymbol{k}_T^4 \left(g_T^{ij} g_T^{kl} + g_T^{ik} g_T^{jl} + g_T^{il} g_T^{jk} \right), \end{aligned} \tag{9}$$

satisfying

$$g_{Tij} k_T^{ij} = g_{Tij} k_T^{ijk} = g_{Tij} k_T^{ijkl} = 0. \tag{10}$$

From this decomposition of the Lorentz structures it follows that the functions involved in the parametrization are twist-2 TMDs *of definite rank*. This has the advantage that there is a one-to-one correspondence between the functions in the transverse coordinate $\boldsymbol{b}_T$ space and the $\boldsymbol{k}_T$ space, which turns out to be a very important feature when the TMD evolution equations have to be considered [6].

Separation into different polarizations gives:

$$\Gamma = \Gamma_U + \Gamma_L + \Gamma_T + \Gamma_{LL} + \Gamma_{LT} + \Gamma_{TT}, \tag{11}$$

where:

$$\Gamma_U^{ij}(x,\boldsymbol{k}_T) = \frac{x}{2}\left[-g_T^{ij}\, f_1\left(x,\boldsymbol{k}_T^2\right) + \frac{k_T^{ij}}{M^2}\, h_1^\perp\left(x,\boldsymbol{k}_T^2\right)\right], \tag{12}$$

$$\Gamma_L^{ij}(x,\boldsymbol{k}_T) = \frac{x}{2}\left[i\epsilon_T^{ij} S_L\, g_1\left(x,\boldsymbol{k}_T^2\right) + \frac{\epsilon_{T\,\alpha}^{\{i}k_T^{j\}\alpha}S_L}{2M^2}\, h_{1L}^\perp\left(x,\boldsymbol{k}_T^2\right)\right], \tag{13}$$

$$\begin{aligned}\Gamma_T^{ij}(x,\boldsymbol{k}_T) = \frac{x}{2}\Bigg[&-\frac{g_T^{ij}\epsilon_T^{S_T k_T}}{M}\, f_{1T}^\perp\left(x,\boldsymbol{k}_T^2\right) + \frac{i\epsilon_T^{ij}\boldsymbol{k}_T\cdot\boldsymbol{S}_T}{M}\, g_{1T}\left(x,\boldsymbol{k}_T^2\right)\\ &-\frac{\epsilon_T^{k_T\{i}S_T^{j\}} + \epsilon_T^{S_T\{i}k_T^{j\}}}{4M}\, h_1\left(x,\boldsymbol{k}_T^2\right) - \frac{\epsilon_{T\,\alpha}^{\{i}k_T^{j\}\alpha S_T}}{2M^3}\, h_{1T}^\perp\left(x,\boldsymbol{k}_T^2\right)\Bigg],\end{aligned} \tag{14}$$

$$\Gamma_{LL}^{ij}(x,\boldsymbol{k}_T) = \frac{x}{2}\left[-g_T^{ij} S_{LL}\, f_{1LL}\left(x,\boldsymbol{k}_T^2\right) + \frac{k_T^{ij}S_{LL}}{M^2}\, h_{1LL}^\perp\left(x,\boldsymbol{k}_T^2\right)\right], \tag{15}$$

$$\begin{aligned}\Gamma_{LT}^{ij}(x,\boldsymbol{k}_T) = \frac{x}{2}\Bigg[&-\frac{g_T^{ij}\boldsymbol{k}_T\cdot\boldsymbol{S}_{LT}}{M}\, f_{1LT}\left(x,\boldsymbol{k}_T^2\right) + \frac{i\epsilon_T^{ij}\epsilon_T^{S_{LT}k_T}}{M}\, g_{1LT}\left(x,\boldsymbol{k}_T^2\right)\\ &+\frac{S_{LT}^{\{i}k_T^{j\}}}{M}\, h_{1LT}\left(x,\boldsymbol{k}_T^2\right) + \frac{k_T^{ij\alpha}S_{LT\,\alpha}}{M^3}\, h_{1LT}^\perp\left(x,\boldsymbol{k}_T^2\right)\Bigg],\end{aligned} \tag{16}$$

$$\begin{aligned}\Gamma_{TT}^{ij}(x,\boldsymbol{k}_T) = \frac{x}{2}\Bigg[&-\frac{g_T^{ij}k_T^{\alpha\beta}S_{TT\,\alpha\beta}}{M^2}\, f_{1TT}\left(x,\boldsymbol{k}_T^2\right) + \frac{i\epsilon_T^{ij}\epsilon_{T\,\gamma}^{\beta}k_T^{\gamma\alpha}S_{TT\,\alpha\beta}}{M^2}\, g_{1TT}\left(x,\boldsymbol{k}_T^2\right)\\ &+S_{TT}^{ij}\, h_{1TT}\left(x,\boldsymbol{k}_T^2\right) + \frac{S_{TT\,\alpha}^{\{i}k_T^{j\}\alpha}}{M^2}\, h_{1TT}^\perp\left(x,\boldsymbol{k}_T^2\right)\\ &+\frac{k_T^{ij\alpha\beta}S_{TT\,\alpha\beta}}{M^4}\, h_{1TT}^{\perp\perp}\left(x,\boldsymbol{k}_T^2\right)\Bigg].\end{aligned} \tag{17}$$

Throughout the paper, the dependence of functions on the gauge link is understood and we define $\boldsymbol{k}_T\cdot\boldsymbol{S}_T = -k_T\cdot S_T$. The transverse antisymmetric tensor is defined as $\epsilon_T^{\mu\nu} \equiv \epsilon^{Pn\mu\nu}$, with nonzero components $\epsilon_T^{12} = -\epsilon_T^{21} = 1$. The functions $h_{1L}^\perp$, $f_{1T}^\perp$, h_1, and $h_{1T}^\perp$ are odd under naive time reversal transformatios (T-odd). The functions $h_{1LL}^\perp$, f_{1LT}, h_{1LT}, $h_{1LT}^\perp$, f_{1TT}, $h_{1TT}^\perp$, $h_{1TT}^{\perp\perp}$ are T-even whereas g_{1LT} and g_{1TT} are T-odd and they have been first introduced in [3]. The function f_{1LL} was also called b_1 in the quark case [7], and h_{1TT} shows up in the structure function $\Delta(x, Q^2)$ discussed in Ref. [8] and is called $\Delta_2 G(x)$ in [9]. They are rank-0 functions which survive in the collinear case, i.e. upon integration over transverse momenta.

3 Positivity Bounds on the Gluon Distributions in Spin-1 Targets

As already done in [10] one can impose positivity constraints to the hadronic tensor and find a probabilistic interpretation for some of the distribution functions. Bounds on the TMDs have been studied in [1] for gluons in spin-$\frac{1}{2}$ hadrons and in [7,11] for quarks in hadrons with spin up to 1.

The correlator Γ^{ij} in (6) is a 2×2 matrix in the transverse gluon polarizations. On the other hand, the spin states of a spin-1 particle are described by the density matrix, which can be written in terms of the independent components of the vector S and the tensor T. In fact one needs a Cartesian basis of 3×3 matrices, consisting of the identity matrix I, the three-dimensional (generalized) Pauli matrices Σ^i, and the five bilinear combinations $\Sigma^{ij} \equiv \frac{1}{2}(\Sigma^i\Sigma^j + \Sigma^j\Sigma^i) - \frac{2}{3}I\delta^{ij}$. The density matrix (in the rest frame) takes the form:

$$\rho = \frac{1}{3}\left(I + \frac{3}{2}S^i\Sigma^i + 3T^{ij}\Sigma^{ij}\right), \tag{18}$$

$$\rho = \begin{pmatrix} \frac{S_L}{2} + \frac{S_{LL}}{3} + \frac{1}{3} & \frac{S_{LTx}-iS_{LTy}}{2\sqrt{2}} + \frac{S_{Tx}-iS_{Ty}}{2\sqrt{2}} & \frac{1}{2}(S_{TTxx} - iS_{TTxy}) \\ \frac{S_{LTx}+iS_{LTy}}{2\sqrt{2}} + \frac{S_{Tx}+iS_{Ty}}{2\sqrt{2}} & \frac{1}{3} - \frac{2SLL}{3} & \frac{-S_{LTx}+iS_{LTy}}{2\sqrt{2}} + \frac{S_{Tx}-iS_{Ty}}{2\sqrt{2}} \\ \frac{1}{2}(S_{TTxx} + iS_{TTxy}) & \frac{-S_{LTx}-iS_{LTy}}{2\sqrt{2}} + \frac{S_{Tx}+iS_{Ty}}{2\sqrt{2}} & -\frac{SL}{2} + \frac{SLL}{3} + \frac{1}{3} \end{pmatrix}. \tag{19}$$

This allows one to write the correlator in a form which explicitly singles out the hadron spin, namely:

$$\Gamma^{ij}(x, \boldsymbol{k}_\perp; P, S, T) = \sum_{s,s'} \rho_{s',s}(S,T)\Gamma^{ij}_{s,s'}(x, \boldsymbol{k}_\perp; P) \tag{20}$$

The matrix $\Gamma^{ij}_{s,s'}$ in the right hand of (20) is a 6×6 matrix in the gluon $\otimes$ hadron spin space:

$$\Gamma^{ij}_{s,s'} = \begin{pmatrix} A & B \\ B^\dagger & C \end{pmatrix} \tag{21}$$

where A, B and C are 3×3 blocks whose complete expressions will be given elsewhere. From symmetry considerations, it follows that the block C is the charge-conjugated of the block A and the off-diagonal block are hermitian-conjugate (see Ref. [3] for more details on the hermiticity, time-reversal and charge-conjugation properties of the gluon correlator). We give here the explicit form of the matrix (21) in the collinear case where the blocks are:

$$A = \begin{pmatrix} f_1 + \frac{f_{1LL}}{2} - g_1 & 0 & 0 \\ 0 & f_1 - f_{1LL} & 0 \\ 0 & 0 & f_1 + \frac{f_{1LL}}{2} + g_1 \end{pmatrix}, \quad B = \begin{pmatrix} 0 & 0 & 0 \\ 0 & 0 & 0 \\ -2h_{1TT} & 0 & 0 \end{pmatrix}$$
$$C = \begin{pmatrix} f_1 + \frac{f_{1LL}}{2} + g_1 & 0 & 0 \\ 0 & f_1 - f_{1LL} & 0 \\ 0 & 0 & f_1 + \frac{f_{1LL}}{2} - g_1 \end{pmatrix} \tag{22}$$

We note that $\Gamma^{ij}_{s,s'}$ is a positive semi-definite matrix, i.e. for each vector in the combined gluon $\otimes$ target spin space it satisfies $v^\dagger\Gamma v \geq 0$.

$$\begin{aligned} \Gamma^{ij}_{s,s'}(x; P) &= \int \frac{d\xi\cdot P}{2\pi} e^{ik\cdot\xi} \langle P, S, T|\, F^{ni}(0)F^{nj}(\xi)\,|P, S, T\rangle\Big|_{\xi\cdot n=\xi_T=0} \\ &= \sum_m \langle P_m|\, F^{ni}(0)\,|P, s\rangle^* \langle P_m|\, F^{nj}(0)\,|P, s'\rangle\, \delta(P_m\cdot n - (1-x)P\cdot n) \end{aligned} \tag{23}$$

This is true for the unintegrated version of Γ as well (see [1,5]). The matrix (21) is positive semi-definite if its eigenvalues are all ≥ 0. Together with the requirement for all the diagonal elements to be positive, we find the following bounds for the integrated gluon distribution functions:

$$\begin{aligned} &f_1 \geq f_{1LL}, \\ &|g_1| \leq f_1 + \frac{f_{1LL}}{2} \leq \frac{3}{2}f_1 \\ &|h_{1TT}| \leq \frac{1}{2}\left(f_1 + \frac{f_{1LL}}{2} + g_1\right). \end{aligned} \tag{24}$$

Including transverse momenta produces many more relations on the functions which will be shown in a future work.

4 Conclusions

We have presented the parametrization of the gluon–gluon correlator in terms of TMDs and we have considered targets with spin up to 1. Some of the new functions presented in [3] survive under integration over transverse momenta and we derived the bounds fullfilled by these collinear distributions.

Acknowledgements This work was done in collaboration with D. Boer, T. van Daal, P. J. Mulders, A. Signori and Y. Zhou. The research was financially supported by the European Research Council (ERC) under the program QWORK (Contract No. 320389) and the "Nederlandse Organisatie voor Wetenschappelijk Onderzoek (NWO)".

References

1. P.J. Mulders, J. Rodrigues, Transverse momentum dependence in gluon distribution and fragmentation functions. Phys. Rev. D **63**, 094021 (2001). [arXiv:hep-ph/0009343]
2. S. Meissner, A. Metz, K. Goeke, Relations between generalized and transverse momentum dependent parton distributions. Phys. Rev. D **76**, 034002 (2007). [arXiv:hep-ph/0703176 [HEP-PH]]
3. D. Boer, S. Cotogno, T. van Daal, P.J. Mulders, A. Signori, Y.J. Zhou, Gluon and Wilson loop TMDs for hadrons of spin $\leq$ 1. JHEP **1610**, 013 (2016). [arXiv:1607.01654 [hep-ph]]
4. E. Leader, Spin in particle physics. Camb. Monogr. Part. Phys. Nucl. Phys. Cosmol. **15**, 1 (2011)
5. A. Bacchetta, P.J. Mulders, Deep inelastic leptoproduction of spin-one hadrons. Phys. Rev. D **62**, 114004 (2000). [arXiv:hep-ph/0007120]
6. A. Signori, Flavor and evolution effects in TMD phenomenology: manifestation of Hadron structure in high-energy scattering processes (2016), https://inspirehep.net/record/1493030/files/Thesis-2016-Signori.pdf. Accessed 14 Feb 2017
7. A. Bacchetta, Probing the transverse spin of quarks in deep inelastic scattering. arXiv:hep-ph/0212025
8. R.L. Jaffe, A. Manohar, Nuclear gluonometry. Phys. Lett. B **223**, 218 (1989)
9. X. Artru, M. Mekhfi, Transversely polarized parton densities, their evolution and their measurement. Z. Phys. C **45**, 669 (1990)
10. W. Lu, J.J. Yang, Bounding the Levelt–Mulders asymmetry from the positivity of the hadronic tensor and Callan–Gross relation. Z. Phys. C **73**, 689 (1997). [arXiv:hep-ph/9601255]
11. A. Bacchetta, M. Boglione, A. Henneman, P.J. Mulders, Bounds on transverse momentum dependent distribution and fragmentation functions. Phys. Rev. Lett. **85**, 712 (2000). [arXiv:hep-ph/9912490]

Few-Body Syst (2017) 58:84
DOI 10.1007/s00601-017-1237-y

L. Mantovani · A. Bacchetta · B. Pasquini · X. Xiong

The Gauge-Field Propagator in Light-Cone Gauge: Which is the Correct One?

Received: 17 November 2016 / Accepted: 17 January 2017 / Published online: 13 February 2017

Abstract In the literature one can find two different expressions for the gauge-field propagator in light-cone gauge, containing the sum of three rather than two terms. The question of which of the two is the correct one has been a subject of debate. We propose a solution to this question by evaluating one-loop level processes in QED, both in the covariant approach in the light-cone gauge and in the light-front time-ordered perturbation theory (TOPT) approach, proving the equivalence between the two formulations of the theory. The form of the propagator turns out to be crucial in the proof, in particular as concerns its relation with the diagrams containing instantaneously propagating photons and instantaneous interactions. We show that the diagrams in light-front TOPT with instantaneous photons can be recovered in the covariant approach starting from the propagators with only two terms. Our proof of the equivalence clarifies which form should be used for the gauge-field propagator in the covariant approach. This result naturally applies to the QCD case also.

1 Introduction

The proof of the equivalence between time-ordered perturbation theory (TOPT) in light-front quantization and the covariant approach with Feynman diagrams has been a subject of several works in the literature, ranging from scalar ϕ^3 theory [1], to scalar QED [2] and standard QED [3]. In a gauge theory a complication arises concerning the gauge field propagator, whose analytical expression in the covariant approach depends on the gauge choice. This aspect is strictly related to the presence, in the light-front TOPT formalism for QED, of diagrams containing instantaneously propagating particles, and hence also instantaneous photons.

This article belongs to the Topical Collection "Light Cone 2016".

This research has received funding from the European Research Council (ERC) under the European Union's Horizon 2020 research and innovation programme (Grant Agreement Nos. 647981, 3DSPIN). This work was supported through the McCartor Fund Fellowship program under the auspices of International Light-Cone Advisory Committee (ILCAC).

L. Mantovani (✉) · A. Bacchetta · B. Pasquini
Dipartimento di Fisica, Università degli Studi di Pavia, Pavia, Italy
E-mail: luca.mantovani@pv.infn.it

A. Bacchetta
E-mail: alessandro.bacchetta@unipv.it

B. Pasquini
E-mail: barbara.pasquini@unipv.it

L. Mantovani · A. Bacchetta · B. Pasquini · X. Xiong
INFN, Sezione di Pavia, 27100 Pavia, Italy
E-mail: x.xiong@fz-juelich.de

The usual expression for the photon propagator $\mathcal{D}^{\mu\nu}(q)$ in light-cone gauge, calculated by considering only the transverse degrees of freedom of the photon polarization vectors as the physical ones [4], contains the sum of three terms:

$$\mathcal{D}_T^{\mu\nu}(q) = \frac{-i}{q^2}\left(g^{\mu\nu} - \frac{q^\mu n^\nu + q^\nu n^\mu}{q^+} + q^2 \frac{n^\mu n^\nu}{\left(q^+\right)^2}\right). \tag{1}$$

The same expression can be recovered by starting from the free photon Lagrangian and adding to it a gauge-fixing term of the form $\left(n_\mu A^\mu\right)^2 + \left(\partial_\mu A^\mu\right)^2$ [5], where $n^\mu A_\mu = 0$ and $\partial_\mu A^\mu = 0$ correspond respectively to the light-cone gauge and the Lorenz-gauge conditions. However the choice of a non-covariant gauge (such as the light-cone gauge) also causes modifications in the interaction Hamiltonian. Since the contribution coming from the third term of the propagator is, in any physical process, exactly canceled by a new instantaneous interaction appearing in the Hamiltonian, one often encounters a different expression for $\mathcal{D}^{\mu\nu}(q)$, where the third term is dropped:

$$\mathcal{D}_2^{\mu\nu}(q) = \frac{-i}{q^2}\left(g^{\mu\nu} - \frac{q^\mu n^\nu + q^\nu n^\mu}{q^+}\right). \tag{2}$$

This last result is again recovered from the free photon Lagrangian, but this time adding only the light-cone gauge fixing term $\left(n_\mu A^\mu\right)^2$, which in any case is not enough to fix the gauge completely.

As a consequence, both expressions for the propagator can be used effectively, if one also takes into account the instantaneous interaction properly. This has been a source of confusion and misinterpretations; we therefore reconsidered the proof of the equivalence between TOPT and covariant approach at one-loop level in QED, where the role of the photon propagator is crucial. In Ref. [6] we enlightened some subtleties and misleading statement in the literature concerning this topic, and here we review our main findings.

The rest of this work is organized as follows: in Sect. 2 we introduce some of the basics about QED in light-front TOPT and describe the interaction Hamiltonian in light-cone gauge; in Sect. 3 we discuss the equivalence between TOPT and covariant approach for one-loop level processes in QED, focusing in particular on the vertex correction; we finally collect our conclusions in Sect. 4.

2 QED Hamiltonian on the Light Front

We would like to recall the basics of light-front TOPT, following Ref. [7]. We introduce light-cone coordinates $a^\mu = \left(a^+, a^-, \mathbf{a}_\perp\right)$, defining $a^+ = \left(a^0 + a^3\right)/\sqrt{2}$, $a^- = \left(a^0 - a^3\right)/\sqrt{2}$, and $\mathbf{a}_\perp = (a^1, a^2)$. Starting from the QED Lagrangian density

$$\mathscr{L}_{QED} = \bar{\psi}(i\partial\!\!\!/ - m)\psi - \frac{1}{4}F^{\mu\nu}F_{\mu\nu} + e\bar{\psi}\gamma^\mu\psi A_\mu, \tag{3}$$

if we use the equations of motion in light-cone coordinates and the light-cone gauge condition $A^+ = 0$, the corresponding Hamiltonian reads

$$H = \int d^2\mathbf{x}_\perp dx^- \, T^{+-} = H_0 + V_1 + V_2 + V_3. \tag{4}$$

In Eq. (4), the V_1 term describes the standard QED three-point vertex, while V_2 and V_3 are peculiar of the light-front formalism since they represent vertices containing the exchange of an instantaneous fermion and photon, respectively.

Let us now focus on the covariant approach. The Feynman rules for evaluating diagrams can be directly derived from the Lagrangian (3), regardless of the choice of instant-form rather than light-cone coordinates. If we want to compare it to the TOPT approach, however, we need to work in the light-cone gauge, and consequently use the corresponding expression for the photon propagator $\mathcal{D}^{\mu\nu}(q)$. From the free theory for the photon and considering only the transverse degrees of freedom as the physical ones, one finds $\mathcal{D}^{\mu\nu}(q) = \mathcal{D}_T^{\mu\nu}(q)$ (see Eq. (1)). Nevertheless the use of the noncovariant light-cone gauge also causes, independently of a particular choice of coordinates, a modification in the interaction Hamiltonian. In the interacting theory, indeed, one can use the equations of motion to reduce once again the physical degrees of freedom of the photon only to the transverse ones. As a result, the interaction Hamiltonian will be rewritten as:

$$H = H_0 + V + V_I, \tag{5}$$

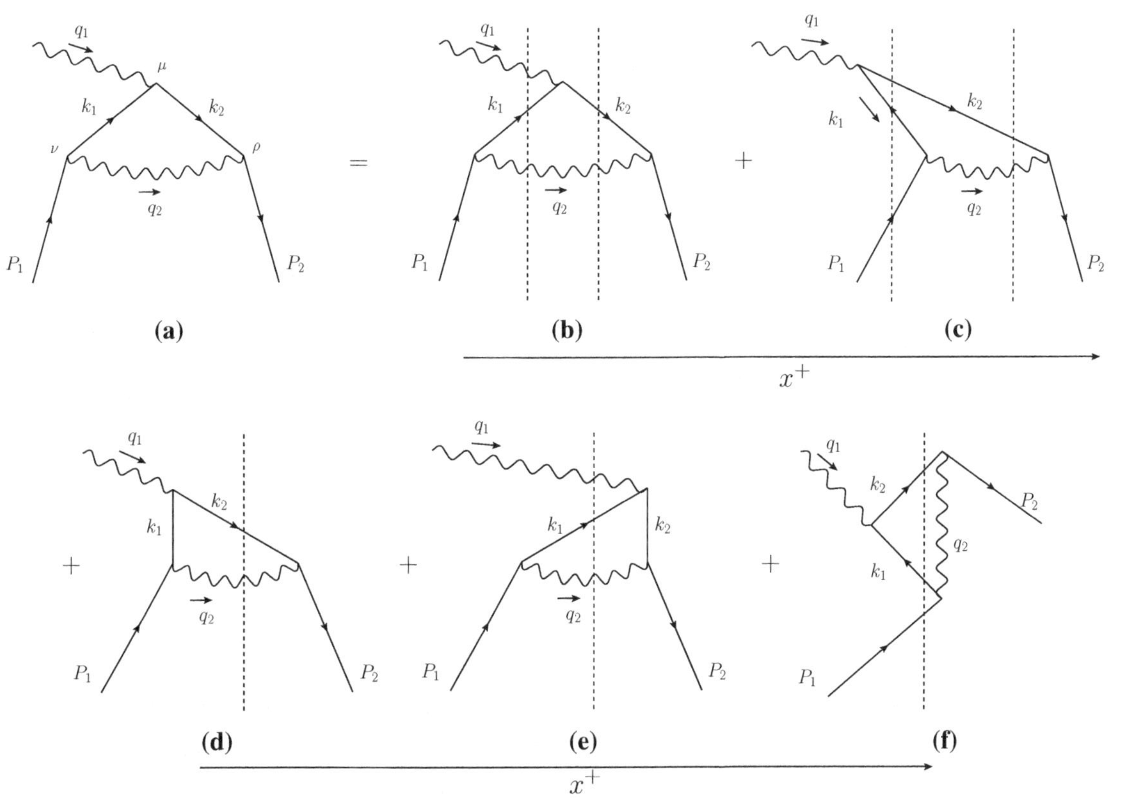

Fig. 1 QED triangle diagram in light-front TOPT. The covariant diagram (**a**) results into the sum of the five time-ordered diagrams (**b**–**f**). Vertical dashed lines are at fixed light-cone time

where V is again the usual QED three-point vertex, while V_I describes an instantaneous four-fermion interaction. It can be shown that, in all physical processes, the contribution from diagrams containing this instantaneous interaction is exactly compensated for by the contribution from the third term of the photon propagator in Eq. (1) [4]. We remark again that the appearance of the V_I term is due only to the choice of the gauge, and it is a common feature of every noncovariant gauge: the same cancellation happens also in the case of the Coulomb gauge, using instant-form coordinates, as discussed e.g. in [2]. Due to the above-described cancellation, it will be therefore equivalent to take as the starting point for the covariant calculation either the three-term propagator, including also the diagrams containing instantaneous interactions, or the two-term propagator (2) neglecting the V_I term in the Hamiltonian at the same time.

3 One-Loop Processes in QED

In this section, we discuss the comparison between TOPT and covariant approach for all one-loop processes in QED; we will focus, in particular, on the vertex correction described by the QED triangle diagram of Fig. 1(a). We evaluate the current associated to this process first in the language of light-front TOPT, and then considering the projection onto the light-front of its covariant expression in the light-cone gauge. In this way, we show the equivalence between the two approaches. The same argument was used by Sawicki in Ref. [1], in the framework of scalar ϕ^3 theory, to prove that the limit of infinite momentum frame is not needed to connect the projection of the current onto the light-front and onto the instant-form surfaces. As in the scalar case, out of the possible six different time-ordered diagrams that appear in the instant-form TOPT, only two (namely diagrams (b) and (c) of Fig. 1) survive when switching to the light-front case, due to the impossibility of having vacuum fluctuations. The complication that arises in QED is the fact that new contributions containing instantaneously propagating particles arise from the interaction Hamiltonian (4); as a result, also the three diagrams (d) to (f) in the bottom side of Fig. 1 will contribute to the current.

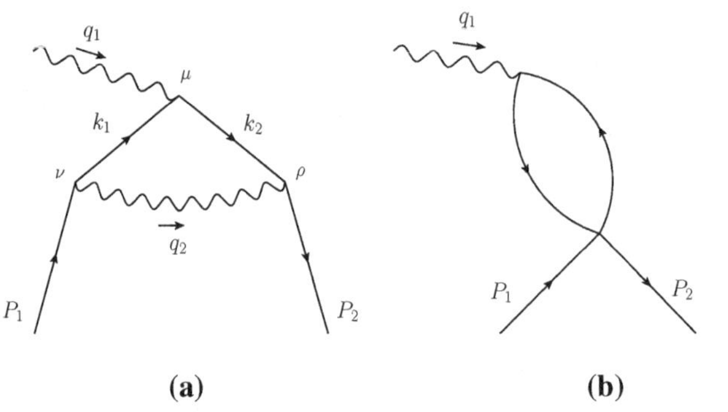

Fig. 2 Vertex correction at one-loop order in covariant QED: **a** triangle diagram; **b** swordfish diagram containing the vertex V_I in Eq. (5)

We take diagram (b) of Fig. 1 as an example. Its amplitude in light-front TOPT is given by:

$$\Lambda_{(b)}^{TOPT} = (2\pi)^3 \delta\left(P_1^+ + q_1^+ - P_2^+\right) \delta^{(2)}\left(\mathbf{P_1}_\perp + \mathbf{q_1}_\perp - \mathbf{P_2}_\perp\right) (-ie)\, \varepsilon_\mu(q_1) J_{(b)}^\mu \,. \tag{6}$$

We can set the symmetric notation for the momenta, defining $q_1 \equiv \Delta$, $P_{1,2} = P \mp \frac{\Delta}{2}$, $k_{1,2} = k \mp \frac{\Delta}{2}$, $q_2 = P - k$, along with the longitudinal momentum fraction $x = k^+/P^+$ and the skewedness parameter $\xi = \Delta^+/2P^+$. The current associated to this diagram, appearing in (6), will then results into:

$$J_{(b)}^\mu = -\frac{e^2 P^+}{(2\pi)^3} \int_\xi^1 dx \int d^2\mathbf{k}_\perp \bar{u}(P_2) \frac{\gamma^\rho \left(\tilde{\not{k}}_2 + m\right) \gamma^\mu \left(\tilde{\not{k}}_1 + m\right) \gamma^\nu}{\left(2P^+\right)^3 (1-x)\left(x^2 - \xi^2\right)\left(\kappa_3^- - \kappa_1^-\right)\left(\kappa_3^- - \kappa_2^-\right)} d_{\nu\rho}\left(\tilde{q}_2\right) u(P_1), \tag{7}$$

where

$$\kappa_{1,2}^- = \frac{\mathbf{k}_\perp^2 \mp \mathbf{k}_\perp \cdot \boldsymbol{\Delta}_\perp + (1 \mp \xi x)\left(m^2 - \frac{t}{4}\right)}{2P^+(x \mp \xi)}, \quad \kappa_3^- = \frac{\mathbf{k}_\perp^2 + (x-1)\left(m^2 - \frac{t}{4}\right)}{2(x-1)P^+}, \tag{8}$$

and

$$d^{\mu\nu}(q) = -g^{\mu\nu} + \frac{q^\mu n^\nu + q^\nu n^\mu}{q^+}. \tag{9}$$

If we consider instead diagram (c) of Fig. 1, we can define in a similar way the associated current $J_{(c)}^\mu$ which reads:

$$J_{(c)}^\mu = -\frac{e^2 P^+}{(2\pi)^3} \int_{-\xi}^{\xi} dx \int d^2\mathbf{k}_\perp \bar{u}(P_2) \frac{\gamma^\rho \left(\tilde{\not{k}}_2 + m\right) \gamma^\mu \left(\tilde{\not{k}}_1 + m\right) \gamma^\nu}{\left(2P^+\right)^3 (1-x)\left(x^2 - \xi^2\right)\left(\kappa_3^- - \kappa_2^-\right)\left(\kappa_2^- - \kappa_1^-\right)} d_{\nu\rho}\left(\tilde{q}_2\right) u(P_1). \tag{10}$$

The contributions from diagrams (d) to (f) of Fig. 1 can be recovered by taking proper limits of instantaneously propagating particles from diagrams (a) and (b) of Fig. 1; this is equivalent to making the substitutions $(\not{k} + m) \mapsto \gamma^+$ and $d^{\mu\nu}(q)/q^+ \mapsto n^\mu n^\nu/(q^+)^2$ in Eqs. (7) and (10) (for instantaneous fermions and photons, respectively), and taking away at the same time the energy denominator of the intermediate state involving the instantaneous particle.

We now turn to the covariant approach. If we work in the noncovariant light-cone gauge $A^+ = 0$, as mentioned in Sect. 3, in addition to the triangle diagram (a) of Fig. 2 we also have to include the so-called swordfish diagram, containing a four-fermion instantaneous interaction (see diagram (b) in Fig. 2). However, as we mentioned in Sect. 2, we can neglect this diagram if we use, at the same time, the two-term propagator in the expression of the current associated to the triangle diagram. We hence take the following as the starting point for the evaluation of the current in the covariant approach:

$$J_{COV}^\mu = \frac{e^2}{(2\pi)^4} \int d^4k_1\, \bar{u}\left(P_2\right) \gamma^\rho S_F(k_2) \gamma^\mu S_F(k_1) \gamma^\nu\, \mathcal{D}_2^{\nu\rho}(q_2) u(p_1), \tag{11}$$

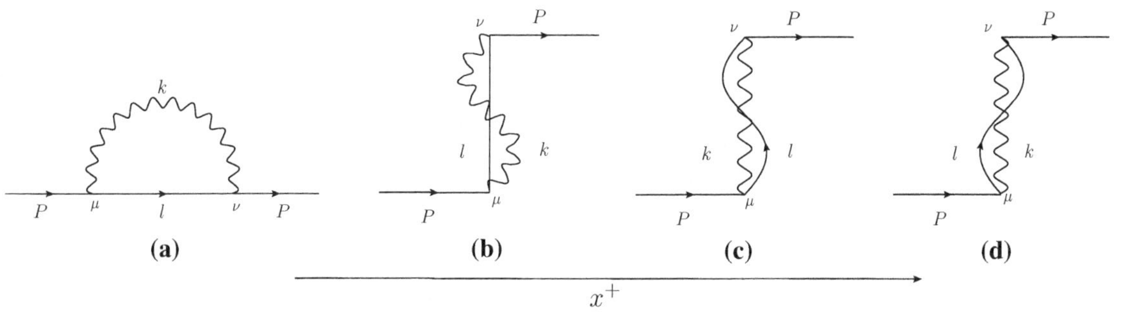

Fig. 3 Contributions to the electron self-energy in light-front TOPT

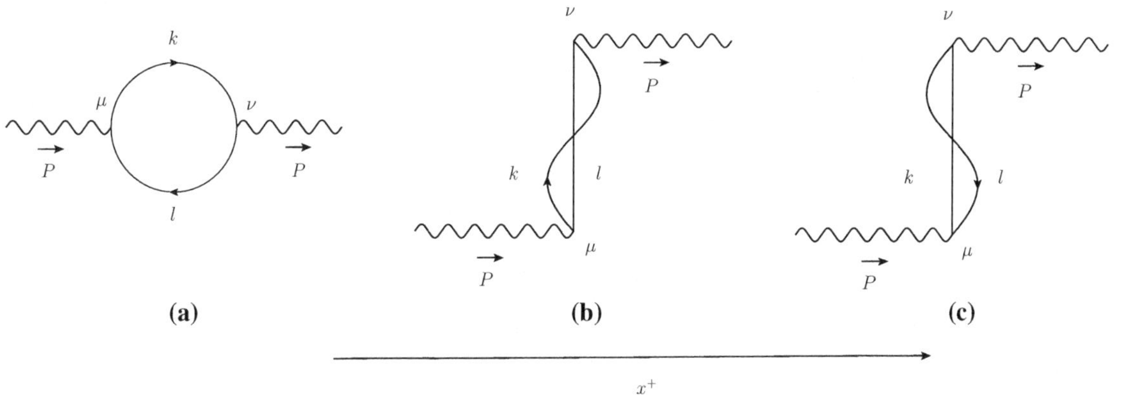

Fig. 4 Contributions to the photon self-energy in light-front TOPT

with $\mathcal{D}_2^{\mu\nu}$ given by Eq. (2). It is useful to introduce the following decomposition of the momentum k of a particle with mass m into its on-shell and off-shell components:

$$k^\mu_{\text{on-shell}} \equiv \tilde{k}^\mu = \left(k^+, \frac{\mathbf{k}_\perp^2 + m^2}{2k^+}, \mathbf{k}_\perp\right), \quad k^\mu_{\text{off-shell}} \equiv \hat{k}^\mu = \left(0, \frac{k^2 - m^2}{2k^+}, \mathbf{0}_\perp\right) = \frac{k^2 - m^2}{2k^+} n^\mu . \tag{12}$$

The propagators appearing in the covariant formula (11) for the current can be then, in turn, systematically split into an on-shell part, and an off-shell part yielding instantaneous propagation:

$$\mathcal{S}_F(k) = \frac{i(\not{k} + m)}{k^2 - m^2 + i\epsilon} = \frac{i\left(\tilde{\not{k}} + m\right)}{k^2 - m^2 + i\epsilon} + \frac{i\gamma^+}{2k^+}, \tag{13a}$$

$$\mathcal{D}_2^{\mu\nu}(q) = \frac{-i}{q^2 + i\epsilon}\left(g^{\mu\nu} - \frac{\tilde{q}^\mu n^\nu + \tilde{q}^\nu n^\mu}{q^+}\right) + \frac{i n^\mu n^\nu}{(q^+)^2} = \frac{i d^{\mu\nu}(\tilde{q})}{q^2 + i\epsilon} + \frac{i n^\mu n^\nu}{(q^+)^2}. \tag{13b}$$

As a result, the current J_μ is in turn split into eight terms. If we turn to light-cone coordinates and integrate over the k^- variable in (11), four out of the eight terms (namely the ones containing more than one off-shell component) will give vanishing contribution to the amplitude; the ones containing only on-shell propagators, instead, will match expressions (7) and (10) of the TOPT approach; finally, the remaining ones will match the TOPT results for diagrams (c) to (e) of Fig. 1.

It is important to notice that the contribution which matches the TOPT diagram with an instantaneous photon comes from the off-shell part of the two-term gauge-field propagator (2); therefore there is no need to take the three-term propagator so as to obtain the equivalence, as claimed in Ref. [3]. This result is obtained simply with the usual method of integration by residues.

It can be shown that the same procedure applies also for the proof of the equivalence for the photon and electron self-energy diagrams in light-front TOPT and covariant approach. The schematic splitting of the propagators into on-shell and off-shell components allows one to easily obtain a one-by-one correspondence between the TOPT diagrams (shown in Figs. 3, 4) and different terms from the splitting of the covariant current projected onto the light-front. Again, for the fermion self-energy, it turns out that the contribution with an instantaneous photon in TOPT is recovered by starting with the two-term propagator (2) in the covariant formula, neglecting at the same time the contact interaction.

4 Conclusions

The present work helps clarifying some subtle statements in the literature, concerning the expression of gauge-field propagator in light-cone gauge.

We re-examined the correspondence between the light-front TOPT approach and the covariant approach in light-cone gauge, for the case of one-loop QED processes. We assumed as the starting point for the covariant calculation the gauge-field propagator containing the sum of three terms, which can be derived from the free equation of motion for the photon in light-cone gauge. However, the use of light-cone gauge also generates a new contribution (namely a contact interaction) in the Hamiltonian. The cancellation between the third term of the propagator and the diagrams containing the contact interaction, which is a generic property of every noncovariant gauge (such as the Coulomb gauge in the instant form) makes it equivalent to use the two-term propagator as an effective one, by neglecting at the same time the contributions from contact terms.

We proved that the diagrams containing instantaneously propagating photons, which are an intrinsic property of TOPT on the light-front, can be generated starting from the two-term propagator and using the standard technique of integration by residue to project the covariant formula onto the light-front. This is in contrast with the existing literature, see e.g. Ref. [3].

This work also helps generalizing the analysis addressed in Ref. [2] for the tree-level scalar QED case, using coordinates which interpolate between light-front form and instant form and in a generic axial gauge $n_\mu A^\mu = 0$. We think that it would be illustrative to complete the treatment of these problems using the interpolating coordinates also for the one-loop QED case.

We finally mention that all the above-listed results naturally translate also to the QCD case of the gluon, since they are related only to the gauge choice and are not in direct connection with the Abelian nature of QED.

References

1. M. Sawicki, Light front limit in a rest frame. Phys. Rev. D **44**, 433–440 (1991)
2. C. Ji, Z. Li, A.T. Suzuki, Electromagnetic gauge field interpolation between the instant form and the front form of the Hamiltonian dynamics. Phys. Rev D **91**, 065020 (2014)
3. A. Misra, S. Warawdekar, Equivalence of covariant and light front QED at one loop level. Phys. Rev. D **71**, 125011 (2005)
4. P.P. Srivastava, S.J. Brodsky, Light front quantized QCD in light cone gauge. Phys. Rev. D **64**, 045006 (2000)
5. A.T. Suzuki, J.H.O. Sales: Light front gauge propagator reexamined (2003). arXiv:hep-th/0304065
6. Mantovani, L., Pasquini, B., Xiong, X., Bacchetta, A.: Revisiting the equivalence of light-front and covariant QED in the light-cone gauge. Phys. Rev. D **94**, 116005 (2016). doi:10.1103/PhysRevD.94.116005
7. D. Mustaki, S. Pinsky, J. Shigemitsu, K. Wilson, Perturbative renormalization of null plane QED. Phys. Rev. D **43**, 3411–3427 (1991)

Few-Body Syst (2017) 58:96
DOI 10.1007/s00601-017-1256-8

Rohini M. Godbole · Abhiram Kaushik · Anuradha Misra · Vaibhav Rawoot · Bipin Sonawane

Heavy Flavour Production as Probe of Gluon Sivers Function

Received: 8 December 2016 / Accepted: 4 February 2017 / Published online: 27 February 2017

Abstract Heavy flavour production like J/ψ and D-meson production in scattering of electrons/unpolarized protons off polarized proton target offer promising probes to investigate gluon Sivers function. In this talk, I will summarize our recent work on transverse single spin asymmetry in J/ψ-production and D-meson production in $pp^\uparrow$ scattering using a generalized parton model approach. We compare predictions obtained using different models of gluon Sivers function within this approach and then, taking into account the transverse momentum dependent evolution of the unpolarized parton distribution functions and gluon Sivers function, we study the effect of evolution on asymmetry.

1 Introduction

Gluon Sivers function (GSF) belongs to a class of transverse momentum dependent parton distribution functions, collectively called TMDs, which are needed to explain the transverse single spin asymmetries (TSSAs) that arise in the scattering of a transversely polarized nucleon off an unpolarized nucleon (or virtual photon) target and are due to orbital motion of quarks and gluons or due to recoil of gluons radiated off the active quarks. TSSA for inclusive process $A^\uparrow + B \to C + X$ is defined as

$$A_N = \frac{d\sigma^\uparrow - d\sigma^\downarrow}{d\sigma^\uparrow + d\sigma^\downarrow} \tag{1}$$

where $d\sigma^\uparrow$ and $d\sigma^\downarrow$ represent the cross section for scattering of a transversely polarized hadron A off an unpolarized hadron (or lepton) B with A being upwards (downwards) polarized with respect to the production

This article belongs to the Topical Collection "Light Cone 2016".

R. M. Godbole · A. Kaushik
Indian Institute of Science, Bangalore, India
E-mail: rohini@chep.iisc.ernet.in

A. Kaushik
E-mail: abhiramb@chep.iisc.ernet.in

A. Misra (✉) · V. Rawoot · B. Sonawane
University of Mumbai, Mumbai, India
E-mail: misra@physics.mu.ac.in

V. Rawoot
E-mail: vaibhavrawoot@gmail.com

B. Sonawane
E-mail: bipin.sonawane@physics.mu.ac.in

plane. The magnitude of these asymmetries observed by HERMES and COMPASS collaborations [1–3] was found to be larger than perturbative QCD (pQCD) predictions based on conventional collinear factorization [4]. One of the two approaches to explain these large asymmetries is based on a generalized factorization formula called transverse momentum dependent (TMD) factorization, in which parton distribution functions (PDFs) of collinear factorization theorem are replaced by TMDs, which take into account the spin and intrinsic transverse momentum dependence of PDFs. Gluon Sivers Function (GSF) is one such TMD, which parametrizes the correlations between the azimuthal distribution of an unpolarized parton and the spin of its transversely polarized parent hadron [5,6]:

$$\Delta^N f_{a/p^\uparrow}(x, k_{\perp a}) \equiv \hat{f}_{a/p^\uparrow}(x, \mathbf{k}_\perp) - \hat{f}_{a/p^\downarrow}(x, \mathbf{k}_\perp) \tag{2}$$

Quark Sivers functions have been studied extensively and their parameterizations have been extracted from SIDIS experimental data from HERMES, COMPASS and JLab experiments. However, not much information is available about GSF so far. Recently, the first rough estimates of GSF have been obtained by D'Alesio et al. [7] by fitting to midrapidity data on $pp^\uparrow \to \pi^0 X$ taken by PHENIX collaboration at RHIC and, by using the quark Sivers parameters extracted earlier from SIDIS data. One now needs to identify processes for which predictions using these fits can be obtained and compared with data.

Heavy quark and quarkonium systems are natural probes to study the GSF as the production is sensitive to intrinsic transverse momentum especially at low momentum. SSAs in low virtuality electroproduction of J/ψ have been estimated by us using a generalization of color evaporation model of quarkonium production with TMDPDF's [8–10]. SSA in D Meson production in $pp^\uparrow$ scattering, which can be used as a clean probe of Gluon Sivers Function [11] were calculated using a Generalized Parton Model approach with two extreme values of GSF-zero and maximum. It was shown that the gluon contribution dominates over quark contribution at RHIC energy and asymmetries up to 25% were predicted with saturated GSF.

Recent measurement of asymmetry in J/ψ production at PHENIX experiment with polarized proton beam suggests the need for a phenomenological study of TSSA in heavy flavor production—both open and closed-as probes of GSF. Here, I summarize our recent work on transverse single spin asymmetry in the processes $pp^\uparrow \to J/\psi + X$ and $pp^\uparrow \to D + X$, where we have estimated TSSA using various parametrizations of GSF presently available and have also studied the effect of TMD evolution of Sivers function on TSSA.

In Sect. 2, we present different parameterizations of GSF that we have used. In Sect. 3, we have summarized our work in Ref. [12] and in Sect. 3, we present our preliminary results for TSSA in $pp^\uparrow \to J/\psi + X$. Here, we also compare our estimates with the recent PHENIX data. In Sect. 4, we present a summary of our results [12,13].

2 Gluon Sivers Function

The standard gaussian form is used for the unpolarized TMD PDF

$$f_{i/p}(x, k_\perp; Q) = f_{i/p}(x, Q)\frac{1}{\pi\langle k_\perp^2\rangle}e^{-k_\perp^2/\langle k_\perp^2\rangle} \tag{3}$$

with $\langle k_\perp^2\rangle = 0.25\ \mathrm{GeV}^2$ and $i = q, g$. For the Sivers function, we use the parametrization [7]

$$\Delta^N f_{i/p^\uparrow}(x, k_\perp; Q) = 2\mathcal{N}_i(x) f_{i/p}(x, Q) h(k_\perp)\frac{e^{-k_\perp^2/\langle k_\perp^2\rangle}}{\pi\langle k_\perp^2\rangle} \tag{4}$$

with

$$\mathcal{N}_i(x) = N_i x^{\alpha_i}(1-x)^{\beta_i}\frac{(\alpha_i+\beta_i)^{\alpha_i+\beta_i}}{\alpha_i^{\alpha_i}\beta_i^{\beta_i}}, \qquad h(k_\perp) = \sqrt{2e}\frac{k_\perp}{M_1}e^{-k_\perp^2/M_1^2} \tag{5}$$

and

$$h(k_\perp)\frac{e^{-k_\perp^2/\langle k_\perp^2\rangle}}{\pi\langle k_\perp^2\rangle} = \frac{\sqrt{2e}}{\pi}\sqrt{\frac{1-\rho}{\rho}}k_\perp\frac{e^{-k_\perp^2/\rho\langle k_\perp^2\rangle}}{\langle k_\perp^2\rangle^{3/2}}, \tag{6}$$

where $N_i, \alpha_i, \beta_i, M_1$ and ρ are the best fit parameter.

Table 1 SIDIS parameters for GSF and best fit parameters for u and d quark used in BV parametrization

DMP-SIDIS 1	$N_g = 0.65$	$\alpha_g = 2.8$	$\beta_g = 2.8$	$\rho = 0.687$	$\langle k_\perp^2 \rangle = 0.25\ \text{GeV}^2$
DMP-SIDIS 2	$N_g = 0.05$	$\alpha_g = 0.8$	$\beta_g = 1.4$	$\rho = 0.576$	
u quark	$N_u = 0.4$	$\alpha_u = 0.35$	$\beta_u = 0.26$	$\langle k_\perp^2 \rangle = 0.25\ \text{GeV}^2$	$M_1^2 = 0.19$
d quark	$N_d = -0.97$	$\alpha_d = 0.44$	$\beta_d = 0.90$		

We estimate asymmetries using the recently fitted GSF parameters given in Table 1.These fits have been obtained by fitting $pp^\uparrow \to \pi^0 + X$ asymmetry data at RHIC [7,14]. We present estimates for SSA in D meson production and J/ψ production using these parameters, which we will call DMP fits following Ref. [12]. In case of J/ψ production, we also compare our predictions with those obtained using parameters we have used in our earlier work, which we call BV parameters as those were fitted using a model given by Boer and Vogelsang [15]. This parametrization is expressed as

$$\mathcal{N}_g(x) = \frac{\mathcal{N}_u + \mathcal{N}_d}{2} \quad (BV - a) \quad and \quad \mathcal{N}_g(x) = \mathcal{N}_d \quad (BV - b). \tag{7}$$

The difference between DMP-SIDIS1 and DMP-SIDIS2 parameters is that the former were fitted using a parametrization of quark Sivers fucntion which took into account only u and d quark flavors while for the latter, s quark contribution was also included. The best fits parameter for up and down valance quark Sivers function we have used are given in Ref. [16]. We also study the effect of QCD evolution of TMD PDF's using the formalism of Ref. [17]. For this study we use the BV parameters only for both DGLAP and TMD evolution cases, since the DMP fits took into account DGLAP evolution only.

3 Transverse Single Spin Asymmetry in $pp^\uparrow \to DX$

We estimate SSA in D Meson production in $pp^\uparrow$ scattering using the Generalized Parton Model approach. The numerator and denominator of asymmetry are given by [11],

$$\begin{aligned}
&\frac{E_D\, d\sigma^{p^\uparrow p \to DX}}{d^3\mathbf{p_D}} - \frac{E_D\, d\sigma^{p^\downarrow p \to DX}}{d^3\mathbf{p_D}} \\
&= \int dx_a\, dx_b\, dz\, d^2\mathbf{k}_{\perp a}\, d^2\mathbf{k}_{\perp b}\, d^3\mathbf{k}_D\, \delta\left(\mathbf{k}_D \cdot \hat{\mathbf{p}}_c\right)\, \delta\left(\hat{s} + \hat{t} + \hat{u} - 2m_c^2\right)\, \mathcal{C}\left(x_a, x_b, z, \mathbf{k}_D\right) \\
&\left\{ \sum_q \left[\Delta^N f_{q/p^\uparrow}\left(x_a, \mathbf{k}_{\perp a}\right)\, f_{\bar{q}/p}\left(x_b, \mathbf{k}_{\perp b}\right)\, \frac{d\hat{\sigma}^{q\bar{q} \to c\bar{c}}}{d\hat{t}}(x_a, x_b, \mathbf{k}_{\perp a}, \mathbf{k}_{\perp b}, \mathbf{k}_D)\, D_{D/c}(z, \mathbf{k}_D) \right] \right. \\
&\left. + \left[\Delta^N f_{g/p^\uparrow}(x_a, \mathbf{k}_{\perp a})\, f_{g/p}(x_b, \mathbf{k}_{\perp b})\, \frac{d\hat{\sigma}^{gg \to c\bar{c}}}{d\hat{t}}(x_a, x_b, \mathbf{k}_{\perp a}, \mathbf{k}_{\perp b}, \mathbf{k}_D)\, D_{D/c}(z, \mathbf{k}_D) \right] \right\} \\
&\frac{E_D\, d\sigma^{p^\uparrow p \to DX}}{d^3\mathbf{p_D}} + \frac{E_D\, d\sigma^{p^\downarrow p \to DX}}{d^3\mathbf{p_D}} \\
&= 2 \int dx_a\, dx_b\, dz\, d^2\mathbf{k}_{\perp a}\, d^2\mathbf{k}_{\perp b}\, d^3\mathbf{k}_D\, \delta(\mathbf{k}_D \cdot \hat{\mathbf{p}}_c)\, \delta\left(\hat{s} + \hat{t} + \hat{u} - 2m_c^2\right)\, \mathcal{C}(x_a, x_b, z, \mathbf{k}_D) \\
&\times \left\{ \sum_q \left[\hat{f}_{q/p}(x_a, \mathbf{k}_{\perp a})\, \hat{f}_{\bar{q}/p}(x_b, \mathbf{k}_{\perp b})\, \frac{d\hat{\sigma}^{q\bar{q} \to c\bar{c}}}{d\hat{t}}(x_a, x_b, \mathbf{k}_{\perp a}, \mathbf{k}_{\perp b}, \mathbf{k}_D)\, \hat{D}_{D/c}(z, \mathbf{k}_D) \right] \right. \\
&\left. + \left[\hat{f}_{g/p}(x_a, \mathbf{k}_{\perp a})\, \hat{f}_{g/p}(x_b, \mathbf{k}_{\perp b})\, \frac{d\hat{\sigma}^{gg \to c\bar{c}}}{d\hat{t}}(x_a, x_b, \mathbf{k}_{\perp a}, \mathbf{k}_{\perp b}, \mathbf{k}_D)\, \hat{D}_{D/c}(z, \mathbf{k}_D) \right] \right\}
\end{aligned} \tag{8}$$

respectively, where all the symbols have usual meaning [12].

We present our estimates of asymmetry in D meson production at $\sqrt{s} = 200$ GeV using DMP fits given in Table 1. Estimates of asymmetry at $\sqrt{s} = 115$ GeV and $\sqrt{s} = 500$ GeV can be found in Ref. [12]. In Fig. 2, we have presented comparison of predictions using DGLAP evolved and TMD evolved GSF using BV parameters [15].

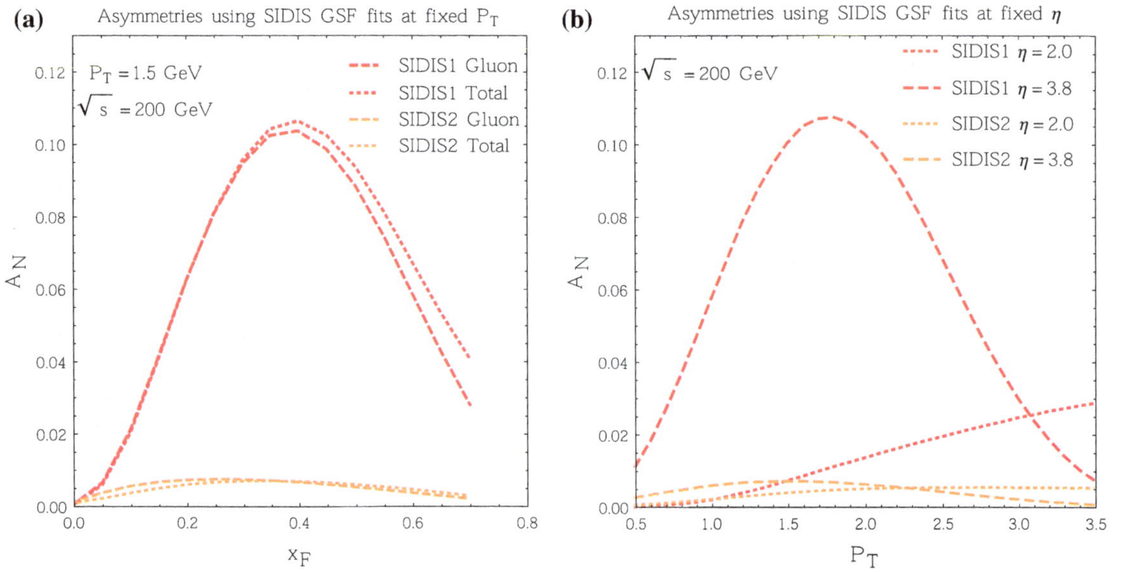

Fig. 1 Asymmetries in D-meson production using DMP fits, **a** x_F-distribution, **b** p_T-distribution using DMP SIDIS1 and DMP SIDIS 2 fits [12]

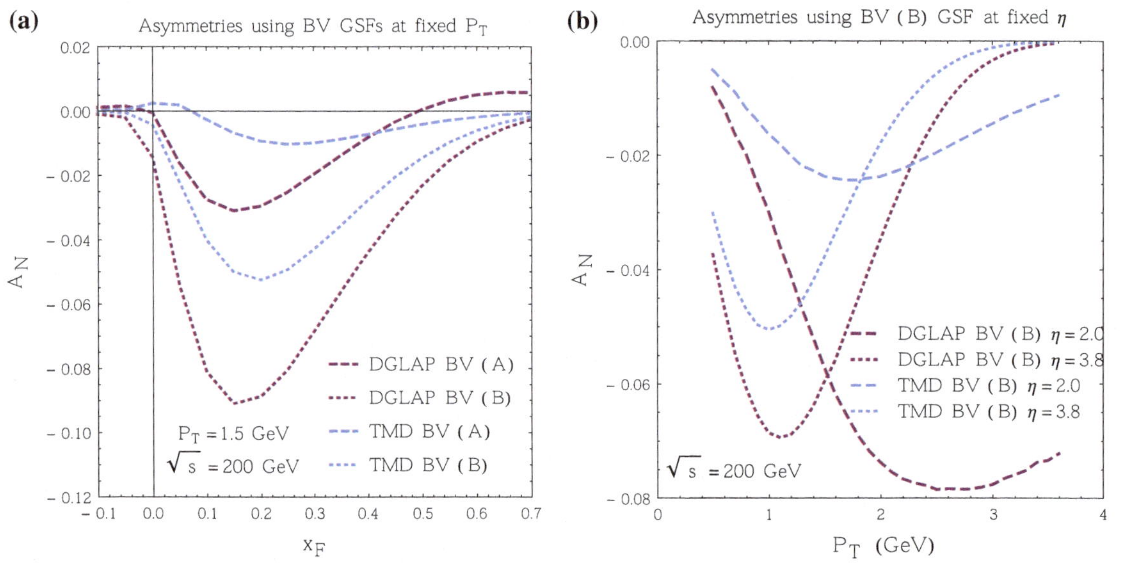

Fig. 2 Comparison of asymmetries in D-meson production using BV parameters for GSF fitted using DGLAP evolution and TMD evolution, **a** x_F-distribution, **b** p_T-distribution [12]

4 Transverse Single Spin Asymmetry in $pp^\uparrow \to J/\psi X$

In this section, we present our estimates of TSSA in the process $p + p^\uparrow \to J/\psi + X$ obtained using a generalized parton model approach and the color evaporation model (CEM) of J/ψ production [13]:

$$\sigma^{p+p\to J/\psi+X} = F_{J/\psi} \int_{4m_c^2}^{4m_D^2} dM_{c\bar{c}}^2 \int dx_a d^2k_{\perp a} dx_b d^2k_{\perp b} f_{g/p^\uparrow}(x_a, k_{\perp a}) f_{g/p}(x_b, k_{\perp b}) \frac{d\hat{\sigma}^{gg\to c\bar{c}}}{dM_{c\bar{c}}^2} \tag{9}$$

In this generalization of CEM, the numerator of the asymmetry in $p + p^\uparrow \to J/\psi + X$ is parameterized in terms of the Sivers function

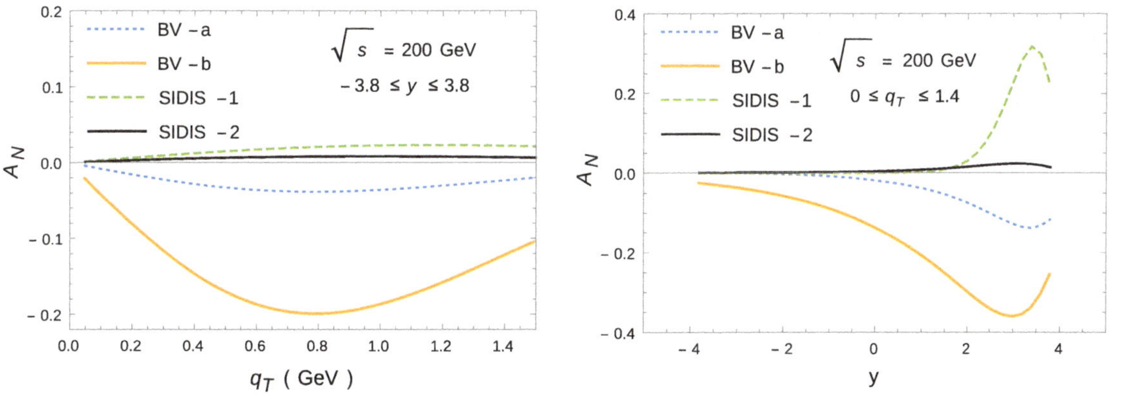

Fig. 3 Predictions of asymmetry in $p^{\uparrow} + p \rightarrow J/\psi + X$ at PHENIX ($\sqrt{s} = 200\,\text{GeV}$). q_T and y integration ranges are $0 \leq q_T \leq 1.4\,\text{GeV}$ and $-3.8 \leq y \leq 3.8$ respectively. Left panel and right panel are plots of q_T and y distribution of asymmetry respectively [13]

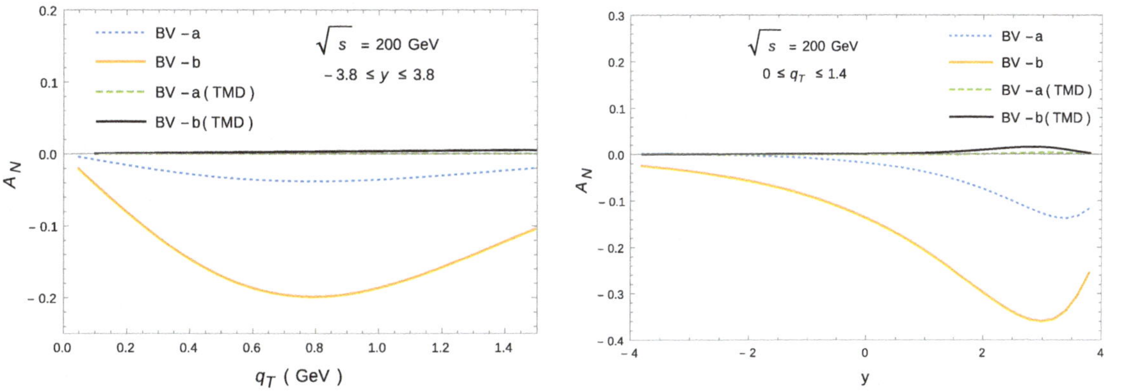

Fig. 4 Predictions of asymmetry with DGLAP and TMD evolution in $p^{\uparrow} + p \rightarrow J/\psi + X$ at $\sqrt{s} = 200\,\text{GeV}$. q_T and y integration ranges are $0 \leq q_T \leq 1.4\,\text{GeV}$ and $-3.8 \leq y \leq 3.8$. Left panel and right panel are plots of q_T and y distribution of asymmetry respectively [13]

$$\frac{d^4\sigma^{\uparrow}}{dydM^2d^2\mathbf{q}_T} - \frac{d^4\sigma^{\downarrow}}{dydM^2d^2\mathbf{q}_T} = \frac{1}{s}\int [d^2\mathbf{k}_{\perp a} d^2\mathbf{k}_{\perp b}] \Delta^N f_{g/p^{\uparrow}}(x_a, \mathbf{k}_{\perp a}) f_{g/p}(x_b, \mathbf{k}_{\perp b}) \times \delta^2(\mathbf{k}_{\perp a} + \mathbf{k}_{\perp b} - \mathbf{q}_T)\hat{\sigma}_0^{gg\rightarrow c\bar{c}}(M^2) \tag{10}$$

The denominator of the asymmetry is

$$\frac{d^4\sigma^{\uparrow}}{dydM^2d^2\mathbf{q}_T} + \frac{d^4\sigma^{\downarrow}}{dydM^2d^2\mathbf{q}_T} = \frac{2}{s}\int [d^2\mathbf{k}_{\perp a} d^2\mathbf{k}_{\perp b}] \Big[f_{g/p}(x_a, \mathbf{k}_{\perp a}) f_{g/p}(x_2, \mathbf{k}_{\perp b}) \delta^2(\mathbf{k}_{\perp a} + \mathbf{k}_{\perp b} - \mathbf{q}_T)\hat{\sigma}_0^{gg\rightarrow c\bar{c}}(M^2) + f_{q/p}(x_a, \mathbf{k}_{\perp a}) f_{\bar{q}/p}(x_b, \mathbf{k}_{\perp b}) \delta^2(\mathbf{k}_{\perp a} + \mathbf{k}_{\perp b} - \mathbf{q}_T)\hat{\sigma}_0^{q\bar{q}\rightarrow c\bar{c}}(M^2) \Big] \tag{11}$$

where, in Eq. (10), we have neglected the contribution of $q\bar{q}$ subprocess as it is negligible compared to gg subprocess. In Fig. 3, we present our estimates of asymmetries calculated using the DMP SIDIS1 and SIDIS2 fits at $\sqrt{s} = 200$ GeV. Estimates of asymmetry at $\sqrt{s} = 115$ GeV and 500 GeV can be found in Ref. [13]. To investigate the effect of TMD evolution on asymmetry, we have also calculated asymmetries using BV parameters and the best fit parameters of the quark Sivers function (used in our earlier work on SSA in $e + p^{\uparrow} \rightarrow J/\psi + X$ and $p + p^{\uparrow} \rightarrow D + X$). In Fig. 4, we compare the predictions with BV parameters obtained using DGLAP evolved and TMD evolved Sivers functions to assess the effect of TMD evolution on asymmetries. We compare our estimates with the PHENIX data [14] in Fig. 5. We have given estimates

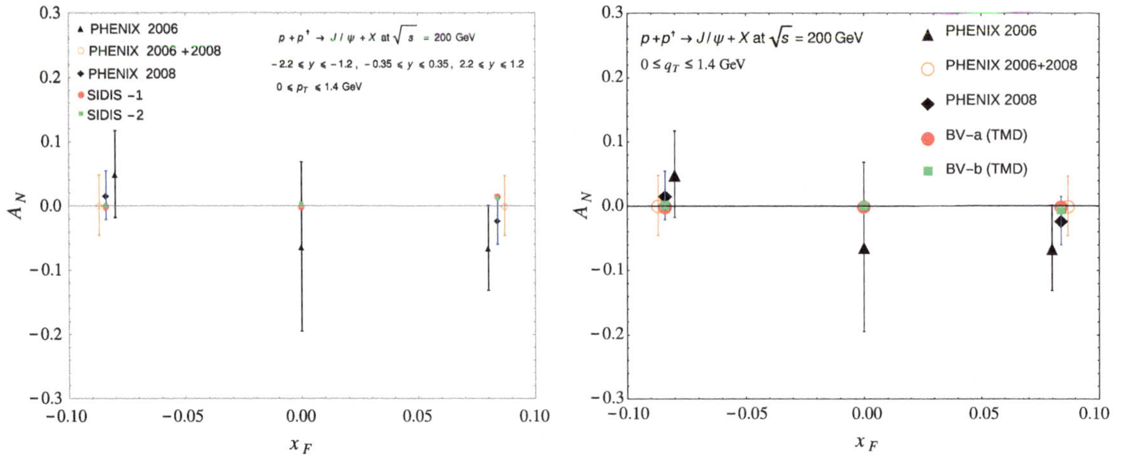

Fig. 5 Plots of asymmetry in $p^{\uparrow} + p \rightarrow J/\psi + X$ at PHENIX experiment ($\sqrt{s} = 200\,\text{GeV}$). Region of q_T integration is $0 \leq q_T \leq 1.4\,\text{GeV}$. Asymmetry is in backward ($-2.2 \leq y \leq -1.2$), mid ($-0.35 < y < 0.35$) and forward ($1.2 \leq y \leq 2.2$) rapidity regions. In left panel we have compared asymmetry obtained using DMP SIDIS-2 and DMP SIDIS-1 parameters with PHENIX data. In right panel we have compared asymmetry obtained using BV-a and BV-b parametrization with TMD evolution with PHENIX data [13]

in three rapidity regions to compare with PHENIX experimental data. The backward and forward rapidity regions correspond to ranges $-2.2 \leq y \leq -1.2$ and $2.2 \leq y \leq 1.2$ respectively and the mid rapidity region correspond to range $-0.35 \leq y \leq 0.35$.

5 Summary

We have presented estimates of TSSA in D Meson production and J/ψ production in $pp^{\uparrow}$ scattering at $\sqrt{s} = 200$ GeV. In case of D-Meson production, symmetries are found to be substantial (up to 10%), while in case of J/ψ production, asymmetries are found to be consistent with the almost zero result of PHENIX experiment. In both csases, the contribution of gluon Sivers function is found to dominate over the quark contribution and the effect of TMD evolution is found to be reduction in the asymmetry estimates. Our comparison of estimated asymmetry with PHENIX data shows that asymmetry with DMP SIDIS-2 and DMP SIDIS-1 parameter set is within the experimental uncertainties of PHENIX data. We have also verified that results obtained using BV-a parameterization is well within experimental uncertainties of PHENIX data, while the estimates of asymmetry with BV-b is not in agreement with PHENIX data [13]. However, after taking into account TMD evolution, both the parameter sets lead to asymmetries that agree with near zero result of experiment.

Acknowledgements A. M and B. S. would like to thank DST, India for financial support under the Project No. EMR/2014/0000486. AM would like to thank organizers of LC2016 for partial financial support.

References

1. A. Airapentian, et al. (HERMES Collboration), Observation of single spin asymmetry in semi-inclusive pion electroproduction. Phys. Rev. Lett. **84**, 4047 (2000)
2. A. Airapentian, et al. (HERMES Collboration), Single spin asymmetry in electroproduction of neutral pion in semi-inclusive deep inelastic scattering. Phys. Rev. D **64**, 097101 (2001)
3. V. Alexakhin (COMPASS Collboration), First measurement of transverse spin asymmetry in semi-inclusive deep inelastic scattering. Phys. Rev. Lett. **94**, 202002 (2005)
4. U. D'alesio, F. Murgia, Azimuthal and single spin asymmetry in hard scattering processes. Proj. Part. Nucl. Phys. **61**, 394 (2001)
5. D. Sivers, Single spin production asymmetries from the hard scattering of point-like constituents. Phys. Rev. D **41**, 83 (1990)
6. D. Sivers, Hard scattering scaling laws for single spin production asymmetries. Phys. Rev. D **43**, 261 (1991)
7. U. D'alesio, F. Murgia, C. Pisano, Towards a first estimate of gluon Sivers function from A_N data in pp collisions at RHIC. JHEP **1509**, 119 (2015)

8. R.M. Godbole, A. Misra, A. Mukherjee, V.S. Rawoot, Sivers effect and transverse single spin asymmetry in $e + p^{\uparrow} \rightarrow e + J/\psi + X$. Phys. Rev. D **85**, 094013 (2012)
9. R.M. Godbole, A. Misra, A. Mukherjee, V.S. Rawoot, Transverse single spin asymmetry in $e + p^{\uparrow} \rightarrow e + J/\psi + X$ and transverse momentum dependent evolution of Sivers function. Phys. Rev. D **88**, 014029 (2013)
10. R.M. Godbole, A. Kaushik, A. Misra, V.S. Rawoot, Transverse single spin asymmetry in $e + p^{\uparrow} \rightarrow e + J/\psi + X$ and Q^2 evolution of Sivers function. Phys. Rev. D **91**, 014005 (2015)
11. M. Anselmino, M. Boglione, U. D'alesio, E. Leader, F. Murgia, Accessing Sivers gluon distribution via transverse single spin asymmetry in $p + p^{\uparrow} \rightarrow D + X$ process at RHIC. Phys. Rev. D **70**, 074025 (2004)
12. R.M. Godbole, A. Kaushik, A. Misra, Transverse single spin asymmetry in $p + p^{\uparrow} \rightarrow D + X$. Phys. Rev. D **94**,114022 (2016). arXiv:1606.01818
13. R.M. Godbole, A. Kaushik, A. Misra, V.S. Rawoot, B. Sonawane, Transverse single spin asymmetry in $p+p^{\uparrow} \rightarrow J/psi+X$. In progress (2016)
14. A. Adare, et al. [PHENIX Collaboration], Measurement of transverse-single-spin asymmetries for midrapidity and forward-rapidity production of hadrons in polarized p+p collisions at $\sqrt{s}$ =200 and 62.4 GeV. Phys. Rev. D **90**(1), 012006 (2014) [arXiv:1312.1995 [hep-ex]]
15. D. Boer, W. Vogelsang, Asymmetric jet corrections in $pp^{\uparrow}$ scattering. Phys. Rev. D **69**, 094025 (2004)
16. M. Anselmino, M. Boglione, S. Mellis, Strategy towards the extraction of the Sivers function with TMD evolution. Phys. Rev. D **86**, 014028 (2012)
17. M.G. Echvarria, A. Idilbi, Z.B. Kang, I. Vitev, QCD Evolution of the Sivers Asymmetry. Phys. Rev. D **89**,074013 (2014)

Few-Body Syst (2017) 58:42
DOI 10.1007/s00601-016-1209-7

Chueng-Ryong Ji

Interpolation Between the Instant Form and the Front Form of Relativistic Dynamics

Received: 15 November 2016 / Accepted: 28 December 2016 / Published online: 19 January 2017

Abstract The instant form and the front form of relativistic dynamics proposed by Dirac (Rev Mod Phys 21:392–399, 1949) can be linked by introducing an interpolating angle. The instant form dynamics (IFD) provides a traditional approach evolved from non-relativistic dynamics and makes a close contact with Euclidean space developing temperature-dependent quantum field theory, lattice QCD, etc. The front form dynamics, now known as the light-front dynamics (LFD), however, works strictly in Minkowski space and provides useful frameworks to study deep inelastic scattering, parton distribution functions, deeply virtual Compton scattering, generalized parton distributions, etc. We present the recent development of the interpolation between IFD and LFD and discuss ramifications from this development.

1 Introduction

When particle systems have characteristic momenta which are of the same order or even much larger than the masses of the particles involved, it is part of nature that a relativistic treatment is called for in order to describe those systems properly. In particular, relativistic effects are most essential to describe the low-lying hadron systems in terms of strongly interacting quarks/antiquarks and gluons in quantum chromodynamics (QCD). For the study of relativistic particle systems, Dirac [1] proposed three different forms of the relativistic Hamiltonian dynamics in 1949: i.e. the instant ($x^0 = 0$), front ($x^+ = (x^0 + x^3)/\sqrt{2} = 0$), and point ($x_\mu x^\mu = a^2 > 0, x^0 > 0$) forms. The instant form dynamics (IFD) of quantum field theories is based on the usual equal time $t = x^0$ quantization (units such that $c = 1$ are taken here) and provides a traditional approach evolved from non-relativistic dynamics. The IFD makes a close contact with Euclidean space, developing temperature-dependent quantum field theory, lattice QCD, etc. The equal light-front time $\tau \equiv (t + z/c)/\sqrt{2} = x^+$ quantization yields the front form dynamics, nowadays more commonly called light-front dynamics (LFD) and provides a progressive approach for the study of relativistic dynamics. The LFD works strictly in Minkowski space, developing useful frameworks for the analyses of deep inelastic scattering (DIS), parton distribution functions (PDFs), deeply virtual Compton scattering (DVCS), generalized parton distributions (GPDs), etc. The quantization in the point form ($x^\mu x_\mu = a^2 > 0, x^0 > 0$) is called radial quantization and this quantization procedure has been much used in string theory and conformal field theories [2] as well as in hadron physics [3–5]. Among these three forms of relativistic dynamics proposed by Dirac, the LFD carries the largest number (seven) of the kinematic (or interaction independent) generators leaving the least number (three) of the dynamic (or interaction dependent) generators while both the IFD and the point form dynamics carry six kinematic and

This article belongs to the Topical Collection "Light Cone 2016".
This Work Was Supported by the US DOE Contract No. DE-FG02-03ER41260.

C.-R. Ji (✉)
Department of Physics, North Carolina State University, Box 8202, Raleigh, NC 27695-8202, USA
E-mail: crji@ncsu.edu

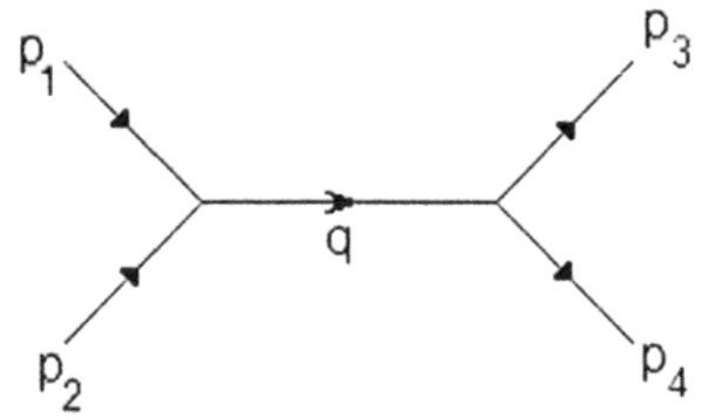

Fig. 1 Scattering amplitude of spinless particles

four dynamic generators within the total ten Poincaré generators. Indeed, the maximum number of kinematic generators allowed in any form of relativistic dynamics is seven and the LFD is the only one which possesses this maximum number of kinematic generators. Effectively, the LFD maximizes the effectiveness in describing hadrons by saving a lot of dynamical efforts in obtaining the QCD solutions that reflect the full Poincaré symmetries.

In this presentation, we interpolate between the IFD and the LFD to link the LFD with the IFD which has been the traditional approach in relativistic dynamics and discuss the key advantageous feature of the LFD in hadron physics. In the next section, Sect. 2, we first interpolate the scattering amplitudes between the IFD and the LFD and discuss the maximization of the number of kinematic generators in LFD, presenting the whole landscape of each time-ordered scattering amplitudes in the space to display the interpolation and the longitudinal boost. This exposition of whole landscape reveals the appearance of the light—front zero—modes. In Sect. 3, we discuss the role of light-front zero-modes in obtaining the model-independent leading nonanalytic behavior in chiral dynamics and its application to the phenomenology of the flavor asymmetry in proton sea. Conclusion and future directions follow in Sect. 4.

2 Interpolating Scattering Amplitudes

The IFD and the LFD can be interpolated [6–10] by an interpolation angle between the ordinary time t and the light-front time τ. Introducing the interpolating scattering amplitude [8–10] that links the corresponding time-ordered amplitudes between the two forms of dynamics, we discussed the physical meaning of the kinematic transformations in contrast to the dynamic transformations by means of checking the invariance of each individual time-ordered amplitude for an arbitrary interpolation angle. Using this interpolation method, we were able to also trace a dramatic change of the longitudinal boost's characteristics, namely, from "dynamic"in IFD to "kinematic" in LFD. This provides a distinct advantage to LFD in hadron phenomenology.

To trace the forms of relativistic quantum field theory between IFD and LFD, we begin by adopting the following convention of the space-time coordinates to define the interpolation angle [6–10]:

$$\begin{bmatrix} x^{\widehat{+}} \\ x^{\widehat{-}} \end{bmatrix} = \begin{bmatrix} \cos\delta & \sin\delta \\ \sin\delta & -\cos\delta \end{bmatrix} \begin{bmatrix} x^0 \\ x^3 \end{bmatrix}, \tag{1}$$

in which the interpolation angle is allowed to run from 0 through 45°, $0 \leq \delta \leq \frac{\pi}{4}$. All the indices with the wide-hat notation signify the variables with the interpolation angle δ. For the limit $\delta \to 0$ we have $x^{\widehat{+}} = x^0$ and $x^{\widehat{-}} = -x^3$ so that we recover usual space-time coordinates although the z-axis is inverted while for the other extreme limit, $\delta \to \frac{\pi}{4}$ we have $x^{\widehat{\pm}} = (x^0 \pm x^3)/\sqrt{2} \equiv x^{\pm}$ which leads to the standard light-front coordinates. Of course, the same interpolation applies to the momentum variables.

For the simplest possible illustration of applying the interpolation to relativistic quantum field theory [8], we first discuss the scattering amplitude of two spin-less particles, i.e., the analogue of the well known QED annihilation/production process $e^+e^- \to \mu^+\mu^-$ in a toy ϕ^3 model theory, as depicted in Fig. 1. As shown in Refs. [9, 10], we have extended our discussion to the gauge bosons and the fermion spinors introducing the interpolating helicity of polarization vectors and spinors. However, for the simplicity of our discussion in this presentation, we don't involve spins and any other degrees of freedom but just discuss the main point with the fundamental degrees of freedom, i.e., particle momenta. Modulo inessential factors including the square of the coupling constant, the lowest order tree-level Feynman diagram shown in Fig. 1 is proportional to the propagator of the intermediate particle: $\Sigma = 1/(s - m^2)$, where $s = (p_1 + p_2)^2$ is the Mandelstam variable, which is invariant under any Poincaré transformations, and m is the mass of the intermediate boson. Of course,

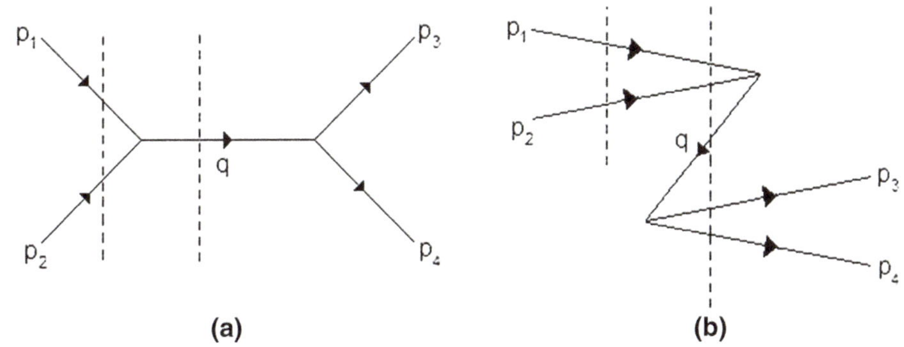

Fig. 2 Time-ordered amplitudes for the Feynman amplitude depicted in Fig. 1

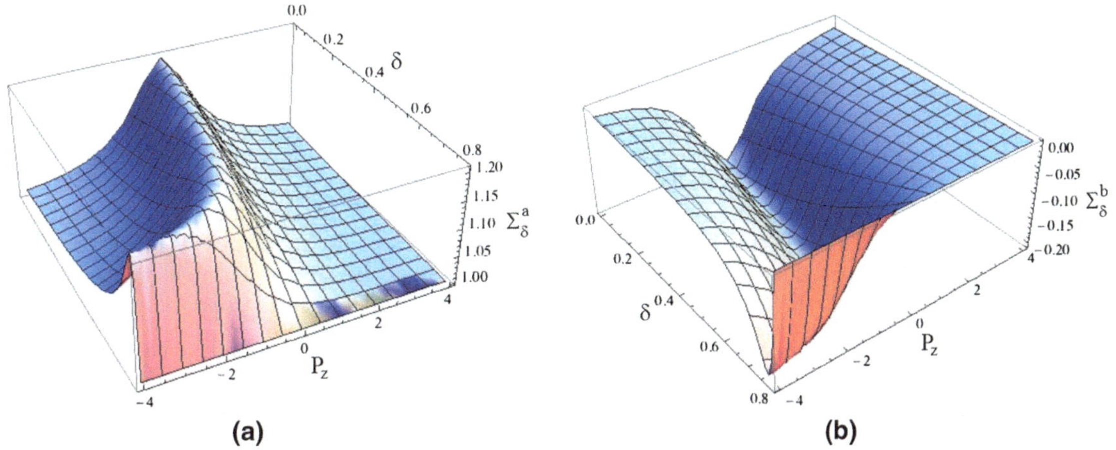

Fig. 3 Interpolating amplitudes

the physical process can take place only above the threshold $s > 4M^2$, where M is the mass of the final particle and anti-particle that are produced, e.g. like the muon mass in the $e^+e^- \rightarrow \mu^+\mu^-$ scattering process. The corresponding time-ordered amplitudes for an arbitrary interpolation angle δ are graphically represented in Fig. 2a, b and respectively given by

$$\Sigma_\delta^a = \frac{1}{2q_{\hat{+}}}\left(\frac{\mathbb{C}}{p_1^{\hat{+}} + p_2^{\hat{+}} - q^{\hat{+}}}\right), \quad \Sigma_\delta^b = -\frac{1}{2q_{\hat{+}}}\left(\frac{\mathbb{C}}{p_1^{\hat{+}} + p_2^{\hat{+}} + q^{\hat{+}}}\right), \tag{2}$$

where $\mathbb{C} = \cos 2\delta$. We note that the individual time-ordered amplitudes are invariant only under the transformations given by the kinematic generators, i.e. they are not invariant under the transformations given by the dynamic generators, although the sum of the time-ordered amplitudes, e.g. the Feynman diagram shown in Fig. 1, is invariant under all ten Poincaré generators. This discussion clarifies the physical meaning of the kinematic generators vs. the dynamic generators.

Figure 3 shows Σ_δ^a and Σ_δ^b as functions of the initial particle's total momentum $(\mathbf{p}_1 + \mathbf{p}_2) \cdot \hat{z} = P_z$ while $(\mathbf{p}_1 + \mathbf{p}_2) \cdot \hat{x} = 0$ and $(\mathbf{p}_1 + \mathbf{p}_2) \cdot \hat{y} = 0$, and the interpolation angle δ. The ranges of δ and P_z are taken as $0 \leq \delta < \frac{\pi}{4}$ and $-4 \leq P_z \leq 4$ in some unit of energy, e.g. GeV, respectively. For illustrative purpose, we took $s = 2$ and $m = 1$ using the same energy unit.

As clearly visible in Fig. 3, the contributions from Σ_δ^a and Σ_δ^b are such that the sum of them yields a constant, independent of P_z and δ. For $\delta = 0$, Σ_δ^a and Σ_δ^b are the maximal and minimal, respectively, at $P_z = 0$. For $\delta = \frac{\pi}{4}$, Σ_δ^a is the whole answer and $\Sigma_\delta^b = 0$. The individual time-ordered amplitudes, $\Sigma_{\delta=\pi/4}^a$ and $\Sigma_{\delta=\pi/4}^b$, are now separately independent of P_z or invariant under the longitudinal boost. This shows clearly that the longitudinal boost becomes the kinematic generator in LFD and makes the number of kinematic generators get maximized in LFD.

We now pay attention to the apparent J-shaped ridge in Σ_δ^a matched by a similar valley in Σ_δ^b. For positive values of momentum, $P_z > 0$, the amplitudes Σ_δ^a and Σ_δ^b show a smooth behaviour, while for negative values of P_z we observe the presence of the J-shaped curve. We find that these J-shaped features are given by the function $P_z = -\sqrt{\frac{s(1-\mathbb{C})}{2\mathbb{C}}}$. An interesting point to observe in this J-shaped curve for negative values of momentum P_z is that it is stable as it is independent of the mass and does not disappear as the momentum goes to the negative infinity. Thus, if the limit $\delta \to \frac{\pi}{4}$ is taken in the exact amplitude with P_z given by the J-shaped curve, i.e. $P_z = -\sqrt{\frac{s(1-\mathbb{C})}{2\mathbb{C}}} \stackrel{\mathbb{C}\to 0}{\longrightarrow} -\infty$, then the connected contribution to the current arising from the vacuum $\Sigma^b_{\delta\to\frac{\pi}{4}}$ does not vanish but remains as a nonzero constant, i.e., $-\frac{1}{2m(\sqrt{s}+m)} = -\frac{1}{2(\sqrt{2}+1)} \approx -0.207$, although this nonzero constant (i.e. the minimum of $\Sigma^b_{\delta\to\frac{\pi}{4}}$) is cancelled by the overshoot in the maximum of $\Sigma^a_{\delta\to\frac{\pi}{4}}$, given by $\frac{1}{2m(\sqrt{s}-m)} = \frac{1}{2(\sqrt{2}-1)} \approx 1.207$ to yield the total amplitude $\frac{1}{s-m^2} = 1$.

Here, one may notice that the limits $P_z \to -\infty$ and $\delta \to \frac{\pi}{4}$ do not commutate. However, this fact should not be construed as an ambiguity of the suggested interpolation procedure, which is completely independent of the order of taking these limits. Rather, one may understand the non-commutativity as a possible issue for the common folklore of "equivalence of IFD and LFD in the infinite momentum frame (IMF)". Our work qualifies this notion, since it applies in the limit of $P_z \to \infty$, but requires a great caution in the limit $P_z \to -\infty$, i.e. $P^+ = (P^0 + P_z)/\sqrt{2} \to 0$ limit. One should note that, as $P^+ = p_1^+ + p_2^+ \to 0$, both $p_1^+ \to 0$ and $p_2^+ \to 0$, i.e. the light-front plus (+) momenta of all the particles involved in the scattering become zero. Thus, only the light-front zero-modes contribute to the scattering amplitude in this limit. This is remarkable because the limits from the IMF [11] of IFD to LFD are clearly asymmetric between $P^+ \to \infty$ and $P^+ \to 0$, while the IMF of IFD itself is entirely symmetric between $P_z \to \infty$ and $P_z \to -\infty$. In the interpolation we discuss here, the treacherous point is related to the particular way the limit $P_z \to -\infty$ is taken. If it is taken along the J-shaped curve, i.e., $P_z = -\sqrt{\frac{s(1-\mathbb{C})}{2\mathbb{C}}} \stackrel{\mathbb{C}\to 0}{\longrightarrow} -\infty$, then the result $\Sigma^b_{\delta=\frac{\pi}{4}} = -\frac{1}{2m(\sqrt{s}+m)} \neq 0$ is valid, but if δ and P_z are not so correlated that the longitudinal boost invariance of the LFD is manifest then $\Sigma^b_{\delta=\frac{\pi}{4}} = 0$. No matter how the limits turn out, one should note that the longitudinal boost is kinematic in LFD and the invariance of each time-ordered amplitude must hold under the transformations given by the kinematic generators. The sum of the two amplitudes $\Sigma^a_{\delta=\frac{\pi}{4}} + \Sigma^b_{\delta=\frac{\pi}{4}}$ remains of course invariant as it should be.

Although we discussed here the simplest possible scattering amplitude of two spinless particles in a toy ϕ^3 model theory, it clarified the confusion in the prevailing notion of the equivalence between the infinite momentum frame and the LFD. It also alerts the caution in handling the contribution from the light-front zero-modes [12]. Much more details of our work on interpolating the gauge fields and the fermion spinors have been presented in Refs. [9,10], respectively. In particular, we discussed the link between the Coulomb gauge in IFD and the light-front gauge in LFD [9] and the chiral representation of the helicity spinors interpolating between the IFD and the LFD [10].

3 Leading Nonanalytic Behavior in Chiral Dynamics

While the light-front zero-modes appear treacherous as discussed above, they play a vital role in guiding the hadron phenomenology to fulfill the first principles in hadron physics. As an example, we discuss here the leading nonanalytic (LNA) behavior determined by the infrared properties of chiral loops (characteristic of Goldstone boson loops in chiral perturbation theory [13]). The LNA behavior in chiral dynamics is model-independent [14–16], and places the physics of the pion cloud on a firm ground in QCD.

The pion cloud of the nucleon plays a vital role in understanding the nucleon's long-range structure as it is known from the approximate chiral symmetry of QCD. It provides important corrections to static nucleon properties, such as the mass, magnetic moment and axial charge, and significantly influences its electric and magnetic charge distributions (for a review see Ref. [17]). This will also have important consequences for correctly implementing chiral extrapolation of lattice QCD data on parton distribution moments and subsequently on other quantities, such as GPDs [18,19], as well as on understanding the pion distributions in the nucleon in a well-defined theoretical framework. In addition to the traditional studies of elastic form factors and parton distributions, there has also been a great deal of interest in GPDs. As with the ordinary PDFs, the physical interpretation of GPDs in terms of probability distributions is most natural in the LFD, and one knows that here

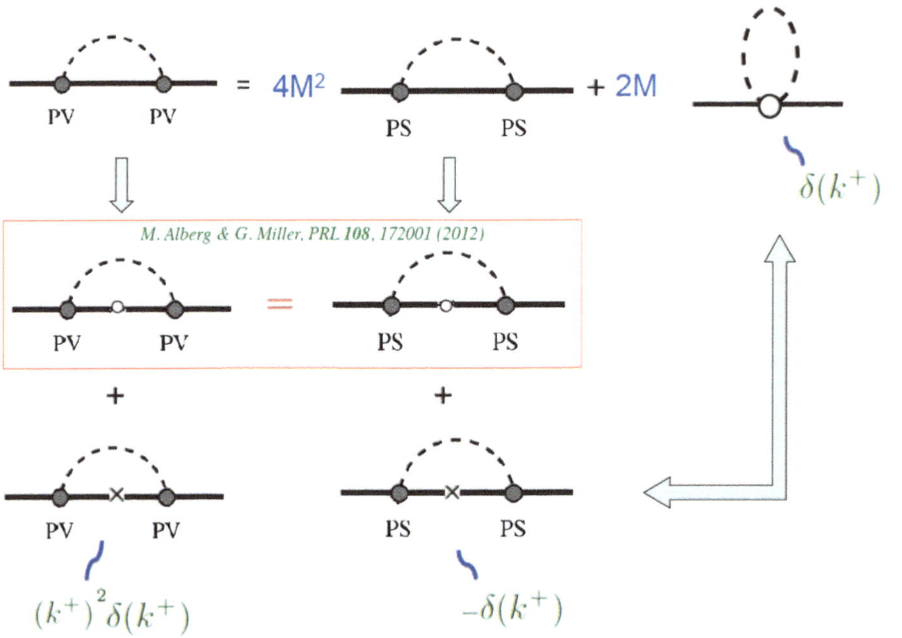

Fig. 4 The equivalence between the pseudovector (PV) coupling and the pseudoscalar (PS) coupling in chiral dynamics. The chiral partner in the PS theory corresponds to the two-pion coupling denoted by the *large white blob* in the upper rightmost (tadpole type) diagram. The *two parallel downward arrows* represent the equality of the fermion propagator given by Eq. (3). The equivalence in the "on-shell" contribution between PV and PS coupling theories is boxed with Ref. [34]. The cancellation between the two light-front zero-mode contributions in the PS coupling is denoted by the *double-arrow* pointing to those two contributions, i.e. the chiral partner contribution and the instantaneous contribution

too chiral corrections can be very important. It is timely, therefore, to address the question of how to provide a consistent derivation of the chiral corrections to all of these observables in LFD.

We have recently applied the LFD methodology to the case of pion loops and their effects on the $\bar{d} - \bar{u}$ asymmetry in the proton [20]. The observation of the $\bar{d} - \bar{u}$ flavor asymmetry in the light quark sea of the proton [21–24] has been one of the seminal results in hadron physics of the past two decades, leading to a major reevaluation of our understanding of the quark structure of the nucleon. In particular, the measurement revealed the importance of 5-quark Fock state components of the nucleon's light-front wave function, and the crucial role played by chiral symmetry breaking and the pion cloud in describing the nucleon's internal structure. An important development in establishing a formal link between models of $\bar{d} - \bar{u}$ and QCD came with the realization that the moments of PDFs could be formally expanded in chiral perturbation theory (χPT) in terms of power series in the pion mass squared, m_π^2. In particular, the LNA contributions were found to depend on the (model-independent) long range structure of the pion cloud, with a characteristic $m_\pi^2 \log m_\pi^2$ dependence [13], whose coefficient is given by physical constants [15]. This idea was then applied to the chiral extrapolation of lattice QCD moments of the nonsinglet $u - d$ distribution in order to reconcile the lattice data at large pion masses with experiment [14]. We recently refined the computation [25] of the $\bar{d} - \bar{u}$ flavor asymmetry in the proton sea with a complementary effort to reveal the dynamics of pion exchange in high-energy processes and generalized our approach to the SU(3) sector [26,27].

Initial calculations of pion loop effects were performed in the context of the "Sullivan" process [28], using pseudoscalar (PS) pion–nucleon coupling, which involves only the pion "rainbow" diagram. Analysis within the χPT, which implements the chirally symmetric pseudovector (PV) coupling, reveals differences in the off-shell behavior of the loops [29], as well as additional pion bubble terms [15,16] which contribute only at the light-front zero-mode $k^+ = 0$ [30]. The equivalence between the PS and PV theories was recently discussed in detail in Refs. [30,31]. This equivalence between the PV and PS theories may be summarized in Fig. 4. In this description, the first line shows the equivalence between PV and PS for which it is crucial to include the contribution from the chiral partner that corresponds to the two-pion contribution, *á la* $\sigma \bar{N} N \sim \pi\pi \bar{N} N$, depicted as the white blob of the vertex in the tadpole diagram shown in first line of Fig. 4. This equality can be verified straightforwardly using the Dirac equation for the external nucleons [32].

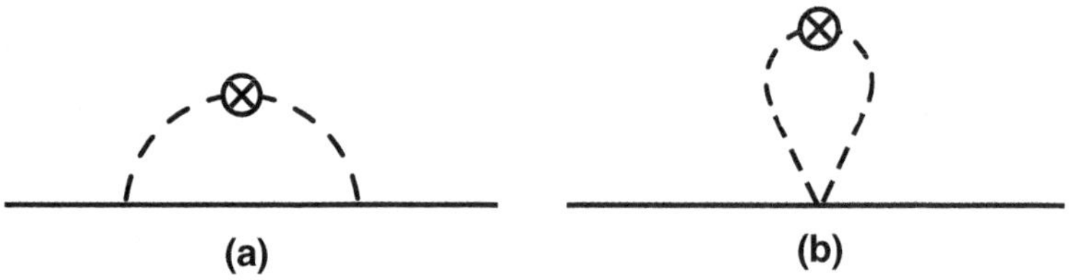

Fig. 5 Contributions to the pion light-cone momentum distributions in the nucleon, from **a** the pion rainbow, and **b** the pion bubble diagrams

Moreover, as depicted in the second and third lines of Fig. 4, the fermion propagator can be split into the so-called "on-shell" contribution plus the LF instantaneous contribution [33]: i.e.

$$\frac{1}{\not{q} - m} = \frac{\sum_s u(q,s)\bar{u}(q,s)}{q^2 - m^2} + \frac{\gamma^+}{2q^+}, \tag{3}$$

where q is the four-momentum of the propagating fermion. We see here that both the instantaneous contribution and the tadpole diagram contribution are the light-front zero-mode contributions proportional to $\delta(k^+)$. As shown in Fig. 4, these two zero-mode contributions (tadpole type diagram and instantaneous contribution) cancels each other in the PS theory while the instantaneous contribution in the PV theory vanishes by itself due to the derivative coupling. Thus, one may effectively get the equivalence between the "on-shell"contributions of the PV and PS theories in the self-energy calculation [34]. This also confirms the model-independence of the LNA behavior in the nucleon self-energy regardless whether the PV theory or the PS theory is used. However, it is still crucial to keep track of the light-front zero-modes as their finite nonzero contribution exists in other process such as the vertex correction [29]. We have indeed verified the chiral symmetry of QCD in the $\bar{d} - \bar{u}$ asymmetry in the proton [20] by examining the LNA behavior of the lowest moment $\overline{D} - \overline{U} \equiv \int_0^1 dx(\bar{d} - \bar{u})$. The LNA behavior of $\overline{D} - \overline{U}$ is given by

$$\begin{aligned}(\overline{D} - \overline{U})_{\text{LNA}} &= \frac{3g_A^2 + 1}{2(4\pi f_\pi)^2} m_\pi^2 \log m_\pi^2 \\ &= \frac{4g_A^2 + (1 - g_A^2)}{2(4\pi f_\pi)^2} m_\pi^2 \log m_\pi^2,\end{aligned} \tag{4}$$

where $g_A = 1.267$ is the axial charge and $f_\pi = 93$ MeV is the pion decay constant. In the second line of this equation, Eq. (4), the model-independent LNA coefficient $3g_A^2$ is explicitly split into the "on-shell" contribution $4g_A^2$ and the light-front zero-mode (δ-function) contribution $-g_A^2$. This result in the second line of Eq. (4) resolved [30,31] the factor 4/3 difference between the "Sullivan" process calculation [13] and the χPT calculation [15,16], which has been puzzling for more than a decade. In fact, one can now realize that the absence of the δ-function contributions in the PS version of the rainbow diagrams leads to the coefficient of the leading nonanalytic term in the chiral expansion of the quark distribution function moments that differs by a factor 4/3 from the PV result. Thus, in this resolution of the lingering factor difference, it is evident that δ-function contribution from the light-front zero-mode has been crucial.

The model-independence of the LNA behavior in chiral dynamics is so useful that one may regard it as a sort of global positioning system "GPS" in the practice of chiral phenomenology. With this guidance, we recently refined the computation [25] of the $\bar{d} - \bar{u}$ flavor asymmetry in the proton sea with a complementary effort to reveal the dynamics of pion exchange in high-energy processes and generalized our approach to the SU(3) sector [26,27]. This complementary effort has been the study of leading neutron production in semi-inclusive DIS on the proton, in which a forward moving neutron is produced in coincidence with the scattered lepton in the high-energy reaction $ep \to enX$. Several dedicated experiments at the ep collider HERA [35–37] have collected high-precision data on the spectrum of leading neutrons carrying a large fraction of the proton's energy. Moreover, the upcoming tagged DIS (TDIS) experiment at Jefferson Lab [38] plans to take data on the production of leading protons from an effective neutron target in the reaction $en \to epX$, which, in analogy with the HERA leading neutron leptoproduction, can be described through the exchange of a π^-. In the proposed experiment, the effective neutron target will be prepared by tagging spectator protons with momenta between 60 MeV and 400 MeV at backward kinematics in the DIS of the electron from a deuteron nucleus, using the same technique that was developed for the measurement of the neutron structure function in the BONuS experiment at Jefferson Lab [39].

In particular, we have recently addressed [25] the question of whether one can reduce the model dependence of the pion structure function F_2^π extracted from the HERA data [35–37] by using additional constraints from the data on the SU(2) flavor asymmetry $\bar{d} - \bar{u}$, especially those from the E866 Drell–Yan experiment [24]. Because the E866 data are at relatively high x values compared with the HERA measurements, within the pion exchange framework they are sensitive to the pion PDFs at large x_π, where these are well determined from pion–nucleon Drell–Yan data [40–42]. The main variable in describing the $\bar{d} - \bar{u}$ asymmetry is therefore the pion distribution function in the nucleon. In our study [25], we used methodology adopted from global PDF analysis [43,44] to simultaneously fit both the HERA leading neutron and E866 $\bar{d} - \bar{u}$ asymmetry data. This study allowed us to refine the analysis of the $\bar{d} - \bar{u}$ asymmetry and at the same time discuss a possible prediction of the future tagged DIS experiments in JLab.

4 Conclusion and Future Directions

The LFD is not just formal but consequential in the analysis of physical observables. The dramatic change of the longitudinal boost's character from dynamic to kinematic in LFD is a great advantage in saving dynamic efforts to obtain the QCD solutions that reflect the full Poincaré symmetries. While the light-front zero-mode is treacherous, it plays a vital role in resolving theoretical issues as we have discussed the lingering LNA factor difference in chiral dynamics which has been persistent more than a decade. The model-independence of the LNA behavior in chiral dynamics provides a good guidance in chiral phenomenology. More recently, we generalized this approach to the SU(3) sector [26,27]. We performed a comprehensive analysis of the strange-antistrange parton distribution function (PDF) asymmetry in the proton in the framework of chiral effective theory, including the full set of lowest-order kaon loop diagrams with off-shell and contact interactions, in addition to the usual on-shell contributions previously discussed in the literature. We identified the presence of δ-function contributions to the $\bar{s}$ PDF at $x = 0$, with a corresponding valencelike component of the s-quark PDF at larger x, which allows greater flexibility for the shape of $s - \bar{s}$. Expanding the moments of the PDFs in terms of the pseudoscalar kaon mass, we computed the leading nonanalytic behavior of the number and momentum integrals of the s and $\bar{s}$ distributions, consistent with the chiral symmetry of QCD. We discussed the implications of our results for the understanding of the NuTeV anomaly [45] and for the phenomenology of strange-quark PDFs in global QCD analysis. However, the quantitative analysis of the light-front zero-mode contribution in the hadron phenomenology is still open for the progress.

Acknowledgements This work was in collaboration with a number of colleagues. In particular, I am indebted to Wally Melnitchouk, Tony Thomas, Ping Wang, Xuangong Wang, Yusupujiang Salamu, Ben Bakker, Ho-Meoyng Choi, Ziyue Li, Bailing Ma, Murat An and Alfredo Suzuki. I also thank Teresa Peña and Alfred Stadler of this stimulating workshop.

References

1. P.A.M. Dirac, Rev. Mod. Phys. **21**, 392–399 (1949)
2. S. Fubini, A.J. Hanson, R. Jackiw, Phys. Rev. D **7**, 1732 (1973)
3. LYa. Glozman, W. Plessas, K. Varga, R.F. Wagenbrunn, Phys. Rev. D **58**, 094030 (1998)
4. R.F. Wagenbrunn, S. Boffi, W. Klink, W. Plessas, M. Radici, Phys. Lett. B **511**, 33 (2001)
5. R.F.T. Melde, K. Berger, L. Canton, W. Plessas, R.F. Wagenbrunn, Phys. Rev. D **76**, 074020 (2007)
6. K. Hornbostel, Phys. Rev. D **45**, 3781–3801 (1992)
7. C.-R. Ji, C. Mitchell, Phys. Rev. D **64**, 085013 (2001)
8. C.-R. Ji, A.T. Suzuki, Phys. Rev. D **87**, 065015 (2013)
9. C.-R. Ji, Z. Li, A.T. Suzuki, Phys. Rev. D **91**, 065020 (2015)
10. Z. Li, M. An, C.-R. Ji, Phys. Red. D **92**, 105014 (2015)
11. S. Weinberg, Phys. Rev. **150**, 1313 (1966)
12. B.L.G. Bakker, M.A. DeWitt, C.-R. Ji, Y. Mishchenko, Phys. Rev. D **72**, 076005 (2005)
13. A.W. Thomas, W. Melnitchouk, F.M. Steffens, Phys. Rev. Lett. **85**, 2892 (2000)
14. W. Detmold, W. Melnitchouk, J.W. Negele, D.B. Renner, A.W. Thomas, Phys. Rev. Lett. **87**, 172001 (2001)
15. J.W. Chen, X. Ji, Phys. Lett. B **523**, 107 (2001)
16. D. Arndt, M.J. Savage, Nucl. Phys. A **697**, 429 (2002)
17. A.W. Thomas, Adv. Nucl. Phys. **13**, 1–137 (1984)
18. X. Ji, Ann. Rev. Nucl. Part Sci. **54**, 413–450 (2004)
19. A.V. Belitsky, A.V. Radyushkin, Phys. Rep. **418**, 1–387 (2005)

20. Y. Salamu, C.-R. Ji, W. Melnitchouk, P. Wang, Phys. Rev. Lett. **114**, 122001 (2015)
21. M. Arneodo et al., Phys. Rev. D **50**, 1 (1994)
22. K. Ackerstaff et al., Phys. Rev. Lett. **81**, 5519 (1998)
23. A. Baldit et al., Phys. Lett. B **332**, 244 (1994)
24. R.S. Towell et al., Phys. Rev. D **64**, 052002 (2001)
25. J.R. McKenney, N. Sato, W. Melnitchouk, C.-R. Ji, Phys. Rev. D **93**, 054011 (2016)
26. X. Wang, C.-R. Ji, W. Melnitchouk, Y. Salamu, A.W. Thomas, P. Wang, Phys. Lett. B **762**, 52 (2016)
27. X. Wang, C.-R. Ji, W. Melnitchouk, Y. Salamu, A.W. Thomas, P. Wang, Phys. Rev. D **94**,094035 (2016)
28. J.D. Sullivan, Phys. Rev. D **5**, 1732 (1972)
29. C.-R. Ji, W. Melnitchouk, A.W. Thomas, Phys. Rev. Lett. **110**, 179101 (2013)
30. M. Burkardt, K.S. Hendricks, C.-R. Ji, W. Melnitchouk, A.W. Thomas, Phys. Rev. D **87**, 056009 (2013)
31. C.-R. Ji, W. Melnitchouk, A.W. Thomas, Phys. Rev. D **88**, 076005 (2013)
32. C.-R. Ji, W. Melnitchouk, A.W. Thomas, Phys. Rev. D **80**, 054018 (2009)
33. S.-J. Chang, T.-M. Yan, Phys. Rev. D **7**, 1147 (1973)
34. M. Alberg, G. Miller, Phys. Rev. Lett. **108**, 172001 (2012)
35. S. Chekanov et al., Nucl. Phys. B **637**, 3 (2002)
36. F.D. Aaron et al., Eur. Phys. J. C **68**, 381 (2010)
37. S. Chekanov et al., Nucl. Phys. B **776**, 1 (2007)
38. C. Keppel, B. Wojtsekhowski, P. King, D. Dutta and J. Annand, as spokespersons: Measurement of Tagged Deep-Inelastic Scattering (TDIS), Jefferson Lab experiment PR12-15-006
39. N. Baillie et al., Phys. Rev. Lett. **108**, 142001 (2012)
40. J. Badier et al., Z. Phys. C **18**, 281 (1983)
41. B. Betev et al., Z. Phys. C **28**, 15 (1985)
42. J.S. Conway et al., Phys. Rev. D **39**, 92 (1989)
43. P. Jimenez-Delgado, W. Melnitchouk, J.F. Owens, J. Phys. G **40**, 093102 (2013)
44. S. Forte, G. Watt, Ann. Rev. Nucl. Part Sci. **63**, 291 (2013)
45. G.P. Zeller et al., Phys. Rev. Lett. **88**, 091802 (2002)

Few-Body Syst (2017) 58:39
DOI 10.1007/s00601-017-1215-4

Franz Gross

The CST: Its Achievements and Its Connection to the Light Cone

Received: 16 December 2016 / Accepted: 2 January 2017 / Published online: 19 January 2017

Abstract I review the applications of the Covariant Spectator Theory (CST) since its inception in 1969. Applications discussed here include calculations of NN scattering, 3N bound states, electromagnetic form factors of few-nucleon systems, and the recent successes in describing the dynamical generation of quark mass and the meson spectrum using a chirially invariant quark–antiquark interaction that includes confinement. The common origin of the Light Cone technique and the CST, which dates back to the 1970's, will be discussed.

1 Background

1.1 Context: Two Ways to Approach Non-perturbative Calculations

If the goal is to solve QCD non-perturbatively, and to treat the one and two-body problems consistently, then we can identify two major approaches:

- Approach #1 Start from the Dyson–Schwinger (DS) equation for the one-body self energy, and generalize to the two-body problem (c.f. Fig. 1), or
- Approach #2 Start from the two-body Bethe–Salpeter (BS) equation, and construct the one-body equation (c.f. Fig. 2).

If we had a theory where the vertices or kernels could be calculated exactly to all orders, the two approaches would be equivalent, but if the vertices or kernels are modeled using phenomenological expressions, each of them has disadvantages. The disadvantage of approach #1 (which starts from the one-body DS equation) is that its un-symmetric structure (only one of the vertices can be dressed if double counting is to be avoided) makes it less than straightforward how to extend it to the two-body sector without missing the dressing on the second vertex (new interaction terms are required). The disadvantage of approach #2 which starts from the two-body BS equation is that double counting may be introduced when the self-energy is calculated using the phenomenological two-body kernel. Spontaneous chiral symmetry breaking can be realized in both approaches, so this feature does not distinguish between them. The CST starts for the BS equation.

This article belongs to the Topical Collection "Light Cone 2016".

Many thanks to my collaborators who have worked with me over the years. Those who have contributed to the discussion covered in this mini-review include: Alfred Stadler, Teresa Peña, Elmar Biernat, J.W. Van Orden, D.O.Riska, Joe Milana, Cetin Savkli, M. Uzzo, Carl Carlson, John Tjon, Yohanes Surya, Gilberto Ramalho, and Sofia Leitão.

F. Gross (✉)
Jefferson Laboratory, 12000 Jefferson Avenue, Newport News, VA 23606, USA
E-mail: gross@jlab.org

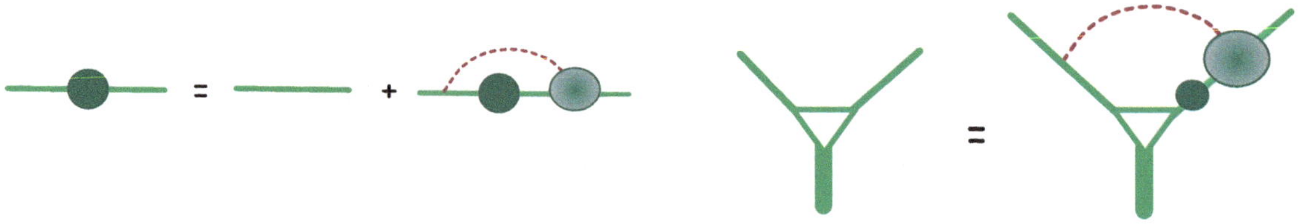

Fig. 1 (*Left panel*) one-body Dyson–Schwinger equation. (*Right panel*) attempt to generalize the one-body DS equation to describe two-body bound states

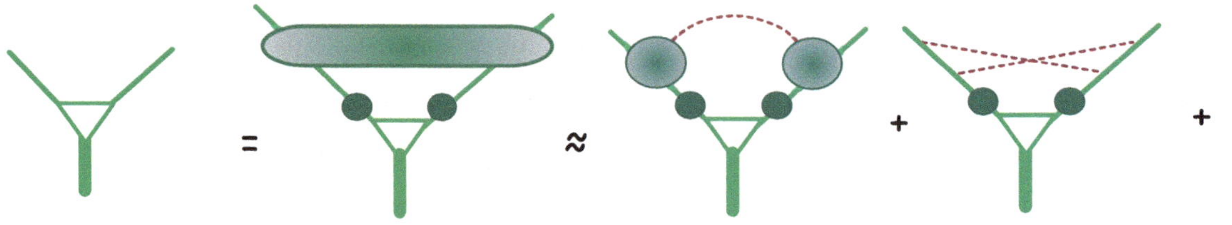

Fig. 2 Two-body Bethe–Salpter equation with ladder and crossed ladder approximations

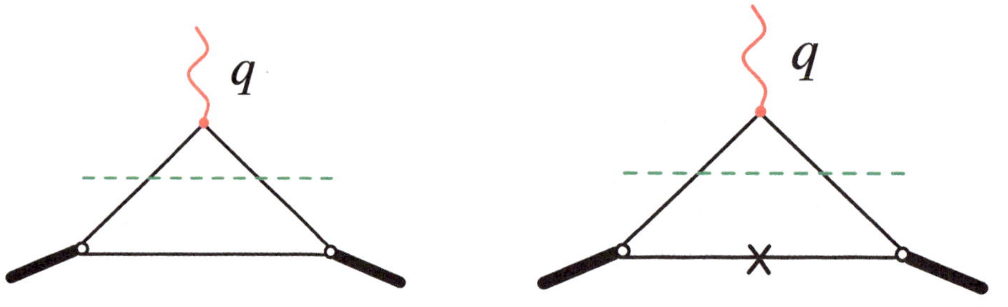

Fig. 3 (*Left panel*) triangle diagram showing the normal two-body $N\bar{N}$ "cut" and (*right panel*) the anomalous cut where the "spectator" is on-shell

1.2 Origin of the CST: From 1964

Historical perspective: In the early 1960's, field theory was not considered useful for non-perturbative calculations, and quarks and QCD had not yet been discovered. Strong interaction physics was the study of nucleon and meson (mostly pion) interactions, and dispersion theory (together with generalized unitary) was the only non-pertubative technique available. (With the emergence of QCD and the development of dynamical quark models, dispersion theory fell into disfavor for a period of time. It has recently been "rediscovered" as a useful tool.)

It was in this context that CST was developed. At that time I calculated the deuteron form factors using dispersion theory [1,2]. Each of the three form factors could be written in the form

$$F(Q^2) = \frac{1}{\pi} \int_{s_0}^{\infty} ds \, \frac{\mathrm{Im} F(s)}{s + Q^2} \tag{1}$$

where $Q^2 = -q^2$ with q the four-momentum transferred by the virtual photon. The threshold s_0 is normally determined by the threshold for the physical production of the intermediate $N\bar{N}$ system (at $s_0 = 4m^2 \simeq 181.4m_\pi^2$) that couples the photon to the deuteron-antideuteron final state (see the left panel of Fig. 3). This large threshold gives a very flat form factor that describes a very small deuteron, and is not in agreement with our picture of the deuteron as a weakly bound, extended system. To obtain the correct dispersion result for an extended deuteron with a mass $M_d \lesssim 2m$, one treats M_d as a complex variable and analytically continues the result for small M_d to large M_d, being careful to stay on the same sheet. A careful study of this continuation [3] shows that when $M_d > \sqrt{2}m$, the dispersion integral must be deformed with a new threshold that migrates to a smaller value. If the deuteron binding energy E_B is small, the new threshold is at $s_0 = 16mE_B \simeq 1.73m_\pi^2$, in complete agreement with our non-relativistic intuition for the behavior of a form factor of an extended, weakly bound system. Even more remarkable is the fact that the imaginary part of the form factor in the anomalous

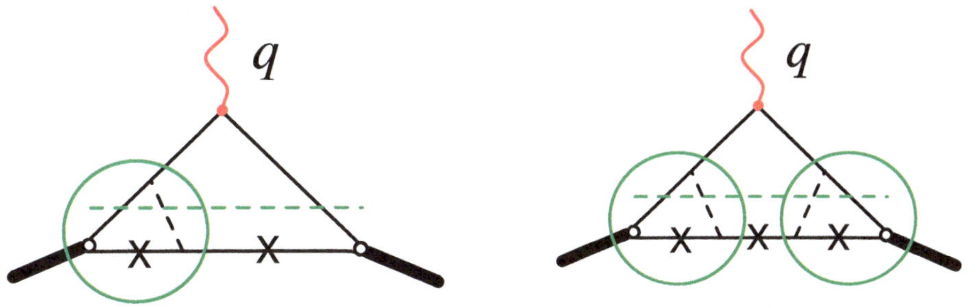

Fig. 4 (*Left panel*) anomalous region for the triangle diagram with a one-pion exchange (OPE) correction to the left vertex and (*right panel*) the anomalous region with OPE corrections at both vertices

Table 1 Thresholds s_0 for the pion exchange diagrams shown in Fig. 4

Diagram	Normal threshold	Anomalous threshold
One pion (left panel)	$(2m+\pi)^2 \simeq 209.3m_\pi^2$	$\simeq 11m_\pi^2$
Two pions (right panel)	$(2m+2\pi)^2 \simeq 239.2m_\pi^2$	$\simeq 28m_\pi^2$

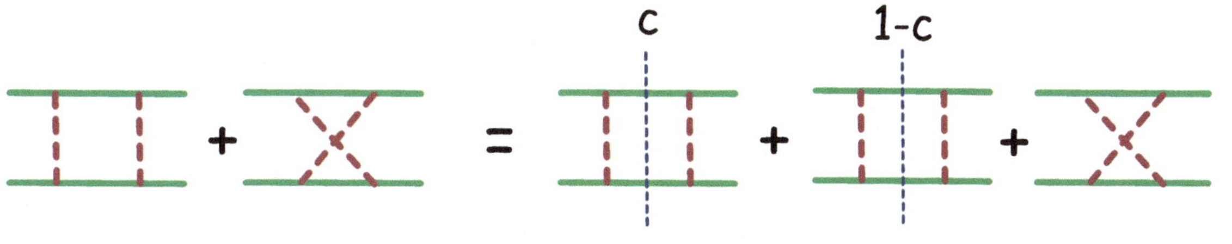

Fig. 5 The ladder and crossed ladders to 4th order. The ladder can be separated into two parts represented by the restriction "c" on the two-body propagator, and its remainder, "$1-c$"

region (where $s_0 < s < 4m^2$) is given by the imaginary part of the triangle diagram with the *spectator on mass-shell*, illustrated in the right panel of Fig. 3.

When pion exchanges at the vertices are included, the cut structure becomes more complicated, and the anomalous thresholds and anomalous regions show a pattern illustrated in Fig. 4. The thresholds for the one and two pion exchanges shown in Fig. 4 are given in Table 1.

Note that these contributions can be summed by calculating a dressed dpn vertex with one nucleon off-shell and the other nucleon restricted to its mass shell. When these contributions are summed to all orders, one has the spectator equation—this is the origin of the idea!

1.3 Quick Derivation of the CST for Two-Body Scattering

Start with the two-body Bethe–Salpeter equation for the scattering amplitude

$$\mathcal{M}(p,p';P) = \mathcal{V}(p,p';P) - i\int \frac{d^4k}{(2\pi)^4}\mathcal{V}(p,k;P)S(k)S(P-k)\mathcal{M}(k,p';P) \tag{2}$$

where $\mathcal{V}$ is the scattering kernel, S the dressed one body propagator, p, p', k are four-momenta of particle 1, and P the total four momentum. Unfortunately, the kernel is unknown and must be modeled. A common choice is to calculate the kernel from the infinite sum of all ladder and crossed ladder exchanges, illustrated to fourth order in Fig. 5. If the restriction "c" is chosen carefully, its remainder ("$1-c$") and the crossed box will be very small. If this cancellation works to all orders, then the two-body equation using the restricted two-body propagator with a one boson exchange (OBE) kernel will give a very good approximation to the BS equation using the full *generalized ladder* (ladder plus crossed ladder) exchanges.

It is commonly assumed that the choice $c = 1$ (the BS in ladder approximation) is very good, but the crossed ladders are not small an this is not generally correct. In some cases, it can be proved that choosing c to restrict one particle to its mass-shell (the heavier one if they have unequal masses) is more efficient [3,4], and in particular, this restriction is *exact* for scalar theories when the heavy (on shell particle) mass approaches infinity. This remarkable result provides a strong motivation for use of the CST.

1.4 Comparison of the Two-Body CST with the Two-Body Light-Cone Equation

I close this background section by showing how the CST and the light cone equation are related. To see the connection, look at the 4th order box restricted to one of the two particles being on shell:

$$\mathcal{V}_4 = i\int \frac{d^4k}{(2\pi)^4}\frac{\mathcal{V}(p,k;P)\mathcal{V}(k,p';P)}{D(k)D(P-k)} \to -\int \frac{d^3k}{(2\pi)^3}\frac{\mathcal{V}(p,\hat{k};P)\mathcal{V}(\hat{k},p';P)}{2E_kD(P-\hat{k})} = \mathcal{V}_4^{\rm CST} \tag{3}$$

where $D(k) = m^2 - k^2 - i\epsilon$, $\hat{k} = \{E_k, \mathbf{k}\}$, $E_k = \sqrt{m^2+\mathbf{k}^2}$, and the decomposition is done in the rest frame $P = \{W, \mathbf{0}\}$, so that $D(P-\hat{k}) = W(2E_k - W)$.

The fourth order light cone result is computed in a similar way – by integrating over one of the components of k. Specifically, one decomposes k into light cone variables using $k_\pm = k_0 \pm k_z$, so that

$$\begin{aligned} k &= \{k_+, k_-, \mathbf{k}_\perp\} \\ P &= \{W, W, \mathbf{0}\} \end{aligned} \tag{4}$$

and integrates over k_+ keeping only the poles from the two particle propagators. Introducing the momentum fraction x defined through $k_- = xP_- = xW$, gives

$$\begin{aligned} \mathcal{V}_4 &= iW\int_{-\infty}^{\infty} dx \int \frac{dk_+d^2k_\perp}{2(2\pi)^4}\frac{\mathcal{V}(p,k;P)\mathcal{V}(k,p';P)}{(E_\perp^2 - Wxk_+ - i\epsilon)(E_\perp^2 - W(1-x)(W-k_+) - i\epsilon)} \\ &\to -\int_0^1 dx\int \frac{d^2k_\perp}{2(2\pi)^3}\frac{\mathcal{V}(p,k_{\rm LC};P)\mathcal{V}(k_{\rm LC},p';P)}{m^2+\mathbf{k}_\perp^2 - x(1-x)W^2} = \mathcal{V}_4^{\rm LC} \end{aligned} \tag{5}$$

where $k_{\rm LC} = \{\frac{m^2+\mathbf{k}_\perp^2}{xW}, xW, \mathbf{k}_\perp\}$. Note the famous result that the integral over x is restricted to the interval $[0, 1]$. This happens because outside of this interval the two poles from the propagator lie in the same half of the complex k_- plane, giving zero when the integral over k_- is evaluated using the Cauchy theorem.

At first the two results (3) and (5) appear to be unrelated, but introducing the variable y and the transformations

$$\begin{aligned} y &= \frac{E_k - k_z}{W} = \frac{\hat{k}_-}{P_-} \qquad \hat{k} = k'_{\rm LC} = k_{\rm LC}(x\to y) \\ W(2E_k - W) &= \frac{1}{y}\left[m^2 + \mathbf{k}_\perp^2 - y(1-y)W^2\right] \end{aligned} \tag{6}$$

were are able to transform $\mathcal{V}_4^{\rm CST}$ into

$$\mathcal{V}_4^{\rm CST} = -\int_0^\infty dy\int \frac{d^2k_\perp}{2(2\pi)^3}\frac{\mathcal{V}(p,k'_{\rm LC};P)\mathcal{V}(k'_{\rm LC},p';P)}{m^2+\mathbf{k}_\perp^2 - y(1-y)W^2} \tag{7}$$

Note that if we now change $y \to x$, (7) is *identical* to (5) except that the range of the x integration is now $[0, \infty)$ instead of $[0, 1]$ (a profound difference, of course). This similarity showing the close connection between CST and LC has been noticed by several people over many years.

2 Applications of the CST

The CST has been used to study many systems:

- Few-quark systems (discussed in Sect. 2.2).
- Few-nucleon systems (discussed in Sect. 2.1).
- Atomic systems (Brodsky and Lepage [5] and Edies et al. [6]).

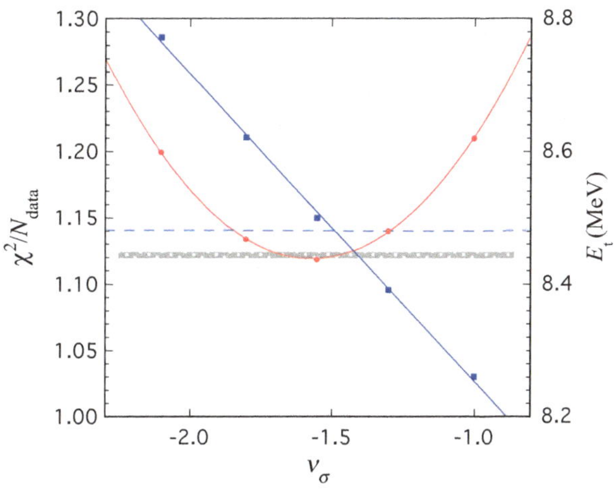

Fig. 6 Variation of χ^2 from the fit to the two-body NN data (*red line*) and the triton binding energy E_t (*blue line*), with the off-shell coupling ν_σ defined in Eq. (8). This figure is for model WJC-2 discovered in 2008, but is qualitatively similar to the result first discovered in 1997

- Inelastic scattering (Ford et al. [7]).
- Nucleon-nucleus scattering (Maung et al. [8]).
- Nuclear matter (Celenza and Shaken [9]).
- Pion-nucleon scattering (Sura and Gross [10]).
- Relativistic explanation of the EMC Effect (Liuti and Gross [11]).

In this talk I will only have time to briefly review the first two topics.

2.1 Applications of CST to Few-Nucleon Systems

A few key papers in the use of CST for the study of few-nucleon systems are:

- 1969 (Gross) First CST paper deriving the cancellation of ladder and crossed ladder exchanges for scalar and electromagnetic interactions [4].
- 1987 (Gross and Riska) Derivation of a two-body current that is conserved, even if the interaction contains phenomenological form factors [12].
- 1995 (Van Orden et al.) Calculation of deuteron form factors using the first realistic CST wave functions [13].
- 1997 (Stadler and Gross) Observation that CST OBE models require *no* three-body forces. Demonstration that the fits to the NN data and the triton binding energy are very sensitive to the strength, ν_σ, of the off-shell coupling of the σ meson, and that the value of ν_σ that gives the best fit to the NN data also gives the correct triton binding energy, E_t (see more detailed discussion below) [14].
- 1997 (Kvinikhidze and Blankleider) Three-body CST currents fixed by the gauging method [15].
- 2004 (Gross et al.) For CST OBE models, a conserved three-body interaction current is derived that uses the same one-body and two-body interaction currents needed for the two-body system [16].
- 2008 (Gross and Stadler) The CST OBE model (WJC-2) is found that gives a precise fit to np scattering data ($\chi^2/N_{\text{data}} \sim 1.1$) and requires only 15 parameters from the exchange of the photon and only 6 strongly interacting bosons $b[I(J^P)]$: $\sigma[0(0^+)]$, $\eta[0(0^-)]$, $\pi[1(0^-)]$, $\delta[1(0^+)]$, $\omega[0(1^-)]$, and $\rho[1(1^-)]$. These more accurate models show the same correlation between the quality of fit to the NN data and the three-body binding energy, first observed in 1997 [17].
- 2014 (Gross) derivation of a unique *isoscalar* two-body interaction current [18].

There is only space to focus on one of the most remarkable results of these CST OBE studies, illustrated in Fig. 6. The off-shell sigma coupling constant appears in the generalized σNN coupling

$$\Lambda^\sigma(p, p') = g_\sigma \mathbf{1} - \nu_\sigma[\Theta(p) + \Theta(p')] \tag{8}$$

where $p(p')$ is the four-momenta of the outgoing (incoming) nucleon and $\Theta(p) = (m - \not{p})/(2m)$ is the off-shell nucleon projection operator. Note that the interaction is *independent* of ν_σ if the nucleon is *on-shell*.

2.2 Applications of CST to Few-Quark Systems

The major results for the applications of CST to few-quark systems are:

- 1991–4 (Gross and Milana) Initial studies of the meson spectrum with a relativistic generalization of confinement (for $L = 0$ states only) [19].
- 1999 (Uzzo and Gross) Study of the stability of vector and scalar interactions using various equations [20].
- 2001 (Savkli and Gross) Implications of the relativistic treatment of confinement [21].
- 2008—(Ramalho et al.) Phenomenological models of hadronic form factors and static moments [22].
- 2014—(Biernat et al.) Self energies and chiral symmetry breaking in a conjugation invariant CST model [23–26].
- 2017—(Leitão et al.) Meson spectrum with $L \neq 0$ and spin (see the talk elsewhere in these proceedings) [27].

In this section I will highlight a few of these results. I start by emphasizing that here are two different ways to model confinement in QCD:

1. Require that the quark propagator have no poles on the real axis, *or*
2. Allow quark propagators to have poles on the real axis (i.e. quarks can be on shell), but construct an interaction that is zero when two or more quarks are on-shell (so that no physical scattering is possible)

The nonrelativistic and the relativistic CST approaches both assume the latter, while the well known Dyson Schwinger approach assumes the former. It may turn out that while the language used by the two approaches is quite different, they give equivalent results when properly carried out to all orders; future study may resolve these issues.

To understand the CST approach to confinement, first consider a screened definition of the linear confining potential in coordinate space

$$\tilde{V}_L(r) = -\lim_{\epsilon\to 0}\frac{\sigma}{\epsilon}(e^{-\epsilon r} - 1) = \sigma r\,. \tag{9}$$

Transforming this to momentum space gives

$$V_L(\mathbf{q}) = \int d^3r e^{i\mathbf{q}\cdot\mathbf{r}} = -8\pi\sigma \lim_{\epsilon\to 0}\left[\frac{1}{(\mathbf{q}^2+\epsilon^2)^2} - \delta(\mathbf{q})\int d^3q'\frac{1}{(\mathbf{q}'^2+\epsilon^2)^2}\right] \tag{10}$$

Note that this satisfies the important constraint

$$0 = \int d^3q\, V_L(\mathbf{q}) \to \tilde{V}_L(0) \tag{11}$$

and leads immediately to the bound state momentum space Schrödiger equation for linear confinement

$$\left[\frac{\mathrm{p}^2}{2m_R} - E\right]\Psi(\mathbf{p}) = -\sigma\int\frac{d^3k}{\pi^2}\,\frac{\Psi(\mathbf{k}) - \Psi(\mathbf{p})}{(\mathbf{p}-\mathbf{k})^4}. \tag{12}$$

This equation is regular and has been solved numerically, giving solutions identical to the well known coordinate space solutions. Modifications are required for $L \neq 0$ states [27].

These result are easily generalized for use with the one-channel CST equations. Labeling the momenta as in Eq. (2), it is convenient to generalize (10), using the on-shell particle 1 four-momentum in place of the nonrelativistic three-momentum, giving

$$V_L(\hat{p}-\hat{k}) = -8\pi\sigma\lim_{\epsilon\to 0}\left\{\frac{1}{[(\hat{p}-\hat{k})^2-\epsilon^2]^2} - 2E_k\delta(\mathbf{p}-\mathbf{k})\int\frac{d^3k'}{2E_{k'}}\frac{1}{[(\hat{p}-\hat{k}')^2-\epsilon^2]^2}\right\}. \tag{13}$$

Note that this reduces to (10) in the limit when the mass $m \to \infty$, and that it satisfies the relativistically invariant constraint

$$0 = \int d^3k\,\frac{m}{E_k}V_L(\hat{p}-\hat{k})\,, \tag{14}$$

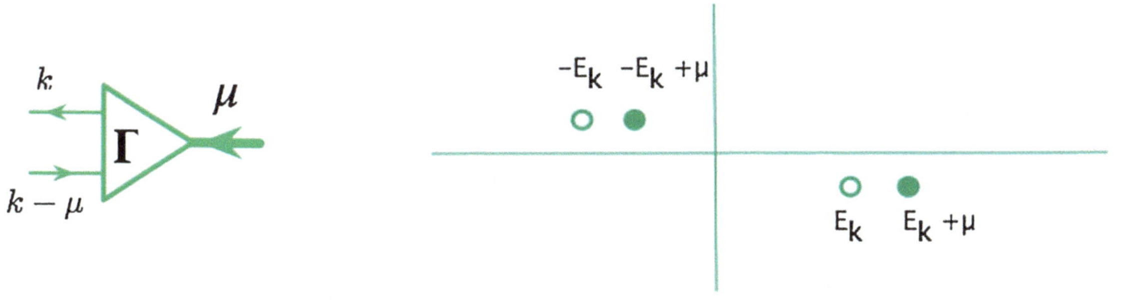

Fig. 7 (*Left panel*) the vertex function Γ with bound state mass μ and the momenta of the particles labeled (here μ is a shorthand notation for the four-momentum $P = \{\mu, \mathbf{0}\}$). (*Right panel*) the location of the particle poles (16) in the complex k_0 plane

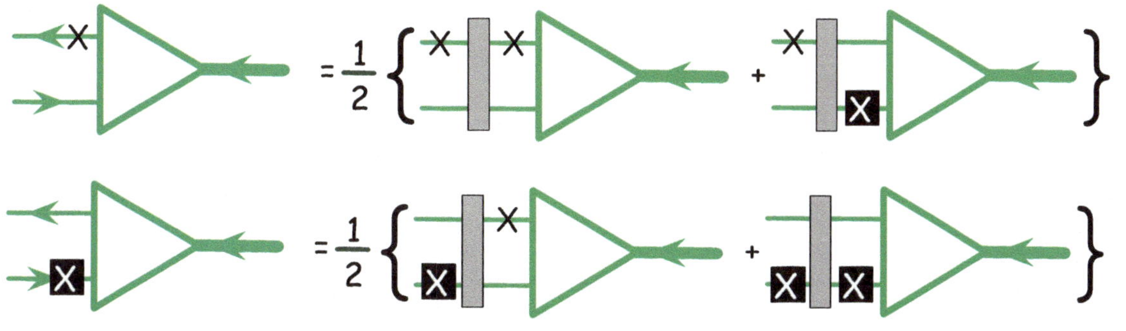

Fig. 8 The two-channel CST equation with exact charge conjugation symmetry. The particles marked with a *times* are on their positive energy mass shell and those with an *times* inside of a *black square* on their negative energy mass shell (see the discussion in the text). The *triangles* represent the vertex functions Γ

which reduces to (11) in the nonrelativistic limit. The covariant one channel CST equation becomes

$$[m^2 - (P - \hat{p})^2]\Psi(\mathbf{p}) = M_b(2E_p - M_b)\Psi(\mathbf{p}) = -2m\sigma \int \frac{d^3k}{\pi^2} \frac{m}{E_k} \frac{\Psi(\mathbf{k}) - \Psi(\mathbf{p})}{(\hat{p} - \hat{k})^4}. \tag{15}$$

These equations provide a convenient relativistic generalization of the linear confining potential, and their solutions have been shown to give a good qualitative description of the low lying $L = 0$ states.

To study $q\bar{q}$ states which exhibit charge conjugation ($\mathcal{C}$) symmetric properties, it is necessary to generalize the one channel CST equation to two channels, and if the bound state mass (represented by μ in this discussion, so that in the rest system $P = \{\mu, \mathbf{0}\}$) is small, it must be extended to include all four possible channels. To understand how these coupled channel equations arise, examine the location of the four poles in the two-body propagator in the complex k_0 plane. In the rest system they are at:

$$\begin{aligned} S(k)^{-1}S(P-k)^{-1} &= (m^2 - k^2 - i\epsilon)(m^2 - (P-k)^2 - i\epsilon) \\ &= (E_k - k_0 - i\epsilon)(E_k + k_0 - i\epsilon)(E_k - \mu + k_0 - i\epsilon)(E_k + \mu - k_0 - i\epsilon) \end{aligned} \tag{16}$$

where the four-momenta of outgoing particle 1 is $k_1 = k$, the four-momenta of the the incoming particle 2 is $k_2 = k - P$ (or $-k_2 = P - k$ when reinterpreted as an outgoing antiparticle), as illustrated in the right panel of Fig. 7, and for simplicity I have used propagators for scalar particles since the pole structure is independent of spin. The location of these poles in the complex k_0 plane is shown in the left panel of Fig. 7. If μ is large and the system is not $\mathcal{C}$ invariant, it is sufficient to keep only the pole at $k_0 = E_k$ in the lower half plane, which restricts particle 1 to its positive energy mass shell. This is the one channel case I have discussed above.

When the system exhibits $\mathcal{C}$ invariance, it is necessary to construct CST equations that preserve this invariance. Since $\mathcal{C}$ transforms particles to antiparticles, the vertex function for a $q\bar{q}$ state is transformed from $\Gamma(k_1, k_2) = \Gamma(k, P)$ to $\Gamma(-k_2, -k_1) = \Gamma(P - k, P)$ (where, for simplicity, I write the result only for scalar particles). Hence, when particle 1 is on-shell the vertex function is $\Gamma(\hat{k}, P)$, and under $\mathcal{C}$ it is transformed to $\Gamma(P - \hat{k}, P)$. This amplitude has $k_2 = -\hat{k}$, and so corresponds to restricting the incoming particle 2 to is *negative* energy mass shell, and is *not* identical to $\Gamma(\hat{k}, P)$. The coupled equations shown in Fig. 8 couple these two different amplitudes together, ensuring that the system is $\mathcal{C}$ invariant. Operationally, the equations are obtained by averaging the contributions from the pole at $k_0 = E_k$ in the lower half plane and the contribution from the pole at $k_0 = -E_k + \mu$ in the upper half plane (hence a factor of $\frac{1}{2}$ is needed). Finally, if $\mu \to 0$, it is clear that none of the poles can be ignored, and in this limit a four channel equation is needed. For a detailed discussion, see Ref. [23].

Using the four channel equations, it has been shown what conditions must be met for spontaneous chiral symmetry breaking to occur [23]. If the kernel is a combination of scalar and vector interactions, so that it has the form $\mathcal{V} = V_S \mathbf{1} \otimes \mathbf{1} + V_V \gamma^\mu \otimes \gamma_\mu$, and the quark self energy is decomposed into $\Sigma(k) = A(k^2) + \not{k}\, B(k^2)$ (with $A_0 \equiv A(m^2)$, etc.), spontaneous chiral symmetry breaking can only occur if two conditions are simultaneously satisfied

$$1 = \frac{1 + m_0/A_0}{4m(1 - B_0)^2} \int \frac{d^3k}{(2\pi)^2} \frac{m}{E_k} [V_V(\hat{p}, \hat{k}) + V_V(\hat{p}, -\hat{k}) + V_S(\hat{p}, \hat{k}) + V_S(\hat{p}, -\hat{k})]$$

$$1 = \frac{1}{4m(1 - B_0)^2} \int \frac{d^3k}{(2\pi)^2} \frac{m}{E_k} [V_V(\hat{p}, \hat{k}) + V_V(\hat{p}, -\hat{k}) - V_S(\hat{p}, \hat{k}) - V_S(\hat{p}, -\hat{k})]\,. \qquad (17)$$

The first condition is a consequence of the requirement that the mass function $M(k^2)$ have a solution for a spontaneously generated mass m (so that $M(m^2) = m$) and the second must be satisfied if the four-channel CST equations are to have a solution for a pion with zero mass. If the bare quark mass m_0 is zero, the equations are identical (signaling spontaneous chiral symmetry breaking) *provided* the constraint (14) is satisfied by the scalar part of the kernel. Hence, contrary to the normal intuition, a scalar interaction is *consistent* with the requirements of spontaneous chiral symmetry breaking provided it is *purely confining*.

3 Summary and Outlook

In conclusion, the CST

- is an efficient reorganization of the Feynman series,
- is manifestly covariant,
- has a smooth nonrelativistic limit which is easily interpreted
- provides a successful and efficient dynamical description of
  - few-quark systems including spontaneous chiral symmetry breaking, quark self energies, and understanding of the meson spectrum, and
  - few-nucleon systems including the ability to provide a precise fit to the np data base and the triton binding energy without the need for three-body forces .

Shortcomings of the CST (not discussed in this talk) include

- difficulties in extending it to new systems, where unexpected singularities can sometimes pose an obstacle,
- absence of a regularization scheme other than that based on the use of form factors, and
- the lack of a clear way to drive its equations from first principals.

Acknowledgements This work was partially support by Jefferson Science Associates, LLC, under U.S. DOE Contract No. DE-AC05-06OR23177.

References

1. F. Gross, Relativistic calculation of the deuteron electromagnetic form factor I. Phys. Rev. **134**, B405 (1964)
2. F. Gross, Relativistic calculation of the deuteron electromagnetic form factor II. Phys. Rev. **136**, B140 (1965)
3. F. Gross, *Relativistic Quantum Mechanics and Field Theory* (Wiley, New York, 1993)
4. F. Gross, Three dimensional covariant integral equations for low energy systems. Phys. Rev. **186**, 1448 (1969)
5. S.J. Brodsky, G.P. Lepage, Recent advances in bound state quantum electrodynamics. In: Fourth International Colloquium on Advanced Computing Methods in Theoretical Physics, Saint Maximin, France, March 21–23, 1977 (1977)
6. M.I. Edies et al., Theory of light hydrogenlike atoms. Phys. Rep. **342**, 63–261 (2001)
7. W.P. Ford, S. Jeschonnek, J.W. Van Orden, Momentum distributions for ^{2}H(e, e'p). Phys. Rev. C **90**, 064006 (2014)
8. K.M. Maung et al., Relativistic proton nucleus scattering and one boson exchange models. Phys. Rev. C **43**, 1378–1388 (1991)

9. L.S. Celenza, C. Shaken, *Relativistic Nuclear Physics: Theories of Structure and Scattering* (World Scientific Publishers, Singapore, 1986)
10. Y. Surya, F. Gross, Unitary, gauge invariant, relativistic resonance model for pion photoproduction. Phys. Rev. C **53**, 2422 (1996)
11. S. Liuti, F. Gross, Extraction of the ratio of the neutron to proton structure functions from deep inelastic scattering. Phys. Lett. **356B**, 157 (1995)
12. F. Gross, D.O. Riska, Current conservation and interaction currents in relativistic meson theories. Phys. Rev. C **36**, 1928 (1987)
13. J.W. Van Orden, N. Devine, F. Gross, Elastic electron scattering from the deuteron using the gross equation. Phys. Rev. Lett. **75**, 4369 (1995)
14. A. Stadler, F. Gross, Relativistic calculation of the triton binding energy and its implications. Phys. Rev. Lett. **78**, 26 (1997)
15. A.N. Kvinikhidze, B. Blankleider, Gauging the three nucleon spectator equation. Phys. Rev. C **56**, 2973–2986 (1997)
16. F. Gross, A. Stadler, M.T. Peña, Relativistic calculation of the triton binding energy and its implications. Phys. Rev. C **69**, 034007 (2004)
17. F. Gross, A. Stadler, Covariant spectator theory of np scattering: phase shifts obtained from precision fits to data below 350-MeV. Phys. Rev. C **78**, 014005 (2008)
18. F. Gross, Covariant Spectator Theory of np scattering: isoscalar interaction currents. Phys. Rev. C **89**, 064001
19. F. Gross, J. Milana, Goldstone pion and other mesons using a scalar confining interaction. Phys. Rev. D **50**, 3332 (1994)
20. M. Uzzo, F. Gross, Stability of the spectator, dirac, and salpeter equations for mesons. Phys. Rev. C **59**, 1009 (1999)
21. C. Savkli, F. Gross, Quark–antiquark bound states in the relativistic spectator formalism. Phys. Rev. C **63**, 035208 (2001)
22. F. Gross, G. Ramalho, M.T. Peña, Spin and angular momentum in the nucleon. Phys. Rev. D **85**, 093006 (2012)
23. E. Biernat, F. Gross, M.T. Peña, A. Stadler, Confinement, quark mass functions, and spontaneous chiral symmetry breaking in Minkowski space. Phys. Rev. D **89**, 016005 (2014)
24. E. Biernat, F. Gross, M.T. Peña, A. Stadler, Pion electromagnetic form factor in the Covariant Spectator Theory. Phys. Rev. D **89**, 016006 (2014)
25. E. Biernat et al., Chiral symmetry and $\pi\pi$ scattering in the Covariant Spectator Theory. Phys. Rev. D **90**, 096008 (2014)
26. E. Biernat, F. Gross, M.T. Peña, A. Stadler, Charge-conjugation symmetric complete impulse approximation for the pion electromagnetic form factor in the Covariant Spectator Theory. Phys. Rev. D **92**, 076011 (2015)
27. S. Leitão, A. Stadler, M.T. Peña, E. Biernat, Covariant Spectator Theory of heavy-light and heavy mesons and the predictive power of covariant interaction kernels. Phys. Lett. B **764**, 38–41 (2017)

Few-Body Syst (2017) 58:109
DOI 10.1007/s00601-017-1266-6

Yang Li

Quarkonium Wave Functions on the Light Front

Received: 5 December 2016 / Accepted: 13 February 2017 / Published online: 11 March 2017

Abstract We study heavy quarkonium within the light-front Hamiltonian formalism. Our effective Hamiltonian is based on the holographic QCD confining potential and the one-gluon exchange interaction with a running coupling. The obtained spectra are compared with experimental measurements. We present a set of light-front wave functions, which exhibit rich structure and are consistent with the nonrelativistic picture. Finally, we use the wave functions to compute the charge and mass radii.

1 Introduction

The light-front Hamiltonian approach is poised as a unique tool to understand the non-perturbative dynamics of quantum field theory in Minkowski space-time [1]. The Hamiltonian formalism provides frame-independent wave functions that are essential for understanding the underlying structure of relativistic bound state systems. The obtained light-front wave functions (LFWFs) can be used to compute hadronic observables and distributions defined in the infinite momentum frame (IMF), and to study exclusive processes in the deep inelastic scattering (DIS), which are otherwise not easily accessible in other methods [2]. Therefore, the light-front Hamiltonian approach is complementary to Lagrangian methods formulated in Euclidean space, e.g., lattice gauge theory.

Recently we proposed a phenomenological light-front potential model for mesons based on light-front holography [3]. We applied this model to charmonium and bottomonium, and solved the system in a basis function representation. The obtained mass spectra, form factors and decay constants are compared with experiments and other established methods. The LFWFs are also used to study exclusive vector meson productions in DIS [4]. Very recently, similar work has been done within the Dyson-Schwinger/Bethe-Salpeter approach [5] and the covariant spectator theory (CST) [6].

In this paper, we propose an improvement of the model in Ref. [3] by incorporating the evolution of the coupling based on perturbative QCD (pQCD). We will show that the mass spectra are improved. Another aim of this paper is to display the LFWFs, which reveal first-hand information of the hadrons, and also help to visualize relativistic bound states on the light front. Heavy quarkonium is an ideal system to explore LFWFs, as the characteristic heavy quark velocities are small [$v \sim (0.1\text{–}0.3)$] compared to the speed of light and the familiar nonrelativistic quantum mechanical language and intuition may be applied.

This article belongs to the Topical Collection "Light Cone 2016".

Y. Li (✉)
Department of Physics and Astronomy, Iowa State University, Ames, IA 50011, USA
E-mail: leeyoung@iastate.edu

2 Formalism

The effective Hamiltonian for the model reads [3],

$$H_{\rm eff} = \frac{\mathbf{k}_\perp^2 + m_q^2}{x} + \frac{\mathbf{k}_\perp^2 + m_{\bar{q}}^2}{1-x} + \kappa^4 \boldsymbol{\zeta}_\perp^2 - \frac{\kappa^4}{(m_q + m_{\bar{q}})^2}\partial_x\big(x(1-x)\partial_x\big)$$
$$- \frac{C_F 4\pi\alpha_s(Q^2)}{Q^2}\bar{u}_{s'}(k')\gamma_\mu u_s(k)\bar{v}_{\bar{s}}(\bar{k})\gamma^\mu v_{\bar{s}'}(\bar{k}'). \quad (1)$$

where $Q^2 = -\bar{q}^2 = -(1/2)(k'-k)^2 - (1/2)(\bar{k}'-\bar{k})^2$ is the average 4-momentum squared carried by the exchanged gluon. The confining term $\kappa^4\boldsymbol{\zeta}_\perp^2$ comes from the "soft-wall" light-front holography, κ is the strength of the confinement, and $\boldsymbol{\zeta}_\perp \equiv \sqrt{x(1-x)}\mathbf{r}_\perp$ is Brodsky and de Téramond's holographic variable [7]. $\boldsymbol{\zeta}_\perp$ and x are taken as the independent variables, $\partial_x f(x, \boldsymbol{\zeta}_\perp) = \partial f(x, \boldsymbol{\zeta}_\perp)/\partial x|_{\zeta_\perp}$. $C_F = (N_c^2-1)/(2N_c) = 4/3$ is the color factor. m_q ($m_{\bar{q}}$) is the mass of the quark (anti-quark).

The longitudinal confinement comes from completing the transverse holographic confinement in the finite mass case and is consistent with the pQCD asymptotics of the distribution amplitude $x^\beta(1-x)^\alpha$. The longitudinal confining strength is fixed in the nonrelativistic limit hence no additional parameter is introduced.

In Ref. [3], the strong coupling α_s is assumed fixed for each system, say, charmonium, whereas for different systems, the values of α_s are different and are related through the pQCD evolution. While such treatment is reasonable (indeed traditional, see e.g. [8]), as the mass difference between states in the same system, e.g. J/ψ and $\psi(2S)$, is small compared to their respective masses, nevertheless, the inclusion of the running coupling implements important QCD physics. Another motivation is to improve the rotational symmetry and the hyperfine structure. In the effective one-gluon exchange kernel, [9–11], a non-covariant UV counterterm is introduced to regularize the UV divergence appearing in the box diagram [9,12], which further spoils the rotational symmetry as will be seen in the mass spectra below. The introduction of the running coupling changes the UV asymptotics of the one-gluon exchange kernel.[1] As a result, the counterterm is not necessary anymore. While rotational symmetry is not fully restored, as we will see, the breaking of rotational symmetry is moderated, and the hyperfine structure is significantly improved.

In the effective one-gluon exchange interaction Eq. (1), the 4-momentum squared of the exchanged gluon Q^2 naturally appears. The running coupling can be introduced as a function of Q^2:

$$\alpha_s(Q^2) = \frac{\alpha_s(M_Z^2)}{1 + \alpha_s(M_Z^2)\beta_0 \ln(\mu_{\rm IR}^2 + Q^2)/(\mu_{\rm IR}^2 + M_Z^2)}. \quad (2)$$

where $N_f = 4$ for charmonium and $N_f = 5$ for bottomonium. Here $\mu_{\rm IR}$ is an IR cutoff introduced to avoid the pQCD IR catastrophe. Similar forms are used in the literature, e.g. Ref. [13]. In practice, we choose $\alpha_s(0) = 0.6$. We find the spectrum is not sensitive for the choice of $\alpha_s(0)$ within the range $0.4 \le \alpha_s(0) \le 0.8$.

The state vectors, hence the LFWFs, are obtained from diagonalizing the Light Cone Hamiltonian operator $\hat{H}_{\rm LC} \equiv P^+\hat{P}^- - \mathbf{P}_\perp^2$:

$$H_{\rm LC}|\psi_h(P^+, \mathbf{P}_\perp, j, m_j)\rangle = M_h^2|\psi_h(P^+, \mathbf{P}_\perp, j, m_j)\rangle. \quad (3)$$

The state vectors can be represented in the Fock space:

$$|\psi_h(P^+, \mathbf{P}_\perp, j, m_j)\rangle = \sum_{s,\bar{s}}\int_0^1 \frac{{\rm d}x}{2x(1-x)}\int \frac{{\rm d}^2\mathbf{k}_\perp}{(2\pi)^3}\psi_{s\bar{s}/h}^{(m_j)}(\mathbf{k}_\perp, x)$$
$$\times \frac{1}{\sqrt{N_c}}\sum_{i=1}^{N_c} b_{si}^\dagger(xP^+, \mathbf{k}_\perp + x\mathbf{P}_\perp)d_{\bar{s}i}^\dagger((1-x)P^+, -\mathbf{k}_\perp + (1-x)\mathbf{P}_\perp)|0\rangle. \quad (4)$$

Here j and m_j are the intrinsic spin and the spin projection of the particle. The coefficient of the expansion, $\psi_{s\bar{s}/h}^{(m_j)}(\mathbf{k}_\perp, x)$, is the (two-body) LFWF, which is normalized as,

$$\sum_{s,\bar{s}}\int_0^1 \frac{{\rm d}x}{2x(1-x)}\int \frac{{\rm d}^2\mathbf{k}_\perp}{(2\pi)^3}\psi_{s\bar{s}/h'}^{(m_j')*}(\mathbf{k}_\perp, x)\psi_{s\bar{s}/h}^{(m_j)}(\mathbf{k}_\perp, x) = \delta_{hh'}\delta_{m_j, m_j'}. \quad (5)$$

[1] This is not to say, however, that now the UV asymptotics of the one-gluon exchange kernel is consistent with the perturbation theory. But the UV asymptotics is not easily studied in the basis representation and the investigation of the UV asymptotics is beyond the scope of this paper.

Table 1 Summary of the model parameters

	N_f	μ_g (GeV)	κ (GeV)	m_q (GeV)	$\langle M_{\rm cal} - M_{\rm pdg}\rangle$	$\langle \delta M_{m_j}\rangle$	# states	$N_{\max} = L_{\max}$
$c\bar{c}$	4	0.02	0.985	1.570	41 MeV	15 MeV	8	8
			0.979	1.587	32	21		16
			0.972	1.596	31	17		24
			0.966	1.603	31	17		32
$b\bar{b}$	5	0.02	1.387	4.894	48 MeV	6 MeV	14	8
			1.392	4.899	41	6		16
			1.390	4.901	39	6		24
			1.389	4.902	38	8		32

μ_g is the gluon mass, used to regularize the integrable Coulomb singularity. $\langle M_{\rm cal} - M_{\rm pdg}\rangle$ represents the r.m.s. deviation of the masses from the Particle Data Group values for states below the threshold. $\langle \delta M_{m_j}\rangle$ represents the r.m.s. value of the mass spreads for the same state with different magnetic projection m_j's

It is useful to introduce LFWFs in the transverse coordinate space:

$$\widetilde{\psi}_{s\bar{s}}(\mathbf{r}_\perp, x) \equiv \frac{1}{\sqrt{x(1-x)}} \int \frac{\mathrm{d}^2 k_\perp}{(2\pi)^2} e^{i\mathbf{k}_\perp \cdot \mathbf{r}_\perp} \psi_{s\bar{s}}(\mathbf{k}_\perp, x). \tag{6}$$

Following Ref. [3], the effective Hamiltonian Eq. (1) was solved in a basis function approach. The LFWFs then are represented as,

$$\begin{aligned} \psi_{ss'/h}(\mathbf{k}_\perp, x) &= \sum_{n,m,l} \psi_h(n,m,l,s,s')\, \phi_{nm}(\mathbf{k}_\perp/\sqrt{x(1-x)})\chi_l(x); \\ \widetilde{\psi}_{ss'/h}(\mathbf{r}_\perp, x) &= \sqrt{x(1-x)} \sum_{n,m,l} \psi_h(n,m,l,s,s')\, \widetilde{\phi}_{nm}(\sqrt{x(1-x)}\mathbf{r}_\perp)\chi_l(x). \end{aligned} \tag{7}$$

Here the coefficients $\psi_h(n,m,l,s,s')$ are obtained from diagonalization. Basis function ϕ_{nm} and χ_l are the solutions of the Hamiltonian without the one-gluon exchange interaction, i.e., the kinetic energy plus the confinement terms. Specifically, $\widetilde{\phi}_{nm}(\boldsymbol{\zeta}_\perp)$ is the harmonic oscillator function in the holographic variables. $\chi_\ell(x)$ is the Jacobi polynomials weighed by $x^{\frac{\beta}{2}}(1-x)^{\frac{\alpha}{2}}$, where $\alpha = 2m_{\bar{q}}(m_q + m_{\bar{q}})/\kappa^2$, $\beta = 2m_q(m_q + m_{\bar{q}})/\kappa^2$.

In practical calculations, the basis is truncated and LFWFs have a corresponding finite set of terms in Eq. (7). Observables are evaluated from finite-dimensional representations and, in principle, have to be extrapolated to the complete basis. Following Ref. [3], the basis are truncated according to:

$$2n + |m| + 1 \le N_{\max}, \quad 0 \le l \le L_{\max}. \tag{8}$$

3 Results

The model parameters are summarized in Table 1. In particular, the confining strength κ and the effective quark mass m_q are obtained from fitting to the PDG mass spectrum below the continuum thresholds. The reconstructed charmonium and bottomonium spectra are shown in Fig. 1. The r.m.s. deviation of the masses from the PDG values are 31 MeV and 38 MeV for charmonia and bottomonia below the thresholds, respectively. These r.m.s. deviations are significantly reduced ($\sim$50% for charmonium, $\sim$25% for bottomonium) from the fixed coupling results reported in Ref. [3]. Our spectroscopy is competitive with those obtained from other methods, e.g. [5,6,14,15]. Not only are the mass eigevalues improved, their dependence on m_j, an indicator for the violation of the rotational symmetry, is also weakened. For example, the mean mass spreads, i.e. the mean heights of the boxes in Fig. 1, are 17 MeV and 8 MeV for charmonia and bottomonia below the thresholds, respectively.

Wave functions provide full information of the system. Despite of their significance in hadron physics, LFWFs, especially those of the excited states, are rarely found in the literature (cf. [16–18]). We present representative quarkonium LFWFs in Figs. 2, 3 and 4.

Figure 2 shows the coordinate space LFWFs for vector meson J/ψ. For each spin alignment, the orbital angular momentum projection $m_\ell = \lambda - s_1 - s_2$ is definite ($\lambda \equiv m_j$). Hereafter, we drop the global phases $\exp(i m_\ell \theta)$ related to the orbital angular momentum, while retaining the relative sign for negative values of $r_\perp$.

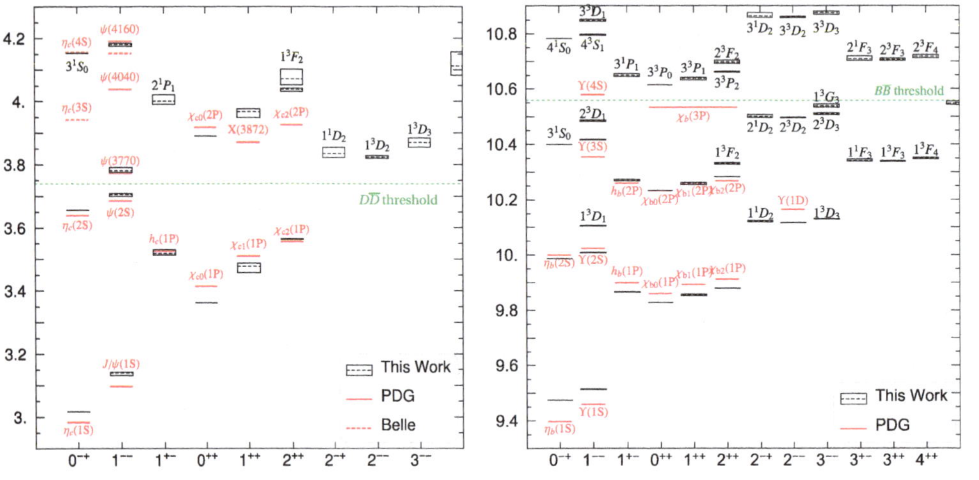

Fig. 1 The reconstructed charmonium (*left*) and bottomonium (*right*) spectra at $N_{\max} = L_{\max} = 32$. The horizontal axes marks the quantum numbers J^{PC}. The vertical axes marks the mass eigenvalues in GeV. The parameters are listed in Table 1. The r.m.s. deviation of the masses from the PDG values are 31 and 38 MeV for charmonium and bottomonium, respectively. The mean mass spreads, i.e. the mean heights of the boxes, are 17 and 8 MeV for charmonium and bottomonium, respectively

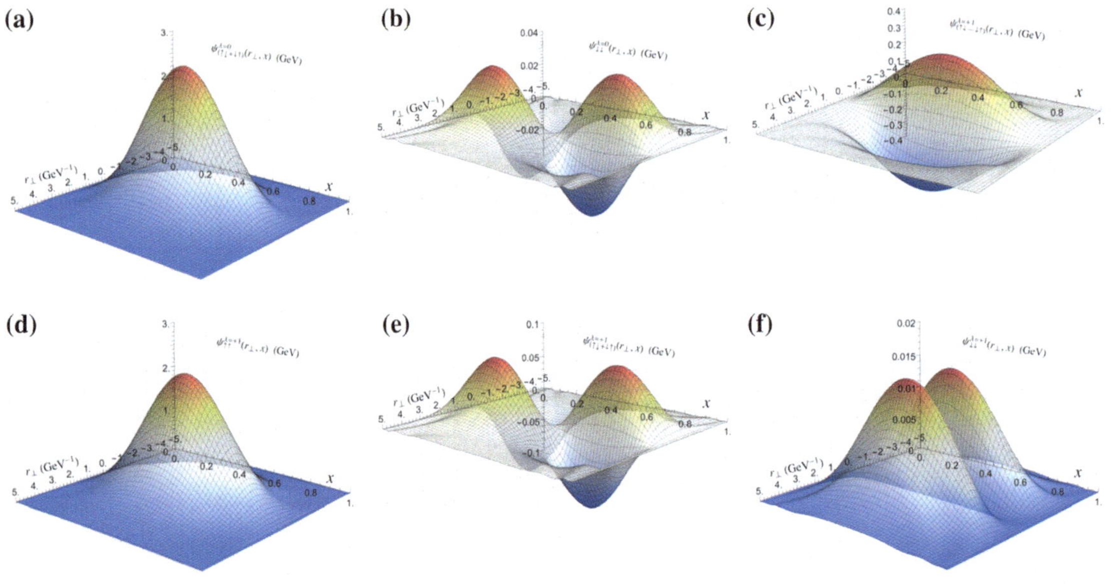

Fig. 2 (Color online) The J/ψ LFWFs plotted in 3D for polarizations $\lambda = 0$ (**a**, **b**), and $\lambda = +1$ (**c**–**f**), for various spin alignments. **a** $\widetilde{\psi}^{\lambda=0}_{\uparrow\downarrow+\downarrow\uparrow}(\mathbf{r}_\perp, x)$, **b** $\widetilde{\psi}^{\lambda=0}_{\downarrow\downarrow}(\mathbf{r}_\perp, x)$, **c** $\widetilde{\psi}^{\lambda=+1}_{\uparrow\downarrow-\downarrow\uparrow}(\mathbf{r}_\perp, x)$, **d** $\widetilde{\psi}^{\lambda=+1}_{\uparrow\uparrow}(\mathbf{r}_\perp, x)$, **e** $\widetilde{\psi}^{\lambda=+1}_{\uparrow\downarrow+\downarrow\uparrow}(\mathbf{r}_\perp, x)$, **f** $\widetilde{\psi}^{\lambda=+1}_{\downarrow\downarrow}(\mathbf{r}_\perp, x)$

It is apparent that $\psi^{\lambda=0}_{\uparrow\downarrow+\downarrow\uparrow}$ (Fig. 2a) and $\psi^{\lambda=+1}_{\uparrow\uparrow}$ (Fig. 2d) are the leading LFWFs for polarization $\lambda = 0$ and $\lambda = +1$, respectively. And they are almost identical in magnitude. Thus the nonrelativistic picture emerges that J/ψ is an S-wave in direct product with a triplet spin configuration. Subleading LFWFs also exist (Figs. 2b, c, e, f) due to relativity. For example, $\psi^{\lambda=+1}_{\downarrow\downarrow}$ (Fig. 2f) resembles a D-wave function ($\ell = 2, m_\ell = 2$), as a result of the S-D mixing.

Figure 3 compares the leading components of the vector meson J/ψ with its "radial" excitation $\psi(2S)$ and "angular" excitation $\psi(1D)$. Rich details emerge in the excited states. The $\psi(2S)$ LFWFs show nodes in both the transverse radial direction ($r_\perp$) and the longitudinal direction (x), consistent with the nonrelativistic interpretation, where the radial excitation is homogeneous in all 3 directions. LFWFs on the top panel (bottom panel) are with the polarization $\lambda = 0$ ($\lambda = +1$). While the leading components of a S-wave vector meson, e.g., J/ψ and $\psi(2S)$, for the all polarizations are almost identical (Fig. 3a vs. d and b vs. e), those of the D-wave vector meson $\psi(1D)$ are dramatically different (Fig. 3c vs. f). However, this is consistent with the nonrelativistic picture that the wave function is dominated by the D-wave ($\ell = 2$). It is easy to see that the leading components

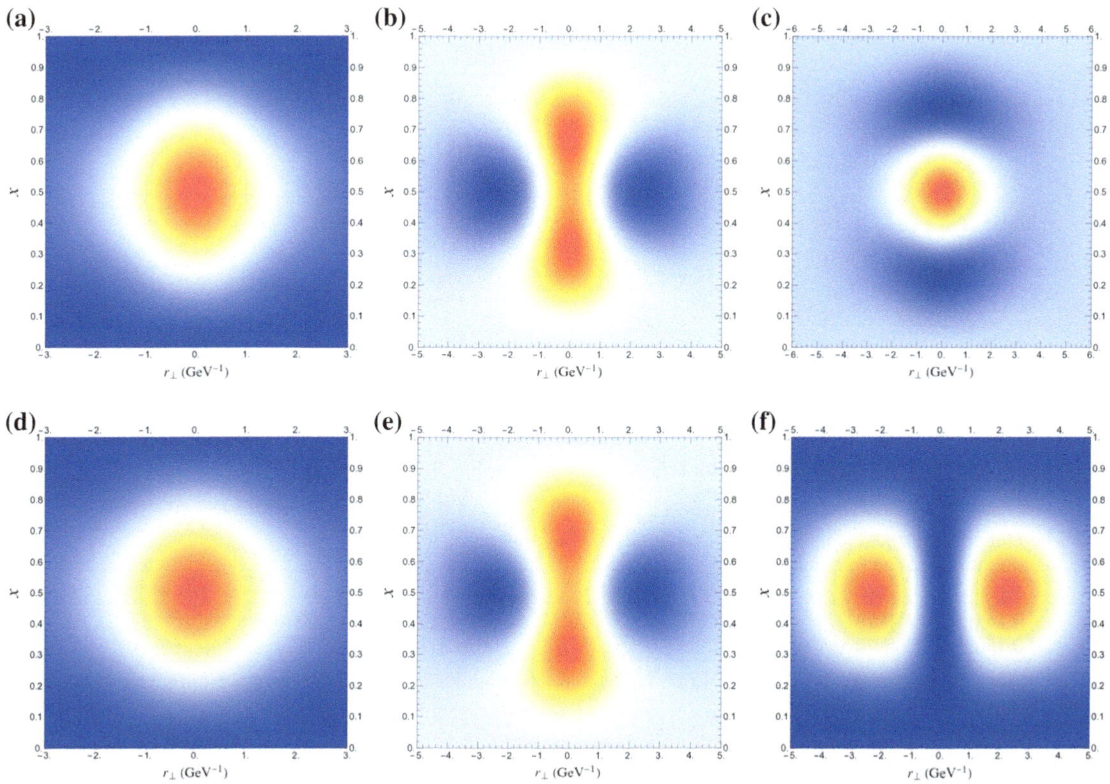

Fig. 3 (Color online) Leading components of the vector meson J/ψ and its radial excitation $\psi(2S)$ and angular excitation $\psi(1D)$ for polarization $\lambda = 0$ (**a–c**) and $\lambda = +1$ (**d–f**). The color scheme is the same as Fig. 2. **a** J/ψ: $\widetilde{\psi}^{\lambda=0}_{\uparrow\downarrow+\downarrow\uparrow}(\mathbf{r}_\perp, x)$, **b** $\psi(2S)$: $\widetilde{\psi}^{\lambda=0}_{\uparrow\downarrow+\downarrow\uparrow}(\mathbf{r}_\perp, x)$, **c** $\psi(1D)$: $\widetilde{\psi}^{\lambda=0}_{\uparrow\downarrow+\downarrow\uparrow}(\mathbf{r}_\perp, x)$, **d** J/ψ: $\widetilde{\psi}^{\lambda=+1}_{\uparrow\uparrow}(\mathbf{r}_\perp, x)$, **e** $\psi(2S)$: $\widetilde{\psi}^{\lambda=+1}_{\uparrow\uparrow}(\mathbf{r}_\perp, x)$, **f** $\psi(1D)$: $\widetilde{\psi}^{\lambda=+1}_{\downarrow\downarrow}(\mathbf{r}_\perp, x)$

of $\psi(1D)$ are $\widetilde{\psi}^{\lambda=0}_{\uparrow\downarrow+\downarrow\uparrow}(\mathbf{r}_\perp, x)$ $(m_\ell = 0)$ and $\widetilde{\psi}^{\lambda=+1}_{\downarrow\downarrow}(\mathbf{r}_\perp, x)$ $(m_\ell = 2)$. The apparent difference between the two polarizations in fact resembles the difference between the D-wave spherical harmonics $Y_2^0(\hat{r})$ and $Y_2^2(\hat{r})$.

Figure 4 presents the LFWFs of $\eta_b(3S)$. The wave functions for this highly excited state reveal complicated inner structure. The leading components $\psi^{\lambda=0}_{\uparrow\downarrow-\downarrow\uparrow}(\hat{r}_\perp, x)$ has additional nodes in both the transverse and the longitudinal directions compared with 2S-wave states and is presumably dominated by a $3S$-wave. The subleading LFWF $\psi^{\lambda=0}_{\uparrow\downarrow-\downarrow\uparrow}(\mathbf{r}_\perp, x)$ also exhibits muliple nodes and may contain both radial and angular excitations.

Another way of visualizing relativistic bound states, in connection with the experiments, is to use the impact parameter generalized parton distributions (GPDs) [19]. These quantities are closely related to the LFWFs [20]. In fact, they are just the squared LFWFs in the coordinate space, inserting gauge links as necessary. The impact parameter GPDs for a similar system—positronium—have been shown in Ref. [21].

4 Charge and Mass Radii

We use the LFWFs to study the the root-mean squared (r.m.s.) charge and the mass radii. In relativistic dynamics, the r.m.s. radii are defined from the corresponding form factors (electromagnetic form factor F_c and the gravitational form factor F_g) which in turn are related to the hadron matrix elements of conserved currents (electromagnetic current J^μ and the energy-momentum tensor $T^{\mu\nu}$):

$$\langle r_c^2 \rangle \equiv -6 \frac{\partial}{\partial q^2} F_c(q^2)\big|_{q\to 0}, \qquad \langle r_m^2 \rangle \equiv -6 \frac{\partial}{\partial q^2} F_g(q^2)\big|_{q\to 0}. \tag{9}$$

For the charge form factor, we only couple the probing photon to the quark. In light-front dynamics, it can be shown (see "Appendix") that the r.m.s. radii are related to the impact parameter $\mathbf{b}_\perp \equiv (1-x)\mathbf{r}_\perp$ [19] and the holographic variable $\boldsymbol{\zeta}_\perp \equiv \sqrt{x(1-x)}\mathbf{r}_\perp$ [7]:

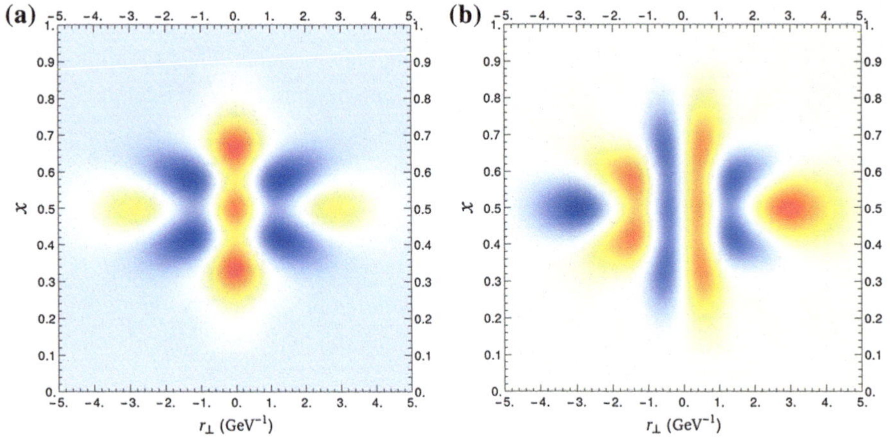

Fig. 4 (Color online) LFWFs of $\eta_b(3S)$. The color scheme is the same as Fig. 2. **a** $\widetilde{\psi}^{\lambda=0}_{\uparrow\downarrow-\downarrow\uparrow}(\mathbf{r}_\perp, x)$, **b** $\widetilde{\psi}^{\lambda=0}_{\downarrow\downarrow}(\mathbf{r}_\perp, x)$

Table 2 The r.m.s. charge and the mass radii of charmonia and bottomonia

(fm)	η_c	χ_{c0}	$\eta_c(2S)$	η_b	χ_{b0}	$\eta_b(2S)$	$\chi_{b0}(2P)$	$\eta_b(3S)$
Charge	0.165(7)	0.231(7)	0.3468(8)	0.109(1)	0.175(3)	0.2235(4)	0.267(3)	0.306(2)
Mass	0.150(6)	0.209(6)	0.332(2)	0.107(4)	0.171(1)	0.221(2)	0.263(1)	0.302(2)

Results are extrapolated from $N_{\max} = 8, 16, 24, 32$. Numerical uncertainties are estimated from the difference between the extrapolated values and the $N_{\max} = 32$ values

$$\langle r_c^2 \rangle = \frac{3}{2} \sum_{s,s'} \int_0^1 \frac{\mathrm{d}x}{4\pi} \int \mathrm{d}^2 r_\perp \, (1-x)^2 \mathbf{r}_\perp^2 \, \widetilde{\psi}^{(\lambda)*}_{ss'/h}(\mathbf{r}_\perp, x) \widetilde{\psi}^{(\lambda)}_{ss'/h}(\mathbf{r}_\perp, x) \equiv \frac{3}{2} \langle \mathbf{b}_\perp^2 \rangle; \tag{10}$$

$$\langle r_m^2 \rangle = \frac{3}{2} \sum_{s,s'} \int_0^1 \frac{\mathrm{d}x}{4\pi} \int \mathrm{d}^2 r_\perp \, x(1-x) \mathbf{r}_\perp^2 \, \widetilde{\psi}^{(\lambda)*}_{ss'/h}(\mathbf{r}_\perp, x) \widetilde{\psi}^{(\lambda)}_{ss'/h}(\mathbf{r}_\perp, x) \equiv \frac{3}{2} \langle \boldsymbol{\zeta}_\perp^2 \rangle. \tag{11}$$

The results are summarized in Table 2. Note that the difference between $\langle r_c^2 \rangle$ and $\langle r_m^2 \rangle$ is a relativistic effect.

5 Summary and Outlook

The light-front Hamiltonian formalism provides a natural framework to understand strongly interacting relativistic bound states. We demonstrate this in a model that implements a holographic confinement and a one-gluon exchange interaction in the heavy quarkonium systems. We update the previous results by incorporating the evolution of the strong interaction coupling constant. The resulting spectroscopy is compared favorably with the experiments. We present the wave functions, which bridge the nonrelativistic picture and the quantum field theoretical treatment of heavy quarkonium. They offer a unique way to visualize and to understand the structure of the relativistic bound states.

Our model can be applied to other mesons, e.g., heavy-light and light mesons, which is in progress. Another important improvement is to include the self-energy correction [23,24], which will allow us to go beyond the light-front potential model treatment and to build a consistent effective light-front model for hadrons.

There are several promising ways under development to go beyond the current phenomenological model. The first one is the systematic Fock sector expansion with a non-perturbative sector dependent renormalization [25,26]. The second one is the full configuration interaction simulation, with or without the coupled cluster technique, of the light-front QCD Hamiltonian within the basis function approach [27,28]. The third line of investigation is to use the effective Hamiltonians obtained from a similarity renormalization group with a systematic perturbative expansion [29,30]. Last but not least, progress in the holographic approach to QCD may also provide new insights. The current model can be linked to all these lines of investigation. The physics learned from the current model, and the techniques employed in our model, may be helpful to explore those more ambitious approaches.

Acknowledgements I wish to thank the organizers of the Light Cone 2016 conference. I acknowledge valuable discussions with J. P. Vary, P. Maris, K. Tuchin, S. J. Brodsky, S. D. Głazek, X. Zhao, G. Chen, M. Gomez-Rocha, and S. Leitão. I also want to thank the hospitality of the High Energy Nuclear Theory Group at the Institute of Modern Physics, Chinese Academy of Sciences, Lanzhou, China, where the work is being completed. This work was supported in part by the Department of Energy under Grant Nos. DE-FG02-87ER40371 and DESC0008485 (SciDAC-3/NUCLEI). Computational resources were provided by the National Energy Research Supercomputer Center (NERSC), which is supported by the Office of Science of the U.S. Department of Energy under Contract No. DE-AC02-05CH11231.

Appendix: Derivation of Eqs. (10–11)

We used the identity:

$$\nabla^2_{q_\perp} = \frac{\partial^2}{\partial q_\perp^2} + \frac{d-1}{q_\perp}\frac{\partial}{\partial q_\perp} - \frac{1}{q_\perp^2}L_z^2 = 4t\frac{\partial^2}{\partial t^2} + 2d\frac{\partial}{\partial t} - \frac{1}{t}L_z^2 \tag{12}$$

where $d = 2$ is the spatial dimension, and $t \equiv q_\perp^2$. The angular momentum operator $L_z = i\partial/\partial\phi$. The two-dimensional Laplacian has a different coefficient from the three-dimensional one, which results the factor 3/2. At $t \to 0$, the first term $4t\partial^2/\partial t^2$ vanishes. The form factor does not have angle dependence, so, the third term vanishes as well. Then, at $t \to 0$,

$$\langle r^2 \rangle = -6\frac{\partial}{\partial t}F(t)\Big|_{t\to 0} = -\frac{3}{2}\nabla^2_{q_\perp} I_{\lambda\lambda'}(\mathbf{q}_\perp)\Big|_{q_\perp \to 0}, \tag{13}$$

where $I_{\lambda\lambda'}(\mathbf{q}_\perp)$ is the helicity amplitude. In the LFWF representation [22],

$$I^{\text{em}}_{\lambda\lambda'}(\mathbf{q}_\perp) = \sum_{s,s'} \int_0^1 \frac{\mathrm{d}x}{2x(1-x)} \int \frac{\mathrm{d}^2 k_\perp}{(2\pi)^3} \psi^{(\lambda')*}_{ss'/h}(\mathbf{k}_\perp + (1-x)\mathbf{q}_\perp, x)\psi^{(\lambda)}_{ss'/h}(\mathbf{k}_\perp, x), \tag{14}$$

$$\begin{aligned} I^{\text{gr}}_{\lambda\lambda'}(\mathbf{q}_\perp) = &\sum_{s,s'} \int_0^1 \frac{\mathrm{d}x}{2x(1-x)} \int \frac{\mathrm{d}^2 k_\perp}{(2\pi)^3} \Big[x\, \psi^{(\lambda')*}_{ss'/h}(\mathbf{k}_\perp + (1-x)\mathbf{q}_\perp, x)\psi^{(\lambda)}_{ss'/h}(\mathbf{k}_\perp, x) \\ &+ (1-x)\, \psi^{(\lambda')*}_{ss'/h}(\mathbf{k}_\perp - x\mathbf{q}_\perp, x)\psi^{(\lambda)}_{ss'/h}(\mathbf{k}_\perp, x)\Big]. \end{aligned} \tag{15}$$

Equations (10–11) immediately follow.

References

1. B.L.G. Bakker et al., Nucl. Phys. Proc. Suppl. **251–252**, 165 (2014)
2. G.P. Lepage, S.J. Brodsky, Phys. Rev. D **22**, 2157 (1980)
3. Yang Li, P. Maris, X. Zhao, J.P. Vary, Phys. Lett. B **758**, 118 (2016)
4. G. Chen, Y. Li, P. Maris, K. Tuchin, J.P. Vary, arXiv:1610.04945 [nucl-th]
5. T. Hilger, C. Popovici, M. Gomez-Rocha, A. Krassnigg, Phys. Rev. D **91**, 034013 (2015)
6. S. Leitão, A. Stadler, M.T. Peña, E.P. Biernat, Phys. Lett. B **764**, 38 (2017)
7. S.J. Brodsky, G.F. de Teramond, H.G. Dosch, J. Erlich, Phys. Rep. **584**, 1 (2015)
8. D. Griffiths, Introduction to Elementary Particles, 2nd revised ed. (Wiley, New York). ISBN: 978-3-527-40601-2
9. M. Krautgärtner, H.C. Pauli, F. Wölz, Phys. Rev. D **45**, 3755 (1992)
10. H. Lamm, R.F. Lebed, J. Phys. G **41**, 125003 (2014)
11. P. Wiecki, Yang Li, X. Zhao, P. Maris, J.P. Vary, Phys. Rev. D **91**, 105009 (2015)
12. M. Mangin-Brinet, J. Carbonell, V.A. Karmanov, Phys. Rev. C **68**, 055203 (2003)
13. D. Atkinson, P.W. Johnson, Phys. Rev. D **37**, 2296 (1988)
14. H.W. Crater, J. Schiermeyer, Phys. Rev. D **82**, 094020 (2010)
15. J.R. Spence, J.P. Vary (**in preparation**)
16. S.D. Głazek, A. Harindranath, S.S. Pinsky, J. Shigemitsu, K.G. Wilson, Phys. Rev. D **47**, 1599 (1993)
17. U. Trittmann, H.-C. Pauli, MPI H-V4-1997, arXiv:hep-th/9704215
18. D. Chakrabarti, A. Harindranath, Phys. Rev. D **64**, 105002 (2001)
19. M. Burkardt, Phys. Rev. D **62**, 071503 (2000), Erratum: ibid. **66**, 119903 (2002)
20. M. Diehl, Phys. Rep. **388**, 41 (2003)
21. L. Adhikari, Yang Li, X. Zhao, P. Maris, J.P. Vary, A.A. El-Hady, Phys. Rev. C **93**, 055202 (2016)
22. S.J. Brodsky, D.S. Hwang, B. Ma, I. Schmidt, Nucl. Phys. B **593**, 311 (2001)
23. V.A. Karmanov, J.-F. Mathiot, A.V. Smirnov, Phys. Rev. D **69**, 045009 (2004)
24. V.A. Karmanov, J.-F. Mathiot, A.V. Smirnov, Phys. Rev. D **75**, 045012 (2007)

25. V.A. Karmanov, J.-F. Mathiot, A.V. Smirnov, Phys. Rev. D **77**, 085028 (2008)
26. Yang Li, V.A. Karmanov, P. Maris, J.P. Vary, Phys. Lett. B **748**, 278 (2015)
27. J.P. Vary et al., Phys. Rev. C **81**, 035205 (2010)
28. S.S. Chabysheva, J.R. Hiller, Phys. Lett. B **711**, 417 (2012)
29. M. Gómez-Rocha, S.D. Głazek, Phys. Rev. D **92**, 065005 (2015)
30. E.L. Gubankova, F. Wegner, Phys. Rev. D **58**, 025012 (1998)

Few-Body Syst (2017) 58:91
DOI 10.1007/s00601-017-1251-0

Sofia Leitão · Alfred Stadler · M. T. Peña · Elmar P. Biernat

Application of the Covariant Spectator Theory to the Study of Heavy and Heavy-Light Mesons

Received: 18 December 2016 / Accepted: 27 January 2017 / Published online: 16 February 2017

Abstract As an application of the Covariant Spectator Theory (CST) we calculate the spectrum of heavy-light and heavy-heavy mesons using covariant versions of a linear confining potential, a one-gluon exchange, and a constant interaction. The CST equations possess the correct one-body limit and are therefore well-suited to describe mesons in which one quark is much heavier than the other. We find a good fit to the mass spectrum of heavy-light and heavy-heavy mesons with just three parameters (apart from the quark masses). Remarkably, the fit parameters are nearly unchanged when we fit to experimental pseudoscalar states only or to the whole spectrum. Because pseudoscalar states are insensitive to spin-orbit interactions and do not determine spin-spin interactions separately from central interactions, this result suggests that it is the covariance of the kernel that correctly predicts the spin-dependent quark-antiquark interactions.

1 Introduction

At low energies and large distances quarks and gluons interact strongly, and therefore the standard perturbative methods cannot be used for a meaningful description. In order to study the most striking features of the strong interaction, dynamical chiral-symmetry breaking and color confinement, one needs to employ nonperturbative tecniques.

In this work we concentrate on mesons that can be described as strongly-bound states of one quark and one antiquark. The physics of mesons is a very active field of research, especially due to the vast amount of data currently being collected in experimental facilities such as the LHC, BaBaR, Belle, and CLEO. In the near future, exciting results are also expected from Jefferson Lab (GlueX) and FAIR (PANDA). Theoretical predictions are therefore important not only to guide the identification of new states—some of them with exotic non-$q\bar{q}$ content [1]—but also to calculate other observables, such as form factors, decay rates, etc., important for the study of the structure of mesons.

In the recent years, lattice QCD approaches have made impressive progress, providing us with a large amount of valuable predictions for physical observables [2]. At the same time, non-perturbative continuum methods (for a good review see Ref. [3]) have attracted attention with their potential of providing a deeper understanding of QCD in the infrared regime from information that cannot be extracted from lattice data alone. Very recently, Hamiltonian approaches with a phenomenological confinement obtained from light-front

This article belongs to the Topical Collection "Light Cone 2016".

S. Leitão (✉)· A. Stadler · M. T. Peña · E. P. Biernat
CFTP, Instituto Superior Técnico, Universidade de Lisboa, Av. Rovisco Pais 1, 1049-001 Lisboa, Portugal
E-mail: sofia.leitao@tecnico.ulisboa.pt

A. Stadler
Departamento de Física da Universidade de Évora, 7000-671 Évora, Portugal

holographic QCD [4] and renormalization-group procedures for effective particles [5] have been used to study heavy quarkonium.

Our approach, the Covariant Spectator Theory (CST) [6], is close in spirit to the Dyson–Schwinger/ Bethe–Salpeter (DSBS) formalism [7,8]. They both aim at a quantum-field-theoretical description where the one- and two-body dynamics are treated self-consistently. Unlike DSBS, CST works directly in Minkowski space. In addition, the two-body CST equation sums the infinite series of all ladder and crossed-ladder exchange diagrams more efficiently than the ladder Bethe–Salpeter equation (BSE), due to important cancellations that occur when the mass of one of the constituent particles becomes large. It has been proven that, in a scalar theory and in the limit of the heavy mass tending to infinity, these cancellations even become exact, and that the CST equation therefore gives the exact result (for more details, see Ref. [9]).

Further virtues of the CST worth mentioning include:

- Meson wave functions are given in terms of covariant vertex functions which have simple transformation properties under Lorentz boosts.
- CST equations are manifestly covariant, but nevertheless require only three-dimensional loop integrations.
- The two-body CST equation reduces in the one-body limit to the Dirac equation, and in the nonrelativistic limit to the Schödinger equation.

This paper is organized as follows: In Sect. 2 we introduce the formalism, in Sect. 3 we present the results and discussion, and in Sect. 4 we conclude with a summary and an outlook.

2 Formalism

In order to derive the CST set of equations we start with the BSE for the quark-antiquark vertex function $\Gamma_{BS}(p_1, p_2)$ with an irreducible interaction kernel $\mathcal{V}(p, k; P)$, where P is the total four-momentum, and p and k are the external and internal relative four-momenta, respectively. The BSE is given by

$$\Gamma_{BS}(p_1, p_2) = i \int \frac{d^4k}{(2\pi)^4} \mathcal{V}(p, k; P) S_1(k_1)\, \Gamma_{BS}(k_1, k_2)\, S_2(k_2)\,, \tag{1}$$

where $S_i(k_i)$ is the dressed quark propagator depending on the individual four-momentum k_i of quark i. In the CST, the heavier quark, say quark 1 with mass m_1, is on-mass-shell. This yields the CST equation for the vertex function Γ_{1CS}, where "1CS" or "1CSE" stands for one-channel spectator equation [10]. More specifically, the 1CSE results from the BSE by keeping in the k_0-contour integration only the contribution from the residue of the positive-energy pole of the quark 1 propagator. When all quark pole contributions are included in the k_0-contour integration this leads to a coupled set of CST equations, depicted diagrammatically in Fig. 1. For the heavy and heavy-light systems the 1CSE is a good approximation [11], as it retains the most important properties of the complete set of CST equations, i.e. manifest covariance, cluster separability, and the correct one-body and nonrelativistic limits. It is also a good approximation for equal-mass particles, as long as the bound-state mass is large and of the order of the sum of the quark masses. However, a property the 1CSE does not maintain, in general, is charge-conjugation symmetry. Therefore, states calculated with the 1CSE are not expected to be C-parity eigenstates. In principle, this problem is easily remedied by using the set of two-channel CST equations inside the dashed rectangle of Fig. 1 instead.

The 1CSE reads

$$\Gamma_{1CS}(\hat{p}_1, p_2) = -\int \frac{d^3k}{(2\pi)^3} \frac{m_1}{E_{1k}} \sum_K V_K(\hat{p}_1, \hat{k}_1) \Theta_1^{K(\mu)} \frac{m_1 + \hat{\not{k}}_1}{2m_1} \Gamma_{1CS}(\hat{k}_1, k_2) \frac{m_2 + \not{k}_2}{m_2^2 - k_2^2 - i\epsilon} \Theta^K_{2(\mu)}\,, \tag{2}$$

where $\Theta_i^{K(\mu)} = \mathbf{1}_i, \gamma_i^5$, or γ_i^μ; $V_K(\hat{p}_1, \hat{k}_1)$ describes the momentum dependence of the kernel labelled K, m_i is the mass of quark i, and $E_{ik} \equiv (m_i^2 + \mathbf{k}^2)^{1/2}$. A "ˆ" over a four-momentum indicates that it is on-mass-shell.

The kernel employed in our calculations with the 1CSE consists of a covariant generalization of the linear (L) confining potential used in [12], a color Coulomb (Coul), and a constant (C) interaction:

$$\mathcal{V} \equiv \sum_K V_K \Theta_1^{K(\mu)} \otimes \Theta^K_{2(\mu)} = \left[(1-y)\left(\mathbf{1}_1 \otimes \mathbf{1}_2 + \gamma_1^5 \otimes \gamma_2^5\right) - y\,\gamma_1^\mu \otimes \gamma_{\mu 2}\right] V_{\rm L} - \gamma_1^\mu \otimes \gamma_{\mu 2}\,[V_{\rm Coul} + V_{\rm C}]\,. \tag{3}$$

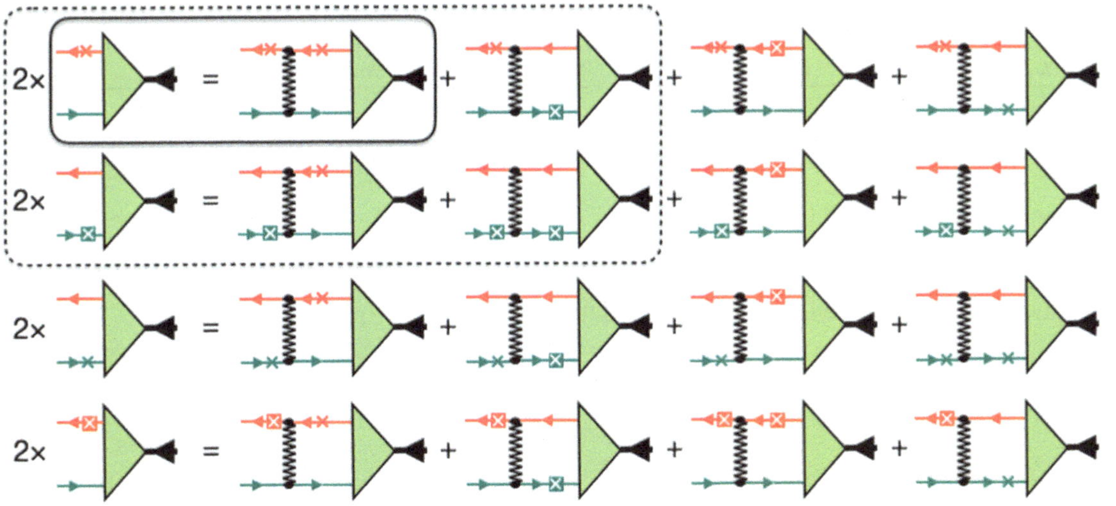

Fig. 1 The set of the four-channel CST equations (4CSE). The *solid rectangle* indicates the one-channel CST equation (1CSE) used in this work, the *dashed rectangle* a two-channel extension with charge-conjugation symmetry. Crosses on *quark lines* indicate that only the positive-energy pole contribution of the propagator is kept, light crosses in a *dark square* refer to the negative-energy pole contribution

The mixing parameter y allows to dial between a scalar-plus-pseudoscalar structure, which preserves chiral symmetry as shown in [13], and a vector structure, while leaving the nonrelativistic limit unchanged. The precise Lorentz structure of the confining interaction is not known, and by fitting the y parameter from the mesonic mass spectra, further information can be gained. Early results favored pure scalar-plus-pseudoscalar confinement, and therefore we set $y = 0$ in this work. The momentum-dependent structures of the interaction kernel are

$$V_{\mathrm{L}}(p,k) = -8\sigma\pi\left[\left(\frac{1}{q^4}-\frac{1}{\Lambda^4+q^4}\right)-\frac{E_{1p}}{m_1}(2\pi)^3\delta^3(\mathbf{q})\int\frac{d^3k'}{(2\pi)^3}\frac{m_1}{E_{1k'}}\left(\frac{1}{q'^4}-\frac{1}{\Lambda^4+q'^4}\right)\right], \quad (4)$$

$$V_{\mathrm{Coul}}(p,k) = -4\pi\alpha_s\left(\frac{1}{q^2}-\frac{1}{q^2-\Lambda^2}\right), \quad V_{\mathrm{C}}(p,k) = (2\pi)^3\frac{E_{1k}}{m_1}C\delta^3(\mathbf{q}), \quad (5)$$

where $q^{(\prime)} = p - k^{(\prime)}$.

The three coupling strengths, σ, α_s, and C, are free parameters of the model. An analysis of the asymptotic behaviour for large momenta k shows that we need to regularize the kernel. We use Pauli–Villars regularization for both the linear and the Coulomb parts, which yields one additional parameter, the cut-off parameter Λ. The results turn out not to be very sensitive to the choice of Λ and we set $\Lambda = 2m_1$.

Next we expand both the projector and the propagator of Eq. (2) in terms of u^ρ-spinors ($\rho = \pm$), defined as

$$u_i^+(p,\lambda) = \sqrt{\frac{E_{ip}+m_i}{2m_i}}\begin{pmatrix}1\\ \frac{\sigma\cdot\mathbf{p}}{E_{ip}+m_i}\end{pmatrix}\otimes\chi_\lambda(\hat{p}), \qquad u_i^-(p,\lambda) = \sqrt{\frac{E_{ip}+m_i}{2m_i}}\begin{pmatrix}-\frac{\sigma\cdot\mathbf{p}}{E_{ip}+m_i}\\ 1\end{pmatrix}\otimes\chi_\lambda(\hat{p}), \quad (6)$$

where χ_λ are two-component spinors. Introducing the notation

$$\Theta_{i,\lambda\lambda'}^{\rho\rho' K(\mu)}(\mathbf{p},\mathbf{k}) \equiv \bar{u}_i^\rho(\mathbf{p},\lambda)\Theta^{K(\mu)}u_i^{\rho'}(\mathbf{k},\lambda'), \qquad \Gamma_{\lambda\lambda'}^{+\rho'}(p) \equiv \bar{u}_1^+(\mathbf{p},\lambda)\Gamma(p)u_2^{\rho'}(\mathbf{p},\lambda'), \quad (7)$$

for the spinor matrix elements of the interaction vertices and of the vertex function, respectively, we obtain

$$\Gamma_{\lambda\lambda'}^{+\rho'}(p) = -\int\frac{d^3k}{(2\pi)^3}\frac{m_1}{E_{1k}}\frac{m_2}{E_{2k}}\sum_{\rho\lambda_1\lambda_2}\sum_K V_K(p,k)\Theta_{1,\lambda\lambda_1}^{++K(\mu)}(\mathbf{p},\mathbf{k})\Gamma_{\lambda_1\lambda_2}^{+\rho}(k)\frac{\rho}{E_{2k}-\rho k_{20}}\Theta_{2,\lambda_2\lambda' K(\mu)}^{\rho\rho'}(\mathbf{k},\mathbf{p}). \quad (8)$$

Table 1 Wave function components for the mesons considered in this work

J^P	$K_1^-(\hat{\mathbf{p}})$	Wave	$K_2^-(\hat{\mathbf{p}})$	Wave	$K_1^+(\hat{\mathbf{p}})$	Wave	$K_2^+(\hat{\mathbf{p}})$	Wave
0^-	$\mathbf{1}$	S	–	–	$\sigma\cdot\hat{\mathbf{p}}$	P	–	–
0^+	$\sigma\cdot\hat{\mathbf{p}}$	P	–	–	$\mathbf{1}$	S	–	–
1^-	$\sigma\cdot\hat{\xi}$	S	$\frac{1}{\sqrt{2}}\left(3\xi\cdot\hat{\mathbf{p}}\sigma\cdot\hat{\mathbf{p}}-\sigma\cdot\hat{\xi}\right)$	D	$\sqrt{3}\xi\cdot\hat{\mathbf{p}}$	P_s	$\sqrt{\frac{3}{2}}\left(\sigma\cdot\hat{\xi}\sigma\cdot\hat{\mathbf{p}}-\xi\cdot\hat{\mathbf{p}}\right)$	P_t
1^+	$\sqrt{3}\xi\cdot\hat{\mathbf{p}}$	P_s	$\sqrt{\frac{3}{2}}\left(\sigma\cdot\hat{\xi}\sigma\cdot\hat{\mathbf{p}}-\xi\cdot\hat{\mathbf{p}}\right)$	P_t	$\sigma\cdot\hat{\xi}$	S	$\frac{1}{\sqrt{2}}\left(3\xi\cdot\hat{\mathbf{p}}\sigma\cdot\hat{\mathbf{p}}-\sigma\cdot\hat{\xi}\right)$	D

Multiplying Eq. (8) from the left by $\bar{u}_1^+(\mathbf{p},\lambda)$ and from the right by $u_2^{\rho'}(\mathbf{p},\lambda')$ yields

$$\rho'(E_{2p}-\rho' p_{20})\sqrt{\frac{m_1m_2}{E_{1p}E_{2p}}}\rho'\frac{\Gamma^{+\rho'}_{\lambda\lambda'}(p)}{E_{2p}-\rho' p_{20}} = -\int\frac{d^3k}{(2\pi)^3}\sqrt{\frac{m_1m_2}{E_{1k}E_{2k}}}\sqrt{\frac{m_1m_2}{E_{1p}E_{2p}}}$$
$$\times\sum_{\rho\lambda_1\lambda_2}\sum_K V_K(p,k)\Theta^{++K(\mu)}_{1,\lambda\lambda_1}(\mathbf{p},\mathbf{k})\sqrt{\frac{m_1m_2}{E_{1k}E_{2k}}}\Gamma^{+\rho}_{\lambda_1\lambda_2}(k)\frac{\rho}{E_{2k}-\rho k_{20}}\Theta^{\rho\rho'}_{2,\lambda_2\lambda' K(\mu)}(\mathbf{k},\mathbf{p})\,. \quad (9)$$

Introducing the CST wave functions when quark 1 is on-shell,

$$\Psi^{+\rho}_{1,\lambda_1\lambda_2}(k)\equiv\sqrt{\frac{m_1m_2}{E_{1k}E_{2k}}}\frac{\rho}{E_{2k}-\rho(E_{1k}-\mu)}\Gamma^{+\rho}_{\lambda_1\lambda_2}(k), \quad (10)$$

we can finally cast Eq. (2) into the form

$$(\rho' E_{2p}-E_{1p}+\mu)\Psi^{+\rho'}_{1,\lambda\lambda'}(p) = -\int\frac{d^3k}{(2\pi)^3}N(p,k)\sum_{\rho\lambda_1\lambda_2}\sum_K V_K(p,k)\Theta^{++K(\mu)}_{1,\lambda\lambda_1}(\mathbf{p},\mathbf{k})$$
$$\times\Psi^{+\rho}_{1,\lambda_1\lambda_2}(k)\Theta^{\rho\rho'}_{2,\lambda_2\lambda' K(\mu)}(\mathbf{k},\mathbf{p}) \quad (11)$$

where $N(p,k)=m_1m_2/\sqrt{E_{1k}E_{2k}E_{1p}E_{2p}}$. The CST wave functions can be written in terms of two-component spinors χ_λ and $K_j^\rho(\hat{\mathbf{p}})$ operators, which are 2×2 matrices that depend on the total angular momentum and the parity of the meson under study (in this work we consider $J^P=0^\pm,1^\pm$),

$$\Psi^{+\rho}_{1,\lambda\lambda'}(p)=\sum_j\psi_j^\rho(p)\chi_\lambda^\dagger(\hat{\mathbf{p}})\,K_j^\rho(\hat{\mathbf{p}})\,\chi_{\lambda'}(\hat{\mathbf{p}}). \quad (12)$$

Table 1 lists the $K_j^\rho(\hat{\mathbf{p}})$ used in this work. The main advantage of using this basis for the wave function is that it explicitly displays its orbital-angular-momentum content and thus enables us to determine the spectroscopic identity of our solutions, which is indispensable when comparing to the measured states. Our wave functions contain relativistic components not present in nonrelativistic solutions. For instance, the S-waves of our pseudoscalar states couple to small P-waves (with opposite intrinsic parity) that vanish in the nonrelativistic limit, whereas, for vector mesons, coupled S- and D-waves are accompanied by relativistic spin-singlet and spin-triplet P-waves, denoted P_s and P_t, respectively.

3 Results and Discussion

In this work we present two models: model P1 was fitted to the masses of pseudoscalar states only, whereas model PSV1 was fitted to the masses of pseudoscalar, scalar, and vector mesons. The parameters of the models are listed in Table 2. The constituent quark masses were first determined in preliminary calculations and then held fixed in the final fits of σ, α_s, and C.

Our results for $b\bar{b}$, $b\bar{c}$, $b\bar{s}$, $b\bar{q}$, $c\bar{c}$, $c\bar{s}$, $c\bar{q}$ states (q stands for a u or a d quark) are given in Table 3. The calculated masses are very close to the experimental data, with an rms difference of roughly 30 MeV for both models P1 and PSV1. These differences are comparable with those reported in Ref. [16] and, more recently, in Ref. [4], and give us confidence that our approach is appropriate for the systems under study.

Table 2 Parameters of models P1 and PSV1. Both models use the quark masses $m_b = 4.892$ GeV, $m_c = 1.600$ GeV, $m_s = 0.448$ GeV, and $m_u = m_d \equiv m_q = 0.346$ GeV. Shown are also the number of states used in the fits, N_{fit}, and the respective rms differences between the model predictions and the data

Model	σ [GeV^2]	α_s	C [GeV]	N_{fit}	Rms difference [GeV]
P1	0.2493	0.3643	0.3491	9	0.036
PSV1	0.2247	0.3614	0.3377	25	0.031

Shown are also the number of states used in the fits, N_{fit}, and the respective rms differences between the model predictions and the data

Table 3 Comparison of the masses of experimental meson states (Exp) with at with least one b or c quark and $J^P = 0^\pm, 1^\pm$ to the theoretical predictions of models P1 and PSV1

	$J^P = 0^-$			$J^P = 1^-$			$J^P = 0^+$			$J^P = 1^+$		
	Exp	P1	PSV1	Exp	P1	PSV1	Exp	P1	PSV1	Exp	P1	PSV1
$b\bar{b}$	$9.398^{\triangle,\square}$	9.386	9.415	$9.460^{\square}$	9.470	9.487	$9.859^{\square}$	9.856	9.850	9.896	9.886	9.875
	$9.999^{\triangle,\square}$	9.982	9.968	$10.02^{\square}$	10.02	10.00	$10.23^{\square}$	10.25	10.22	10.26	9.890	9.879
	10.30	10.37	10.33	10.15(?)	10.16	10.13	–	10.57	10.52	10.51	10.27	10.24
	–	10.68	10.63	$10.36^{\square}$	10.40	10.35	–	10.86	10.80	–	10.28	10.24
	–	10.96	10.89	–	10.49	10.44	–	11.13	11.03	–	10.60	10.54
	–	11.28	11.16	$10.58^{\square}$	10.71	10.65	-	11.48	11.32	–	10.60	10.54
$b\bar{c}$	$6.275^{\triangle,\square}$	6.302	6.319	–	6.394	6.397	–	6.745	6.730	–	6.777	6.757
	6.842	6.888	6.865	–	6.941	6.912	–	7.161	7.121	–	6.777	6.758
	–	7.293	7.246	–	7.057	7.019	–	7.505	7.445	–	7.191	7.146
$b\bar{s}$	$5.367^{\triangle,\square}$	5.362	5.367	$5.415^{\square}$	5.442	5.436	–	5.784	5.763	5.829	5.796	5.770
	–	5.938	5.910	–	5.993	5.957	–	6.208	6.163	–	5.811	5.785
	–	6.349	6.297	–	6.093	6.051	–	6.559	6.495	–	6.234	6.184
$b\bar{q}$	$5.279^{\triangle,\square}$	5.288	5.293	$5.325^{\square}$	5.366	5.360	–	5.709	5.688	5.726	5.716	5.690
	–	5.864	5.835	–	5.918	5.882	–	6.132	6.087	–	5.735	5.708
	–	6.274	6.221	–	6.017	5.974	–	6.483	6.418	–	6.157	6.106
$c\bar{c}$	$2.984^{\triangle,\square}$	3.009	3.030	$3.097^{\square}$	3.110	3.120	$3.415^{\square}$	3.424	3.424	3.518	3.461	3.454
	$3.639^{\triangle,\square}$	3.647	3.627	$3.686^{\square}$	3.702	3.677	3.918	3.930	3.894	–	3.474	3.465
	–	4.123	4.073	$3.773^{\square}$	3.784	3.756	–	4.355	4.291	–	3.950	3.911
$c\bar{s}$	$1.968^{\triangle,\square}$	1.944	1.966	$2.112^{\square}$	2.107	2.109	$2.318^{\square}$	2.399	2.396	2.459	2.434	2.422
	–	2.612	2.591	–	2.697	2.667	–	2.910	2.872	2.535	2.458	2.444
	–	3.100	3.048	–	2.769	2.737	–	3.340	3.274	–	2.934	2.893
$c\bar{q}$	$1.867^{\triangle,\square}$	1.858	1.881	$2.009^{\square}$	2.029	2.030	$2.318^{\square}$	2.319	2.316	2.421	2.351	2.339
	–	2.529	2.507	–	2.617	2.587	–	2.828	2.790	–	2.377	2.362
	–	3.016	2.964	–	2.687	2.655	–	3.257	3.191	–	2.852	2.810

The $\triangle$ and $\square$ symbols indicate the states used in the fits of models P1 and PSV1, respectively. The states with no symbol assigned are pure predictions. The experimental values for axial-vector mesons in $b\bar{b}$ and $c\bar{c}$ are averages over the two possible charge-conjugation parities. All masses are given in units of GeV. There is weak evidence (at the level of 1.8 σ) that the $\Upsilon(1D)$ (10.15 GeV, marked with "?") has been seen [14,15]

The parameters of the two models are very similar, as Table 2 shows. This is remarkable, because pseudoscalar states (the only ones used in the fit of model P1) are essentially pure S-waves and are not sensitive to spin-orbit and tensor forces. Moreover, in pseudoscalar mesons the spin-spin forces act only in spin-singlet states and therefore cannot be separated from the central forces. This means that no information about the spin-dependent interactions was used to constrain the kernel of model P1. On the other hand, the scalar and vector states to which model PSV1 was fitted *are* sensitive to the spin-dependent interactions and do constrain its parameters. The fact that both fits give essentially the same parameters and rms differences to the data means that the spin-dependent interactions are *correctly predicted* by the covariant kernel of model P1.

We have also calculated the wave functions needed to study the structure of mesons and to calculate form factors and decay rates. In Figs. 2–5, the ground-state wave function components $\psi(p)$, defined as in Eq. (12) and calculated with model P1, are depicted for pseudoscalar, scalar, vector and axial-vector mesons. These components are normalized as

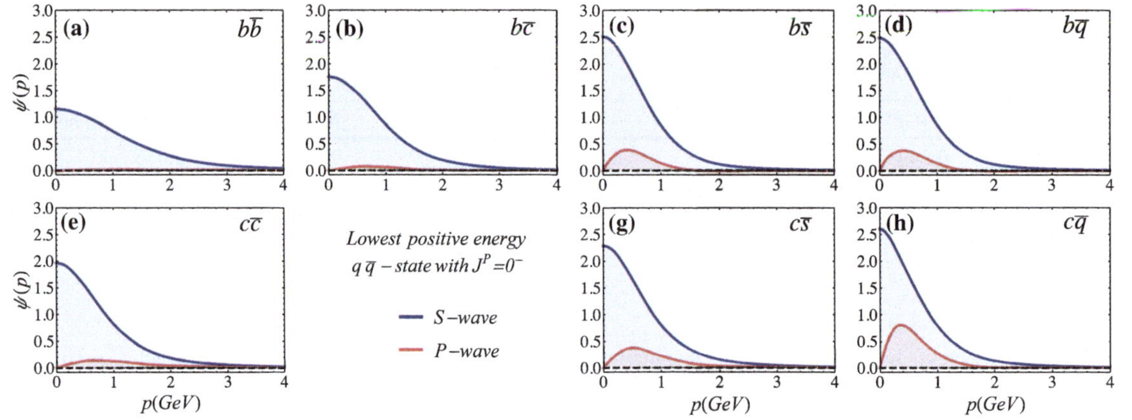

Fig. 2 Model P1 ground-state wave functions for pseudoscalar mesons ($J^P = 0^-$)

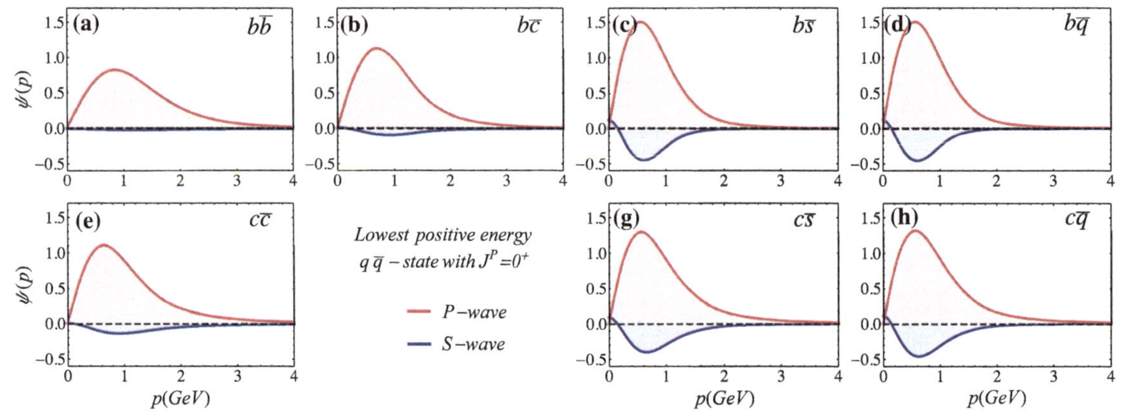

Fig. 3 Model P1 ground-state wave functions for scalar mesons ($J^P = 0^+$)

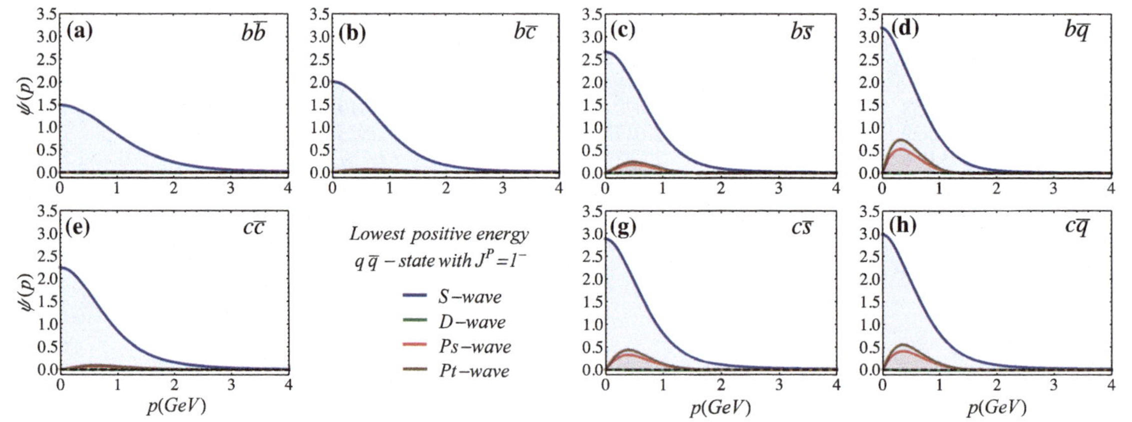

Fig. 4 Model P1 ground-state wave functions for vector mesons ($J^P = 1^-$)

$$\int dp\, p^2 \left[\psi_S^2(p) + \psi_D^2(p)\right] = 1\,, \qquad \text{for } J^P = 0^{\pm}\,, \tag{13}$$

$$\int dp\, p^2 \left[\psi_S^2(p) + \psi_D^2(p) + \psi_{P_s}^2(p) + \psi_{P_t}^2(p)\right] = 1\,, \qquad \text{for } J^P = 1^{\pm}\,. \tag{14}$$

Tables 4–6 list the probabilities of the different wave function components. As expected, the relativistic components are tiny in heavy mesons, but become larger as the constituent quark masses decrease, reaching almost 10 % for $c\bar{q}$ mesons, the lightest systems considered here. The momentum-space wave functions tend to be

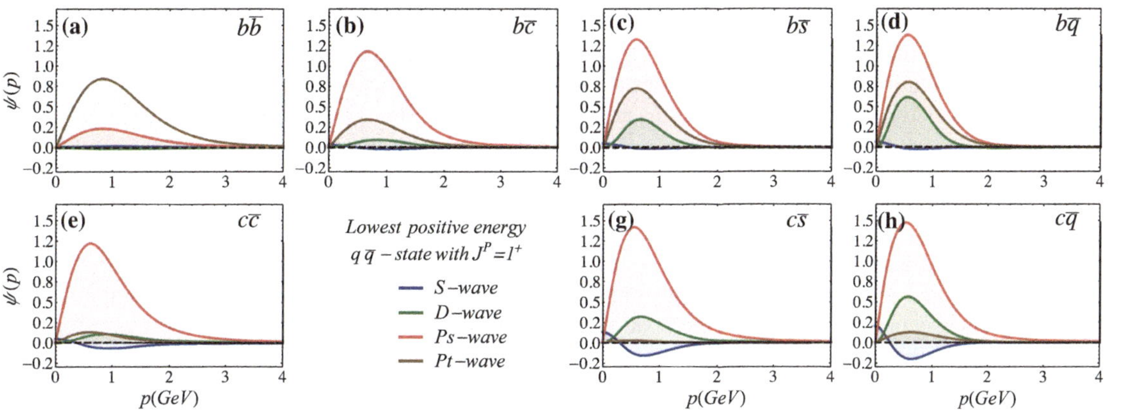

Fig. 5 Model P1 ground-state predictions for axial-vector mesons $J^P = 1^+$

Table 4 Probabilities of S- and P-wave components for pseudoscalar mesons ($J^P = 0^-$), left panel, and scalar mesons ($J^P = 0^+$), right panel

Meson	Fig. 2	S-wave (%)	P-wave (%)	Meson	Fig. 3	S-wave (%)	P-wave (%)
$b\bar{b}$	(a)	99.5	0.528	$b\bar{b}$	(a)	0.203	99.8
$b\bar{c}$	(b)	99.4	0.586	$b\bar{c}$	(b)	0.855	99.1
$b\bar{s}$	(c)	96.8	3.19	$b\bar{s}$	(c)	6.77	93.2
$b\bar{q}$	(d)	95.7	4.34	$b\bar{q}$	(d)	6.04	94.0
$c\bar{c}$	(e)	96.3	3.71	$c\bar{c}$	(e)	2.97	97.0
$c\bar{s}$	(g)	92.0	8.04	$c\bar{s}$	(g)	6.38	93.6
$c\bar{q}$	(h)	90.8	9.15	$c\bar{q}$	(h)	9.24	90.8

Table 5 Probabilities of S-, D-, P_s- and P_t-wave components for vector mesons ($J^P = 1^-$)

Meson	Fig. 4	S-wave (%)	D-wave (%)	P_s-wave (%)	P_t-wave (%)
$b\bar{b}$	(a)	99.9	0.0130	0.0186	0.0477
$b\bar{c}$	(b)	99.6	0.0118	0.144	0.196
$b\bar{s}$	(c)	97.6	0.00862	0.858	1.49
$b\bar{q}$	(d)	93.5	0.00783	2.30	4.22
$c\bar{c}$	(e)	98.9	0.0124	0.397	0.731
$c\bar{s}$	(g)	95.3	0.0309	1.87	2.78
$c\bar{q}$	(h)	94.0	0.0354	2.38	3.61

Table 6 Probabilities of S-, D-, P_s- and P_t-wave components for axial-vector mesons ($J^P = 1^+$)

Meson	Fig. 5	S-wave (%)	D-wave (%)	P_s-wave (%)	P_t-wave (%)
$b\bar{b}$	(a)	0.0559	0.0399	7.54	92.4
$b\bar{c}$	(b)	0.0262	0.670	89.3	9.97
$b\bar{s}$	(c)	0.0134	5.93	71.9	22.1
$b\bar{q}$	(d)	0.0111	7.34	69.7	23.0
$c\bar{c}$	(e)	0.578	1.66	97.0	0.750
$c\bar{s}$	(g)	1.14	6.39	92.4	0.0954
$c\bar{q}$	(h)	1.15	7.90	90.5	0.397

more spread out for heavier states, and as one decreases the mass of the lighter quark, the wave functions concentrate more in the region of low momenta.

4 Summary and Outlook

In this work we report on recent progress in our studies of heavy and heavy-light mesons in the framework of the CST. Our kernel consists of covariant versions of a linear confining potential with Lorentz-scalar coupling, in combination with Lorentz-vector one-gluon-exchange plus a constant interaction kernel. When fitting a small number of global adjustable parameters we obtain an accurate description of the mass spectrum, which gives us confidence that our goal of a global description of all $q\bar{q}$ mesons may indeed be feasible.

From the observation that a fit to the masses of mesons whose wave functions are essentially pure S-waves leads to almost exactly the same parameters as another fit to meson states whose orbital angular momenta are not restricted, we conclude that the covariance of our interaction kernel accurately predicts the spin-dependent quark-antiquark interactions.

We also compared the orbital-angular-momentum components of the various meson wave functions and found that they behave as expected: the lighter the constituent quark masses of the quark-antiquark system, the larger become the probabilities of the wave function components of purely relativistic origin.

For the near future we plan to include also heavy and heavy-light tensor mesons in our calculations with the one-channel spectator equation. The next step is then to extend our model to the light-meson sector, which requires the use of the more complicated four-channel equations. In addition, with the covariant meson wave functions we will compute other observables such as decay rates, and study the structure of mesons in more detail by calculating electroweak elastic and transition form factors.

Acknowledgements This work was supported by the Portuguese Fundação para a Ciência e a Tecnologia (FCT) under contracts SFRH/BD/92637/2013, SFRH/BPD/100578/2014, and UID/FIS/0777/2013.

References

1. Belle Collaboration, S.-K. Choi et al., Phys. Rev. Lett. **91**, 262001 (2003)
2. E. Klempt, A. Zaitsev, Phys. Rep. **454**, 1 (2007)
3. N. Brambilla et al., Eur. Phys. J. C **74**, 2981 (2014)
4. Y. Li, P. Maris, X. Zhao, J.P. Vary, Phys. Lett. B **758**, 118–124 (2016)
5. M. Gomez, S. Glazek, Phys. Rev. D **92**, 065005 (2015)
6. F. Gross, Phys. Rev. **186**, 1448 (1969)
7. T. Hilger, C. Popovici, M. Gomez-Rocha, A. Krassnigg, Phys. Rev. D **91**, 034013 (2015)
8. G. Eichmann, H. Sanchis-Alepuz, R. Williams, R. Alkofer, C.S. Fischer, Prog. Part. Nucl. Phys. **91**, 1 (2016)
9. A. Stadler, F. Gross, Few-Body Syst. **49**, 91110 (2011)
10. E.P. Biernat, F. Gross, M.T. Peña, A. Stadler, Phys. Rev. D **89**, 016005 (2014)
11. S. Leitão, A. Stadler, M.T. Peña, E.P. Biernat, Phys. Lett. B **764**, 38 (2017)
12. S. Leitão, A. Stadler, M.T. Peña, E.P. Biernat, Phys. Rev. D **90**, 096003 (2014)
13. E.P. Biernat, M.T. Peña, J.E. Ribeiro, A. Stadler, F. Gross, Phys. Rev. D **90**, 096008 (2014)
14. C.L.E.O. Collaboration, G. Bonvicini et al., Phys. Rev. D **70**, 032001 (2004)
15. B.A.B.A.R. Collaboration, P. del Amo Sanchez et al., Phys. Rev. D **82**, 111102 (2010)
16. J.R. Spence, J.P. Vary, Phys. Rev. C **47**, 1282 (1993)

Few-Body Syst (2017) 58:81
DOI 10.1007/s00601-016-1200-3

Gernot Eichmann

More About the Light Baryon Spectrum

Received: 30 November 2016 / Accepted: 23 December 2016 / Published online: 8 February 2017

Abstract We discuss the light baryon spectrum obtained from a recent quark–diquark calculation, implementing non-pointlike diquarks that are self-consistently calculated from their Bethe–Salpeter equations. We examine the orbital angular momentum content in the baryons' rest frame and highlight the fact that baryons carry all possible values of L compatible with their spin, without the restriction $P = (-1)^L$ which is only valid nonrelativistically. We furthermore investigate the meaning of complex conjugate eigenvalues of Bethe–Salpeter equations, their possible connection with 'anomalous' states, and we propose a method to eliminate them from the spectrum.

1 Introduction

The light baryon spectrum has traditionally been a benchmark of our theoretical understanding of Quantum Chromodynamics (QCD), and it continues to be an exciting area of research in light of new advances with photo- and electroproduction experiments [1–4]. In recent years lattice QCD has successfully charted the low-lying hadron spectrum, but lattice determinations of baryon resonances are still in their early stages [5–8]. Recent progress has also been made using covariant bound-state equations [9–11]. Here the starting point is the covariant three-body Faddeev equation for baryons [12], whose kernel contains all possible quark–gluon interactions that are encoded in QCD's n-point functions. The latter can be determined by their Dyson–Schwinger equations, which are the quantum equations of motions of QCD and therefore exact. Due to the complexity of the problem most studies so far have employed comparatively simple kernels which, in particular, do not implement decay mechanisms: the hadrons obtained so far correspond to poles on the real axis in the associated scattering amplitudes. On the other hand, the absence of meson–baryon interactions that would shift these poles into the complex plane can also be viewed as an advantage, as it provides insight into the 'quark core' of a baryon that is stripped of its meson cloud.

A possibility to simplify the three-body problem is to introduce diquarks. There is a wide range of theoretical quark–diquark studies including quantum-mechanical models [1,13], light-front holographic QCD [14], or quantum-field theoretical approaches within QCD where diquarks appear as intermediate structures in the quark–quark correlation function [9,11]. The original motivation for diquarks has been the problem of 'missing resonances' in the baryon spectrum which, despite predictions from the quark model, had not been observed in πN scattering experiments. The addition of new states extracted from photo- and electroproduction experiments to the PDG [15] has ruled out early pointlike diquark models. On the other hand, sophisticated quark–diquark approaches derived from the three-body Faddeev equation produce a rich baryon spectrum that resembles the

This article belongs to the Topical Collection "Light Cone 2016".

G. Eichmann (✉)
Institut für Theoretische Physik, Justus-Liebig-Universität Giessen, 35392 Giessen, Germany
E-mail: gernot.eichmann@theo.physik.uni-giessen.de

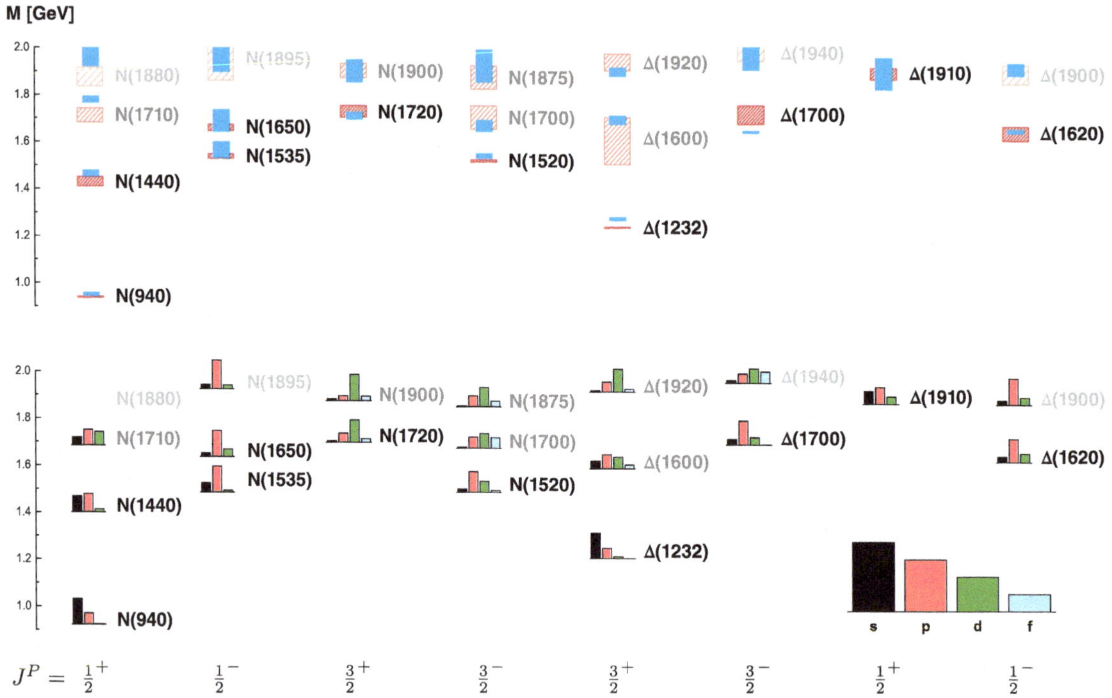

Fig. 1 Light baryon spectrum in the quark–diquark approach with reduced pseudoscalar and vector diquarks. *Top* The calculated masses (*solid boxes, blue*) are compared to the PDG values (*hatched boxes, red*) [15]. *Bottom* Orbital angular momentum contributions (in %) for the individual states; all *bars* sum up to 100%

experimental situation [16]. It is therefore desirable to get a closer look at the properties of the baryons obtained in such a framework.

2 Light Baryons and Their Orbital Angular Momentum

In Ref. [16] an effort was made to calculate the light baryon spectrum both from the covariant three-body Faddeev equation and its quark–diquark simplification, using the same underlying ingredients. The quarks and diquarks were not modeled but calculated from their Dyson–Schwinger and Bethe–Salpeter equations including their full covariant tensor decomposition. The only input is a common rainbow-ladder interaction which models the qq and $q\bar{q}$ kernel by an effective gluon exchange.

As a result, the three-quark and quark–diquark results essentially agree with each other, at least for the low-lying states that are accessible in both cases. The rainbow-ladder truncation performs well in the $N({}^1\!/{}_2^+)$ and $\Delta({}^3\!/{}_2^+)$ channels but the remaining channels typically come out too low. In the quark–diquark system this behavior can be attributed to the underlying diquark properties: the 'good' states are dominated by scalar and axialvector diquarks, whose meson counterparts (the pion and ρ meson) are stable beyond rainbow-ladder. The pseudoscalar and vector diquarks, on the other hand, are bound too strongly and this affects the remaining channels. The strategy employed in Ref. [16] was to reduce the strength in these 'bad' diquark channels by attaching a common constant to their Bethe–Salpeter kernels, which was adjusted by the $\rho - a_1$ splitting in the meson sector. As a result, the 'bad' diquarks together with their (scalar and axialvector) meson counterparts acquire a higher mass and thereby resemble the outcome of beyond-rainbow-ladder calculations. The resulting baryon spectrum is shown in Fig. 1 and agrees very well with the experimental spectrum, which suggests that a three-body calculation beyond rainbow-ladder should produce similar results.

The resulting baryon Faddeev amplitudes carry a rich tensorial structure which can be organized into eigenstates of spin and orbital angular momentum in the baryon's rest frame [11,18]. The orbital angular momentum content of a baryon can be quantified from the contributions to its Bethe–Salpeter normalization (see Eq. (15) below). The resulting s, p, d and f-wave components (corresponding to $L = 0 \dots 3$) are shown in Fig. 1:

– the low-lying states in the $N(\frac{1}{2}^+)$, $\Delta(\frac{3}{2}^+)$ and $\Delta(\frac{1}{2}^+)$ channels carry sizeable $L = 0$ components;

- the $N(\frac{1}{2}^-)$, $N(\frac{3}{2}^-)$, $\Delta(\frac{3}{2}^-)$, $\Delta(\frac{1}{2}^-)$ channels are typically dominated by $L = 1$;
- the $N(\frac{3}{2}^+)$ channel is dominated by $L = 2$.

In principle this is also expected from the nonrelativistic quark model, where the $SU(6) \times O(3)$ classification together with the relation $P = (-1)^L$ entails that the nucleon and Δ belong to the 56-plet with $L = 0$ and positive parity, followed by the 70-plet with $L = 1$ and negative parity, and so on. Relativistically, however, these baryons carry all possible L values allowed by $|L - S| \leq J \leq L + S$, without the restriction $P = (-1)^L$: only J and P are good quantum numbers but L and S are not, and therefore a relation between parity and orbital angular momentum cannot hold in general. Figure 1 shows that the nucleon and $\Delta(1232)$ have nonvanishing p-wave components, the $N(1535)$ has s waves, and the p waves appear to be even dominant for the Roper. The subleading partial waves can have important consequences for form factors, for example in the $N\gamma \to \Delta$ transition [17]. These results clearly demonstrate that light baryons are relativistic objects and Poincaré invariance is a necessary ingredient in their realistic description.

3 Complex Conjugate Eigenvalues?

A point noted in Ref. [16] is that some states in the calculated spectrum correspond to complex conjugate eigenvalues of the quark–diquark BSE. Their imaginary parts are small and thus they may simply be numerical artifacts, but from a conceptual point of view their appearance is not completely satisfactory. In fact, complex conjugate eigenvalues are a typical feature of BSEs for unequal-mass systems. They occur in scalar theories [19, 20], where they can be associated with the interference of 'normal' and 'anomalous' states [21], but for example also in heavy-light mesons [22–24]. As we will see below, complex conjugate eigenvalues may even appear in equal-mass systems and therefore a closer inspection is desirable.

To provide a basis for the following discussion, let us quickly collect some basic formulas (more details can be found in [11]). The generic form of a homogeneous BSE reads

$$\mathbf{\Gamma}(p, P) = \int \frac{d^4 p'}{(2\pi)^4} K(p, p', P)\, G(p', P)\, \mathbf{\Gamma}(p', P)\,, \tag{1}$$

where $\mathbf{\Gamma}(p, P)$ is the Bethe–Salpeter amplitude, K is the kernel, and G is the disconnected product of propagators as illustrated in Fig. 2. P is the total momentum ($P^2 = -M^2$) and p, p' are the external and internal relative momenta. The simplest example is a scalar theory with massive constituent particles of masses $m_{1,2}$ and $m = (m_1 + m_2)/2$, which are bound by a scalar exchange particle with mass μ and coupling strength c:

$$K(p, p', P) = \frac{(4\pi m)^2 c}{(p - p')^2 + \mu^2}\,, \qquad G(p, P) = \frac{1}{p_+^2 + m_1^2}\,\frac{1}{p_-^2 + m_2^2}\,, \qquad p_\pm = p + \frac{\varepsilon \pm 1}{2} P. \tag{2}$$

For $\mu \to 0$ this becomes the well-known Wick-Cutkosky model [25,26] which admits analytic solutions. $\varepsilon \in [-1, 1]$ is an arbitrary momentum partitioning parameter and 'ideal momentum partitioning' corresponds to $\varepsilon = (m_1 - m_2)/(m_1 + m_2)$. Here the amplitude $\mathbf{\Gamma}(p, P)$ is Lorentz invariant and therefore only depends on the variables p^2, $p \cdot P$ and $P^2 = -M^2$. In practice it is convenient to factor out the angular dependencies in $z = \hat{p} \cdot \hat{p}$, for example with Chebyshev polynomials of the second kind,

$$\mathbf{\Gamma}(p, P) = \sum_m \phi_m(p^2)\, U_m(z)\,, \qquad \frac{2}{\pi} \int dz \sqrt{1 - z^2}\, U_m^*(z)\, U_n(z) = \delta_{mn}\,, \tag{3}$$

which leads to an equation for the Chebyshev moments ($x = p^2$ and $x' = p'^2$):

$$\phi_m(x) = \sum_{nl} \int dx'\, K_{mn}(x, x')\, G_{nl}(x')\, \phi_l(x'). \tag{4}$$

The resulting 'kernel' and 'propagator matrix' are given by (cf. Eq. (3.44) in [11])

$$\begin{aligned} K_{mn}(x, x') &= \frac{2}{\pi} \int dz \int dz' \sqrt{1 - z^2} \sqrt{1 - z'^2}\, U_m^*(z)\, U_n(z') \int \frac{dy}{(2\pi)^3}\, K(p, p', P)\,, \\ G_{mn}(x) &= \frac{2}{\pi} \frac{x}{2} \int dz \sqrt{1 - z^2}\, U_m^*(z)\, U_n(z)\, G(p, P). \end{aligned} \tag{5}$$

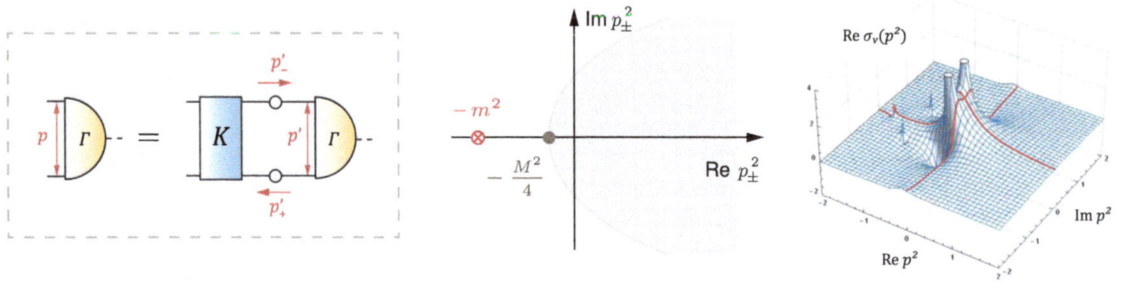

Fig. 2 *Left* Generic form of a two-body Bethe–Salpeter equation. *Center* Parabolic domain of the propagator that is integrated over in the equation. *Right* Solution for the quark propagator in a rainbow-ladder truncation. Figures adapted from Ref. [11]

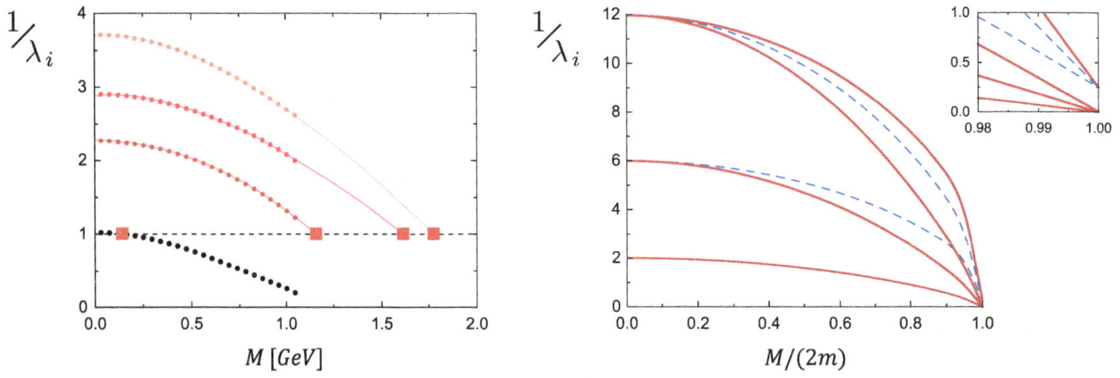

Fig. 3 *Left* Eigenvalue spectrum for the pion channel with $J^{PC} = 0^{-+}$ obtained in rainbow-ladder. The *boxes* show the intersections $\lambda_i = 1$ which correspond to the physical pion and its radial excitations. *Right* Eigenvalue spectrum for the Wick-Cutkosky model. *Solid* (*red*) *curves* are the solutions with positive 'C parity', *dashed* (*blue*) *curves* those with negative C parity (see text)

Upon discretizing also the radial variable x, the equation can be cast in the form

$$K\,G\,\phi_i = \lambda_i\,\phi_i\,, \tag{6}$$

which generally holds for any homogeneous BSE. K and G are typically large matrices. For a two-body system their dimensions are of the order $\sim 10^3 \times 10^3$, depending on the number of grid points and invariant tensors (one for the scalar system, four for the pion, ...) whereas for a three-body problem they become giant matrices of order $\gtrsim 10^6 \times 10^6$. The momentum dependencies in Eq. (1) are encoded in the matrix dimensions of K and G, except for the total momentum $P^2 = -M^2$ which remains an external variable. The equation is then solved for different values of M, which results in an eigenvalue spectrum $\lambda_i(M)$ of KG and corresponding eigenvectors $\phi_i(M)$. At the intersections $\lambda_i = 1$ one recovers the onshell Bethe–Salpeter amplitudes together with their masses, $P^2 = -M_i^2$.

An exemplary result is shown in the left panel of Fig. 3 for a pseudoscalar light quark–antiquark system with $J^{PC} = 0^{-+}$, calculated in rainbow-ladder. The corresponding BSE is given in Eq. (17) below. The first eigenvalue λ_0 corresponds to the pion, the next one to its first radial excitation (experimentally, the $\pi(1300)$), and so on. The manifestation of spontaneous chiral symmetry breaking is visible in the plot: for massless quarks the eigenvalue for the ground state would start at $\lambda_0(M = 0) = 1$, so the pion becomes massless. In any case, it turns out that complex conjugate eigenvalues can occur even in the pion spectrum (see Fig. 4): the first few eigenvalues are real, but in the higher-lying spectrum one encounters complex conjugate ones. How can one interpret them?

In unequal-mass systems complex conjugate eigenvalues signal the loss of charge-conjugation invariance. This is related with the occurrence of anomalous states which appear for example in the Wick-Cutkosky model [25–27]. The right panel in Fig. 3 shows the eigenvalue spectrum of the Wick-Cutkosky model for equal constituent masses $m_1 = m_2 = m$ and $\mu = 0$. In this case there are no complex conjugate eigenvalues. In contrast to the case of the pion, the coupling c in Eq. (2) is a free parameter and choosing a different value does not affect the propagators. We used $c = 1$, but since a different choice only vertically rescales the eigenvalue spectrum one may as well replace the condition $1/\lambda_i = 1$ for bound states by $1/\lambda_i = c$.

In the Wick-Cutkosky model the constituent propagators have real mass poles at $p_\pm^2 = -m^2$. The propagators in the BSE are sampled within complex parabolas, cf. Fig. 2, from where one infers the condition $-m^2 < -M^2/4 \Rightarrow M < 2m$. This leads to a threshold, where the eigenvalues diverge at $M = 2m$ and the inverse eigenvalues vanish—except some that do not, which defines the anomalous states. Since they produce a vanishing binding energy ($M - 2m = 0$) for a nonvanishing coupling ($c = 1/4$), these states are unphysical. Anomalous states can be associated with excitations in relative time, which amounts to the distinction $\mathbf{\Gamma}(p, P) = \pm\mathbf{\Gamma}(-p, P)$ where the Bethe–Salpeter amplitudes are either even or odd in the angular variable z and thus correspond to even or odd Chebyshev moments.

The analogous symmetry for mesons is charge-conjugation invariance:

$$\bar{\mathbf{\Gamma}}(p, P) = \pm\mathbf{\Gamma}(p, -P), \qquad \bar{\mathbf{\Gamma}}(p, P) := C\,\mathbf{\Gamma}(-p, -P)^T\,C^T, \qquad C = \gamma^4\gamma^2, \tag{7}$$

which also leads to a distinction between even and odd Chebyshev moments. Equal-mass quarkonia are invariant under charge conjugation and their BSEs decouple for these two types of solutions, so they correspond to different physical particles with opposite C parities: $J^{PC} = 0^{-+}$ for the pion and its radial excitations, and $J^{PC} = 0^{--}$ for members of the 'exotic' channel. The latter do not have a non-relativistic limit due to the non-relativistic relations $P = (-1)^{L+1}$ and $C = (-1)^{L+S}$. Relativistically these constraints no longer hold, for the same reasons as discussed in Sect. 2, and therefore exotic quantum numbers are in general not forbidden as $q\bar{q}$ states. The distinction between the pion and its exotic partner then amounts to nothing more than, for instance, the distinction between the a_1 (1^{++}) and the b_1 (1^{+-}) meson.

In the scalar system charge-conjugation invariance can be easily verified: flipping the signs of p and p' in Eq. (1) leads back to the same equation for $\mathbf{\Gamma}(-p, P)$ because $K(-p, -p', P) = K(p, p', P)$ and, for equal constituent masses, $G(-p, P) = G(p, P)$. One can then form symmetric and antisymmetric combinations whose equations decouple. Without constraining the symmetry properties in advance, the eigenvalue spectrum is a superposition of both types of solutions. In the meson spectrum of Fig. 3 we already removed the 0^{--} eigenvalues, whereas in the Wick-Cutkosky spectrum we display both solutions: the solid (dashed) curves are those with positive (negative) 'C parity'. One can see that there is no 1:1 correspondence between the 'wrong' C parity and anomalous states. It has also been argued that anomalous states can indeed survive three-dimensional reductions and that the non-relativistic spectrum contains remnants of these states [28]. Hence, the remaining characteristic feature is that their eigenvalues remain finite at threshold. Ref. [21] concluded that the anomalous states in the Wick-Cutkosky model only appear for values of the coupling that clash with the renormalization of the underlying quantum field theory. In Fig. 3 one can see that they only appear in the physical mass spectrum if $c \geq 1/4$; the consistent renormalization (which involves solving Dyson–Schwinger equations for the propagators) restricts c to tiny values so they can never produce physical bound states and thus do not contribute poles to the S-matrix. In this sense anomalous states can be viewed as an artifact of the inconsistent combination of a ladder truncation with the usual tree-level propagators.

What Fig. 3 also makes clear is that the definition of anomalous states in the above sense only applies to systems without confinement. The rainbow-ladder solution for the quark propagator produces complex conjugate poles (see Fig. 2). Although this may not be the final word on the analytic structure of the quark propagator, it still poses a sufficient criterion for quark confinement due to positivity violations [29]. Consequently, the meson eigenvalues in Fig. 3 do not diverge at $M = 2m \sim 1$ GeV, where m is defined from the parabola that passes through the first complex conjugate pole pair (in analogy to the center graph in Fig. 2). Hence, there is no threshold and the mesons do not decay into free quarks, which also allows one to extrapolate their eigenvalues beyond the singularity limit.

In summary it appears that neither of the existing definitions of anomalous states is truly applicable to QCD. QCD exhibits confinement and does not have quark thresholds, and it is a relativistic quantum field theory that may produce bound states without a nonrelativistic limit. Perhaps the only remaining characteristic of anomalous states is in terms of eigenvalues of KG that never reach $\lambda_i = 1$ and thus do not produce poles in the S-matrix. Whether the appearance of such eigenvalues in BSE solutions is truly a problem or not remains an open question; but even if so, it is not necessarily an inherent flaw of the theory because they may equally be truncation artifacts. Another possible criterion for unphysical states concerns the signs in spectral functions [30] which are determined by the slopes of the eigenvalues λ_i; see the comment after Eq. (14) below.

Let us now return to the question of complex eigenvalues. The typical situation in a BSE is that K and G are both Hermitian but $KG \neq (KG)^\dagger$ because K and G do not commute. For unequal mass systems charge-conjugation invariance is lost, and the remnants of the positive- and negative C-parity solutions may start to interfere with each other. As discussed in [21], this is the underlying mechanism that can lead to pairs of complex conjugate eigenvalues. On the other hand, complex eigenvalues may also occur in the excitation spectrum of equal-mass bound states (although not in the scalar system). Can one find a criterion that enforces real eigenvalues?

Equation (6) is a generalized eigenvalue problem since it can be cast in the form

$$K\,[G\,\phi_i] = G^{-1}\lambda_i\,[G\,\phi_i]. \tag{8}$$

If G^{-1} is Hermitian and positive definite, one can employ a Cholesky decomposition $G^{-1} = LL^\dagger$ to transform the equation into

$$C\,[L^\dagger G\,\phi_i] = \lambda_i\,[L^\dagger G\,\phi_i]\,, \qquad C = L^{-1}K\,(L^\dagger)^{-1}\,, \tag{9}$$

where $C = C^\dagger$ if $K = K^\dagger$. Since the equation has the same eigenvalues λ_i, they must indeed be real and the eigenvectors orthogonal:

$$\phi_i^\dagger\,G^\dagger LL^\dagger G\,\phi_j = \phi_i^\dagger\,G\,\phi_j = \delta_{ij}. \tag{10}$$

One should emphasize that the λ_i and ϕ_i are defined for any $P^2 = -M^2$, not only at the onshell points $P^2 = -M_i^2$, and their orthogonality holds for each value of M separately (but in general not between states with different M). From the spectral decomposition of C we then further conclude:

$$K = \sum_i \lambda_i\,\phi_i\,\phi_i^\dagger\,, \qquad K^{-1} = \sum_i \frac{1}{\lambda_i}\,\varphi_i\,\varphi_i^\dagger\,, \qquad G = \sum_i \varphi_i\,\varphi_i^\dagger\,, \qquad G^{-1} = \sum_i \phi_i\,\phi_i^\dagger\,, \tag{11}$$

where $\varphi_i = G\,\phi_i$ is the vector corresponding to the Bethe–Salpeter wave function. Moreover, if we project the full n-body Green function $\mathbf{G}$ and its connected, amputated part (the $\mathbf{T}$-matrix) onto the Bethe–Salpeter eigenbasis for fixed J^{PC} we obtain the relations

$$\begin{array}{l} \mathbf{G} = G + G\,K\,\mathbf{G} \\ \mathbf{T} = K + K\,G\,\mathbf{T} \end{array} \quad \Rightarrow \quad \mathbf{G} = \sum_i \frac{\varphi_i\,\varphi_i^\dagger}{1-\lambda_i}\,, \qquad \mathbf{T} = \sum_i \lambda_i\,\frac{\phi_i\,\phi_i^\dagger}{1-\lambda_i} \tag{12}$$

which illustrate that each bound state ($\lambda_i = 1$ at $P^2 = -M_i^2$) leads to a pole in $\mathbf{G}$ and $\mathbf{T}$. They also allow us to derive the proper *normalization* from the residues at the pole positions:

$$\lambda_i = 1 + (P^2 + M_i^2)\,\frac{d\lambda_i}{dP^2}\bigg|_{P^2=-M_i^2} + \ldots\,, \tag{13}$$

so that the onshell Bethe–Salpeter amplitudes and wave functions pick up normalization factors g_i:

$$\mathbf{G} \xrightarrow{P^2\to -M_i^2} g_i^2\,\frac{\varphi_i\,\varphi_i^\dagger}{P^2+M_i^2}\,, \qquad g_i^2 = \left[-\frac{d\lambda_i}{dP^2}\right]^{-1}_{P^2=-M_i^2}. \tag{14}$$

Therefore, g_i^2 is always positive as long as $1/\lambda_i(M)$ approaches 1 from above as in Fig. 3—the opposite behavior would be another signature of an unphysical state. Equation (14) is equivalent to the standard Bethe–Salpeter normalization condition which can be written in the form [27]

$$-g_i^2\,\frac{d\lambda_i}{dP^2}\,\phi_i^\dagger\,G\,\phi_i\,\bigg|_{P^2=-M_i^2} \stackrel{!}{=} 1. \tag{15}$$

Finally, a similar decomposition holds for the vertex function ψ, defined by $G\,\psi = \mathbf{G}\,\psi_0$, which satisfies the inhomogeneous BSE $\psi = \psi_0 + K\,G\,\psi$:

$$\psi = \sum_i \frac{\alpha_i\,\phi_i}{1-\lambda_i} \xrightarrow{P^2\to -M_i^2} g_i^2\,\frac{\alpha_i\,\phi_i}{P^2+M_i^2}\,, \qquad \alpha_i = \phi_i^\dagger\,G\,\psi_0. \tag{16}$$

The previous relations (real eigenvalues, orthogonal eigenvectors, etc.) hold as long as $K = K^\dagger$, $G = G^\dagger$, and G is positive. In fact, the positivity of G is only a sufficient but not a necessary criterion because it can be relaxed in several ways: we could have equally employed a Cholesky decomposition for K, which means that either G *or* K should be positive; and if G^{-1} had only negative eigenvalues, we could still write $G^{-1} = -LL^\dagger$ to arrive at a Hermitian matrix $(-C) = (-C)^\dagger$. In any case, the occurrence of complex conjugate eigenvalues of KG signals that one or more of these criteria may be violated. In the following we will therefore review these assumptions in more detail.

Symmetry: The symmetry properties $K = K^T$, $G = G^T$ are not difficult to satisfy. The symmetry of G is already ensured by the BSE itself; for example, in Eq. (5) the condition $G_{mn}(x) = G_{nm}(x)$ is trivial because $z \in [-1, 1]$ and the Chebyshev polynomials are real on that domain. The symmetry of $K_{mn}(x, x') = K_{nm}(x', x)$, on the other hand, leads to $K(p, p', P) = K(p', p, P)$ which can be implemented by a symmetrized kernel. Pictorially it refers to the mirror symmetry of K in Fig. 2 around the vertical axis. The symmetry properties carry over to other types of BSEs, for example for $J = 0$ mesons as fermion–antifermion systems:

$$[\mathbf{\Gamma}(p, P)]_{\alpha\beta} = \int \frac{d^4 p'}{(2\pi)^4} \, [K(p, p', P)]_{\alpha\gamma,\beta\delta} \left[S(p'_+)\, \mathbf{\Gamma}(p', P)\, S(p'_-)\right]_{\gamma\delta}. \tag{17}$$

In this case the equation can be transformed into a set of Lorentz-invariant integral equations by projecting the amplitude onto its Dirac basis elements (see Sect. 3.4 in Ref. [11] for explicit examples):

$$[\mathbf{\Gamma}(p, P)]_{\alpha\beta} = \sum_i f_i(p^2, z, P^2 = -M^2)\, [\tau_i(p, P)]_{\alpha\beta}\,, \qquad \frac{1}{4}\,\text{Tr}\{\overline{\tau}_i\, \tau_j\} = \delta_{ij}\,, \tag{18}$$

where we used an orthonormal basis and charge conjugation was defined in Eq. (7). The resulting propagator matrix is again symmetric,

$$\begin{aligned} G_{ij}(p, P) &= \frac{1}{4}\,\text{Tr}\{\overline{\tau}_i(p, P)\, S_A(p_+)\, \tau_j(p, P)\, S_B(p_-)\} = \frac{1}{4}\,\text{Tr}\{S_B^T(p_-)\, \tau_j^T(p, P)\, S_A^T(p_+)\, \overline{\tau}_i^T(p, P)\} \\ &= \frac{1}{4}\,\text{Tr}\{\overline{\tau}_j(-p, -P)\, S_A(-p_+)\, \tau_i(-p, -P)\, S_B(-p_-)\} = G_{ji}(-p, -P) = G_{ji}(p, P)\,, \end{aligned} \tag{19}$$

and the same is true for its Chebyshev moments. In the last step we exploited the Lorentz invariance of $G_{ij}(p, P)$, which depends on p^2, z and $P^2 = -M^2$ only. Note that we only employed the *definition* of charge conjugation but not charge-conjugation invariance itself, so the symmetry property also holds for heavy-light systems as highlighted by the subscripts A and B on the propagators.

Reality: that K and G are real is less obvious and in general also not the case. In the presence of thresholds the eigenvalues can become complex as the masses are shifted into the complex plane; for instance by an explicit decay mechanism in the kernel K, or potentially already in G if a threshold in the propagators is crossed. In any case, the threshold locations are known in advance and below them the eigenvalues of KG should be real. In the left panel of Fig. 3 one expects real eigenvalues for all values of M because a rainbow-ladder gluon exchange does not provide a decay mechanism for mesons and the quarks have complex conjugate poles, whereas for the Wick-Cutkosky model the eigenvalues should be real and positive below the threshold at $M = 2m$.

Still, even in the absence of thresholds K and G are not necessarily real. For example, depending on the τ_i in Eq. (18) the dressing functions f_i can be either real or imaginary; or in unequal-mass systems the propagator product $G(p, P)$ becomes complex. In these cases, however, one can recast Eq. (6) in the form

$$(h\, K\, h)\, (h^\dagger G\, h^\dagger)\, (h\, \phi_i) = \lambda_i\, (h\, \phi_i)\,, \tag{20}$$

where the vectors $h\,\phi_i$ are real and h is a diagonal matrix whose entries are either 1 or i. As a consequence, $h\,K\,h$ and $h^\dagger G\, h^\dagger$ are real and symmetric matrices. The same reasoning applies for unequal-mass systems: with the Chebyshev expansion (3), odd Chebyshev moments are imaginary and therefore $K_{mn}(x)$ and $G_{mn}(x)$ are only symmetric and not real, but combinations such as $i^m G_{mn}(x)\, i^n$ are both real and symmetric. (It is customary to include a factor i^m directly in the Chebyshev expansion, but this is still not enough because in that case K_{mn} and G_{mn} are real but not symmetric.) In general, K and G are not Hermitian but only *pseudo-*Hermitian because they satisfy $K^\dagger = h^2 K\, h^2$ and $G^\dagger = h^2 G\, h^2$, where h^2 has diagonal entries ± 1. In these cases, however, one can redefine $h\,K\,h \to K$, $h^\dagger G\, h^\dagger \to G$ and $h\,\phi_i \to \phi_i$ to arrive at real and symmetric matrices K and G, which leads back to Eq. (6).

Positivity: Neither of these alterations have changed the original equation (6). The leftmost plot in Fig. 4 shows again the pion spectrum, but now also including the eigenvalues for the higher-lying states. We have ensured in the calculation that K and G are strictly real and symmetric and yet we occasionally encounter complex conjugate eigenvalues which split into real branches. We are therefore left with the third option, namely that G is not positive. To investigate this, we calculated the complete set of eigenvalues η_i and eigenvectors ξ_i

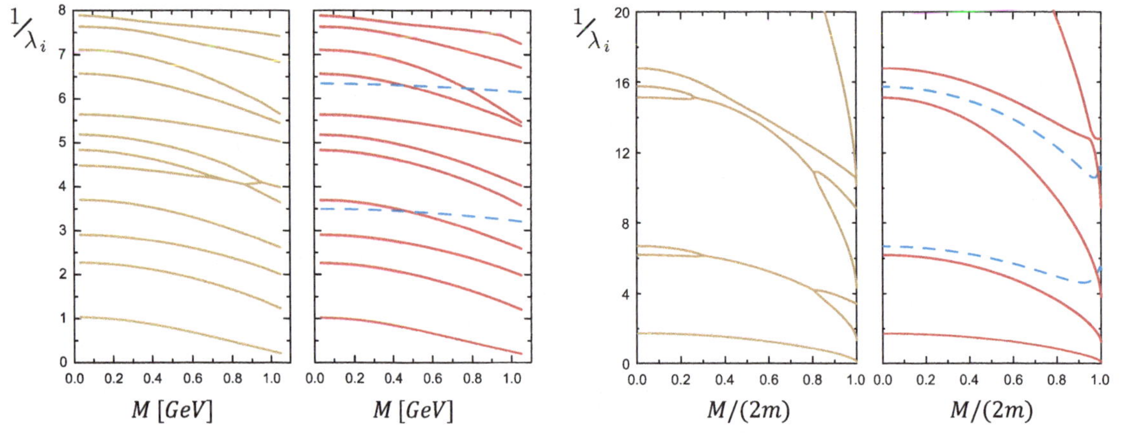

Fig. 4 *Left panels* Eigenvalue spectrum for the pion before and after modifying the propagator matrix. The *solid* (*red*) *lines* are obtained by removing the negative eigenvalues of G and the *dashed* (*blue*) *lines* by removing its positive eigenvalues. *Right panels* Analogous results for the Wick-Cutkosky model with unequal masses ($m_1/m_2 = 4$) and a nonvanishing exchange mass ($\mu = m_2$)

of G and reconstructed the propagator matrix from its spectral decomposition, where we retained its positive eigenvalues only:

$$G\,\xi_i = \eta_i\,\xi_i\,, \qquad \rightarrow \qquad G' = \sum_{\eta_i>0} \eta_i\,\xi_i\,\xi_i^\dagger\,. \tag{21}$$

We then insert the new propagator matrix into the original equation (6),

$$K\,G'\,\phi_i' = \lambda_i'\,\phi_i'\,, \tag{22}$$

and find that the new eigenvalues λ_i' are indeed all real.

It turns out that for equal-mass mesons removing the negative (positive) eigenvalues of G also removes the negative (positive) C-parity states in the spectrum. This could have equally been achieved by restricting to even or odd Chebyshev moments; however, the procedure accomplishes more than that. Fig. 4 for the pion is already restricted to positive C-parity states. In the second panel we show the new eigenvalue spectrum obtained by removing either the negative eigenvalues of G (solid, red) or its positive eigenvalues (dashed, blue). Removing the negative eigenvalues of G does not change the low-lying spectrum, but it *does* remove the complex eigenvalues in the higher-lying spectrum. Hence, all eigenvalues become real and the eigenvectors are automatically orthogonal in the sense of Eq. (10). The 'spurious' eigenvalues resemble the anomalous states in the Wick-Cutkosky model in the sense that they show much less curvature, which implies that they may not produce physical states.

To test the effect on an unequal-mass system, we investigated the Wick-Cutkosky model with $m_1/m_2 = 4$ and a nonvanishing exchange mass $\mu = m_2$. The third panel in Fig. 4 displays the original eigenvalue spectrum, where the splitting of complex conjugate eigenvalue pairs into two real branches is clearly visible (see also Fig. 5 in Ref. [21]). In the right panel we show the new eigenvalue spectrum. The previous complex conjugate pairs are now encompassed by two real eigenvalues, where the smaller one bends to zero as $M \rightarrow 2m$ and the larger one turns upwards and therefore does not generate a bound state. Interestingly, the procedure also leads to avoided level crossings (visible close to the threshold), which can switch the nature of 'normal' and 'anomalous' states. The 'normal' eigenvalues display sharp curvatures very close to the threshold which are sensitive to the numerical resolution. Since this makes the spectral reconstruction (21) numerically expensive we cannot yet resolve whether they truly vanish at the threshold, although it appears to be likely. We also note that in the unequal-mass system the restriction of G to its positive or negative spectrum does not restrict the resulting Bethe–Salpeter amplitudes to even or odd Chebyshev moments because they still contain both of them.

4 Summary

We discussed the light baryon spectrum in a quark–diquark approach where the quarks and diquarks are calculated from their Dyson–Schwinger and Bethe–Salpeter equations in QCD. The baryons' rest-frame orbital angular momentum contributions show clear traces of the non-relativistic quark-model, but due to Poincaré covariance their Faddeev amplitudes contain all possible values of L and such relativistic effects are already important for the nucleon, the Δ, and the Roper resonance.

We furthermore investigated the nature of complex conjugate eigenvalues that can appear in the excitation spectrum of Bethe–Salpeter equations. Our findings support the conjecture of Ref. [21] that complex eigenvalues are generated by the interference of 'normal' and 'anomalous' solutions. Although it is not clear whether a rigorous criterion for anomalous states exists in QCD, one may associate them with eigenvalues that do not produce poles in the S-matrix. We find that a restriction of the propagator matrix to its positive (or negative) spectrum disentangles the complex conjugate eigenvalues of the Bethe–Salpeter equation, so they become real and exhibit the characteristics of normal (or anomalous) solutions. It will be interesting to apply this strategy to the spectra of heavy-light mesons and, in particular, to the light baryon spectrum.

Acknowledgements I am grateful for interactions with C. S. Fischer, S.-X. Qin, H. Sanchis-Alepuz, and J. Segovia. This work was funded by DFG Project No. FI 970/11-1 and by the DFG collaborative research center TR 16.

References

1. E. Klempt, J.M. Richard, Baryon spectroscopy. Rev. Mod. Phys. **82**, 1095 (2010)
2. L. Tiator, D. Drechsel, S.S. Kamalov, M. Vanderhaeghen, Electromagnetic excitation of nucleon resonances. Eur. Phys. J. Special Top. **198**, 141 (2011)
3. I.G. Aznauryan et al., Studies of nucleon resonance structure in exclusive meson electroproduction. Int. J. Mod. Phys. E **22**, 1330015 (2013)
4. V. Crede, W. Roberts, Progress towards understanding baryon resonances. Rep. Prog. Phys. **76**, 076301 (2013)
5. M. Gockeler et al., Scattering phases for meson and baryon resonances on general moving-frame lattices. Phys. Rev. D **86**, 094513 (2012)
6. D.J. Wilson et al., Coupled $\pi\pi$, KK^- scattering in P-wave and the ρ resonance from lattice QCD. Phys. Rev. D **92**(9), 094502 (2015)
7. C.B. Lang, L. Leskovec, M. Padmanath, S. Prelovsek, Pion-nucleon scattering in the Roper channel from lattice QCD. arXiv:1610.01422 [hep-lat]
8. J.J. Wu, H. Kamano, T.-S.H. Lee, D.B. Leinweber, A.W. Thomas, Nucleon resonance structure in the finite volume of lattice QCD. arXiv:1611.05970 [hep-lat]
9. A. Bashir et al., Collective perspective on advances in Dyson-Schwinger Equation QCD. Commun. Theor. Phys. **58**, 79 (2012)
10. J. Segovia et al., Completing the picture of the Roper resonance. Phys. Rev. Lett. **115**(17), 171801 (2015)
11. G. Eichmann et al., Baryons as relativistic three-quark bound states. Prog. Part. Nucl. Phys. **91**, 1 (2016)
12. G. Eichmann, R. Alkofer, A. Krassnigg, D. Nicmorus, Nucleon mass from a covariant three-quark Faddeev equation. Phys. Rev. Lett. **104**, 201601 (2010)
13. M. Anselmino et al., Diquarks. Rev. Mod. Phys. **65**, 1199 (1993)
14. S.J. Brodsky, G.F. de Teramond, H.G. Dosch, J. Erlich, Light-Front Holographic QCD and Emerging Confinement. Phys. Rep. **584**, 1 (2015)
15. C. Patrignani et al., Review of Particle Physics. Chin. Phys. C **40**(10), 100001 (2016). (**PDG Collaboration**)
16. G. Eichmann, C.S. Fischer, H. Sanchis-Alepuz, On light baryons and their excitations. Phys. Rev. D **94**, 094033 (2016)
17. G. Eichmann, D. Nicmorus, Nucleon to Delta electromagnetic transition in the Dyson-Schwinger approach. Phys. Rev. D **85**, 093004 (2012)
18. M. Oettel, G. Hellstern, R. Alkofer, H. Reinhardt, Octet and decuplet baryons in a covariant and confining diquark-quark model. Phys. Rev. C **58**, 2459 (1998)
19. W.B. Kaufmann, Numerical solutions of the bethe-salpeter equation. Phys. Rev. **187**, 2051 (1969)
20. N. Seto, I. Fukui, Scalar scalar ladder model in the unequal mass case. 1: Numerical evidence for the complex eigenvalues. Prog. Theor. Phys. **89**, 205 (1993)
21. S. Ahlig, R. Alkofer, (In) consistencies in the relativistic description of excited states in the Bethe-Salpeter equation. Ann. Phys. **275**, 113 (1999)
22. E. Rojas, B. El-Bennich, J.P.B.C. de Melo, Exciting flavored bound states. Phys. Rev. D **90**, 074025 (2014)
23. B. El-Bennich, G. Krein, E. Rojas, F.E. Serna, Excited hadrons and the analytical structure of bound-state interaction kernels. Few Body Syst. **57**(10), 955 (2016)
24. T. Hilger, A. Krassnigg, Charming quasi-exotic open-flavor mesons. arXiv:1611.04334 [hep-ph]
25. G.C. Wick, Properties of Bethe-Salpeter Wave Functions. Phys. Rev. **96**, 1124 (1954)
26. R.E. Cutkosky, Solutions of a Bethe-Salpeter equations. Phys. Rev. **96**, 1135 (1954)
27. N. Nakanishi, A General survey of the theory of the Bethe-Salpeter equation. Prog. Theor. Phys. Suppl. **43**, 1 (1969)
28. J. Bijtebier, Bethe-Salpeter equation: 3-D reductions, heavy mass limits and abnormal solutions. Nucl. Phys. A **623**, 498 (1997)
29. R. Alkofer, L. von Smekal, The Infrared behavior of QCD Green's functions: Confinement dynamical symmetry breaking, and hadrons as relativistic bound states. Phys. Rep. **353**, 281 (2001)
30. M.S. Bhagwat et al., Schwinger functions and light-quark bound states. Few Body Syst. **40**, 209 (2007)

Few-Body Syst (2017) 58:118
DOI 10.1007/s00601-017-1286-2

Giovanni Salmè · Wayne de Paula · Tobias Frederico · Michele Viviani

Two-Fermion Bethe–Salpeter Equation in Minkowski Space: The Nakanishi Way

Received: 18 January 2017 / Accepted: 8 March 2017 / Published online: 23 March 2017

Abstract The possibility of solving the Bethe–Salpeter equation in Minkowski space, even for fermionic systems, is becoming actual, through the applications of well-known tools: (1) the Nakanishi integral representation of the Bethe–Salpeter amplitude and (2) the light-front projection onto the null-plane. The theoretical background and some preliminary calculations are illustrated, in order to show the potentiality and the wide range of application of the method.

1 Introduction

To achieve a fully covariant description of two-fermion systems, directly in Minkowski space, represents a challenging issue, that nowadays can be considered quite conceivable at least for analytic interaction kernels. As a matter of fact, the description can be carried out within the non perturbative framework given by the Bethe–Salpeter equation(BSE) [1] (see also Ref. [2] for a general introduction) and exploiting new approaches based on (1) the so-called Nakanishi integral representation (NIR) [3,4] of the BS amplitude (BSA) and (2) the light-front (LF) machinery. For instance, to work directly in the LF environment has a clear advantage in hadron physics, since one is immediately ready to calculate the relevant LF momentum distributions to be adopted in the investigation of several processes. However, it is worth reminding that to determine from the BSA, directly in Minkowski space, the LF momentum distributions (indeed, to be used in many areas, besides hadron physics) is the Holy Grail for both fundamental approaches, like the lattice calculations (though in Euclidean space) and phenomenological studies. As it is well-known, the fermionic nature of the interacting constituents produces difficulties that are far from trivial, but in our approach, where the NIR plays a pivotal role in the elaboration of the strategy for getting actual solutions of BSE, can be put under control.

This article belongs to the Topical Collection "Light Cone 2016".

G. Salmè (✉)
INFN, Sezione di Roma, Rome, Italy
E-mail: salmeg@roma1.infn.it

W. de Paula · T. Frederico
Departamento de Física, ITA, São José dos Campos, São Paulo, Brazil
E-mail: wayne@ita.br

T. Frederico
E-mail: tobias@ita.br

M. Viviani
INFN Sezione di Pisa, Pisa, Italy
E-mail: michele.viviani@pi.infn.it

2 The BSE in a Nutshell

We briefly recall the main path to BSE (for an extended review see, e.g., Ref. [2]), simplifying to the case of two scalars. One has to start with the 4-point Green's Function, given by

$$G(x_1, x_2; y_1, y_2) =< 0 \,|\, T\{\phi_1(x_1)\phi_2(x_2)\phi_1^+(y_1)\phi_2^+(y_2)\} \,|\, 0 > . \tag{1}$$

It fulfills an integral equation (see Fig. 1) that symbolically reads as follows

$$G = G_0 \,+\, G_0\,\mathcal{I}\,G \tag{2}$$

where $\mathcal{I}$ is the interaction kernel given by the infinite sum of irreducible Feynman graphs. In Fig. 2 some examples are given for the simple case of a ϕ^3 theory (see Ref. [5] for the caveats about such a model).

It should be pointed out that the iteration of the integral equation gives all the expected contributions. If one inserts a complete Fock basis in $G(x_1, x_2; y_1, y_2)$, then the bound state contribution (assuming only one non degenerate bound state for the sake of simplicity) appears as a pole in the Fourier space, i.e.

$$G_B(k, q; p_B, \beta) \simeq \frac{i}{(2\pi)^{-4}} \frac{\phi(k; p_B, \beta)\, \bar{\phi}(k; p_B, \beta)}{2\omega_B(p_0 - \omega_B + i\epsilon)} \tag{3}$$

where β is the set of possible quantum numbers, $\omega_B = \sqrt{M_B^2 + |\mathbf{p}|^2}$, $\phi(k; p_B, \beta)$ is BSA for a bound state, in momentum space. In configuration space, the BSA reads $\langle 0|T\{\phi_1(x_1)\phi_2(x_2)\}|p_B\,\beta\rangle$.

Close to the bound-state pole, i.e. $p_0 \to \omega_B$ the 4-point Green's function can be approximated by $G \simeq G_B + \mathit{regular\ terms}$ and one gets the homogeneous BSE, that holds for a bound state. In conclusion, the integral equation determining the BSA for a bound system, for simplicity without self-energy and vertex corrections, is given by

$$\phi(k; p_B, \beta) = G_0(k; p_B, \beta) \int d^4q'\, \mathcal{I}(k, q'; p_B)\, \phi(q'; p_B, \beta) \tag{4}$$

where for a two-scalar system the two-body free-propagator is

$$G_0 = \frac{i}{(\frac{p_B}{2} + k)^2 - m^2 + i\epsilon} \; \frac{i}{(\frac{p_B}{2} - k)^2 - m^2 + i\epsilon} \tag{5}$$

Notably, $\mathcal{I}(k, q'; p_B)$, the irreducible kernel in BSE, is the same one meets in the integral equation for the 4-point Green's function [cf Eq. (2)].

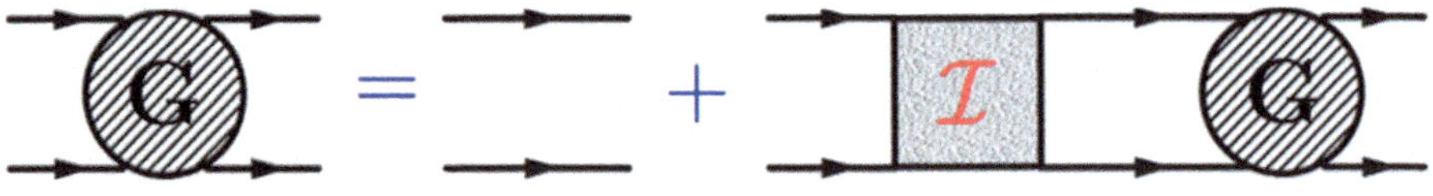

Fig. 1 Pictorial representation of the integral equation fulfilled by the 4-point Green's function

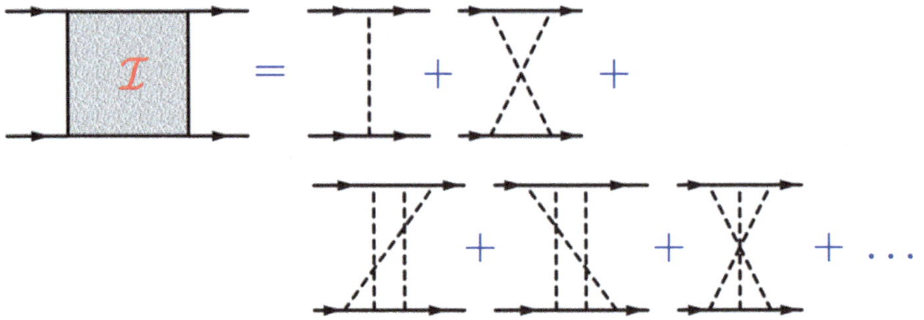

Fig. 2 First contributions to the interaction kernel in the integral equation for the 4-leg Green's function [cf Eq. (2)]

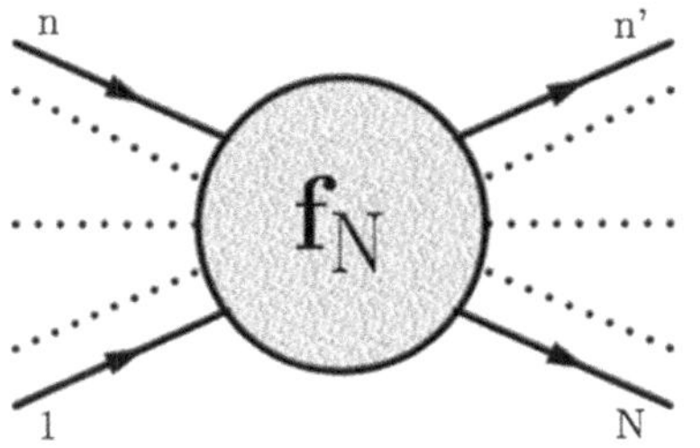

Fig. 3 Pictorial representation of a N-leg transition amplitude. Notice that the legs can correspond to on-mass-shell particles or to off-mass-shell ones, depending on the actual process under scrutiny

3 Nakanishi Integral Representation of N-Leg Transition Amplitudes

In the sixties, Nakanishi [3,4] proposed an integral representation for N-leg transition amplitudes, based on the parametric formula of the Feynman diagrams (Fig. 3). In a scalar theory, for N external legs, a generic contribution to the transition amplitude is given by

$$f_{\mathcal{G}_{nk}}(p_1, p_2, \ldots, p_N) \propto \prod_{r=1}^{k} \int d^4q_r \frac{1}{(\ell_1^2 - m_1^2)(\ell_2^2 - m_2^2) \ldots (\ell_n^2 - m_n^2)} \tag{6}$$

where n propagators and k loops are present (k is the number of integration variables). It is important to notice that the external momenta are N and does not change from one diagram to another, all belonging to the infinite set contributing to the given process one is investigating.

Nakanishi proposed a compact and elegant expression for the full N-leg amplitude

$$f^N(s) = \sum_{\mathcal{G}_{nk}} f_{\mathcal{G}_{nk}}(s)$$

where $s \equiv \{s_h\}$ is the set of all the independent scalar products one can construct from the N external momenta. The main ingredient for obtaining the Nakanishi integral representation, once the Feynman parametric representation for the amplitudes is adopted, is the following identity

$$1 \doteq \prod_h \int_0^1 dz_h \delta\left(z_h - \frac{\eta_h(\boldsymbol{\alpha})}{\beta}\right) \int_0^\infty d\gamma\ \delta\left(\gamma - \sum_l \frac{\alpha_l m_l^2}{\beta}\right) \tag{7}$$

where $\{\eta_h\}$ are suitable combinations of the Feynman parameters $\boldsymbol{\alpha} \equiv \{\alpha_i\}$, and $\beta = \sum_h \eta_h(\boldsymbol{\alpha})$ [3,4]. By integrating by parts $n - 2k - 1$ times one gets

$$f_{\mathcal{G}_{nk}}(s) \propto \prod_h \int_0^1 dz_h \int_0^\infty d\gamma \frac{\delta(1 - \sum_h z_h)\, \tilde{\phi}_{\mathcal{G}}(z, \gamma)}{(\gamma - \sum_h z_h s_h)} \tag{8}$$

where $\tilde{\phi}_{\mathcal{G}_{nk}}(z, \gamma)$ is a proper function (indeed a distribution). By adopting the Nakanishi change of variables one is able to move the dependence upon the details of the diagram, i.e. (n, k), from the denominator to the numerator. More notably, after performing the change of variables any diagram $\mathcal{G}_{nk}$ leads to a contribution $f_{\mathcal{G}_{nk}}(s)$ with just the same formal expression for the denominator, as given in Eq. (8). Eventually, this allows one to formally carry on the sum over the infinite set of Feynman diagrams contributing to a given N-leg transition amplitude $f^N(s)$.

As a matter of fact, the full N-leg transition amplitude can be formally written as

$$f^N(s) = \sum_{\mathcal{G}_{nk}} f_{\mathcal{G}_{nk}}(s) \propto \prod_h \int_0^1 dz_h \int_0^\infty d\gamma \frac{\delta(1 - \sum_h z_h)\, \phi_N(z, \gamma)}{(\gamma - \sum_h z_h s_h)} \tag{9}$$

where $\phi_N(z, \gamma) = \sum_{\mathcal{G}_{nk}} \tilde{\phi}_{\mathcal{G}_{nk}}(z, \gamma)$, is called the Nakanishi weight function for the N-leg transition amplitude.

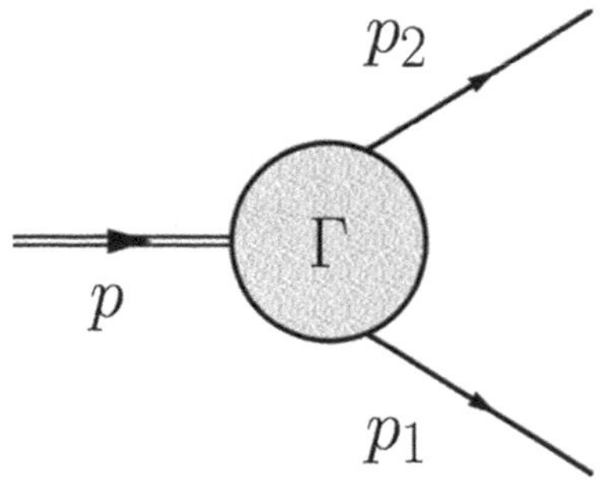

Fig. 4 The pictorial representation of the BSA, with p an on-mass-shell 4-momentum

Such an elegant expression can be exploited for obtaining the 3-leg transition amplitude or vertex function, that will be the main quantity to be tentatively adopted within the BS framework for describing a two-body interacting system, as discussed in what follows. The vertex function reads

$$f_3(s) = \int_0^1 dz \int_0^\infty d\gamma \frac{\phi_3(z,\gamma)}{\gamma - \frac{p^2}{4} - k^2 - zk \cdot p - i\epsilon} \tag{10}$$

with $p = p_1 + p_2$ and $k = (p_1 - p_2)/2$. Notice that only three independent scalar products can be constructed from the 4-momenta at disposal. In Fig. 4, it is pictorially shown the BSA, that can be obtained from $f_3(s)$ after multiplying the external legs (off-mass-shell) by the corresponding propagators.

We should anticipate that the application of the NIR of the vertex function to the BS framework is only formal. Let us clarify the meaning of this statement, that allows one to better understand the spirit of extending NIR, elaborated within a perturbative framework, such as the one based on Feynman diagrams, to a non perturbative regime, that is compelling for a realistic description of an interacting system (in a bound or scattering states). As shown in Eq. (10), the denominator contains the analytic structure of the amplitude and the numerator is a quantity that in principle one can obtain once a perturbation theory is considered, i.e. when a scattering process is described perturbatively. One can tentatively assume the formal expression (10) is valid also in a non perturbative regime, but taking as arbitrary the function in the numerator and keeping the same analytical structure in the denominator, i.e. the one determined by all the independent scalar products composed by the external 4-momenta. To construct the BSA from Eq. (10) one can simply multiply by the constituent propagators. This step preserves the overall analytic structure shown in Eq. (10) (the external momenta do not change!) but change the power of the denominator (see next section).

Summarizing: for positive energies of the system, one gets perturbatively the expression (10), and assumes that the analytic structure is the same also for a non perturbative scenario, planning to exploit the freedom associated to the numerator. This is taken as an arbitrary function, once we move from the perturbative framework to the non perturbative one. In spirit, this strategy could resemble the familiar steps one carries on when the harmonic oscillator eigen-problem is solved. First one determines the asymptotic behavior of the solutions and then, one realizes that in order to solve the eigenvalue problem it is necessary the presence of the Hermite polynomials. Similar steps are followed for actually solving BSE once the NIR of the BSA is introduced. Noteworthy, elaborating the Nakanishi arguments [3,4], one can always assume NIR for the BSA if the interaction kernel in BSE is analytical.

The homogeneous BSE for two scalars has been numerically solved within NIR framework in Ref. [6] (an early application via the uniqueness theorem of the Nakanishi weight function [4]) and in Refs. [7–9]. It should be pointed out that also the inhomogeneous BSE, suitable for investigating the scattering states, can be studied by using the NIR framework [10,11].

4 Projecting BSE Onto the LF Hyper-Plane $x^+ = 0$

As mentioned in the previous section, the appealing feature of NIR is the explicit appearance of the analytic structure of the vertex function and the presence of some *hidden* freedom, once the weight function is taken as an unknown quantity within a non perturbative context, i.e the BSE in Minkowski space. The basic question to be answered is: do actual solutions of the BSE have the analytic structure exposed by NIR? In what follows, it will be sketched how to reach a quantitative answer to the previous question by exploiting the LF framework (for more details, see Refs. [7,8,10,12,13]).

If one inserts the NIR of the BSA in the homogeneous BSE, one can perform the needed analytic integration obtaining an eigen-equation for the unknown weight function (indeed one gets a generalized eigenvalue problem). It is the integration on the LF variable k^-, also known as LF projection (see Ref. [14] and references quoted therein for a short review of the issue), that allows one to reduce the complexity of the 4D BSE.

Let us illustrate the issue in the simple case of two massive scalars interacting through the exchange of a massive scalar in ladder approximation, within the non explicitly covariant LF framework [8,10] (for the treatment within the explicitly covariant LF framework see Refs. [7,15]). First one assumes a NIR for the BSA, viz

$$\Phi_b(k,p) = \int_0^\infty d\gamma' \int_{-1}^{1} dz' \frac{g_b(\gamma', z', ; \kappa^2)}{\left[\gamma' - \frac{p^2}{4} - k^2 - z'k \cdot p - i\epsilon\right]^3} \tag{11}$$

where $g_b(\gamma', 1-2\xi; \kappa^2)$ is the Nakanishi weight function, to be determined, $p^2 = M^2$ is the 4-momentum of the interacting system and $\kappa^2 = m^2 - M^2/4$ yields the strength of the binding, since m is the mass of the constituents and $M = 2m - B$ with $B \equiv$ binding energy. The *valence component* of the Fock expansion of the two-scalar interacting state is readily obtained by integrating on k^- the BSA, $\Phi_b(k,p)$, and reads

$$\begin{aligned}\psi_{n=2}(\xi, k_\perp) &= \frac{p^+}{\sqrt{2}} \xi\,(1-\xi) \int \frac{dk^-}{2\pi} \Phi_b(k,p) \\ &= \frac{1}{\sqrt{2}} \xi\,(1-\xi) \int_0^\infty d\gamma' \frac{g_b(\gamma', 1-2\xi; \kappa^2)}{[\gamma' + \gamma + \kappa^2 + (2\xi-1)^2 \frac{M^2}{4} - i\epsilon]^2}\end{aligned} \tag{12}$$

where $\gamma \equiv |\mathbf{k}_\perp|^2 \in [0, \infty]$ and $\xi = (1+z)/2 \in [0,1]$. Finally, by applying the same LF projection to both sides of the homogeneous BSE, written as follows

$$\Phi_b(k,p) = G_0(k,p) \int d^4k' \, \mathcal{K}_{BS}(k,k',p)\, \Phi_b(k',p) \tag{13}$$

one gets [10]

$$\int_0^\infty d\gamma' \frac{g_b(\gamma', z; \kappa^2)}{[\gamma' + \gamma + z^2 m^2 + (1-z^2)\kappa^2 - i\epsilon]^2} = \int_0^\infty d\gamma' \int_{-1}^{1} dz' \, V_b^{LF}(\gamma, z; \gamma', z') g_b(\gamma', z'; \kappa^2). \tag{14}$$

with $V_b^{LF}(\gamma, z; \gamma', z')$ a new kernel, fully determined by the irreducible kernel $\mathcal{K}_{BS}(k,k',p)$ present in BSE, (13). The explicit form for the Nakanishi kernel V_b^{LF} in ladder approximation can be found in [7] and in [8], while Ref. [15] presents the cross-ladder case.

In conclusion, once we know an explicit form of the 4D kernel in BSE, (13) ,then we can obtain a generalized eigenvalue problem for determining the Nakanishi weight function $g_b(\gamma', z; \kappa^2)$, as given by Eq. (14). If solutions exist, this indicate that solutions of the BSE in Minkowski space can be written as a NIR. It is worth mentioning that the same approach has been applied to excited states [9], allowing the calculation of the valence momentum distributions for those states, and also to the evaluation of scattering lengths [11], i.e. a first insight in the continuum.

5 Spin Degrees of Freedom and BSE

To introduce spin degrees of freedom in the BSE is a non trivial task, and one has to carefully elaborate the treatment for getting reliable solutions [12,13]. An immediate consequence of the presence of further degrees of freedom is the need of decomposing the BSA in a suitable sum of Dirac structures multiplied by unknown scalar functions, that depend upon the external 4 momenta. Notice that the number of allowed Dirac structures is constrained by parity, total spin and the proper behavior of the BSA under Lorentz transformations. A first big difference between a two-fermion interacting system (but also for a fermion-boson or vector-vector cases) and a two-scalar one is an increase in the number of scalar functions determining the BSA, and consequently the number of Nakanishi weight functions. As a matter of fact, each scalar function that is present in the *macroscopical* expansion of the BSA can be written in terms of a NIR, generalizing the two-scalar case.

The first case that has been investigated is a 0^+ two-fermion system [12,13]. The corresponding BSA contains four scalar functions and can be written as follows

$$\Phi(k, p) = S_1\, \phi_1(k, p) + S_2\, \phi_2(k, p) + S_3\, \phi_3(k, p) + S_4\, \phi_4(k, p) \tag{15}$$

where S_i are the four Dirac structures compatible with the quantum numbers of the 0^+ state. For a 1^+ state the number of S_i doubles. Indeed there is some freedom in the actual choice of the 4×4 S_i matrices, but it is convenient [12] to implement an orthogonality relation between them. In particular, one can choose S_i such that $Tr\{S_i\, S_j\} = \mathcal{N}_i\, \delta_{ij}$. Hence, one gets

$$S_1 = \gamma_5, \quad S_2 = \frac{\not{p}}{M}\, \gamma_5, \quad S_3 = \frac{k \cdot p}{M^3}\not{p}\, \gamma_5 - \frac{1}{M}\not{k}\gamma_5, \, S_4 = \frac{i}{M^2}\sigma^{\mu\nu}\, p_\mu k_\nu\, \gamma_5 \tag{16}$$

The two-fermion BSE reads

$$\Phi(k, p) = \; S(p/2 + k) \int d^4k'\; F^2(k - k')\, i\mathcal{K}(k, k')\Gamma_1\; \Phi(k', p)\; \bar{\Gamma}_2\; S(k - p/2) \tag{17}$$

where $S(q)$ is the Dirac propagator

$$S(q) = i\frac{\not{q} + m}{q^2 - m^2 + i\epsilon} \tag{18}$$

and at each interaction vertex it has to be attached the following form factor

$$F(k - k') = \frac{(\mu^2 - \Lambda^2)}{[(k - k')^2 - \Lambda^2 + i\epsilon]} \tag{19}$$

Finally, the Dirac structure of the interaction vertices depends upon the boson that mediates the interaction in the simple model we are considering. In particular, in Refs. [12,13] three different exchanges have been inserted: (1) a scalar boson, i.e. $\Gamma_1 = \Gamma_2 = 1$; (2) a pseudoscalar boson, i.e $\Gamma_1 = \Gamma_2 = \gamma_5$ and (3) a vector boson, i.e $\Gamma_1 = \Gamma_2 = \gamma^\mu$. The notation $\bar{\Gamma}$ means $C\Gamma^T C^{-1}$ with C the charge conjugation. In Eq. (17), the quantity $\mathcal{K}(k, k')$ is the momentum-dependent part of the interaction kernel, that in ladder approximation is given for a massive scalar (pseudoscalar) by $\mathcal{K}_{S(PS)} = \pm g^2/[(k - k')^2 - \mu^2 + i\epsilon]$, and for a massless vector by $\mathcal{K}_V^{\mu\nu} = g^2\, g^{\mu\nu}/[(k - k')^2 + i\epsilon]$. Each unknown scalar function ϕ_i in Eq. (15) has a well-defined symmetry under the exchange $1 \to 2$, as dictated by the symmetry of both $\Phi(k, p)$ and the matrices S_i. In particular, $\phi_3(k, p) = -\phi_3(-k, p)$, while the others are symmetric. By using a NIR for each ϕ_i, as in Eq. (11), and applying a LF projection one gets

$$\psi_i(\gamma, z) = \int \frac{dk^-}{2\pi}\; \phi_i(k, p) = -\frac{i}{M}\int_0^\infty d\gamma' \frac{g_i(\gamma', z; \kappa^2)}{\left[\gamma + \gamma' + m^2 z^2 + (1 - z^2)\kappa^2 - i\epsilon\right]^2} \tag{20}$$

Finally, applying the LF projection to the rhs of the BSE in (17), one can formally transform the BSE in a coupled-equation system, viz

$$\psi_i(\gamma, z) = g^2 \sum_j \int_{-1}^{1} dz' \int_0^\infty d\gamma'\; g_j(\gamma', z'; \kappa^2)\; \mathcal{L}_{ij}(\gamma, z, \gamma', z'; p) \tag{21}$$

where the matrix $\mathcal{L}_{ij}(\gamma, z, \gamma', z'; p)$ yields the suitable Nakanishi kernel (its explicit expression is given in [13,16]). From Eqs. (20) and (21), one realizes that the coupled-equation system is a generalized eigenvalue problem where the Nakanishi weights $g_j(\gamma', z'; \kappa^2)$ are the eigen-vectors, and the interaction coupling g^2 plays the role of the eigen-value, once the binding B has been fixed. For actual calculations, a suitable orthonormal basis given by the Cartesian product of Laguerre polynomials in γ and Gegenbauer polynomials in z has been used [13,16], as in the case of the two-scalar system [8].

Table 1 Scalar couplings g^2 [they are eigenvalues of Eq. (21)] corresponding to a mass of the exchanged boson $\mu/m = 0.15$ and $\Lambda = 2$ in the vertex form factor

B/m	g^2_{dFSV}(ns)	g^2_{dFSV}(full) [13]	g^2_{CK} [12]
0.01	7.879	7.844	7.813
0.02	10.109	10.040	10.05
0.03	12.041	11.930	11.95
0.04	13.837	13.675	13.69
0.05	15.558	15.336	15.35
0.10	23.745	23.122	23.12
0.20	40.738	38.324	38.32
0.30	61.449	54.195	54.20
0.40	87.303	71.060	71.07
0.50	121.342	88.964	86.95

First column: assigned binding energies. Second column: g^2, obtained without *non singular* term in the kernel matrix. Third column: full calculation with *singular* contributions in $\mathcal{L}_{ij}$ exactly taken into account[13]. Fourth column: results obtained in Ref. [12] by using a smoothing function for treating the singular behavior of the integrals

While in the scalar interacting system the Nakanishi kernel is regular, for the fermionic case [13,16] the kernel matrix $\mathcal{L}_{ij}(\gamma, z, \gamma', z'; p)$ contains singular contributions produced by integrating on k^- the combination of the numerator of the fermionic propagators in the rhs and the operators S_i in $\Phi(k, p)$, on the lhs (notice that the combination is produced when we single out each scalar function to get Eq. (21)). Noteworthy, the non explicitly covariant LF framework allows one to straightforwardly determine the singular contributions to $\mathcal{L}_{ij}$, that have the following general form

$$\mathcal{C}_j = \int_{-\infty}^{\infty} \frac{dk^-}{2\pi} (k^-)^j \; \mathcal{S}(k^-, v, z, z', \gamma, \gamma') \tag{22}$$

with $j = 1, 2, 3$ and $\mathcal{S}(k^-, v, z, z', \gamma, \gamma')$ explicitly calculable [13,16]. For some values of the variables $\{v, z, z'\}$, one can have the worst case $\mathcal{S}(k^-, v, z, z', \gamma, \gamma') \sim 1/[k^-]^2$ for $k^- \to \infty$. Then, one cannot close the arc at ∞ for carrying out the needed analytic integration, but has to deal with singular behavior. Fortunately this kind of LF singularities can be carefully treated [13,16] by exploiting the studies of the field theory in the Infinite Momentum frame, performed in the seventies by T. M. Yan and collaborators (see in particular Ref. [17]). The relevant singular integral has the following expression

$$\mathcal{I}(\beta, y) = \int_{-\infty}^{\infty} \frac{dx}{\left[\beta x - y \mp i\epsilon\right]^2} = \pm \frac{2\pi i \; \delta(\beta)}{\left[-y \mp i\epsilon\right]} \tag{23}$$

Indeed, to fully account the singular behavior encountered in our analysis it is also needed to consider $1/2\partial/\partial\beta\mathcal{I}(\beta)$ that means to deal with $\partial/\partial\beta \; \delta(\beta)$. From the numerical point of view, this does not represent an issue, given the orthonormal basis adopted for expanding the Nakanishi weight functions.

Another positive fact is given by recognizing that in ladder approximation the severity of the singularities, i.e. the power j, depends only upon the constituent propagators and the structure of the BSA. Differently, in the explicit covariant LF framework, the trouble produced by the singular behavior of the relevant integrals needed for solving BSE with NIR was pragmatically fixed by introducing a suitable smoothing function [12].

An example of the numerical results we have obtained, and the quality of the comparisons we have achieved is given in Table 1, where for an exchanged scalar with mass $\mu/m = 0.15$, the values of the coupling constant g^2 corresponding to different values of the binding energy of the two-fermion system are given. The outcomes are obtained by solving the generalized eigenvalue problem in Eq. (21), after discretizing the coupled-equation system by expanding the Nakanishi weight functions $g_i(\gamma, z; \kappa^2)$ on the orthonormal basis above mentioned. The effect generated by the LF singularities can be appreciated by comparing the second column (without singular contributions) and the third one [13] (with singular contributions). In order to have a complete overview of the issue, the fourth column shows the results obtained in Ref. [12] by introducing a smoothing function for taking numerically under control the plague of the singularities. A more exhaustive comparison between the results obtained within the non covariant LF framework and the ones in the explicitly covariant LF approach is presented in Ref. [13]. The very nice agreement between results obtained in Minkowski space is made complete by the favorable comparison, also shown in [13], with the results from Euclidean space calculations [18].

The appealing to get solutions directly in Minkowski space is given by the possibility to evaluate LF distributions, as shown in [13]. There, the dynamical effect of the ladder exchange is illustrated by looking at the tail of the transverse-momentum behavior of the amplitude $\psi_i(\gamma, z)$ that nicely results in agreement with the finding in Ref. [19], where the asymptotic behavior of ψ_i for the pion [cf Eq. (20)] is determined by exploiting a very general counting rule.

6 Conclusions and Perspectives

The technique for solving the fermionic BSE in Minkowski space by using the Nakanishi integral representation of the Bethe–Salpeter amplitude can be now safely applied and extended, since a crucial point related to the treatment of light-front singularities has been singled out and fixed. The rule for dealing with the expected singularities, that in ladder approximation basically depend upon the structure of the BSA and not upon the complexity of the kernel, has been established [13, 16] and a short discussion has been illustrated in the present contribution. To reach a satisfactory treatment of the above mentioned singularities, that allows one to open the possibility of investigating a wide set of interacting systems with spin degrees of freedom, the LF framework plays a basic role due to the well-known advantages in performing analytical integrations. In ladder approximation, after obtaining a manageable form of the eigenvalue problem for the 0^+ system of two fermions, we have solved the coupled-equation system of four integral equations for the Nakanishi weight functions by using a suitable orthonormal basis. Our numerical investigations confirm both the robustness of the Nakanishi Integral Representation for the BSA, valid for any analytical BS kernel, and strongly encourages to extended the technique to other interesting cases: boson-fermion and vector-vector systems.

Calculations are in progress for the LF momentum distributions of the two-fermion system in the valence component, elucidating some formal subtleties.

Acknowledgements W. de Paula and T. Frederico thank the support of the Brazilian Institution CNPq, and G. Salmè acknowledges the support by CAPES.

References

1. E.E. Salpeter, H.A. Bethe, A relativistic equation for bound-state problems. Phys. Rev. **84**, 1232 (1951)
2. N. Nakanishi, A general survey of the theory of the Bethe–Salpeter equation. Prog. Theor. Phys. Suppl. **43**, 1 (1969)
3. N. Nakanishi, Partial-wave Bethe–Salpeter equation. Phys. Rev. **130**, 1230 (1963)
4. N. Nakanishi, *Graph Theory and Feynman Integrals* (Gordon and Breach, New York, 1971)
5. G. Baym, Inconsistency of cubic boson–boson interactions. Phys. Rev. **117**, 886 (1960)
6. K. Kusaka, K. Simpson, A.G. Williams, Solving the Bethe–Salpeter equation for bound states of scalar theories in Minkowski space. Phys. Rev. D **56**, 5071 (1997)
7. V.A. Karmanov, J. Carbonell, Solving Bethe–Salpeter equation in Minkowski space. Eur. Phys. J. A **27**, 1 (2006)
8. T. Frederico, G. Salmè, M. Viviani, Quantitative studies of the homogeneous Bethe–Salpeter equation in Minkowski space. Phys. Rev. D **89**, 016010 (2014)
9. C. Gutierrez, V. Gigante, T. Frederico, G. Salmè, M. Viviani, L. Tomio, Bethe–Salpeter bound-state structure in Minkowski space. Phys. Lett. B **759**, 131 (2016)
10. T. Frederico, G. Salmè, M. Viviani, Two-body scattering states in Minkowski space and the Nakanishi integral representation onto the null plane. Phys. Rev. D **85**, 036009 (2012)
11. T. Frederico, G. Salmè, M. Viviani, Solving the inhomogeneous Bethe–Salpeter equation in Minkowski space: the zero-energy limit. Eur. Phys. J. C **75**, 398 (2015)
12. J. Carbonell, V.A. Karmanov, Solving the Bethe–Salpeter equation for two fermions in Minkowski space. Eur. Phys. J. A **46**, 387 (2010)
13. W. de Paula, T. Frederico, G. Salmè, M. Viviani, Advances in solving the two-fermion homogeneous Bethe–Salpeter equation in Minkowski space. Phys. Rev. D **94**, 071901 (2016). arXiv: 1609.00868
14. T. Frederico, G. Salmè, Projecting the Bethe–Salpeter equation onto the light-front and back: a short review. Few-Body Syst. **49**, 163 (2011)
15. J. Carbonell, V.A. Karmanov, Bethe–Salpeter equation in Minkowski space with cross-ladder kernel. Eur. Phys. J. A **27**, 11 (2006)
16. W. de Paula, T. Frederico, G. Salmè, M. Viviani, in preparation
17. T.M. Yan, Quantum field theories in the infinite-momentum frame. IV. Scattering matrix of vector and dirac fields and perturbation theory. Phys. Rev. D **7**, 1780 (1973)

18. S.M. Dorkin, M. Beyer, S.S. Semikh, L.P. Kaptari, Two-fermion bound states within the Bethe–Salpeter approach. Few-Body Syst. **42**, 1 (2008)
19. X. Ji, J.P. Ma, F. Yuan, Generalized counting rule for hard exclusive processes. Phys. Rev. Lett. **90**, 241601 (2003)

Few-Body Syst (2017) 58:146
DOI 10.1007/s00601-017-1303-5

L'ubomír Martinovič

Quantum Field Theory in Two Dimensions: Light-front Versus Space-like Solutions

Received: 1 December 2016 / Accepted: 11 May 2017 / Published online: 29 June 2017

Abstract A few non-perturbative topics of quantum field theory in D $= 1 + 1$ are studied in both the conventional (SL) and light-front (LF) versions. First, we give a concise review of the recently proposed quantization of the two-dimensional massless LF fields. The LF version of bosonization follows in a simple and natural way including the bosonized form of the Thirring model. As a further application, we demonstrate the closeness of the 2D massless LF quantum fields to conformal field theory (CFT). We calculate several correlation functions including those between the components of the LF energy-momentum tensor and derive the LF version of the Virasoro algebra. Using the Euclidean time variable, we can immediately transform calculated quantities to the (anti)holomorphic form. The results found are in agreement with those from CFT. Finally, we show that the proposed framework provides us with the elements needed for an independent LF study of exactly solvable models. We compute the non-perturbative correlation functions from the exact operator solution of the LF Thirring model and compare it to the analogous results in the SL theory. While the vacuum effects are automatically taken into account in the LF case, the non-trivial vacuum structure has to be incorporated by an explicit diagonalization of the SL Hamiltonians, to obtain the equivalently complete solution.

1 Introduction

It is a prevailing opinion that the light front (LF) form of quantum field theory (QFT) has a great potential, but it is simultaneously considered to be somewhat mysterious, perhaps even incomplete or at least counter-intuitive. Its features that are superior to the conventional ("space-like"—SL) form of QFT, include the minimal number (3) of dynamical Poincaré generators [1], the status of the vacuum state, and a reduced number of independent field components. The most fundamental aspect is the equality of the physical vacuum state to the Fock vacuum, which follows from the positivity and conservation of the LF momentum p^+. Only the field zero modes, carrying $p^+ = 0$, and a narrow set of (symmetry) operators [2,3], can transform the LF Fock vacuum into a more complex object, which however will still be much simpler than its SL counterpart.

What is then the relation between the SL and LF theory? Could it be that the LF version conceptually and technically simplifies the structure of QFT while maintaining ability to correctly represent the complicated relativistic dynamics? The area of 2D solvable relativistic models offers us a very suitable environment to

This article belongs to the Topical Collection "Light Cone 2016".

L. Martinovič (✉)
Institute of Physics, Slovak Academy of Sciences Dúbravská cesta 9, 845 11 Bratislava, Slovakia
E-mail: fyziluma@savba.sk

Present address:
L. Martinovič
Bogoliubov Laboratory of Theoretical Physics, JINR, Joliot-Curie 8, Dubna, Russia 141 980

study these questions [4]. However, for a long time, it was not clear how to correctly quantize 2D massless LF fields which are a prerequisite for exact solutions of these models. It was suspected that some LF massless degrees of freedom cannot be initialized as they propagate along the light-like (characteristic) surfaces.

Recently, a simple and natural way of quantizing the two-dimensional massless LF fields has been suggested [5,6]. In our contribution, we shall first give a brief exposition of this quantization scheme. Its validity will be confirmed by the LF bosonization of the massless fermion field. After that, the closeness of the massless LF quantum fields to conformal field theory (CFT) will be demonstrated by calculating several correlation functions of elementary and composite operators, as well as the LF version of the Virasoro algebra. Finally, the most general operator solution and correlation functions of the LF Thirring model will be discussed.

Throughout this paper, we will use the following LF notation: $x^\mu = (x^+, x^-) = (x^0 + x^1, x^0 - x^1)$. The momentum is designed as k^μ (or p^μ), $k^\mu = (k^+, k^-)$, while $\hat{k}^-$ is the on-shell LF energy:

$$\partial_\pm = \frac{\partial}{\partial x^\pm}, \quad \hat{k} \cdot x = \frac{1}{2}k^+ x^- + \frac{1}{2}\hat{k}^- x^+, \quad k^2 = \mu^2 \Rightarrow \hat{k}^- = \frac{\mu^2}{k^+}. \tag{1}$$

2 Quantization of Free Massless Light-front Fields in 2D

The free massive **LF scalar field** obeys the LF field equation $(4\partial_+\partial_- + \mu^2)\phi(x) = 0$, whose solution is

$$\phi(x) = \int_0^{+\infty} \frac{dk^+}{\sqrt{4\pi k^+}} \left[a(k^+)e^{-\frac{i}{2}k^+x^- - \frac{i}{2}\frac{\mu^2}{k^+}x^+} + a^\dagger(k^+)e^{\frac{i}{2}k^+x^- + \frac{i}{2}\frac{\mu^2}{k^+}x^+}\right], \left[a(k^+), a^\dagger(l^+)\right] = \delta(k^+ - l^+). \tag{2}$$

The vacuum state is defined by means of the annihilation operator as $a(k^+)|0\rangle = 0$. The LF Hamiltonian and momentum operator is given in terms of densities $T^{++} = 4 : \partial_-\phi\partial_-\phi :$ $T^{+-} = \mu^2 : \phi^2 :$

$$P^\nu = \frac{1}{2}\int_{-\infty}^{+\infty} dx^- T^{+\nu}(x) = \int_0^{+\infty} dk^+ \hat{k}^\nu a^\dagger(k^+)a(k^+), \quad \hat{k}^\nu = (k^+, \frac{\mu^2}{k^+}). \tag{3}$$

Setting $\mu = 0$ in the plane-wave factors of (2), we obtain the x^-- dependent massless scalar field right away. But the massless LF Klein-Gordon equation $\partial_+\partial_-\tilde{\phi}(x) = 0$ admits also a piece depending solely on x^+:

$$\tilde{\phi}(x) = \tilde{\phi}(x^-) + \tilde{\phi}(x^+). \tag{4}$$

The latter can be recovered from (2) by the change of variables $k^+ = \frac{\mu^2}{k^-}$. x^+ and x^- interchange their places in (2), and the new Fock operators $\tilde{a}(k^-), \tilde{a}^\dagger(k^-)$ emerge naturally in the massless limit [6]. The following formulae summarize quantization of the massless LF scalar field and its algebra:

$$\tilde{\phi}(x^+) = \int_0^{+\infty} \frac{dk^-}{\sqrt{4\pi k^-}} \left[\tilde{a}(k^-)e^{-\frac{i}{2}k^-x^+} + \tilde{a}^\dagger(k^-)e^{\frac{i}{2}k^-x^+}\right], \quad \left[\tilde{a}(k^-), \tilde{a}^\dagger(l^-)\right] = \delta(k^- - l^-), \tag{5}$$

$$\tilde{\phi}(x^-) = \int_0^{+\infty} \frac{dk^+}{\sqrt{4\pi k^+}} \left[\tilde{a}(k^+)e^{-\frac{i}{2}k^+x^-} + \tilde{a}^\dagger(k^+)e^{\frac{i}{2}k^+x^-}\right], \quad \left[\tilde{a}(k^+), \tilde{a}^\dagger(l^+)\right] = \delta(k^+ - l^+), \tag{6}$$

$$\left[\tilde{\phi}(x^-), \tilde{\phi}(y^-)\right] = -\frac{i}{4}\epsilon(x^- - y^-), \quad \left[\tilde{\phi}(x^+), \tilde{\phi}(y^+)\right] = -\frac{i}{4}\epsilon(x^+ - y^+). \tag{7}$$

One also finds $\left[\tilde{a}(k^+), \tilde{a}^\dagger(l^-)\right] = 0$, i.e. the decoupling of two types of modes. The 2-point functions calculated from the massless fields (5,6) and their $x^\pm$ derivatives exactly coincide with the massless limit of the corresponding two-point functions of the massive field and its derivatives. This is an important consistency check of our quantization scheme. The 2-point functions $\langle 0|\phi(x^\pm)\phi(y^\pm)|0\rangle$ are infrared divergent and have to be regularized by an infrared cutoff. Note that the variables k^+ and k^- actually coincide, in the complete analogy with the SL case $k^0 = |k^1|$. Using the above Fock algebra, one verifies that the dynamical operators

$$P^+ = \int_0^{+\infty} dk^+ k^+ a^\dagger(k^+) a(k^+), \quad P^- = \int_0^{+\infty} dk^+ k^- a^\dagger(k^-) a(k^-) \tag{8}$$

generate the correct Heisenberg equations $2i\partial_\pm\phi(x^\mp) = -[P^\mp, \phi(x^\mp)] = 0$, $2i\partial_\pm\phi(x^\pm) = -[P^\mp, \phi(x^\pm)]$.

The analogous considerations enable us to consistently quantize also the massless **light front fermion field**. The massive two-dimensional field (Dirac) equation $i\gamma^\mu\partial_\mu\psi(x) = m\psi(x)$ decomposes as

$$2i\partial_+\psi_2(x) = m\psi_1(x),\ 2i\partial_-\psi_1(x) = m\psi_2(x) \ \Rightarrow\ \psi_2(x) = \tilde{\psi}_2(x^-),\ \psi_1(x) = \tilde{\psi}_1(x^+),\ \text{if } m = 0. \tag{9}$$

The massless limit of the solution

$$\psi_2(x) = \int_0^{+\infty} \frac{dp^+}{4\pi}\left[b(p^+)e^{-\frac{i}{2}p^+x^- - \frac{i}{2}\frac{m^2}{p^+}x^+} + d^\dagger(p^+)e^{\frac{i}{2}p^+x^- + \frac{i}{2}\frac{m^2}{p^+}x^+}\right], \tag{10}$$

$$\{b(p^+), b^\dagger(q^+)\} = \{d(p^+), d^\dagger(q^+)\} = \delta(p^+ - q^+) \tag{11}$$

of the dynamical equation (9) is straightforward, while the above change of variables is needed for the solution

$$\psi_1(x) = \int_0^{+\infty} \frac{dp^+}{4\pi}\frac{m}{p^+}\left[b(p^+)e^{-\frac{i}{2}p^+x^- - \frac{i}{2}\frac{m^2}{p^+}x^+} - d^\dagger(p^+)e^{\frac{i}{2}p^+x^- + \frac{i}{2}\frac{m^2}{p^+}x^+}\right] \tag{12}$$

of the constraint in (9). The quantum massless representation is then characterized by the following relations:

$$\tilde{\psi}_2(x^-) = \int_0^{+\infty} \frac{dp^+}{\sqrt{4\pi}}\left[\tilde{b}(p^+)e^{-\frac{i}{2}p^+x^-} + \tilde{d}^\dagger(p^+)e^{\frac{i}{2}p^+x^-}\right],\ \{\tilde{b}(p^+), \tilde{b}^\dagger(q^+)\} = \{\tilde{b} \to \tilde{d}\} = \delta(p^+ - q^+),$$

$$\tilde{\psi}_1(x^+) = \int_0^{+\infty} \frac{dp^-}{\sqrt{4\pi}}\left[\tilde{b}(p^-)e^{-\frac{i}{2}p^-x^+} - \tilde{d}^\dagger(p^-)e^{\frac{i}{2}p^-x^+}\right],\ \{\tilde{b}(p^-), \tilde{b}^\dagger(q^-)\} = \{\tilde{b} \to \tilde{d}\} = \delta(p^- - q^-), \tag{13}$$

$$\left\{\tilde{\psi}_1(x^+), \tilde{\psi}_1^\dagger(y^+)\right\} = \delta(x^+ - y^+), \quad \left\{\tilde{\psi}_2(x^-), \tilde{\psi}_2^\dagger(y^-)\right\} = \delta(x^- - y^-). \tag{14}$$

The two kinds of modes decouple: $\{\tilde{b}(p^-), \tilde{b}^\dagger(q^+)\} = \{\tilde{d}(p^-), \tilde{d}^\dagger(q^+)\} = 0$. The two-point functions of the massless components coincide with the massless limits of the corresponding massive two-point function. From the expansions (13), one can construct the bilinear operators (we omit tilde henceforth): the current $j^\mu = (:\psi_1^\dagger\psi_1:, :\psi_2^\dagger\psi_2:)$ and the scalar densities $\psi_2^\dagger\psi_1 \pm \psi_1^\dagger\psi_2$. Thus, the quantum theory of the massless LF fermion field is established. The necessary information is contained in the original massive solutions and no new degrees of freedom have to be introduced by hand [7,8]. The derivation opens the road to the genuine LF solutions of the class of interacting models with massless fermions.

Consistency of the scheme is further confirmed by **LF bosonization**. Bosonization is a remarkable property of the 2D field theory: fermion fields can be represented in terms of boson variables [9,10]. Our derivation of its LF version is based on the natural decomposition of the massless $\phi(x)$ and $\psi(x)$ fields (4), (9).

Consider first $\tilde{\psi}_2(x^-)$. Assume that it can be represented as

$$\varphi_2(x^-) = C : e^{i\alpha\phi(x^-)} : \ = Ce^{i\alpha\phi^{(-)}(x^-)}e^{i\alpha\phi^{(+)}(x^-)}. \tag{15}$$

It is now sufficient to adjust the constants C and α in such a way, that two φ_2 with different arguments anticommute and $\varphi_2(x^-)$, $\varphi_2^\dagger(y^-)$ satisfy the anticommutation relation (14). The first condition fixes α to the value $\hat{\alpha} = 2\sqrt{\pi}$. The second one determines the constant C as $\hat{C} = \left(\frac{\lambda e^{\gamma_E}}{4\pi}\right)^{1/2}$ (λ is the infrared cutoff associated with the massless $D_0^{(+)}$ function [6] and γ_E is the Euler's constant). It follows that the operators $\hat{\varphi}(x^-)$ and the analogously obtained $\hat{\varphi}(x^+)$ represent the bosonized form of the fields $\tilde{\psi}_2(x^-)$ and $\tilde{\psi}_1(x^+)$. Forming their appropriate point-split products, the bosonized vector current is found to be $\hat{j}^\pm(x^\mp) = 2\pi^{-1/2}\partial_\mp\phi(x^\mp)$. It correctly reproduces the Schwinger term in the current-current commutators:

$$\left[\hat{j}^+(x^-), \hat{j}^+(y^-)\right] = \frac{i}{\pi}\partial_x\delta(x^- - y^-), \; \left[\hat{j}^-(x^+), \hat{j}^-(y^+)\right] = \frac{i}{\pi}\partial_x\delta(x^+ - y^+). \tag{16}$$

Similarly, for the scalar densities one obtains in a straightforward way

$$\overline{\psi}(x)\psi(x) = \frac{\lambda e^{\gamma_E}}{4\pi}\cos\left(2\sqrt{\pi}\phi(x)\right), \; \overline{\psi}(x)\gamma^5\psi(x) = i\frac{\lambda e^{\gamma_E}}{4\pi}\sin\left(2\sqrt{\pi}\phi(x)\right). \tag{17}$$

Thus the LF version of bosonization yields the results known from the SL theory.

3 Conformal Properties of the 2D Massless LF Fields

The massless 2D fields exhibit conformal symmetry, whose (anti)holomorphic formulation was developed in [11]. In this section, we calculate a few correlation functions and the Virasoro algebra within our LF formalism. Using the Euclidean time in the obtained results, we find a complete agreement with CFT. Also, a c-number quantum term arises naturally in the commutators between the conformal generators.

Hamiltonian density $T^{+-}(x)$ of the massless scalar field vanishes, as required by conformal symmetry, $TrT^{\mu\nu} = T^{+-} = 0$ (the $\mu = 0$ limit (8) of the massive $P^- \neq 0$, however). The other components are

$$T^{++}(x^-) =: \pi(x^-)\pi(x^-) :, \; T^{--}(x^+) =: \theta(x^+)\theta(x^+) :, \quad \pi(x^-) \equiv 2\partial_-\phi(x^-), \; \theta(x^+) \equiv 2\partial_+\phi(x^+). \tag{18}$$

Note that the LF Hamiltonian (8) can also be obtained as the x^+-integral of the density $T^{--}(x^+)$, analogously to P^+ which is the x^--integral of $T^{++}(x^-)$. Next, we compute a few additional correlation functions

$$\langle 0|\theta(x^+)\theta(y^+)|0\rangle = \frac{1}{\pi}\frac{1}{(x^+ - y^+ - i\delta^+)^2}, \; \langle 0|\pi(x^-)\pi(y^-)|0\rangle = \frac{1}{\pi}\frac{1}{(x^- - y^- - i\delta^-)^2}, \tag{19}$$

as well as those between the components of the energy-momentum tensor,

$$\langle 0|T^{++}(x^-)T^{++}(y^-)|0\rangle = \frac{2}{\pi^2}\frac{1}{(x^- - y^- - i\delta^-)^4}, \; \langle 0|T^{--}(x^+)T^{--}(y^+)|0\rangle = \frac{2}{\pi^2}\frac{1}{(x^+ - y^+ - i\delta^+)^4}. \tag{20}$$

In the holomorphic form of 2D CFT [11,12], the Laurent expansion in the variables

$$z = e^{\frac{2\pi}{L}\zeta}, \; \bar{z} = e^{\frac{2\pi}{L}\bar{\zeta}}, \; \text{where } \zeta = \tau - ix, \; \bar{\zeta} = \tau + ix, \tag{21}$$

is commonly used. It is based on radial quantization with the euclidean time τ, $t \to -i\tau$. We need to reformulate our results for $\phi(x)$ in the form of infinite series to conform with the discrete picture of [11]. Thus, we consider the massive field in a finite box of lenght $2L$ in x^- or $2T$ in x^+ with periodic boundary conditions $\phi(x^+, x^- = -L) = \phi(x^+, x^- = L)$, $\phi(x^+ = -T, x^-) = \phi(x^+ = T, x^-)$. Performing the change of variables and the massless limit as before, we arrive at

$$\phi(x^-) = \phi_0 + \sum_{n=1}^{\infty}\frac{1}{\sqrt{2Lk_n^+}}\left[a_n e^{-\frac{i}{2}k_n^+x^-} + H.c.\right] = \frac{1}{\sqrt{4\pi}}\sum_{n=\pm1,\ldots}\frac{1}{\sqrt{|n|}}a_n e^{-i\frac{\pi}{L}nx^-}, \; a_{-n} \equiv a_n^\dagger, \tag{22}$$

The component $\phi(x^+)$ has the same form as the normal-mode part of (22) with a_n replaced by $\bar{a}_n$. The Fock operators obey $\left[a_m, a_n^\dagger\right] = \left[\bar{a}_m, \bar{a}_n^\dagger\right] = \delta_{m,n}$, $\left[\bar{a}_m, a_n^\dagger\right] = 0$. For $\mu = 0$, $\phi_0 \neq 0$. It reduces however to a constant whose conjugate momentums vanishes. The 2-point functions $D_0^{(+)}$ are evaluated for $L >> 1$ as

$$D_0^{(+)}(z^\pm) = \langle 0|\phi(x^\pm)\phi(y^\pm)|0\rangle = \frac{1}{4\pi}\sum_{n=1}^{\infty}\frac{1}{n}e^{-i\frac{\pi}{L}n(z^\pm - i\epsilon)} \approx \frac{1}{4\pi}\ln\left[\frac{i\pi}{L}(z^\pm - i\epsilon)\right]. \tag{23}$$

L plays the role of the infared regularization parameter. It also introduces the necessary dimension to (23). L drops out of all the other correlation functions (due to the derivatives present). The results match the continuum results (19) - (20). The components of the energy-momentum tensor in the discrete form read

$$T^{++}(x^-) = K \sum_{m,n} \epsilon(m)\epsilon(n)\sqrt{|m||n|} : a_m a_n : e^{-i\frac{\pi}{L}(n+m)x^-}, \; K = -\frac{\pi}{L^2}. \tag{24}$$

$T^{--}(x^+)$ has the same form with a_n replaced by $\bar{a}_n$. Both can be transformed to a "Virasoro form". Indeed, consider the Fourier transform of $T^{++}(x^-)$:

$$T^{++}(x^-) = \frac{1}{4L^2} \sum_{l=0,\pm1,\ldots} L_l e^{-i\frac{\pi}{L}lx^-}, L_l = 2L \int_{-L}^{+L} dx^- e^{i\frac{\pi}{L}lx^-} T^{++}(x^-). \tag{25}$$

Inserting $T^{++}(x^-)$ in the Fock form (24) into (25) gives ($L_0 = 4LP^+$)

$$L_n = -4\pi \sum_{k=\pm1,\ldots} \epsilon(k)\epsilon(n-k)\sqrt{|k||n-k|} a_k a_{n-k}. \tag{26}$$

A calculation based on the commutators below Eq.(22) yields the LF version of the Virasoro algebra, including the c-number term ("anomaly"), not present at the classical level:

$$\left[L_n, L_m\right] = (n-m)L_{n+m} + \frac{c}{12}n(n^2-1)\delta_{n+m,0}, \; c = 1, \tag{27}$$

where c is the "central charge". Taking $T^{--}(x^+)$ in (25) instead of T^{++} generates the algebra (27) with $L_n \to \bar{L}_n$. It follows from $\left[a_n, \bar{a}_m\right] = 0$ that $\left[L_n, \bar{L}_m\right] = 0$.

To give a few details of these calculations, we switch back to the "$a, a^\dagger$" picture:

$$L_n = -\sum_{k=1}^{n-1} \sqrt{k(n-k)} a_k a_{n-k} + 2\sum_{k=n+1}^{\infty} \sqrt{k(k-n)} a^\dagger_{k-n} a_k, \; L^\dagger_n = L_{-|n|}. \tag{28}$$

The "anomaly" comes from the commutator between the first terms:

$$\left[\sum_{l=1}^{m-1} \sqrt{l(m-l)} a_l a_{m-l}, \sum_{k=1}^{n-1} \sqrt{k(n-k)} a^\dagger_k a^\dagger_{n-k}\right] = \sum_{l=1}^{m-1} \sqrt{l(m-l)} \sum_{k=1}^{n-1} \sqrt{k(n-k)}$$
$$\times \left\{\delta_{m-l,k}\delta_{l,n-k} + \delta_{l,k}\delta_{m-l,n-k}\right\} = 2\delta_{m,n} \sum_{l=1}^{m-1} l(m-l) = \frac{1}{3}m(m^2-1)\delta_{m,n}.$$

These results agree with the CFT results after taking into account the different normalization. They can be transformed to the conformal ((anti)holomorphic) form by switching to the variables ζ and $\bar{\zeta}$ (21).

Indeed, with the conventional CFT normalization (factor 2π instead of 4 in $T^{\mu\nu}$) (18), we get (cf. Eq.(19)):

$$\langle 0|\pi(\zeta)\pi(\zeta')|0\rangle = -\frac{1}{(\zeta-\zeta')^2}, \; \langle 0|T(\zeta)T(\zeta')|0\rangle = \frac{c}{2}\frac{1}{(\zeta-\zeta')^4}, \; c = 1. \tag{29}$$

Field expansions (22), analogous to the transition to the conformal field in the usual treatment [11,12], read

$$\phi(\zeta) = \frac{1}{\sqrt{4\pi}} \sum_{n=\pm1,\pm2,\ldots} \frac{1}{\sqrt{|n|}} a_n z^n, \; [a_m, a_n] = \delta_{m+n,0}, \; \phi(\bar{\zeta}) = \phi(\zeta) \text{ with } (\zeta, z, a_n) \to (\bar{\zeta}, \bar{z}, \bar{a}_n). \tag{30}$$

Thus, the results of CFT are derived independently in the LF operator formalism (including the fermion field).

4 Solvable Models in the LF Formulation: The Thirring Model

The quantization of the massless LF fermion field enables one to find independent exact LF operator solutions of a few two-dimensional models. The Thirring model is the most natural starting example. Classical Lagrangian density of the model and the corresponding Euler-Lagrange equation are

$$\mathcal{L} = \frac{i}{2}\overline{\Psi}\gamma^{\mu} \overleftrightarrow{\partial_{\mu}} \Psi + \frac{1}{2}gJ_{\mu}J^{\mu}, \quad i\gamma^{\mu}\partial_{\mu}\Psi = -g\gamma_{\mu}J^{\mu}, \quad J^{\mu}(x) = \overline{\Psi}(x)\gamma^{\mu}\Psi(x). \tag{31}$$

In the LF theory, the field equation decomposes into two equations:

$$2i\partial_{+}\Psi_2 = -gJ^{-}\Psi_2, \;\; 2i\partial_{-}\Psi_1 = -gJ^{+}\Psi_1, \quad J^{+} = 2\Psi_2^{\dagger}\Psi_2, \;\; J^{-} = 2\Psi_1^{\dagger}\Psi_1. \tag{32}$$

Defining the potentials $J(x)$, $\tilde{J}(x)$ and $J(x^{\pm})$ by $J(x) = J(x^{+}) + J(x^{-})$, $\tilde{J}(x) = -(J(x^{+}) - J(x^{-}))$, $J^{+}(x^{-}) = \frac{2}{\sqrt{\pi}}\partial_{-}J(x) = \frac{2}{\sqrt{\pi}}\partial_{-}\tilde{J}(x)$, $J^{-}(x^{+}) = \frac{2}{\sqrt{\pi}}\partial_{+}J(x) = -\frac{2}{\sqrt{\pi}}\partial_{+}\tilde{J}(x)$, the most general solution of Eqs. (32) is

$$\Psi(x) = e^{i\frac{g}{\sqrt{\pi}}\left(\alpha J(x) - \beta\gamma^5\tilde{J}(x)\right)}\psi(x), \quad \alpha + \beta = 1, \;\; J^{\mu}(x) = G(g)j^{\mu}(x). \tag{33}$$

The factor $G(g) = (1 - \frac{g}{2\pi})^{-1}$ results from calculating the interacting current from the solution (33) as the point-split product [5]: the free current gets renormalized by interaction in quantum theory. This aspect was omitted in the original solution [13]. The quantum (normal-ordered) solution in the component form reads

$$\Psi_1(x) = e^{i\lambda G\left((\alpha-\beta)j^{(-)}(x^{+}) + j^{(-)}(x^{-})\right)}\psi_1(x^{+})e^{i\lambda G\left((\alpha-\beta)j^{(+)}(x^{+}) + j^{(+)}(x^{-})\right)}, \tag{34}$$

$$\Psi_2(x) = e^{i\lambda G\left(j^{(-)}(x^{+}) + (\alpha-\beta)j^{(-)}(x^{-})\right)}\psi_2(x^{-})e^{i\lambda G\left(j^{(+)}(x^{+}) + (\alpha-\beta)j^{(+)}(x^{-})\right)}, \quad \lambda \equiv \frac{g}{\sqrt{\pi}}. \tag{35}$$

The expansion of the $\psi(x)$ field is given in (13). We now bosonize the currents by a Fourier transformation:

$$j^{+}(x^{-}) = -\frac{i}{\sqrt{\pi}}\int_0^{\infty}\frac{dk^{+}\,k^{+}}{\sqrt{4\pi k^{+}}}\left[c(k^{+})e^{-\frac{i}{2}k^{+}x^{-}} - c^{\dagger}(k^{+})e^{\frac{i}{2}k^{+}x^{-}}\right], \quad \left[c(k^{+}), c^{\dagger}(l^{+})\right] = \delta(k^{+} - l^{+}),$$

$$\hat{c}(k^{+}) = \int_0^{\infty} ds^{+}\left[b^{\dagger}(s^{+})b(s^{+} + k^{+}) - d^{\dagger}(s^{+})d(s^{+} + k^{+}) + d(p^{+})b(k^{+} - p^{+})\theta(k^{+} - s^{+})\right],$$

$$c(k^{+}) = \frac{i\hat{c}(k^{+})}{2\sqrt{k^{+}}} \tag{36}$$

and analogously for $j^{-}(x^{+})$, where $b(k^{+}), d(k^{+}), c(k^{+}) \to b(k^{-}), d(k^{-}), c(k^{-})$ in the corresponding formulae. With an implicit infrared regularization, the potentials ("integrated currents") are represented as

$$j(x^{-}) = \int_0^{\infty}\frac{dk^{+}}{\sqrt{4\pi k^{+}}}\left[c(k^{+})e^{-\frac{i}{2}k^{+}x^{-}} + c^{\dagger}(k^{+})e^{\frac{i}{2}k^{+}x^{-}}\right], \;\; j(x^{+}) = \int_0^{\infty}\frac{dk^{-}}{\sqrt{4\pi k^{-}}}\left[c(k^{-})e^{-\frac{i}{2}k^{-}x^{+}} + H.c.\right], \tag{37}$$

They are two components of a massless LF scalar field $j(x)$. The interacting potential $J(x) = G(g)j(x)$.

Express now the Lagrangian and Hamiltonian in terms of independent fields, using the solution (33):

$$\mathcal{L} = i\Psi_2 \overleftrightarrow{\partial_{+}} \Psi_2 + i\Psi_1 \overleftrightarrow{\partial_{-}} \Psi_1 + \frac{1}{2}gJ^{+}J^{-} \to \mathcal{L} = i\psi_2 \overleftrightarrow{\partial_{+}} \psi_2 + i\psi_1 \overleftrightarrow{\partial_{-}} \psi_1 - \frac{1}{2}gJ^{+}J^{-}. \tag{38}$$

The Hamiltonian derived from the above Lagrangian acquires a very simple form

$$P^{-} = \frac{1}{2}\int_{-\infty}^{+\infty} dx^{-}T^{+-}(x) = \frac{1}{2}g\int_{-\infty}^{+\infty} dx^{-}J^{+}(x^{-})J^{-}(0) = gG^2(g)Q\tilde{Q}, \;\; Q = \hat{c}(0), \;\; \tilde{Q} = j^{-}(0). \tag{39}$$

The 2-point functions can be calculated from the solutions (35) using the operator identities $Ae^B = e^\lambda e^B A$, if $[A, B] = \lambda A$, λ being a c-number, and $e^A e^B = e^{[A,B]} e^B e^A$, if $[A, B]$ is a c-number. For example, we get

$$\langle 0|\Psi_1(x^+)\Psi_1{}^\dagger(y^+)|0\rangle = \frac{i\mu}{2\pi}\frac{1}{\mu z^+ - i\delta}\frac{1}{(i\mu z^+ + \delta)^{\frac{\frac{g^2}{\pi}G^2(\alpha-\beta)^2 - 2gG(\alpha-\beta)}{4\pi}}}\frac{1}{(i\mu z^- + \delta)^{\frac{g^2}{4\pi^2}}}, \quad z = x - y. \tag{40}$$

Its Fourier transform has no pole indicating absence of asymptotic states [13,14]. It is instructive to compare these results with the SL solution. The main difference rests in the fact that the SL Hamiltonian is not diagonal when expressed in terms of the composite operators $c(k^1)$, $c^\dagger(k^1)$, analogous to (37). After its diagonalization by a unitary operator $U(\tilde{\gamma}) = e^{iS(\tilde{\gamma})}$, where $S(\tilde{\gamma})$ is a hermitian operator with off-diagonal Fock structure [4], the Fock vacuum $|0\rangle$ becomes the true lowest-energy eigenstate. The constant $\tilde{\gamma}$ is fixed at the value ensuring the diagonal form of the Hamiltonian. However, all operators including the solution $\Psi(x)$ have to be simultaneously transformed by U. This step again has not been performed in the original solution [13]. Only with its inclusion one can hope for a match between the correlation functions in the LF and SL form of the model. This question along with the analogous comparison between the two forms of the solutions is under study also for the gradient-coupling model, the Thirring-Wess and the Federbush model (L. Martinovic, unpublished).

Acknowledgements This work was in part done in collaboration with P. Grangé and supported by the Grant VEGA2/0072/2013.

References

1. P.A.M. Dirac, Rev. Mod. Phys. **21**, 392 (1949)
2. L. Martinovic, Phys. Lett. B **509**, 355 (2001)
3. L. Martinovic, J.P. Vary, Phys. Rev. D **64**, 105016 (2001)
4. L. Martinovic, P. Grangé, Phys. Lett. B **724**, 310 (2013)
5. L. Martinovic, P. Grangé, Few Body Syst. **56**, 607 (2015)
6. L. Martinovic, P. Grangé, Few Body Syst. **57**, 565 (2016)
7. G. McCartor, Z. Phys. C **64**, 349 (1994)
8. G. McCartor, S.S. Pinsky, D.G. Robertson, Phys. Rev. D **56**, 1025 (1997)
9. S. Coleman, Phys. Rev. D **11**, 2088 (1975)
10. S. Mandelstam, Phys. Rev. D **11**, 3026 (1975)
11. A.A. Belavin, A.M. Polyakov, A.B. Zamolodchikov, Nucl. Phys. B **241**, 333 (1984)
12. P. Di Francesco, P. Mathieu, D. Senechal, *Conformal Field Theory* (Springer, New York, 1997)
13. B. Klaiber, The Thirring model, in *Lectures in theoretical physics* in theoretical physics, vol XA, (New York, 1968), p. 141
14. E. Abdalla, M.C.B. Abdalla, K.D. Rothe, *Nonperturbative Methods in Two-Dimensional Quantum Field Theory* (World Scientific, Singapore, 1991)

Few-Body Syst (2017) 58:78
DOI 10.1007/s00601-017-1221-6

W. N. Polyzou · Gordon J. Aiello

Euclidean Relativistic Quantum Mechanics: Scattering Asymptotic Conditions

Received: 8 November 2016 / Accepted: 5 January 2017 / Published online: 7 February 2017

Abstract We discuss the formulation of the scattering asymptotic condition in a relativistic quantum theory formulated in terms of reflection positive Euclidean Green functions.

1 Introduction

Schwinger [1] showed that the spectral condition implies that the Green functions of a local quantum field theory have an analytic continuation to Euclidean times. This observation has motivated Euclidean formulations of quantum field theory that have been used successfully in numerical lattice discretizations of the theory. One difficulty with the Euclidean formulation is that the imaginary time leads to problems in formulating scattering problems [2] that involve asymptotic time limits. Osterwalder and Schrader [3] proved a reconstruction theorem that showed how to construct a relativistic quantum field theory from a set of Euclidean invariant Green functions satisfying a condition called reflection positivity. The proof of the reconstruction theorem has three important attributes. First, it is constructive, leading to an explicit representation of the physical Hilbert space and a set of self-adjoint operators on that space satisfying the Poincaré Lie algebra. Second, the locality property is logically independent of the other properties that are needed to construct a relativistic quantum theory. This implies that it should be possible to make approximations or truncations of a local theory that still retain all of the remaining properties of a relativistically invariant quantum theory. Third, the construction of the relativistic quantum theory does not require explicit analytic continuation of the Euclidean times. This paper is based on [4,5] where we argue that given a Hilbert space representation and a set of Poincaré generators on that space that satisfy cluster properties, it should be possible to formulate scattering directly in the Euclidean representation of the Hilbert space.

2 Reconstruction of Quantum Mechanics

A dense set of vectors in the Hilbert space consists of sets of normalizable functions of Euclidean space–time variables,

This article belongs to the Topical Collection "Light Cone 2016".

Supported by the US DOE office of science—award number DE-SC0016457.

W. N. Polyzou (✉)
Department of Physics and Astronomy, The University of Iowa, Iowa city, IA, USA
E-mail: polyzou@uiowa.edu
Tel.: 1-319-335-1856

G. J. Aiello
Department of Applied Mathematics and Computational Sciences, The University of Iowa, Iowa city, IA, USA
E-mail: gordon-aiello@uiowa.edu

$$\mathbf{f} = (f_0, f_1(x_{e11}), f_2(x_{e21}, x_{e22}), \ldots), \tag{1}$$

with support for positive relative Euclidean times, $0 < x^0_{en1} < x^0_{en2} < x^0_{en3} \cdots$.

The Hilbert-space inner product of two such vectors is expressed in terms of a quadratic form involving the Euclidean Green functions, $G^e_n(x_{e1}, \ldots, x_{en})$, by

$$\langle \mathbf{f} | \mathbf{g} \rangle = \tag{2}$$

$$\sum_{mn} \int f^*_m(x_{em1}, \ldots, x_{emm}) G^e_{m+n}(T_e x_{emm}, \ldots, T_e x_{em1}, y_{en1}, \ldots, y_{enn}) \times$$

$$g_n(y_{en1}, \ldots, y_{enn}) d^{4m} x d^{4n} y \tag{3}$$

where T_e is Euclidean time reflection. The collection of Euclidean Green functions are called reflection positive if $\langle \mathbf{f} | \mathbf{f} \rangle \geq 0$ for all $\mathbf{f}$ satisfying the support condition. This represents the physical Hilbert space inner product, even though all of the integrals in (3) are over Euclidean variables and there is no analytic continuation. This inner product has zero norm vectors, so Hilbert space vectors are actually equivalence class of functions $\mathbf{f}$ subject to the equivalence relation $\mathbf{f} \sim \mathbf{g}$ if and only if $\langle \mathbf{f} - \mathbf{g} | \mathbf{f} - \mathbf{g} \rangle = 0$.

The infinitesimal generators of the Poincaré group are represented by the following self-adjoint operators on this space

$$H f_n(x_{en1}, x_{en2}, \ldots, x_{enn}) = \sum_{k=1}^{n} \frac{\partial}{\partial x^0_{enk}} f_n(x_{en1}, x_{en2}, \ldots, x_{enn}) \tag{4}$$

$$\mathbf{P} f_n(x_{en1}, x_{en2}, \ldots, x_{enn}) = -i \sum_{k=1}^{n} \frac{\partial}{\partial \mathbf{x}_{enk}} f_n(x_{en1}, x_{en2}, \ldots, x_{enn}) \tag{5}$$

$$\mathbf{J} f_n(x_{en1}, x_{en2}, \ldots, x_{enn}) = -i \sum_{k=1}^{n} \mathbf{x}_{enk} \times \frac{\partial}{\partial \mathbf{x}_{enk}} f_n(x_{en1}, x_{en2}, \ldots, x_{enn}) \tag{6}$$

$$\mathbf{K} f_n(x_{en1}, x_{en2}, \ldots, x_{enn}) = \sum_{k=1}^{n} \left(\mathbf{x}_{enk} \times \frac{\partial}{\partial x^0_{enk}} - x^0_{enk} \frac{\partial}{\partial \mathbf{x}_{enk}} \right) f_n(x_{en1}, x_{en2}, \ldots, x_{enn}). \tag{7}$$

It is straightforward to show that these operators are Hermetian with respect to the inner product (3) and satisfy the Poincaré commutation relations. They are also self-adjoint by virtue of being generators of unitary one-parameter groups, contractive semigroups, or local symmetric semigroups [6–8] on this Hilbert space. The Hamiltonian H also satisfies a spectral condition as a consequence of reflection positivity [9].

If the Euclidean Green functions satisfy cluster properties,

$$\lim_{\mathbf{a} \to \infty} (G^e_{m+n}(X_{em} + \mathbf{a}, Y_{en}) - G^e_m(X_{em}) G^e_n(Y_{en})) \to 0 \tag{8}$$

$$X_{em} = (x_{em1}, \ldots, x_{emm}) \tag{9}$$

then the Poincaré generators defined by (4–7) also satisfy cluster properties.

3 Formulation of Scattering

Scattering matrix elements, $S_{fi} = \langle \mathbf{f}_+ | \mathbf{f}_- \rangle$, are defined as the inner product of initial and final states at a given time that asymptotically look like systems of non-interacting particles long before and long after a collision. The relation between these states and the corresponding asymptotic states is given by the scattering asymptotic condition, which replaces the initial condition on the initial or final states by the scattering asymptotic conditions

$$\lim_{t \to \pm\infty} \| e^{-iHt} | \mathbf{f}_\pm \rangle - \Phi e^{-iH_{ft}} | \chi \rangle \| = 0. \tag{10}$$

The quantity $|\chi\rangle$ on the right describes free-particle wave packets. For a two-particle final state it has the form

$$\Phi e^{-iH_{f12}t} | \chi_{12} \rangle := \sum_{\mu_1 \mu_2} \int \underbrace{|\phi_{b1}, \mathbf{p}_1, \mu_1, \phi_{b2}, \mathbf{p}_2, \mu_2\rangle}_{\Phi} \tag{11}$$

$$\times \underbrace{e^{-i(E_1(\mathbf{p}_1)+E_2(\mathbf{p}_2))t}}_{e^{-iH_{f12}t}} d\mathbf{p}_1 d\mathbf{p}_2 \underbrace{\chi_1(\mathbf{p}_1,\mu_1)\chi_2(\mathbf{p}_2,\mu_2)}_{wave\ packets} \tag{12}$$

where

$$M_{bi}|\phi_{bi}\rangle = m_{b_i}|\phi_{bi}\rangle \qquad E_i(\mathbf{p}_i) = \sqrt{m_{b_i}^2 + \mathbf{p}_i^2} \tag{13}$$

and m_{b_i} is in the point-spectrum of the subsystem mass operator. It follows that the operator Φ is a mapping from a two-particle asymptotic Hilbert space of square integrable functions of $\mathbf{p}_1, \mu_1, \mathbf{p}_2, \mu_2$, to the physical Hilbert space with inner product (3).

A sufficient condition for the existence of these limits is the Cook condition, which in this notation has the form

$$\int_a^\infty \|(H\Phi - \Phi H_f)e^{\mp iH_f t}|\chi\rangle\| dt < \infty \tag{14}$$

where $(H\Phi - \Phi H_f)$ replaces the short-range potential in the original formulation of Cook's method [10].

This condition can be formulated in the Euclidean representation. Two conditions are needed to establish convergence in the Euclidean case. First, the Green functions need to satisfy cluster properties. Second, the operator Φ must eliminate the contribution from the disconnected parts of the Green function in (10).

To illustrate how this works consider the case of 2–2 scattering. Cluster properties lead to a four-point Green function that can be expressed as the sum of products of two-point functions and a connected fourpoint function:

$$G_4 = G_{2e}G_{2e} + G_{4ec}. \tag{15}$$

In general there will be additional disconnected terms—the analysis below can be easily extended to treat these cases.

In the Euclidean representation the square of integrand of (10) for the case of 2–2 scattering has the form

$$\begin{aligned}
&\|(H\Phi - \Phi H_f)e^{-iH_f t}|\chi_{0\pm}(0)\rangle\|^2 = (\chi_f, e^{iH_f t}(\Phi^\dagger H - H_f\Phi^\dagger), T_e(G_{2e}G_{2e} + G_{4ec})(H\Phi \\
&-\Phi H_f)e^{-iH_{i}t}\chi_i) = \int \chi_1^*(\mathbf{p}_1)\chi_2^*(\mathbf{p}_2)e^{i(E_{m_1}(\mathbf{p}_1)+E_{m_2}(\mathbf{p}_2))t} d\mathbf{p}_1 d\mathbf{p}_2 \\
&\times \left(\frac{\partial}{\partial x_1^0} + \frac{\partial}{\partial x_2^0} - E_{m_1}(\mathbf{p}_1) - E_{m_2}(\mathbf{p}_2)\right)\langle \mathbf{p}_1, \mathbf{p}_2|\Phi^\dagger|x_1, x_2\rangle \\
&\times d^4x_1 d^4x_2 (G_{2e}(T_e x_1, y_1)G_{2e}(T_e x_2, y_2) + G_{4ec}(T_e x_1, T_e x_2, y_2, y_1))d^4y_1 d^4y_2 \\
&\times \left(\frac{\partial}{\partial y_1^0} + \frac{\partial}{\partial y_2^0} - E_{m_1}(\mathbf{p}_1') - E_{m_2}(\mathbf{p}_2')\right)\langle y_1, y_2|\Phi|\mathbf{p}_1', \mathbf{p}_2'\rangle \\
&\times e^{-i(E_{m_1}(\mathbf{p}_1')+E_{m_2}(\mathbf{p}_2'))t}\chi_1(\mathbf{p}_1')\chi_2(\mathbf{p}_2')d\mathbf{p}_1' d\mathbf{p}_2'
\end{aligned} \tag{16}$$

where $E_{m_i}(\mathbf{p}) = \sqrt{m_i^2 + \mathbf{p}^2}$ are the energies of the outgoing particles, and G_{4ec} is the connected part of the Euclidean 4-point Green function.

We expect that the contribution to this expression from the connected Green functions will fall off like $1/t^3$ for large time, as it does in the Minkowski case (see [11, p. 132]). For the disconnected terms it is necessary to study the contribution from two-point Green functions. These can be expressed in terms of the Kallën–Lehmann representation of the Euclidean two-point function:

$$\begin{aligned}
\langle\xi|\psi\rangle &= \int \xi^*(x_e)G_{2e}(T_e x_e - y_e)\psi(y_e)d^4x_e d^4y_e \\
&= \int \frac{d^4p_e d^4x_e d^4y_e}{(2\pi)^4}\xi^*(x_e)\frac{e^{ip_e^0(-x_e^0-y_e^0)+i\mathbf{p}\cdot(\mathbf{x}-\mathbf{y})}\rho(m)}{(p_e^0)^2+\mathbf{p}^2+m^2}\psi(y_e)dm \\
&= \int \underbrace{\xi^*(x_e)\frac{d^4x_e}{(2\pi)^{3/2}}e^{-E_m(\mathbf{p})x_e^0+i\mathbf{p}\cdot\mathbf{x}}}_{\tilde{\xi}^*(\mathbf{p},m)}\frac{\rho(m)dmd\mathbf{p}}{2E_m(\mathbf{p})}\underbrace{e^{-E_m(\mathbf{p})y_e^0-i\mathbf{p}\cdot\mathbf{y}}\frac{d^4y_e}{(2\pi)^{3/2}}\psi(y_e)}_{\tilde{\psi}(\mathbf{p},m)}
\end{aligned} \tag{17}$$

The Euclidean time support condition of the wave functions $\psi(y_e)$ and $\xi^*(x_e)$ allows the p_0 integral to be computed by contour integration.

The subsystem mass operator can be seen to be the Euclidean 4-Laplacian by integration by parts in (17):

$$\int \frac{\rho(m)}{2E_m(\mathbf{p})} e^{-E_m(\mathbf{p})y_e^0 - i\mathbf{p}\cdot\mathbf{y}} \frac{d^4 y_e}{(2\pi)^{3/2}} \nabla^2_{ye} \psi(y_e) \tag{18}$$

$$= \int \frac{\rho(m)}{2E_m(\mathbf{p})} e^{-E_m(\mathbf{p})y_e^0 - i\mathbf{p}\cdot\mathbf{y}} \frac{d^4 y_e}{(2\pi)^{3/2}} m^2 \psi(y_e) \tag{19}$$

The problem is to find subsystem point mass eigenstates $\psi(y_e)$ and $\xi(x_e)$ with the mass of the asymptotic particle that are also consistent with the Euclidean time-support condition.

The spectrum of the subsystem mass operator is identical to the support of the Lehmann weight, $\rho(m)$. The Lehmann weight is assumed to have the form

$$\rho(m) = \sum z_i \delta(m - m_i) + \rho_{ac}(m) \tag{20}$$

where the support of the continuous part, $\rho_{ac}(m)$ of $\rho(m)$ is a half line that starts at the sum of the masses of the lightest possible intermediate particles.

In order to select the desired asymptotic particle it is necessary to choose wave functions $\psi(x_e)$ that only get contributions from the Lehmann weight corresponding to mass of the asymptotic particle. This corresponds to one of the delta functions in (20).

One way to do this, motivated by the method used by Haag and Ruelle [11–13], is to multiply the wave function by a function $h(m^2)$ that is 1 when m^2 is the square of the mass of the asymptotic particle and 0 on the rest of the Lehmann weight, $\rho(m^2)$. Since the square of the cluster mass operator is the Euclidean Laplacian, a wave function of the form

$$\psi(x_e) = h(\nabla^2) g(x_e) \tag{21}$$

has the desired properties provided it satisfies the support condition. Here $g(x_e)$ is another function satisfying the support condition. The simplest choice of $g(x_e)$ satisfying the support condition is a function of the form

$$g(x_e) = \delta(x_e^0 - \tau)\tilde{g}(\mathbf{x}). \tag{22}$$

It is interesting that in spite of the presence of the delta function that enforces the support condition, this is a normalizable vector with respect to the scalar product (3).

The question is whether the support conditions on $h(m^2)$ and $h(\nabla^2)g(x_e)$ are compatible. The concern is that on one hand an infinite series of derivatives, like a translation operator, can change support conditions. On the other hand $h(m^2)$ cannot even be analytic if the Lehmann weight has any continuous spectrum because $h(m^2)$ must vanish on the continuous support of the Lehmann weight.

In the unphysical situation, where the support of the Lehmann weight consists of a finite number of discrete points, a finite-degree polynomial $h(\nabla^2)$ can be constructed that selects the desired mass and automatically preserves the support condition. Also, in the case where the support of the continuous portion of the Lehmann weight is compact, the Weierstrass approximation theorem implies that $h(\nabla^2)$ can be uniformly approximated by a polynomial. While both cases are unphysical, both preserve the support condition of the wave function.

This leads to the question of whether it is possible to make a polynomial approximation to $h(m^2)$ that converges to any continuous function of the half line. It turns out that there is a sufficient condition called the Carleman condition that provides a suitable sufficient (but not necessary) condition.

The concern is that applying $h(\nabla^2)$ to $g(x_e)$ does not automatically preserve the positive Euclidean-time support condition of $g(x_e)$.

The Carleman condition [14,15] gives a sufficient condition

$$\sum_{n=1}^{\infty} \gamma_n^{-\frac{1}{2n}} > \infty \tag{23}$$

on an infinite collection of moments γ_n on the half line

$$\gamma_n = \int_0^{\infty} w(x) dx x^n \tag{24}$$

to uniquely solve for the measure, $w(x)dx$.

When this condition is satisfied, multiplication by x has a unique-self adjoint extension on the Hilbert space with weight $w(x)$ and the orthogonal polynomials with respect to the weight $w(x)$ are complete.

For the Euclidean two-point function the relevant moments are

$$\gamma_n := (\psi, T_e G_{2e}(\nabla^2)^n \psi) = \int_0^\infty w(m) m^{2n}. \tag{25}$$

For the scattering application we actually need the completeness of polynomials in m^2 rather than m which is why the nth moment involves m^{2n}.

For the two-point example above the relevant moments are

$$\gamma_n := \int_0^\infty \frac{e^{-\sqrt{m^2+\mathbf{p}^2}\tau}}{2\sqrt{m^2+\mathbf{p}^2}} \rho(m) m^{2n} dm \tag{26}$$

In order to estimate these moments we assume (see Theorem 6.2.4 in [9]) that the continuous part of the Lehmann weight is polynomially bounded. In this case it is sufficient to replace the moments by

$$\gamma_n \to \gamma_n' = \int_0^\infty \frac{e^{-\sqrt{m^2+\mathbf{p}^2}\tau}}{2\sqrt{m^2+\mathbf{p}^2}} m^{2n+k} dm. \tag{27}$$

where k is a constant (the polynomial bound). With some algebra it is possible to prove the inequality

$$\sum_{n=1}^\infty \gamma_n'^{-\frac{1}{2n}} > \sum_{n=0}^\infty \frac{c}{2n+k-2} > \infty \tag{28}$$

which is sufficient to show that $h(\nabla^2)$ can be approximated by a polynomial, thus ensuring that it is possible to construct point-spectrum mass eigenstates that satisfy the Euclidean time support condition.

Returning to Eq. (16) it is now possible to see how the condition above eliminates the disconnected terms in this equation. The first step is to replace the wave packets $\chi(x_e)$ by $h_n(\nabla^2)g(x_e)$, where h_n is a polynomial approximation to $h(m^2)$. Integrating the Laplacians by parts means that they can be replaced by $h_n(m^2)$, where m^2 is integrated over the support of the Lehmann weight. The completeness of the polynomials in m^2 means that we can make the contribution from the continuous part of the Lehmann weight as small as desired. Integrating the Euclidean time derivatives by parts in the two-point functions leads to replacements of the form $\frac{\partial}{\partial x^0} \to \omega_m(\mathbf{p})$. When this is integrated over m the only term that is picked up is the $m = m_i$ selected by $h(m^2)$. This term cancels the energy factors that come from ΦH_f. What remains is just the contribution from the connected four-point Green function, which is expected to vanish at t gets large.

This shows that it is possible to compute S-matrix elements in the Euclidean representation. As a practical matter e^{iHt} is not easy to compute in the Euclidean representation. There are a number of approaches that can be used given a representation for H. One that we have tested makes use of the fact that given a set of reflection positive Green functions, matrix elements of polynomials in $(e^{-\beta H})$ are easy to calculate using the inner product (3).

To take advantage of this note that the invariance principle [16,17] implies

$$\lim_{t\to\pm\infty} e^{iHt} \Phi e^{-iH_f t} |\chi\rangle = \lim_{n\to\pm\infty} e^{-ine^{-\beta H}} \Phi e^{ine^{-\beta H_f}} |\chi\rangle \tag{29}$$

Because the spectrum of $e^{-\beta H}$ is compact, for any fixed n, $e^{-ine^{-\beta H}}$ can be uniformly approximated by a polynomial in $e^{-\beta H}$. Matrix elements of $e^{-\beta m H}$ simply translate the argument of the Euclidean wave functions by $-\beta m$. In principle β is any positive number; in practice it sets the energy scale of the problem. We have demonstrated in a solvable model that these methods can be applied to calculate sharp momentum cross sections over a wide range of energies [18]. There are many other possibilities given that the reconstruction theorem provides all of the elements necessary to formulate the scattering problem.

References

1. J.S. Schwinger, Proc. Natl. Acad. Sci. USA **44**, 956 (1958)
2. L. Maiani, M. Testa, Phys. Lett. B **245**, 585 (1990)
3. K. Osterwalder, R. Schrader, Commun. Math. Phys. **31**, 83 (1973)
4. G. Aiello, W.N. Polyzou. Phys. Rev. **D93**, 056003 (2016). arXiv:1512.03651
5. W.N. Polyzou, Phys. Rev. **D89**, 076008 (2014). arXiv:1312.03585
6. A. Klein, L. Landau, J. Funct. Anal. **44**, 121 (1981)
7. A. Klein, L. Landau, Commun. Math. Phys **87**, 469 (1983)
8. J. Frohlich, K. Osterwalder, E. Seiler, Ann. Math. **118**, 461 (1983)
9. J. Glimm, A. Jaffe, *Quantum Physics: A functional Integral Point of View* (Springer, Berlin, 1981)
10. J. Cook, J. Math. Phys. **36**, 82 (1957)
11. R. Jost, *The General Theory of Quantized Fields* (AMS, Providence, RI, 1965)
12. R. Haag, Phys. Rev. **112**, 669 (1958)
13. D. Ruelle, Helv. Phys. Acta **35**, 147 (1962)
14. T. Carleman, *Les fonctions quasi analytiques, Collection de Monographies sur la Théorie des Fonctions* (Gauthier-Villars, Paris, 1926)
15. N.I. Akheizer, *The Classical Moment Problem and Some Related Questions in Analysis* (Oliver and Boyd, Edinburgh and London, 1965)
16. T. Kato, *Perturbation Theory for Linear Operators* (Spinger, Berlin, 1966)
17. C. Chandler, A. Gibson, Indiana J. Math. **25**, 443 (1976)
18. P. Kopp, W.N. Polyzou, Phys. Rev. **D85**, 016004 (2012). arXiv:1106.4086

Few-Body Syst (2017) 58:125
DOI 10.1007/s00601-017-1285-3

Kamil Serafin

Relativistic Model of Hamiltonian Renormalization for Bound States and Scattering Amplitudes

Received: 13 January 2017 / Accepted: 9 March 2017 / Published online: 27 March 2017

Abstract We test the renormalization group procedure for effective particles on a model of fermion–scalar interaction based on the Yukawa theory. The model is obtained by truncating the Yukawa theory to just two Fock sectors in the Dirac front form of Hamiltonian dynamics, a fermion, and a fermion and a boson, for the purpose of simple analytic calculation that exhibits steps of the procedure.

1 Introduction

The renormalization group procedure for effective particles (RGPEP) [1] is a tool developed for describing bound-states in QCD [2]. It has been shown that the RGPEP passes the test of producing asymptotic freedom in the front form Hamiltonian of QCD in third-order calculations in expansion in powers of the bare coupling constant [3,4]. Similar calculations for the quark–gluon coupling constant are yet to be done. However, the lowest order required for studying nontrivial effects of nonabelian gauge group of QCD in bound-state dynamics is fourth. In this article, we present a simple Hamiltonian model stemming from Yukawa field theory [5,6], in which we apply the RGPEP in order to verify its utility in fourth-order calculations and dynamics of bound states. The model simplicity allows us to illustrate the properties of an effective theory by a precise example, including bound states. Our analytic results in the simple model are helpful in organizing our thinking about fourth-order derivation of effective QCD, which is needed in calculations of gluon dynamics in heavy-quarkonia, *cf.* [7]. The model we study has been studied before and solved non-perturbatively by Głazek and Perry [5]. Their results were reproduced up to fourth order by Masłowski and Więckowski using similarity renormalization group procedure [6]. Our analysis differs by using the RGPEP with a new generator, which is known to apply well in third-order derivation of effective QCD and can be used in fourth-order. In the next sections we present the model, renormalize it, and study the effective fermion–boson coupling constant.

2 Model Theory

The construction of the model Hamiltonian starts with the Lagrangian of Yukawa theory,

$$\mathcal{L} = \frac{1}{2}(\partial_\mu \phi)^2 - \frac{1}{2}\mu^2\phi^2 + \bar{\psi}(i\not{\partial} - m)\psi - g\phi\bar{\psi}\psi, \tag{1}$$

where ψ is a fermion field and ϕ is a scalar field. From the Lagrangian, we obtain the Euler–Lagrange equations and the stress–energy tensor density $\mathcal{T}^{\mu\nu}$. In the front-form (FF) of dynamics [8], the hypersurface on which

This article belongs to the Topical Collection "Light Cone 2016".

K. Serafin (✉)
Faculty of Physics, University of Warsaw, Pasteura 5, 02-093 Warsaw, Poland
E-mail: Kamil.Serafin@fuw.edu.pl

we quantize the theory is defined by setting $x^+ = x^0 + x^3 = 0$. Variables $x^- = x^0 - x^3$, $x^\perp = (x^1, x^2)$ constitute thc FF "spatial" directions. Only half of the components of the fermion field are independent. These are $\psi_+ = \Lambda_+\psi$, where $\Lambda_\pm = \gamma^0\gamma^\pm/2$ are projection matrices. The part $\psi_- = \Lambda_-\psi$ dynamically depends on ψ_+ and ϕ. The FF energy density is

$$\mathcal{T}^{+-} = -\frac{1}{2}(\partial^\perp\phi)^2 + \frac{1}{2}\mu^2\phi^2 + \psi_+^\dagger(i\alpha^\perp\partial^\perp + \beta m + g\beta\phi)\frac{1}{i\partial^+}(i\alpha^\perp\partial^\perp + \beta m + g\beta\phi)\psi_+, \quad (2)$$

where $1/i\partial^+$ is a result of solving the constraint equation for ψ_-. The quantum Hamiltonian is obtained by replacing the classical fields ψ_+ and ϕ by quantum operators,

$$\psi_+(x) = \Lambda_+ \sum_\sigma \int [p] \left[u_\sigma(p)b_{p\sigma}e^{-ipx} + v_\sigma(p)d_{p\sigma}^\dagger e^{ipx}\right]\Big|_{x^+=0}, \quad (3)$$

$$\phi(x) = \int [k] \left[a_k e^{-ikx} + a_k^\dagger e^{ikx}\right]\Big|_{x^+=0}, \quad (4)$$

where the integration measure is $[k] = \theta(k^+)dk^+d^2k^\perp/16\pi^3k^+$ and the creation and annihilation operators obey canonical commutation, or anticommutation, relations

$$\left[a_p, a_{p'}^\dagger\right] = 2(2\pi)^3 p^+ \delta(p^+ - p'^+)\delta^{(2)}(p^\perp - p'^\perp), \quad (5)$$

$$\left\{b_{p\sigma}, b_{p'\sigma'}^\dagger\right\} = 2(2\pi)^3 p^+ \delta(p^+ - p'^+)\delta^{(2)}(p^\perp - p'^\perp)\delta_{\sigma\sigma'}. \quad (6)$$

We omit the anticommutation relations for antiparticles.

The quantum canonical Hamiltonian, defined as integral of normal ordered product of fields given in Eq. (2), is ill-defined, because it leads to divergent integrals in loop corrections to energies of states. In fact, any estimate of the order of magnitude of the interaction energy gives infinity [9]. Therefore, to properly define the quantum Hamiltonian we need to regulate and renormalize it.

To simplify the renormalization problem we enormously simplify the theory by restricting the Hilbert space of states to one fermion, $|1\rangle = b_{p_1\sigma_1}^\dagger|0\rangle$, and one fermion and one boson, $|2\rangle = b_{p_2\sigma_2}^\dagger a_{k_2}^\dagger|0\rangle$, where $|0\rangle$ is the vacuum state. In such truncated space only three interaction terms are active: creation of boson from a fermion, its Hermitian conjugation, and the so-called seagull term [see Eq. (8) below]. The truncated Hamiltonian still leads to divergences. We are interested in the elements of the RGPEP that are intact in the truncated model.

We regularize the interaction terms by restricting the invariant mass squared of the particles in the ingoing and outgoing states by Λ^2. The regulating function is

$$\theta_2^\Lambda = \theta\left[\Lambda^2 - \mathcal{M}^2(x, \kappa)\right], \quad (7)$$

where θ is the Heaviside theta step function and $\mathcal{M}^2(x, \kappa)$ is the invariant mass squared of the particles in a state $|2\rangle$. The invariant mass squared is a function of relative momenta $x = k_2^+/\mathcal{P}^+$ and $\kappa^\perp = k_2^\perp - x\mathcal{P}^\perp$, where $\mathcal{P}^\mu = k_2^\mu + p_2^\mu$. The model Hamiltonian is

$$H_{\text{model}} = \int_1 p_1^- |1\rangle\langle 1| + \int_2 (p_2^- + k_2^-)|2\rangle\langle 2| + g\int_{21} \theta_2^\Lambda\, \tilde\delta\, \bar u_{\sigma_2}(p_2)u_{\sigma_1}(p_1)|2\rangle\langle 1| + H.c.$$
$$+g^2\int_{22'} \theta_2^\Lambda\theta_{2'}^\Lambda\, \tilde\delta\, \bar u_{\sigma_2}(p_2)\frac{\gamma^+}{2\mathcal{P}^+}u_{\sigma_2'}(p_2')\,|2\rangle\langle 2'| + X_\Lambda, \quad (8)$$

where $\tilde\delta$ denotes the three-momentum conservation Dirac δ-functions and integral symbols contain integrals over momentum variables as well as sums over spins. X_Λ contains any counterterms that are needed for the effective theory not to depend on the cutoff parameter Λ.

3 The RGPEP

Regularized theory is well-defined in the sense that the solutions to the Hamiltonian eigenvalue problem exist for finite Λ. Nevertheless, the eigenvalues and eigenvectors depend badly on the cutoff parameter Λ. In other words, they depend on the number of momentum scales one has to sum over [9]. To eliminate dependence of physical quantities on Λ we introduce the concept of effective particle. Detailed presentation of the RGPEP can be found in Ref. [1].

The effective particle is defined as the state created by the effective creation operator. The latter is produced by a unitary rotation operator $\mathcal{U}_t$ from the initial, bare particle operator:

$$q_t = \mathcal{U}_t \, q_0 \, \mathcal{U}_t^\dagger , \tag{9}$$

where q_0 denotes any of the initial creation and annihilation operators $b_{p\sigma}$, $b_{p\sigma}^\dagger$, a_k, $a_k^\dagger$, and q_t denotes their effective counterparts. t is a scale parameter whose fourth root has interpretation of the size of effective particles. It can assume positive values. The FF vacuum state, $|0\rangle$, is annihilated by annihilation operators irrespective of the value of t. Any state in our truncated Fock space can be constructed using any one of the operator bases: the canonical one at $t = 0$ or effective ones at any value of $t > 0$. In particular, the basis states are related to each other in the following way,

$$|1\rangle_t \; = \; b_{t\, p_1\sigma_1}^\dagger |0\rangle \; = \; \mathcal{U}_t|1\rangle, \quad |2\rangle_t \; = \; b_{t\, p_2\sigma_2}^\dagger a_{t\, k_2}^\dagger |0\rangle \; = \; \mathcal{U}_t|2\rangle. \tag{10}$$

The Hamiltonian of the theory can be expressed with use of either q_0 or q_t,

$$H_t(q_t) = H_0(q_0), \tag{11}$$

where $H_0(q_0)$ means that the Hamiltonian is expressed using operators q_0 and coefficients in front of their products are the ones in the initial theory, while $H_t(q_t)$ is the same Hamiltonian expressed using operators q_t and the coefficients in front of them are functions of t. For technical reasons, we introduce also $H_t(q_0) \equiv \mathcal{H}_t$. In the model,

$$\mathcal{H}_t = \int_{21} \left[\delta\, \widetilde{\mathcal{H}}_t(2;1)\; |2\rangle\langle 1| + H.c. \right] + \int_{22'} \delta\, \widetilde{\mathcal{H}}_t(2;2')\; |2\rangle\langle 2'| + \int_{11'} \delta\, \widetilde{\mathcal{H}}_t(1;1')|1\rangle\langle 1'|. \tag{12}$$

The unitary rotation $\mathcal{U}_t$ and the family of Hamiltonians $\mathcal{H}_t$ are defined through the RGPEP evolution equation

$$\frac{d}{dt}\mathcal{H}_t = \left[[\mathcal{H}_f, \mathcal{H}_{Pt}], \mathcal{H}_t \right], \tag{13}$$

with the initial condition

$$\mathcal{H}_0 = H_0(q_0) = H_{\text{model}}. \tag{14}$$

$\mathcal{H}_f = H_{\text{model}}|_{g=0}$ is the free part of $\mathcal{H}_t$ and $\mathcal{H}_{Pt}$ is the same as $\mathcal{H}_t$ except that every term is multiplied by the square of sum of $+$ momenta of the ingoing particles. Equation (13) implies $\mathcal{U}_t = T \exp\left(-\int_0^t d\tau [\mathcal{H}_f, \mathcal{H}_{P\tau}] \right)$, where T denotes ordering in τ. The double commutator structure of Eq. (13) ensures that the effective particles do not interact unless the difference in free invariant masses between ingoing and outgoing states in the interaction vertex is smaller than $\lambda = t^{-1/4}$. In the lowest order (in g) effective vertex,

$$\widetilde{\mathcal{H}}_t(2;1) = \theta_2^\Lambda \; g \; e^{-t(\mathcal{M}_2^2 - m^2)^2} \, \bar{u}_{\sigma_2}(p_2) u_{\sigma_1}(p_1) + O(g^3). \tag{15}$$

The form factor $e^{-t(\mathcal{M}_2^2 - m^2)^2}$ falls exponentially with the free invariant mass of fermion–boson state effectively preventing the interaction from happening when $\mathcal{M}_2^2 - m^2 \gg t^{-1/2}$. Similar form factor, $e^{-t(\mathcal{M}_2^2 - \mathcal{M}_{2'}^2)^2}$, is present in the $2' \to 2$ vertex.

Because the effective interactions are suppressed by the form factors, we expect that in the effective theory no dependence on $\Lambda \to \infty$ should arise, not just in observables but in all Hamiltonian matrix elements between states of finite kinematical quantum numbers. The initial theory and the effective one are equivalent, and we evaluate the latter from the former. When the former leads to divergences, we have to adjust it so that the effective theory is free from divergences. This is achieved by introducing counterterms in the initial theory. We impose the following prescription for the counterterms in the initial theory:

1. Propose the initial theory.
2. Calculate the effective theory.
3. If any matrix element of the effective theory Hamiltonian is divergent when $\Lambda \to \infty$, then add appropriate counterterm to the initial theory (unique up to the finite part in the Λ-dependent functions), which cancels the divergence.
4. Constrain finite parts of counterterms by available kinematical symmetry requirements.
5. Repeat steps 2–4 until the matrix elements of effective Hamiltonians are free from divergences and obey kinematical symmetries.

Once the counterterms in the initial theory are found, one can solve the dynamical problems in the finite effective theories and adjust the finite parts of the counterterms to data. The last step corresponds to expressing bare constants in terms of observables in perturbative calculations of observables. One can choose freely the RGPEP scale parameter t of an effective theory used to adjust the finite free parameters. Effective theories with different t are related by the RGPEP Eq. (13). One can simplify the description of phenomena of interest by choosing t, which is analogous to procedures known in literature [10].

4 Calculation of Counterterms

In the first order of perturbative expansion in powers of g, the effective Hamiltonian acquires a form factor, see Eq. (15). In the second order, we have effective $2' \to 2$ vertex, which does not require counterterm. We also have the effective mass term,

$$\widetilde{\mathcal{H}}_{t2}(1;1) = \widetilde{\mathcal{H}}_{02}(1;1) + \int \frac{dx d^2\kappa \, \theta_2^\Lambda}{16\pi^3 x(1-x)} \frac{e^{-2t(\mathcal{M}^2-m^2)^2} - 1}{\mathcal{M}^2(x,\kappa) - m^2} \sum_{\sigma_2} \bar{u}_{\sigma_1}(p_1) u_{\sigma_2}(p_2) \bar{u}_{\sigma_2}(p_2) u_{\sigma_1}(p_1), \quad (16)$$

where $\widetilde{\mathcal{H}}_{02}(1;1)$ is the counterterm to be determined.

The part of the integrand multiplied by $e^{-2t(\mathcal{M}^2-m^2)^2}$ falls off to zero quickly for large transverse momentum κ. Therefore, that part depends very little on Λ when $\Lambda \to \infty$. However, the numerator of the integrand contains also 1, which is subtracted from $e^{-2t(\mathcal{M}^2-m^2)^2}$. This 1 gives a part of integral, which diverges when $\Lambda \to \infty$. The product of spinors $\bar{u}_1 u_2 \bar{u}_2 u_1$ behaves like κ^2 for large relative transverse momentum, so does $\mathcal{M}^2$ in the denominator. Therefore, the integration over κ gives the leading term of order Λ^2, which is badly divergent for $\Lambda \to \infty$. To cancel this divergence the counterterm is defined as a term of the same operator structure with a coefficient that cancels the diverging number. Hence,

$$\widetilde{\mathcal{H}}_{02}(1;1) = \omega^2 + 2m^2(\alpha + \beta) + \widetilde{\mathcal{H}}_{02}^{\text{finite}}, \quad (17)$$

where

$$\omega^2 = \int \frac{dx d^2\kappa \, \theta_2^\Lambda}{16\pi^3 x(1-x)} (1-x), \quad (18)$$

$$\alpha = \int \frac{dx d^2\kappa \, \theta_2^\Lambda}{16\pi^3 x(1-x)} \frac{1-x}{\mathcal{M}^2(x,\kappa) - m^2}, \quad (19)$$

$$\beta = \int \frac{dx d^2\kappa \, \theta_2^\Lambda}{16\pi^3 x(1-x)} \frac{1}{\mathcal{M}^2(x,\kappa) - m^2}, \quad (20)$$

and $\widetilde{\mathcal{H}}_{02}^{\text{finite}}$ is the finite part, on which we concentrate in the next paragraph. Division of the counterterm into ω^2, which is quadratically divergent and α and β, which are logarithmically divergent, is dictated by utility of these symbols in the higher order calculations.

To fix the finite part of the mass counterterm we need a physical condition. We demand that the physical fermion state is a solution of the effective Hamiltonian eigenproblem for some t with the FF energy eigenvalue fulfilling the relativistic dispersion relation,

$$H_t |\mathcal{P}\sigma\rangle_{\text{phys},t} = \frac{m_{\text{phys}}^2 + \mathcal{P}^{\perp 2}}{\mathcal{P}^+} |\mathcal{P}\sigma\rangle_{\text{phys},t}, \quad (21)$$

where m_{phys} is the physical mass of the fermion and

$$|\mathcal{P}\sigma\rangle_{\text{phys},t} = \int_1 \mathcal{P}^+\tilde{\delta}\, c^{\sigma}_{\sigma_1 t}|1\rangle_t + \int_2 \mathcal{P}^+\tilde{\delta}\, \phi^{\sigma}_{\sigma_2 t}(x,\kappa)|2\rangle_t. \tag{22}$$

We solve Eq. (21) up to the second order in the expansion in powers of g and find that the solution exists if the following constraints are fulfilled,

$$m = m_{\text{phys}}, \quad \widetilde{\mathcal{H}}^{\text{finite}}_{02} = 0. \tag{23}$$

Moreover, the effective wavefunction of the physical fermion is

$$\phi^{\sigma}_{\sigma_2 t}(x,\kappa) = g\frac{e^{-t(\mathcal{M}_2^2-m^2)^2}}{m^2-\mathcal{M}_2^2}\sum_{\sigma'}\bar{u}_{\sigma_2}(p_2)u_{\sigma'}(\mathcal{P})c^{\sigma}_{\sigma' t} + O(g^3). \tag{24}$$

This result reveals that as we increase t, the contribution of the two-particle component to the physical fermion decreases because of the form factor. In other words, the bigger the size parameter t the more similar the effective fermion to the physical one.

In the third order in g, the effective theory contains only fermion $\to$ fermion–boson vertex (and its Hermitian conjugate). The counterterm, which secures finiteness of the effective third order vertex when $\Lambda \to \infty$ is

$$\widetilde{\mathcal{H}}_{03}(2;1) = m(\alpha+\beta+A)\bar{u}_{\sigma_2}(p_2)\frac{\gamma^+}{2\mathcal{P}^+}u_{\sigma_1}(p_1) + \frac{\alpha+B}{2}\bar{u}_{\sigma_2}(p_2)u_{\sigma_1}(p_1), \tag{25}$$

where A and B are finite parts of the logarithmically divergent functions in front of the two different spinor structures of the counterterm. The counterterm divergence can be absorbed into parameters of the initial theory. The first term on the right hand side of Eq. (25) shifts the mass of the fermion in the fermion sector by $\delta m = g^2 m(\alpha+\beta+A)$. The second term shifts the coupling constant. The fact that the fermion mass may be different in different Fock sectors is a feature of the Tamm–Dancoff-truncated theories [5,11]. The $1 \to 2$ vertex with counterterm divergences absorbed in the initial theory parameters, is

$$\widetilde{\mathcal{H}}_0(2;1) = \theta_2^{\Lambda}\, g_{\Lambda}\, \bar{u}_m(p_2,\sigma_2)u_{m_\Lambda}(p_1,\sigma_1), \tag{26}$$

where u_m means a spinor with mass m, u_{m_Λ} is a spinor with mass m_Λ and

$$g_\Lambda = g + (\alpha+B)g^3 + \cdots, \quad m_\Lambda = m + g^2 m(\alpha+\beta+A) + \cdots. \tag{27}$$

For the effective third-order vertex, see Sect. 5.

The fourth-order calculation of the effective Hamiltonian $\mathcal{H}_t$ reveals the form of the fourth-order counterterms. They are

$$\widetilde{\mathcal{H}}_{04}(2;2') = (\alpha+C)\bar{u}_{\sigma_2}(p_2)\frac{\gamma^+}{2\mathcal{P}^+}u_{\sigma_{2'}}(p_{2'}) \tag{28}$$

for the the seagull interaction vertex and

$$\widetilde{\mathcal{H}}_{04}(1,1') = \delta_{\sigma_1\sigma_{1'}}\left[(\alpha+B)\omega^2 + 2m^2(\alpha+B+A)(\alpha+\beta) + m^2(\alpha+\beta)^2\right] + \widetilde{\mathcal{H}}^{\text{finite}}_{04} \tag{29}$$

for the fermion mass term. C and $\widetilde{\mathcal{H}}^{\text{finite}}_{04}$ are finite parts of the counterterms.

To fix the finite part of the mass counterterm we again use Eq. (21). Solving it to fourth order in g gives us

$$\widetilde{\mathcal{H}}^{\text{finite}}_{04} = 0. \tag{30}$$

The RGPEP respects 7 kinematical symmetries. In order to secure the full Poincare symmetry in the model, we simply demand that the symmetries corresponding to the dynamical symmetry generators are directly visible in the fermion–boson $\to$ fermion–boson scattering amplitude. The fourth-order contribution to the T matrix is

$$g^4\, \theta_{2_o}^{\Lambda}\theta_{2_i}^{\Lambda}\, \bar{u}_{\sigma_{2o}}(p_{2o})\left[\Gamma_1(\mathcal{P}^2)\not{\mathcal{P}} + \Gamma_2(\mathcal{P}^2)m + \Gamma_3(\mathcal{P}^2)\frac{\gamma^+}{2\mathcal{P}^+}\right]u_{\sigma_{2i}}(p_{2i}), \tag{31}$$

where $\mathcal{P}$ is the total four-momentum, evaluated using the physical mass parameters for the boson and fermion. Hence, $\mathcal{P}^2$ is the physical invariant mass squared of the incoming or outgoing particles. Γ_1, Γ_2 and Γ_3 are finite. In particular,

$$\Gamma_3(\mathcal{P}^2) = B - C - \frac{2m^2}{\mathcal{P}^2 - m^2} A. \tag{32}$$

The only term breaking the Lorentz covariance of the scattering amplitude is the one multiplied by Γ_3. Therefore, we demand that $\Gamma_3(\mathcal{P}^2) = 0$. The counterterms securing Lorentz covariance of the scattering amplitude are obtained without introducing functions of momenta by setting

$$A = 0, \quad B = C. \tag{33}$$

B remains unspecified, which allows one to freely choose at which scale coupling constant g is defined.

5 Running of the Effective Hamiltonian Coupling Constant

Beside the prescription for the counterterms, RGPEP provides the family of equivalent effective theories numbered with parameter t. For example, the effective fermion–boson–fermion vertex is

$$\widetilde{\mathcal{H}}_t(2;1) = \theta_2^\Lambda \, \tilde{g}_t\left(\mathcal{M}_2^2\right) \, e^{-t(\mathcal{M}_2^2 - m^2)^2} \cdot \bar{u}_{\sigma_2}(p_2) \left[1 + \delta m_t\left(\mathcal{M}_2^2\right) \frac{\gamma^+}{2\mathcal{P}^+}\right] u_{\sigma_1}(p_1), \tag{34}$$

where $\tilde{g}_t(\mathcal{M}_2^2) = g + B_t(\mathcal{M}_2^2) g^3$. The quantities $B_t(\mathcal{M}_2^2)$ and $\delta m_t(\mathcal{M}_2^2)$ are finite when $\Lambda \to \infty$ but quite complicated. We do not write them explicitly. The square bracket above is similar to the one present in the initial theory, which is interpreted as shifting the mass of the fermion in the fermion sector by δm, *cf.* Eq. (27). In the effective theory, however, δm depends on the free invariant mass $\mathcal{M}_2$ of the outgoing fermion–boson state. For t much smaller than m^{-4} this dependence is negligible.

Almost every element of Eq. (34) depends on the invariant mass $\mathcal{M}_2$. Therefore, to clarify the picture, we define effective coupling constant $g_t = \tilde{g}_t(m^2)$ and rewrite the effective vertex,

$$\widetilde{\mathcal{H}}_t(2;1) = \theta_2^\Lambda \, g_t \, f_t\left(\mathcal{M}_2^2\right) \cdot \bar{u}_{\sigma_2}(p_2) \left[1 + \delta m_t(\mathcal{M}_2^2) \frac{\gamma^+}{2\mathcal{P}^+}\right] u_{\sigma_1}(p_1), \tag{35}$$

where f_t is a new form factor, which contains second order corrections to the exponential form factor of Eq. (15). In this form, one can interpret the effective vertex. First of all, its strength is characterized by the effective coupling constant g_t, which for $t \ll m^{-4}$ depends on t in the following way,

$$g_t = g_{t_0} + \frac{g_{t_0}^3}{32\pi^2} \log \frac{\lambda}{\lambda_0} + \cdots, \quad \lambda \gg m, \tag{36}$$

where $\lambda = t^{-1/4}$ and $\lambda_0 = t_0^{-1/4}$. It is noteworthy that the bare coupling, *cf.* Eq. (27), in the initial Hamiltonian depends in the same way on Λ for $\Lambda \gg m$. This finding is a manifestation of the fact that, in the effective theory, the finite-width vertex form factors assume the role analogous to the regulating role played by the Heaviside θ-functions with cutoff parameter Λ in the initial theory. Therefore, in practice, one can omit θ_2^Λ in the effective-theory Eq. (35). Moreover, higher-order calculations introduce further corrections to the vertex form factors and parameters like δm, because we have freedom of choosing t while in exact calculations no observable depends on t.

6 Concluding Remarks

The counterterms found in the model Hamiltonian with use of the RGPEP in its most recent form agree with the ones found previously using similarity [6]. We also provide the lowest order effective wave function of the physical fermion and an example of the Hamiltonian running coupling constant in the effective theory. An interesting further study of the model would be an exploration of nonperturbative solutions to the RGPEP evolution Eq. (13).

References

1. S.D. Głazek, Perturbative formulae for relativistic interactions of effective particles. Acta Phys. Pol. B **43**, 1843 (2012)
2. K.G. Wilson, T.S. Walhout, A. Harindranath, W.-M. Zhang, R.J. Perry, S.D. Głazek, Nonperturbative QCD: a weak coupling treatment on the light front. Phys. Rev. D **49**, 6720 (1994)
3. S.D. Głazek, Dynamics of effective gluons. Phys. Rev. D **63**, 116006 (2001)
4. M. Gómez-Rocha, S.D. Głazek, Asymptotic freedom in the front-form Hamiltonian for quantum chromodynamics of gluons. Phys. Rev. D **92**, 065005 (2015)
5. S.D. Głazek, R.J. Perry, Special example of relativistic Hamiltonian field theory. Phys. Rev. D **45**, 3740 (1992)
6. T. Masłowski, M. Więckowski, Fourth-order similarity renormalization of a model Hamiltonian. Phys. Rev. D **57**, 4976 (1998)
7. M. Gómez-Rocha, From asymptotic freedom toward heavy quarkonia within the renormalization group procedure for effective particles, in *Contribution to Light Cone 2016 Proceedings*. arXiv:1611.07300 [hep-ph]
8. P.A.M. Dirac, Forms of relativistic dynamics. Rev. Mod. Phys. **21**, 392 (1949)
9. K.G. Wilson, Model Hamiltonians for local quantum field theory. Phys. Rev. **140**, B445 (1965)
10. S.J. Brodsky, G.P. Lepage, P.B. Mackenzie, On the elimination of scale ambiguities in perturbative quantum chromodynamics. Phys. Rev. D **28**, 228 (1983)
11. R.J. Perry, A. Harindranath, K.G. Wilson, Light-front Tamm–Dancoff field theory. Phys. Rev. Lett. **65**, 2959 (1990)

Few-Body Syst (2017) 58:49
DOI 10.1007/s00601-016-1194-x

Stanisław D. Głazek · Arkadiusz P. Trawiński

Effective Particles in Quantum Field Theory

Received: 29 November 2016 / Accepted: 16 December 2016 / Published online: 19 January 2017

Abstract The concept of effective particles is introduced in the Minkowski space-time Hamiltonians in quantum field theory using a new kind of the relativistic renormalization group procedure that does not integrate out high-energy *modes* but instead integrates out the large *changes* of invariant mass. The new procedure is explained using examples of known interactions. Some applications in phenomenology, including processes measurable in colliders, are briefly presented.

1 Introduction

Effective particles can be introduced in the quantum field theory (QFT) in such a way that their size plays the role of a scale parameter in a renormalization group procedure for the corresponding Hamiltonians. The example of QCD is particularly interesting because the effective quarks of large size are expected to correspond to the constituent quarks that are used in classification of hadrons, while point-like quarks and gluons correspond to partons used in description of hadrons in high-energy collisions. But the renormalization group procedure for effective particles (RGPEP) that is used here is not limited to QCD. Its general nature stems form the fact that one does not specify the size of quanta when one expands a quantum field in space into its Fourier modes. Namely, the Heisenberg and Pauli [1,2] formulation of QFT is equally applicable to various particles despite the fact that they have different sizes, see Fig. 1. One can say that the concept of a quantum field operator is blind to the size of individual quanta. In contrast, in the renormalized theories derived using the RGPEP, this free size parameter determines the width of vertex form factors in the Hamiltonian interaction terms for effective particles.

2 Example of the RGPEP Calculation: Asymptotic Freedom in Yang–Mills theory

Canonical quantization of the Yang–Mills (YM) theory begins with the Lagrangian density $\mathcal{L} = -\,\mathrm{tr}\, F^{\mu\nu}F_{\mu\nu}/2$, in which the field strength tensor is $F^{\mu\nu} = \partial^\mu A^\nu - \partial^\nu A^\mu + ig[A^\mu, A^\nu]$. The associated energy-momentum tensor is $\mathcal{T}^{\mu\nu} = -F^{a\mu\alpha}\partial^\nu A^a_\alpha + g^{\mu\nu}F^{a\alpha\beta}F^a_{\alpha\beta}/4$. Consequently, the field four-momentum

This article belongs to the Topical Collection "Light Cone 2016".

Stanisław D. Głazek: On leave of absence from Faculty of Physics, University of Warsaw, Pasteura 5, 02-093 Warszawa, Poland.

S. D. Głazek (✉)
Department of Physics, Yale University, 217 Prospect Street, New Haven, CT 06511-8499, USA
E-mail: Stanislaw.Glazek@yale.edu

A. P. Trawiński
Faculty of Physics, University of Warsaw, Pasteura 5, 02-093 Warszawa, Poland
E-mail: Arkadiusz.Trawinski@fuw.edu.pl

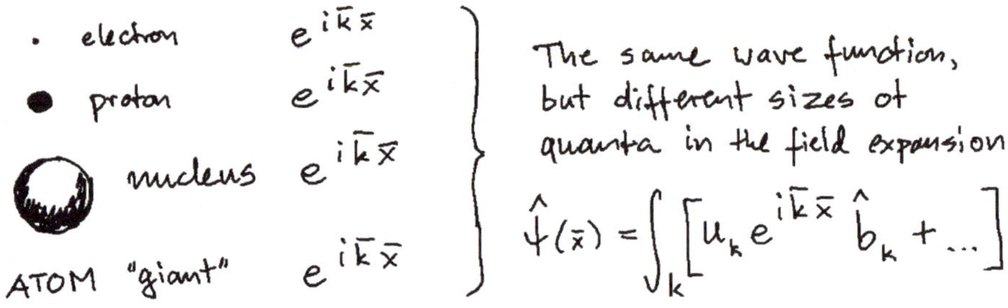

Fig. 1 Quantum field operators are blind to the size of individual quanta

is given by $P^\nu = \int_\Sigma d\sigma_\mu\, T^{\mu\nu}$, where $d\sigma_\mu$ is a measure on the hypersurface Σ in the Minkowski space-time. In the front form (FF) of Hamiltonian dynamics [3], one uses the variables $x^\pm = t \pm z$ and $x^\perp = (x, y)$, with a similar convention for all tensors. The FF Hamiltonian results from integration of $\mathcal{H}_{YM} = T^{+-}$ over the front hypersurface Σ defined by the condition $x^+ = 0$. The canonical FF Hamiltonian density in gauge $A^+ = 0$ contains three terms, $\mathcal{H}_{YM} = \mathcal{H}_{A^2} + \mathcal{H}_{A^3} + \mathcal{H}_{A^4}$. Our example concerns the three-gluon term $\mathcal{H}_{A^3}$, to exhibit asymptotic freedom. Quantization is achieved by replacing the classical gauge-field A by a field operator,

$$\hat{A}^\mu = \sum_{k\sigma c} \left[t^c\, \varepsilon^\mu_{k\sigma}\, a_{k\sigma c}\, e^{-ikx} + t^c\, \varepsilon^{\mu *}_{k\sigma}\, a^\dagger_{k\sigma c}\, e^{ikx} \right]_{on\ \Sigma} \tag{1}$$

where $a_{k\sigma c}$ and $a^\dagger_{k\sigma c}$ are the annihilation and creation operators of gluons with momentum $k = (k^+, k^\perp)$, polarization vectors ε_σ and color matrices t^c that span the algebra of $SU(N_c)$. We consider the number of colors $N_c = 3$. Thus, the starting Hamiltonian is

$$\hat{H}_{YM} = \int_\Sigma \; : \mathcal{H}_{YM}(\hat{A}) : \tag{2}$$

where the double dots indicate normal ordering. Our focus is on the term that comes from $\mathcal{H}_{A^3} = g\, i\partial_\alpha A^a_\beta\, [A^\alpha, A^\beta]^a$, which in the local quantum theory for point-like particles leads to

$$\hat{H}_{A^3} = \int_{x\in\Sigma} \; : g\, i\partial_\alpha \hat{A}^a_\beta(x)\, [\hat{A}^\alpha(x), \hat{A}^\beta(x)]^a : . \tag{3}$$

If the quanta are not point-like, see Fig. 2, the local interaction is merely an approximation, in which the size s is a free parameter. Thus, instead of the quantum size-blind field operator $\hat{A}(x)$, we can introduce an effective particle quantum field $\hat{A}(x, s)$ and the non-local three-gluon interaction term

$$\hat{H}_{A^3,s} = \int_{x_i\in\Sigma} g_s(x_1, x_2, x_3) \; : i\partial_\alpha A^a_\beta(x_1, s)\, [A^\alpha(x_2, s), A^\beta(x_3, s)]^a : . \tag{4}$$

The function $g_s(x_1, x_2, x_3)$ is determined using the RGPEP, except that one works with momentum variables rather than the space-time coordinates. The required elements of the RGPEP can be found in a sequence of Refs. [4–8]. A brief summary of the procedure follows.

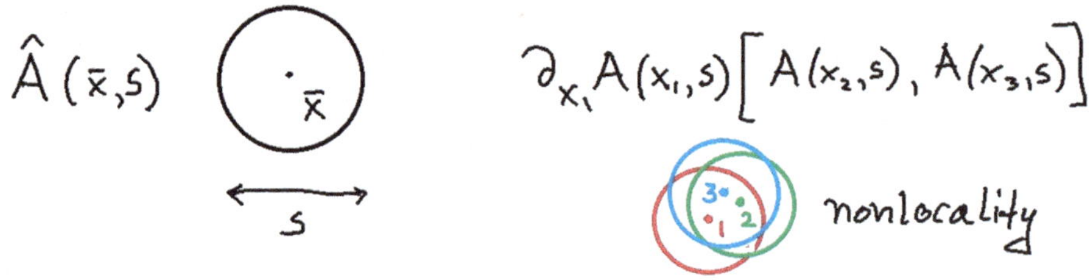

Fig. 2 Nonlocal interaction of quanta of size s can be approximated by a local interaction for wavelengths much greater than s. The effective particle size s is the free scale parameter in the RGPEP

The RGPEP employs equations that connect theories with different values of the effective particle size s. There is an initial condition set at $s = 0$ by canonical QCD. In our example,

$$\hat{H}_{A^3,s=0} = \int_{x \in \Sigma} : g\, i\partial_\alpha \hat{A}^a_\beta(x,0)\, [\hat{A}^\alpha(x,0), \hat{A}^\beta(x,0)]^a : \tag{5}$$

where the field operator is built from creation and annihilation operators for point-like quanta,

$$\hat{A}^\mu(x, s=0) = \sum_{k\sigma c} \left[t^c\, \varepsilon^\mu_{k\sigma}\, a_{k\sigma c,s=0}\, e^{-ikx} + t^c\, \varepsilon^{\mu *}_{k\sigma}\, a^\dagger_{k\sigma c,s=0}\, e^{ikx} \right]_{on\ \Sigma}. \tag{6}$$

Direct evaluation yields

$$\hat{H}_{A^3,s=0} = \sum_{123} \delta_{12.3} \left[g\, Y_{123}\, a^\dagger_{1,0}\, a^\dagger_{2,0}\, a_{3,0} + g\, Y^*_{123}\, a^\dagger_{3,0}\, a_{2,0}\, a_{1,0} \right] \tag{7}$$

where an abbreviated notation is used, such as 1 in place of k_1, σ_1, c_1, the factor $\delta_{12.3}$ takes care of momentum conservation and the coefficients Y_{123} stand for the gluon momentum, spin and color dependent factors implied by Eqs. (5) and (6). This is an example of a term in the general initial Hamiltonian for $s = 0$. The general structure of the latter reads

$$H_0(a_0) = \sum_{n=2}^{\infty} \sum_{i_1, i_2, \ldots, i_n} c_0(i_1, \ldots, i_n)\, a^\dagger_{i_1,0} \cdots a_{i_n,0} \tag{8}$$

where the coefficients c_0 are determined by the regulated canonical theory and counterterms. In the same notation, the Hamiltonian for effective particles of size s reads

$$H_s(a_s) = \sum_{n=2}^{\infty} \sum_{i_1, i_2, \ldots, i_n} c_s(i_1, \ldots, i_n)\, a^\dagger_{i_1,s} \cdots a_{i_n,s} \tag{9}$$

and the coefficients c_s are calculated using the RGPEP. One of these coefficients is $g_s(i_1, i_2, i_3)$ that corresponds to the function $g_s(x_1, x_2, x_3)$ in Eq. (4).

The coefficients c_s in Eq. (9) are derived from the condition that the Hamiltonian is not changed, $H_s(a_s) = H_0(a_0)$, when the canonical quanta with size $s = 0$ are replaced by the effective quanta of size $s > 0$, using a unitary transformation, $a_s = U_s\, a_0\, U_s^\dagger$. One works in a canonical operator basis and calculates the same coefficients c_s using the operator $\mathcal{H}_s = H_s(a_0) = U_s^\dagger H_s(a_s) U_s$ that is determined by the RGPEP equations

$$\mathcal{H}'_s = [\mathcal{G}_s, \mathcal{H}_s] \tag{10}$$

$$\mathcal{G}_s = [\mathcal{H}_f, \tilde{\mathcal{H}}_s] \tag{11}$$

where $\tilde{\mathcal{H}}_s$ differs from $\mathcal{H}_s$ only by multiplication of coefficients c_s by the square of +-momentum carried by the corresponding product of creation or annihilation operators. This multiplication secures boost invariance of the effective theory. In Eqs. (10) and (11), the Hamiltonian is divided into a free part, equal to $\mathcal{H}_0$ for $g = 0$, and the interaction $\mathcal{H}_I$, so that $\mathcal{H}_s = \mathcal{H}_f + \mathcal{H}_I$. The key feature of solutions to these equations, and the reason for using them here, is that the coefficients c_s exponentially suppress interactions that change the invariant mass of effective particles by more than $1/s$, see below. No modes are eliminated from the dynamics. Instead, vertex form factors suppress changes of invariant mass, see below.

Solutions to Eq. (10) define the YM theories *a priori* non-perturbatively. As such, they are not easy to obtain. In contrast, asymptotic freedom is found in the expansion in powers of the coupling constant,

$$\mathcal{H}_s = \mathcal{H}_f + g\mathcal{H}_{1s} + g^2\mathcal{H}_{2s} + g^3\mathcal{H}_{3s} + \cdots . \tag{12}$$

The terms of order g and g^3 yield the result [8]

$$H_{A^3\,(1+3)s} = \sum_{123} \int_{123} \delta_{12.3}\, e^{-s^4 \mathcal{M}^4_{12}} \left[V_{s123}\, a^\dagger_{1,s} a^\dagger_{2,s} a_{3,s} + V^*_{s123}\, a^\dagger_{3,s} a_{2,s} a_{1,s} \right] \tag{13}$$

in which $\mathcal{M}_{12}$ denotes the invariant mass of gluons 1 and 2. We see that the effective three-gluon interaction vertex is softened by the exponential form factor, which suppresses the interaction that involves a pair of gluons with invariant mass $\mathcal{M}_{12}$ the more the larger size s. The calculation shows that in the limit of vanishing relative transverse momentum of gluons $V_{s123} = g_s\, Y_{123}$ and

$$g_s = g_0 + \frac{g_0^3}{48\pi^2} N_c\, 11\, \ln \frac{s}{s_0}. \tag{14}$$

The resulting Hamiltonian function $\beta(s)$ corresponds to the Green's function $\beta(\lambda)$ obtained in Refs. [9,10], where λ denotes the length of a four-dimensional Euclidean momentum. Thus the effective particle formulation of QFT shows how the key feature of four-dimensional calculations of QCD manifests itself in the renormalized Hamiltonians. This way, the effective Hamiltonian approach passes an obligatory test for tackling issues of strong interaction theory. It provides the Minkowskian space-time scale parameter that corresponds to the abstract four-dimensional Euclidean scale parameter: the size of effective particles.

3 Early Examples of RGPEP Insight in Phenomenology and Theory of Particles

The RGPEP provides a conceptual insight into particle dynamics in various systems. For example, in the case of protons, the universal RGPEP parameter s tells us that the low-energy effective theory that can be sought using QCD, to support the quark model classification of hadrons, concerns quarks whose size is comparable with the size of the proton itself. Such large quarks ought to be thought about as built from partons of smaller size s, and the RGPEP provides mathematical tools for the corresponding construction of operators and states in the effective Fock-space basis. Using this insight, one can ask if the configurations of three bulky effective quarks built of small partons can explain ridge-like correlations in high-energy pp collisions observed by the ATLAS and CMS collaborations at LHC [11,12]. It was found [13] that among various simple models, including the gluon string stretched between a quark and a diquark [14], three quarks joined by a three-prong gluon body in the shape of letter Y, or a combination of various shapes and distributions of partons in effective particles, only the Y effective-particle picture of a proton provides eccentricity increasing with multiplicity. The model study [13] was carried out before the data exhibiting such type of behavior of $\sqrt{s} = 13$ TeV pp collisions [11,12] were published. This example illustrates the phenomenological utility of the scale dependent, effective particle picture in QFT.

The RGPEP can be applied to QED in description of hydrogen and muonic hydrogen atoms [15] for the purpose of explanation of the difference between the proton radii extracted from data for level splittings in these atoms [16]. This is a computationally ambitious goal. But the question of principle that RGPEP answers is how to derive the Schrödinger equation from relativistic QFT. In QED, the RGPEP would start from the classical Lagrangian density $\mathcal{L} = -\frac{1}{4}F_{\mu\nu}F^{\mu\nu} + \bar{\psi}(i\partial\!\!\!/ - eA\!\!\!/ - m)\psi$. One treats the electrons and protons as point-like and proceeds to quantization as in Sec. 2 to obtain $\hat{H}$, regularize it, identify counterterms and calculate $\hat{H}_{QED\,s}$. The eigenvalue problem, $\hat{H}_{QED\,s}|\psi_s\rangle = E|\psi_s\rangle$ is identified with the Schrödinger equation for the hydrogen atoms represented as lepton-proton bound states when the size parameter s is sufficiently large to prevent the interactions from inducing the invariant mass changes that exceed the constituent masses. The RGPEP valence constituent picture of atoms is illustrated schematically by

$$\begin{bmatrix} \cdot\cdot \\ \cdot\cdot \\ e\gamma e\bar{e}p\gamma \\ e\gamma p\gamma \\ e\gamma p \\ \boxed{ep} \end{bmatrix} = \begin{bmatrix} \cdot\cdot \\ \cdot\cdot \\ e\gamma e\bar{e} \\ e\gamma\gamma \\ e\gamma \\ e \end{bmatrix} \times \begin{bmatrix} \cdot\cdot \\ \cdot\cdot \\ pe\bar{e} \\ p\gamma\gamma \\ p\gamma \\ p \end{bmatrix} + \cdots = |e_s p_s\rangle + \cdots . \tag{15}$$

The left column represents a hydrogen atom as a superposition of virtual particle components in canonical theory, the size of quanta $s = 0$. The lowest Fock component is put in a frame to indicate the state that provides the first approximation to an atom when one works with canonical QED. Once one introduces the effective quanta, the whole tower of Fock components is rewritten as an effective state of $|e_s p_s\rangle$ plus more effective-particle components, indicated by the three dots. When the size parameter s is increased, the canonical interactions that act in and among all the Fock components are increasingly transformed into a complex interaction that predominantly acts in the component $|e_s p_s\rangle$, because the change of the number of quanta, even massless photons, involves much larger changes of invariant mass than the changes that occur in the component $|e_s p_s\rangle$. For sufficiently large s and for small α, the eigenvalue equation $\hat{H}_{QED\,s}|\psi_s\rangle = E|\psi_s\rangle$ can be reduced to the eigenvalue equation for the component $|e_s p_s\rangle$, which, after including proton electromagnetic form factors, especially the electric one, G_E, takes the familiar Schrödinger form up to terms of formal order α,

$$\frac{\mathbf{p}^2}{2\mu}\,\psi_s(\mathbf{p}) + \int \frac{d^3k}{(2\pi)^3}\,V_s(\mathbf{p},\mathbf{k})\,\psi_s(\mathbf{k}) = -E_B\,\psi_s(\mathbf{p}) \tag{16}$$

where the potential is

$$V_s(\mathbf{p},\mathbf{k}) = e^{-s^4(\mathbf{p}^2-\mathbf{k}^2)^2/c^4}\,\frac{-4\pi\alpha}{\mathbf{q}^2}\,G_E(\mathbf{q}^2). \tag{17}$$

The three-momentum transfer is $\mathbf{q} = \mathbf{p} - \mathbf{k}$. The exponential factor that corrects the Coulomb potential of extended proton, suppresses the interaction that changes the invariant mass of effective electron-proton system by more than $1/s$. Once the size s is selected to secure universal scaling of atomic levels with α, setting $s \sim 1/\sqrt{\mu M}$, where μ is the reduced and M the average constituent mass, this factor introduces a small dependence of the potential on the lepton mass. This minuscule dependence has a divergent perturbative expansion in powers of α when one expresses momenta in units of $\alpha\mu c$. Therefore, it needs to be evaluated numerically. It turns out that it is capable of producing a lepton-mass correction in the extraction of proton radius from atomic levels, of the same magnitude [15] as the variation encountered in the proton radius puzzle [16] when the effective nature of particles appearing in the Schrödinger equation according the RGPEP is not accounted for.

It is worth pointing out that the Schrödinger equation emerges from the RGPEP derivation of effective low-energy Hamiltonian in QFT using the three-momentum variables $\mathbf{p} = (p^\perp, p^z)$,

$$p^\perp = c\left[(1-x)\,p_l^\perp - x\,p_p^\perp\right]/\sqrt{x(1-x)} \tag{18}$$

$$p^z = c\,(m_l + m_p)\,(x-\beta)/\sqrt{x(1-x)} \tag{19}$$

where $\beta = m_l/(m_p + m_l)$, $c = \sqrt{m_l m_p}/(m_l + m_p)$, $x = p_l^+/(p_l^+ + p_p^+)$, while l refers to lepton and p to proton. These variables correspond to the relative momentum variables used in light-front holography [17] for description of hadrons. They appear to be non-relativistic while they are in fact the relative momenta of constituents in a relativistic theory, in which boost invariance is maintained. The correspondence is unlikely to be a meaningless coincidence. The Schrödinger equation is universally valid as an effective theory for atomic physics, despite that it apparently contains no remnants of complexity of relativistic QED. Similarly, the light-front holography [17], motivated by the idea of duality [18], is meant to approximate complex QFT in terms of solutions to relatively simple field equations in AdS [19]. In the case of strong interactions, the same variables are also used in showing that the linear confining potential in the instant form of dynamics [3] corresponds to a quadratic one in the FF [20].

4 Jet Production in Pion-Nucleus Collisions

When a negatively charged pion with energy on the order of TeV collides with a platinum target, as in the E791 Fermilab experiment [21], and diffractively dissociates into two jets, the observed jet distribution is expected to report on the quark-antiquark pion wave function. The collision is illustrated in Fig. 3. Our description of it is based on Ref. [22].

One represents the incoming pion as a bound state of two quarks of constituent size s_c. According to Refs. [19,20], the effective constituent wave function should be a solution of a holographic eigenvalue equation [17] with a quadratic effective potential, which for the ground state is Gaussian, $\psi_\pi \sim \exp[-\mathcal{M}_{q\bar{q}}^2/(2\varkappa^2)]$ where the constituent invariant mass squared is $\mathcal{M}_{q\bar{q}}^2 = (l^{\perp 2} + m_c^2)/[y(1-y)] = 4(\mathbf{l}^2 + m_c^2)$. $\varkappa$ is a parameter of the quadratic effective potential written as $U_{\rm eff} = \varkappa^4 r^2/4$ in terms of the relative distance r between the quark and antiquark, canonically conjugated to the relative momentum variables of Eqs. (18) and (19) according to quantum mechanics. The quark carries fraction y and antiquark fraction $1-y$ of the pion large $+$-momentum. The pion transverse momentum is zero, and the quark carries transverse momentum $l^\perp$ while the antiquark carries $-l^\perp$. The wave function is normalized to 1, assuming that the effective quarks of size s_c saturate the

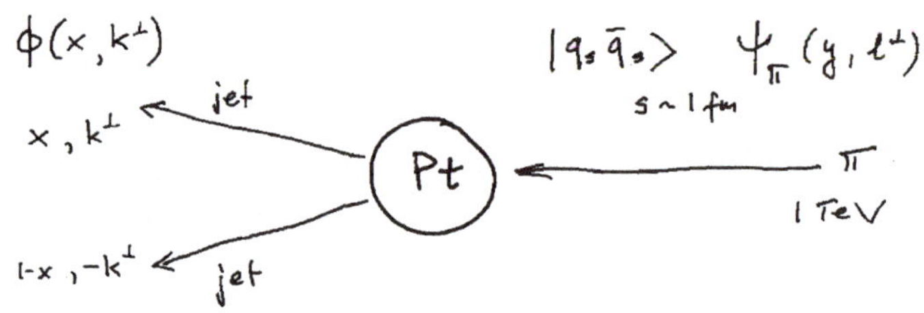

Fig. 3 Pion is split into two quark jets on the gluons in a nucleus that act like a wedge

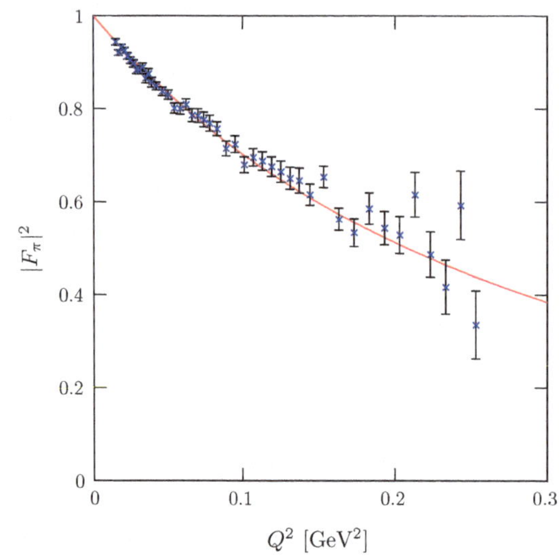

Fig. 4 The pion electromagnetic form factor at low momentum transfers for the wave function model with holographic effective potential $U_{\text{eff}} = \varkappa^4 r^2/4$ in AdS/QCD [17,20]. Data points are from Ref. [26]

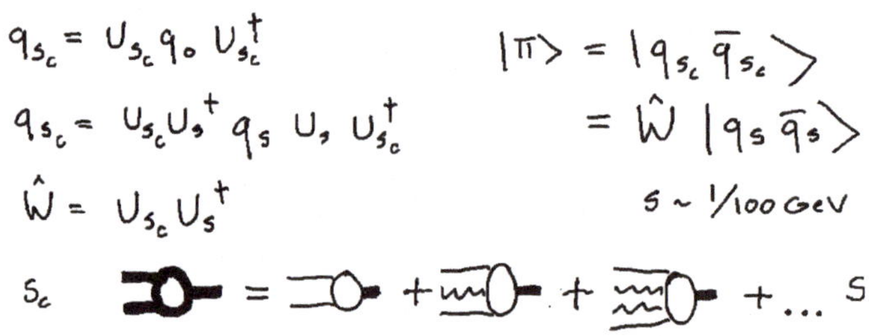

Fig. 5 $|\pi\rangle = |q_{s_c}\bar{q}_{s_c}\rangle$ in terms of quarks and gluons with $1/s \sim 100\,\text{GeV}$

pion state, as it should be the case according to the classification of hadrons in particle data tables [23]. The spin and isospin details are described by multiplying the Gaussian factor by $\bar{d}(p\!\!\!/_\pi + M)\gamma_5 u$ [22]. The wave function parameters are adjusted to reproduce the pion radius and decay constant as well as the Gell-Mann–Okubo formula [24,25]: $m_c \sim 331$ MeV, $\varkappa \sim 436$ MeV and $M \sim -1.9$ GeV. The pion radius squared is ~0.44 fm^2 while data provide 0.45(1) fm^2 [23]. Weak decay constant is $f_\pi \sim 130.7$ MeV, to be compared with the experimental value of 130.4 MeV [23]. The fitted value of $\varkappa$ is about 25% smaller than one could expect on the basis of other models [20]. The pion electromagnetic form factor obtained using these parameters is plotted in Fig. 4.

When the high-energy pion hits the nucleus and its constituents interact with the target, the low-energy wave-function picture for the pion as built just from one constituent quark and one constituent antiquark is not adequate. We need to express the quarks of size s_c by the quarks and gluons of much smaller size s for which we know the nuclear distribution. For example, to use the platinum distribution of gluons in the Bjorken variable x for $Q \sim 100$ GeV [27], we need to use the RGPEP operator U_s to transform the quarks from scale s_c to the quarks of scale $s \sim (100\,\text{GeV})^{-1}$. This is illustrated in Fig. 5. The required formula reads

$$|\pi\rangle = |q_{s_c}\bar{q}_{s_c}\rangle = \hat{W}_{s\,s_c}|q_s\bar{q}_s\rangle \tag{20}$$

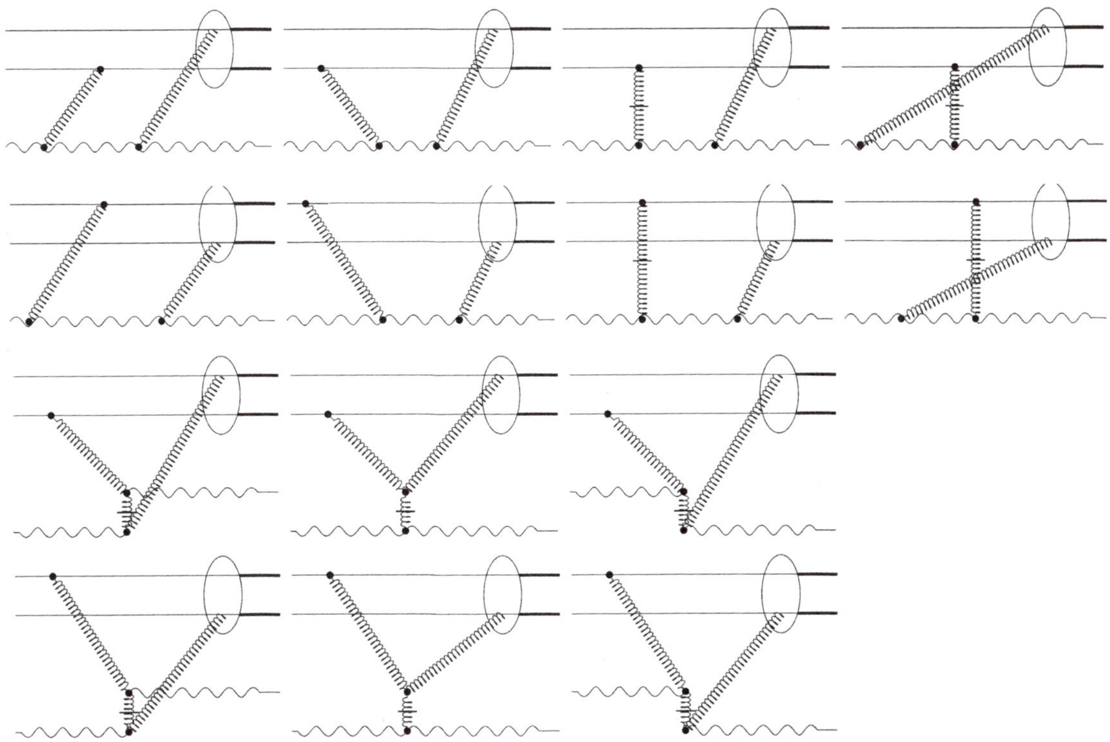

Fig. 6 Examples of interactions that split the pion on gluons in a nucleus according to H_{QCDs} with $1/s \sim 100$ GeV [22]

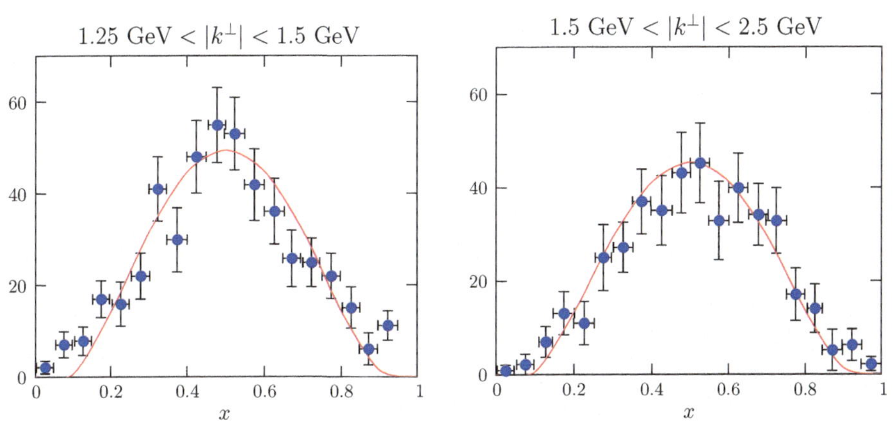

Fig. 7 Jet counts distribution $\phi(x, k^\perp)$, for jets induced by pions impinging on Pt, in two jet-$k^\perp$ bins; data points are from Ref. [21] and theory curves from Ref. [22]

where the state $|q_s\bar{q}_s\rangle$ has the same wave function as $|q_{s_c}\bar{q}_{s_c}\rangle$, $\hat{W} = U_{s_c}U_s^\dagger$ and both operators U are expressed in terms of creators and annihilators of effective particles of size s. The operator $\hat{W}$ can be evaluated using expansion in powers of the coupling constant g_s, which is equivalent to expansion in powers of bare g when one limits the calculation to terms order 1, g and g^2. To this order, the pion quark-antiquark constituent state has components with quark-antiquark, quark-antiquark-gluon and quark-antiquark-gluon-gluon of the small size s. Thus, the calculation of splitting of a pion into two jets amounts to evaluation of action of $\hat{W}$ on the constituent model of pion and calculating scattering amplitude of the resulting components into final quarks, whose momenta are identified with the momenta of the outgoing jets. Examples of the diagrams that contribute are provided in Fig. 6.

The resulting jet counts distributions, $\phi(x, k^\perp)$ in Fig. 3, in comparison with data that are available in two bins of the jet $k^\perp$, are shown in Fig. 7. The absolute normalization factor for the theoretical curves is freely adjusted in both bins, the required coefficients in low to high $k^{\perp}$ bin being of ratio 13/11. The shapes of theoretical curves are in a reasonably good agreement with experimentally observed distributions of jets. However, the calculation does not include effects of propagation of leading quarks through the nucleus and misses description of hadronization. Inclusion of these elements requires development of the RGPEP beyond the current stage.

5 Conclusion

The RGPEP opens new options for developing QFT and applying it in phenomenology of particle, nuclear and atomic physics. We are focused mainly on strong interactions. A critical step that awaits performing in QCD is to derive the Hamiltonian for effective particles with accuracy to fourth power of the coupling constant g. This Hamiltonian is expected to contain so far unknown terms whose structure may shed new light on the form of effective dynamics that applies to hadrons understood in terms of constituent quarks. The simplest theory to start with is QCD of quarks with masses much larger than Λ_{QCD} in the RGPEP scheme. The goal would be to understand the dynamics of effective gluons in colorless objects, with the heavy quarks serving as anchors for the system. Extension to light quarks will require understanding of how many of them participate in the dynamics. This number will be determined by the ratio of Λ_{QCD} to the light quark masses, which is on the order of 100. Such large number explains why the answer is not simple to obtain. It involves understanding of the Hamiltonian mechanism of breaking chiral symmetry and buildup of confinement. The RGPEP does not provide any ready answers to these questions, but it does offer tools for required studies. In particular, the effective coupling constant g_s at suitable values of quark and gluon size s is not known yet and, so far, we do not have any theoretical estimates for the probability of finding large numbers of light quarks in eigenstates of $\hat{H}_{QCD\,s}$ with any value of s that is likely to be effective in describing light hadrons.

Acknowledgements APT acknowledges support of the National Science Center of Poland grant PRELUDIUM under the decision number DEC-2014/15/N/ST2/03451.

References

1. W. Heisenberg, W. Pauli, Zur Quantendynamik der Wellenfelder. Z. Phys. **56**, 1–61 (1929)
2. W. Heisenberg, W. Pauli, Zur Quantentheorie der Wellenfelder II. Z. Phys. **59**, 168–190 (1929)
3. P. Dirac, Forms of relativistic dynamics. Rev. Mod. Phys. **21**, 392–399 (1949)
4. S. Głazek, K. Wilson, Renormalization of Hamiltonians. Phys. Rev. D **48**, 5863–5872 (1993)
5. K. Wilson et al., Nonperturbative QCD: a weak-coupling treatment on the light front. Phys. Rev. D **49**, 6720–6766 (1994)
6. F. Wegner, Flow equations for Hamiltonians. Ann. Phys. **506**, 77–91 (1994)
7. S. Głazek, Perturbative formulae for relativistic interactions of effective particles. Acta Phys. Pol. B **43**, 1843–1862 (2012)
8. M. Gómez-Rocha, S. Głazek, Asymptotic freedom in the front-form hamiltonian for quantum chromodynamics of gluons. Phys. Rev. D **92**, 065005–065019 (2015)
9. D. Gross, F. Wilczek, Ultraviolet behavior of non-Abelian gauge theories. Phys. Rev. Lett. **30**, 1343–1346 (1973)
10. D. Politzer, Reliable perturbative results for strong interactions? Phys. Rev. Lett. **30**, 1346–1349 (1973)
11. ATLAS Collaboration, Observation of long-range elliptic anisotropies in $\sqrt{s} = 13$ and 2.76 TeV pp collisions with the ATLAS detector. Phys. Rev. Lett. **116**, 172301–172320 (2016)
12. C.M.S. Collaboration, Measurement of long-range near-side two-particle angular correlations in pp collisions at $\sqrt{s} = 13$ TeV. Phys. Rev. Lett. **116**, 172302–172319 (2016)
13. P. Kubiczek, S. Głazek, Manifestation of proton structure in ridge-like correlations in high-energy proton–proton collisions. Lith. J. Phys. **55**, 155–161 (2015)
14. J. Bjorken et al., Possible multiparticle ridge-like correlations in very high multiplicity proton–proton collisions. Phys. Lett. B **726**, 344–346 (2013)
15. S. Głazek, Calculation of size for bound-state constituents. Phys. Rev D **90**, 045020–26 (2014)
16. R. Pohl et al., Muonic hydrogen and the proton radius puzzle. Annu. Rev. Nucl. Part. Sci. **63**, 175–204 (2013)
17. S. Brodsky et al., Light-front holographic QCD and emerging confinement. Phys. Rep. **584**, 1–105 (2015)
18. J. Maldacena, The large-N limit of superconformal field theories and supergravity. Int. J. Theor. Phys. **38**, 1113–1133 (1999)
19. S. Głazek, A. Trawiński, Model of the AdS/QFT duality. Phys. Rev. D **88**, 105025–12 (2013)
20. A. Trawiński et al., Effective confining potentials for QCD. Phys. Rev. D **90**, 074017–6 (2014)
21. Fermilab E791 Collaboration, Direct measurement of the pion valence-quark momentum distribution, the pion light-cone wave function squared. Phys. Rev. Lett. **86**, 4768–4772 (2001)

22. A. Trawiński, Hadron light-front wave functions based on AdS/QCD duality. Ph.D. thesis, University of Warsaw, 12 July 2016 (2016)
23. Particle Data Group, The review of particle physics. Chin. Phys. C **40**, 100001 (2016)
24. M. Gell-Mann, The eightfold way: a theory of strong interaction symmetry Report CTSL-20, TID-12608, Synchtron Laboratory, California Institute of Technology, Pasadena (1961)
25. S. Okubo, Note on unitary symmetry in strong interactions. Prog. Theor. Phys. **27**, 949–966 (1962)
26. A. Amendolia, A measurement of the space-like pion electromagnetic form factor. Nucl. Phys. B **277**, 168–196 (1986)
27. J. Pumplin et al., New generation of parton distributions with uncertainties from global QCD analysis. JHEP **0207**, 012 (2002)

Few-Body Syst (2017) 58:65
DOI 10.1007/s00601-017-1218-1

María Gómez-Rocha

From Asymptotic Freedom Toward Heavy Quarkonia Within the Renormalization-Group Procedure for Effective Particles

Received: 22 November 2016 / Accepted: 2 January 2017 / Published online: 30 January 2017

Abstract The renormalization group procedure for effective particles (RGPEP), developed as a nonperturbative tool for constructing bound states in quantum field theories, is applied to QCD. The approach stems from the similarity renormalization group and introduces the concept of effective particles. It has been shown that the RGPEP passes the test of exhibiting asymptotic freedom. We present the running of the Hamiltonian coupling constant with the renormalization-group scale and we summarize the basic elements needed in the formulation of the bound-state problem.

1 Introduction

The renormalization-group procedure for effective particles (RGPEP) has been developed during the last years as a non perturbative tool for constructing bound states in quantum field theory (QFT) [1–5]. It stems from the similarity renormalization group [6,7] and introduces the concept of effective particles in the Fock space [8]. The RGPEP is a Hamiltonian approach and it is formulated in the front form (FF) of relativistic dynamics [9]. The FF possesses several interesting features. For example, in the FF Hamiltonian there are no terms with only creation or only annihilation operators; then, the vacuum state $|0\rangle$ is an eigenstate of the Hamiltonian with eigenvalue 0. Furthermore, from the 10 generators of the Poincaré group, only 3 are dynamical (contain interactions). This is interesting e.g. from the point of view of situations in which one needs to consider bound states in motion.

One of the central problems in describing bound states in QFT is the fact that one needs to deal with an infinite number of degrees of freedom. For example, a quarkonium state in the Fock space can be expressed by an infinite series with the structure

$$|\psi\rangle = |Q\bar{Q}\rangle + |Q\bar{Q}g\rangle + |Q\bar{Q}gg\rangle + \dots, \tag{1}$$

and there is no limit in the number of participating Fock components. The key idea of the RGPEP is that for describing observables characterized by a size scale s, one can derive an effective Hamiltonian Hs, which is written in a scale-dependent operator basis, such that the number of Fock components [cf. Eq. (1)] relevant to the description is sufficiently small for carrying out computations. The parameter s has the meaning of size of the effective particles. When infinitely many Fock components can be neglected and only a few are

This article belongs to the Topical Collection "Light Cone 2016".

Invited talk given at Light Cone 2016, IST, Universidade de Lisboa, Portugal, September 5–8 2016.

M. Gómez-Rocha (✉)
European Centre for Theoretical Studies in Nuclear Physics and Related Areas (ECT*), Strada delle Tabarelle, 286, 38123 Villazzano, TN, Italy
E-mail: gomezr@ectstar.eu

significant, the bound state problem is drastically simplified and one can attempt to seek a numerical solution to the eigenvalue equation. We present the main steps of this method in Sects. 2 and 3.

The notion of effective particles is also used to explain the different behavior of interacting particles at different energy scales. It has been checked that the RGPEP passes the test of asymptotic freedom [5]. It will be shown below how the strength of the running coupling constant depends on the size of the effective particles. Exhibiting asymptotic freedom is a precondition for any approach aiming at describing hadrons using QCD. This issue is addressed in Sect. 4.

In order to start thinking about bound states in QCD we focus on heavy quarkonia. Heavy quarkonium is the simplest bound state in QCD and it provides a fruitful ground for comparing QCD approaches with experimental data [10–23]. A number of effective theories and phenomenological models are based on simplifications appearing due to the very large quark masses as compared with Λ_{QCD}. This condition, and the asymptotic freedom feature will allow us to construct an effective theory using RGPEP. We present the main steps and basic elements of the bound-state problem for heavy quarkonia in Sect. 5.

2 The Initial Hamiltonian

2.1 The QCD Canonical Hamiltonian

We start from the classical Lagrangian of QCD:

$$\mathcal{L} = \bar{\psi}(i\not{D} - m)\psi - \frac{1}{2}\mathrm{tr}F^{\mu\nu}F_{\mu\nu}, \tag{2}$$

where $F^{\mu\nu} = \frac{1}{ig}[D^\mu, D^\nu]$, $D = \partial + igA$, $A = A^a t^a$ and $[t^a, t^b] = if^{abc}t^c$. We use the front form of dynamics and light-front variables (we adopt the conventions of Ref. [24]). The Noether theorem provides the associated energy-momentum tensor which, in the gauge $A^+ = 0$ and with the choice of the initial conditions set on the hypersurface $x^+ = x^0 + x^1 = 0$, leads to the following QCD front-form Hamiltonian

$$H_{QCD} = P^- = \frac{1}{2}\int dx^- d^2x^\perp \mathcal{H}\,|_{x^+=0}, \tag{3}$$

where

$$\mathcal{H} = \mathcal{H}_{\psi^2} + \mathcal{H}_{A^2} + \mathcal{H}_{A^3} + \mathcal{H}_{A^4} + \mathcal{H}_{\psi A \psi} + \mathcal{H}_{\psi AA\psi} + \mathcal{H}_{[\partial AA]^2} + \mathcal{H}_{[\partial AA](\psi\psi)} + \mathcal{H}_{(\psi\psi)^2}, \tag{4}$$

with the expression for every term given in the "Appendix A".

At the quantization surface $x^+ = 0$, the quantum fields are defined by[1]

$$\hat{A}^\mu = \sum_{\sigma c}\int [k]\left[t^c \varepsilon^\mu_{k\sigma}\hat{a}_{k\sigma c}e^{-ikx} + t^c\varepsilon^{\mu *}_{k\sigma}\hat{a}^\dagger_{k\sigma c}e^{ikx}\right]_{x^+=0}, \tag{5}$$

$$\hat{\psi} = \sum_{\sigma c f}\int [k]\left[\chi_c u_{fk\sigma}\hat{b}_{k\sigma cf}e^{-ikx} + \chi_c v_{fk\sigma}\hat{d}^\dagger_{k\sigma cf}e^{ikx}\right]_{x^+=0}, \tag{6}$$

with the shorthand notation for the integration measure $[k] = \theta(k^+)dk^+d^2k^\perp/(16\pi^3k^+)$, the gluon polarization four-vector $\varepsilon^\mu_{k\sigma} = (\varepsilon^+_{k\sigma} = 0, \varepsilon^-_{k\sigma} = 2k^\perp\varepsilon^\perp_\sigma/k^+, \varepsilon^\perp_\sigma)$, and $u_{fk\sigma}$ and $v_{fk\sigma}$ being Dirac spinors. The indices σ and c stand for spin polarization and color, respectively. The replacement in Eq. (4) yields the canonical QCD Hamiltonian, a quantum operator written in terms of creation and annihilation operators (cf. Refs. [18,29] for details in these expressions).

[1] In the sequel we will drop the hats in particle operators, for simplicity.

2.2 Regularization

The canonical Hamiltonian needs regularization. We will follow the same procedure as in Ref. [25]. Every interaction term in the Hamiltonian is labeled by momentum quantum numbers $p^\perp$ and p^+; the total momentum annihilated in a vertex is labeled by $P^\perp$ and P^+; the relative transverse momentum is given by $\kappa^\perp = p^\perp - xP^\perp$, and the longitudinal momentum fraction by $x = p^+/P^+$. In every term in the Hamiltonian, every creation and annihilation operator of any momentum p is multiplied by the regulating function

$$r_{\Delta\delta}(\kappa^\perp, x) = \exp\left(-\kappa^{\perp 2}/\Delta^2\right) x^\delta\, \theta(x). \tag{7}$$

The first factor regulates ultraviolet divergences appearing at large $\kappa^{\perp 2}$. The Gaussian function does not allow the change of any gluon relative transverse momentum $\kappa^\perp$ to exceed the large cutoff Δ. The second factor, depending on the small cutoff parameter δ, regulates divergences appearing due to small-x in denominators.

After the derivation of the effective Hamiltonian (see below), any remaining cutoff-dependent term in the limit $\Delta \to \infty$ and $\delta \to 0$ must be canceled by counterterms. The regularized canonical Hamiltonian plus counterterms, provide the initial condition for the RGPEP equation.

3 Renormalization Group and Effective Particles

The RGPEP introduces the concept of *effective particles* [8]. The initial Hamiltonian H_0, which is written in terms of (bare) creation and annihilation operators of pointlike particles (quark, antiquark or gluons) $q_0^\dagger$ and q_0 can be re-expressed in terms of particle operators of size s. Creation (annihilation) operators of effective particles labeled with s, acting in the Fock space create (annihilate) effective particles of size s. E.g. for gluons:

$$a_0^\dagger|0\rangle = |g\rangle, \quad a_s^\dagger|0\rangle = |g_s\rangle, \quad a_0|0\rangle = a_s|0\rangle = 0. \tag{8}$$

It is also common to use the momentum-width parameter $\lambda = 1/s$. This parameter distinguishes different kinds of gluons according to the rule that: effective gluons of type λ can change their relative motion kinetic energy through a single effective interaction by no more than about λ [25]. It is however more convenient now to illustrate the calculation using the parameter $t = s^4 = 1/\lambda^4$.

In terms of this parameter, bare and effective particle operators are related by the unitary transformation:

$$q_t = \mathcal{U}_t\, q_0\, \mathcal{U}_t^\dagger, \quad \text{with} \quad \mathcal{U}_t = T\exp\left(-\int_0^t d\tau\, \mathcal{G}_t\right), \tag{9}$$

where $\mathcal{G}_t$ is a generator that will be given later. The RGPEP is based on the condition that the effective Hamiltonian cannot depend on the scale parameter t:

$$H_t(q_t) = H_0(q_0). \tag{10}$$

The relation between q_t and q_0 implies that

$$H_t(q_0) = \mathcal{U}_t^\dagger H_0(q_0)\mathcal{U}_s \equiv \mathcal{H}_t. \tag{11}$$

The differentiation of this equation with respect to t leads to the RGPEP equation:

$$\mathcal{H}_t' = [\mathcal{G}_t, \mathcal{H}_t]\,. \tag{12}$$

We consider the generator $\mathcal{G}_t = [\mathcal{H}_f, \mathcal{H}_{Pt}]$ (see "Appendix B" for details on the notation). Other generators are also possible, like the one used in Ref. [25], but they may lead to more complicated expressions.

The initial condition needed for solving Eq. (12) is given by the initial Hamiltonian at $t = 0$ plus counterterms:

$$\mathcal{H}_{t=0} = \mathcal{H}_{\Delta\delta} + \mathrm{CT}_{\Delta\delta}. \tag{13}$$

Although, in principle, Eq. (12) can be solved nonperturbatively, we are not prepared yet to do so in QCD. We use the perturbative formulas presented in Ref. [2] and solve it order by order as given in the "Appendix B".

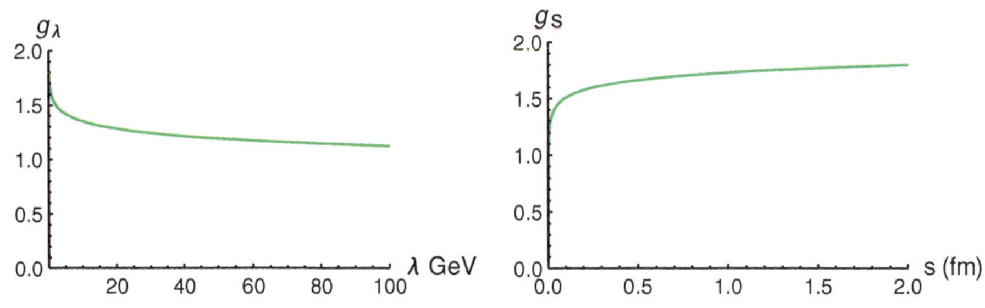

Fig. 1 Running coupling as a function of the momentum-width parameter λ (*left*) and as a function of the size of effective particles *s* (*right*)

4 Asymptotic Freedom

Already at third order, the effective Hamiltonian H_t contains details of the structure of the three-gluon vertex and quark-gluon vertex. We have studied the three-gluon vertex in Ref. [5] and calculated the Hamiltonian running coupling. The result,

$$g_\lambda = g_0 - \frac{g_0^3}{48\pi^2} N_c \, 11 \, \ln \frac{\lambda}{\lambda_0}, \tag{14}$$

leads to the β-function that is familiar to us and is found in Refs. [26,27], if one identifies λ with the momentum scale of external gluon lines in Feynman diagrams,

$$\lambda \frac{d}{d\lambda} g_\lambda = \beta_0 g_\lambda^3, \quad \text{with} \quad \beta_0 = -\frac{11 N_c}{48\pi^2}. \tag{15}$$

The running of the coupling constant with λ and *s* is plotted in Fig. 1. The smaller is the size of the effective particles, the weaker is the interaction between them; this corresponds to the situation of asymptotic freedom. This situation is explained in a visual way using a picture in Ref. [1].

This result suggests a possible direction in which one could start thinking about the equivalence of the Euclidean Greens function calculus and Minkowskian Hamiltonian quantum field operator.

5 Toward Heavy Quarkonium

5.1 A Heavy-Quark Effective Theory

The RGPEP allows one for the construction of effective theories. We will consider a theory of very heavy quarks, with masses much larger than the scale parameter λ (so that $\lambda/m_Q \to 0$), and with only one flavor in the QCD Hamiltonian. In this simplified situation, the coupling constant can still be considered small by means of asymptotic freedom and one can use expansions in powers of g when solving the RGPEP equation. Our hierarchy of scales is sketched in Fig. 2.

5.2 The Bound-State Problem

In our heavy quark effective theory, in which only one (heavy) flavor is present, it is not possible to create quark-antiquark pairs of such large masses out of an intermediate gluon with mass not larger than λ. Hence,

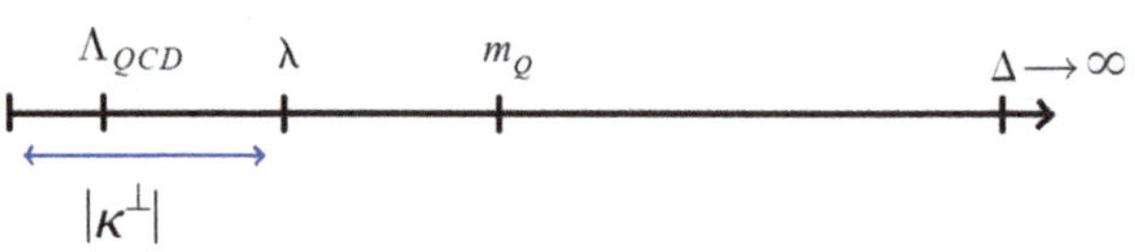

Fig. 2 Hierarchy of scales in the heavy-quark effective theory. Effective particles cannot change their relative momentum $|\kappa^\perp|$ by more than about λ. In turn, this parameter is much smaller than the quark mass m_Q. The ultraviolet parameter Δ is much larger than any other momentum parameter and is sent to infinity

the higher Fock component following to the simple $|Q\bar{Q}\rangle$ sector contain only additional gluons. The minimum number of sectors necessary to account for non-Abelian contributions is three:

$$|\psi\rangle = |Q\bar{Q}\rangle + |Q\bar{Q}\,g\rangle + |Q\bar{Q}\,gg\rangle. \tag{16}$$

Three Fock components in our perturbative RGPEP formalism requires 4th-order calculations.

We follow Ref. [28] and use projection operators to separate Fock components and formulate the eigenvalue equation in terms of the $|Q\bar{Q}\rangle$ state. As distinct from [28], here the operators $\hat{P}_t$ and $\hat{Q}_t = (1 - \hat{P}_t)$ depend on t and act on Fock sectors of particles with size $s = t^{1/4}$:

$$\psi_P \equiv \hat{P}_t|\psi\rangle = |Q\bar{Q}\rangle, \tag{17}$$

$$\psi_Q \equiv \hat{Q}_t|\psi\rangle = |Q\bar{Q}\,g\rangle + |Q\bar{Q}\,gg\rangle, \tag{18}$$

with $\psi = \psi_P + \psi_Q$. States ψ_P and ψ_Q are related by the transformation denoted by $\hat{R}$, $\psi_Q = \hat{R}_t\,\psi_P$. The eigenvalue equation reads[2]

$$H_t(\psi_P + \psi_Q) = E(\psi_P + \psi_Q). \tag{19}$$

Projecting the eigenvalue equation once into the subspace P and once into Q and using $\hat{R}_t$, one can obtain after careful manipulations [28] the effective Hamiltonian for $|Q\bar{Q}\rangle$:

$$H_{\mathrm{eff}\,t}|Q\bar{Q}\rangle = E|Q\bar{Q}\rangle, \tag{20}$$

where

$$H_{\mathrm{eff}\,t} = \frac{1}{\sqrt{P_t + R_t^\dagger R_t}}(P_t + R_t^\dagger)H_t(P_t + R_t)\frac{1}{\sqrt{P_t + R_t^\dagger R_t}}. \tag{21}$$

When one assumes the coupling to be small, like it is in our heavy-quark effective theory, one can use expansions in powers of g also to derive R_t and last $H_{\mathrm{eff}\,t}$. It should be emphasized that perturbation theory is only employed to solve the RGPEP equation and to calculate R_t, but not in solving the eigenvalue problem.

Second-order calculations using a particular generator (different from ours) have been performed, and the spectrum of a large number of quarkonia [18,29] has been obtained. However, details that can provide new hints about deeper aspects of the theory, such as, relevant information related to the confinement mechanism, can appear only starting from fourth order.

Another important feature of the RGPEP is that, since it uses a formulation in the Fock space which includes gluon degrees of freedom explicitly, the method provides an ideal framework for the description of hybrids [30,31].

6 Summary

We have presented the RGPEP as a tool to deal with bound states in QCD using the concept of effective particles. We have given the perturbative formula that solves the RGPEP equation and provides third- and fourth-order renormalized Hamiltonians.

The third-order H_t contains the running of the Hamiltonian coupling constant. We have presented the obtained result and its variation with the scale parameter. The derived running coupling has the same asymptotic-freedom behavior as in the calculus based on renormalized Feynman diagrams. This is important from the point of view of the desired but unknown precise connection between Feynman diagrams for virtual transition amplitudes and the Hamiltonan formalism in the Minkowski spacetime.

Heavy quarkonium is the simplest bound state in QCD and can be studied in the context of an effective theory for heavy quarks within the RGPEP. In order to investigate non-Abelian contributions and thus to include explicitly the running of the coupling in the bound-state problem, at least fourth-order effective Hamiltonians are necessary. Fourth-order calculations are drastically more complicated than third-order ones.

Acknowledgements I am very grateful to the Gary McCartor Fellowship Program for the McCartor travel award, thanks to which I could attend to this conference. I congratulate the LC2016 organizers for the very interesting and enjoyable meeting.

[2] We drop hats again in the operators from now on.

Appendix A: Hamiltonian-Density Terms

The terms in Eq. (4) are

$$\mathcal{H}_{\psi^2} = \frac{1}{2}\bar{\psi}\gamma^+ \frac{-\partial^{\perp 2} + m^2}{i\partial^+}\psi, \tag{22}$$

$$\mathcal{H}_{A^2} = -\frac{1}{2}A^{\perp a}(\partial^\perp)^2 A^{\perp a}, \tag{23}$$

$$\mathcal{H}_{A^3} = g\, i\partial_\alpha A^a_\beta [A^\alpha, A^\beta]^a, \tag{24}$$

$$\mathcal{H}_{A^4} = -\frac{1}{4}g^2\,[A_\alpha, A_\beta]^a[A^\alpha, A^\beta]^a, \tag{25}$$

$$\mathcal{H}_{\psi A \psi} = g\,\bar{\psi}\not{A}\psi, \tag{26}$$

$$\mathcal{H}_{\psi AA\psi} = \frac{1}{2}g^2\,\bar{\psi}\not{A}\frac{\gamma^+}{i\partial^+}\not{A}\psi, \tag{27}$$

$$\mathcal{H}_{[\partial AA]^2} = \frac{1}{2}g^2\,[i\partial^+ A^\perp, A^\perp]^a \frac{1}{(i\partial^+)^2}[i\partial^+ A^\perp, A^\perp]^a, \tag{28}$$

$$\mathcal{H}_{[\partial AA](\psi\psi)} = g^2\,\bar{\psi}\gamma^+ t^a \psi \frac{1}{(i\partial^+)^2}[i\partial^+ A^\perp, A^\perp]^a, \tag{29}$$

$$\mathcal{H}_{(\psi\psi)^2} = \frac{1}{2}g^2\,\bar{\psi}\gamma^+ t^a \psi \frac{1}{(i\partial^+)^2}\bar{\psi}\gamma^+ t^a \psi. \tag{30}$$

Appendix B: Summary of RGPEP Formulas

The RGPEP equation satisfied by the renormalized Hamiltonian $\mathcal{H}_t \equiv H_t(q_0) = \mathcal{U}_t^\dagger H_0(q_0)\mathcal{U}_s$ is:

$$\mathcal{H}'_t = [[\mathcal{H}_f, \mathcal{H}_{Pt}], \mathcal{H}_t] \tag{31}$$

where

$$\mathcal{H}_f = \sum_i p_i^- q^\dagger_{0i} q_{0i}, \quad \text{with} \quad p_i^- = \frac{p_i^{\perp 2}}{p_i^+}, \tag{32}$$

$$\mathcal{H}_t(q_0) = \sum_{n=2}^{\infty} \sum_{i_1, i_2, \ldots, i_n} c_t(i_1, \ldots, i_n)\, q^\dagger_{0i_1} \cdots a_{0i_n}, \tag{33}$$

$$\mathcal{H}_{Pt}(q_0) = \sum_{n=2}^{\infty} \sum_{i_1, i_2, \ldots, i_n} c_t(i_1, \ldots, i_n) \left(\frac{1}{2}\sum_{k=1}^{n} p^+_{i_k}\right)^2 q^\dagger_{0i_1} \cdots q_{0i_n}. \tag{34}$$

The operator $\mathcal{H}_f$ is called the free Hamiltonian, since it is the part of $\mathcal{H}_0(a_0)$ that does not depend on the coupling constant. The index i denotes the quantum numbers of the particle and p_i^- is the free FF energy for the gluon kinematical momentum components p_i^+ and $p_i^\perp$.
In order to solve this equation perturbatively, the Hamiltonian can be expressed as an expansion in powers of the coupling constant g:

$$\mathcal{H}_t = \mathcal{H}_0 + g\mathcal{H}_{t1} + g^2\mathcal{H}_{t2} + g^3\mathcal{H}_{t3} + g^4\mathcal{H}_{t4} + \ldots \tag{35}$$

In an expansion up to fourth-order, the RGPEP equation reads,

$$\begin{aligned} &\mathcal{H}'_0 + g\mathcal{H}_{t1} + g^2\mathcal{H}'_{t2} + g^3\mathcal{H}'_{t3} + g^4\mathcal{H}'_{t4} \\ &= \left[\left[\mathcal{H}_0, \mathcal{H}_0 + g\mathcal{H}_{1Pt} + g^2\mathcal{H}_{2Pt} + g^3\mathcal{H}_{3Pt} + g^4\mathcal{H}_{4Pt}\right], \mathcal{H}_0 + g\mathcal{H}_{t1} + g^2\mathcal{H}_{t2} + g^3\mathcal{H}_{t3} + g^4\mathcal{H}_{t4}\right]. \end{aligned} \tag{36}$$

Collecting terms with the same power of g one can solve the equation order by order, starting from order zero and introducing the result in the next-order equation:

$$\mathcal{H}_0' = 0, \tag{37}$$
$$g\mathcal{H}_{t\,1}' = [[\mathcal{H}_0, g\mathcal{H}_{1Pt}], \mathcal{H}_0], \tag{38}$$
$$g^2\mathcal{H}_{t\,2}' = \left[[\mathcal{H}_0, g^2\mathcal{H}_{2Pt}], \mathcal{H}_0\right] + [[\mathcal{H}_0, g\mathcal{H}_{1Pt}], g\mathcal{H}_{1t}], \tag{39}$$
$$g^3\mathcal{H}_{t\,3}' = \left[[\mathcal{H}_0, g^3\mathcal{H}_{3Pt}], \mathcal{H}_0\right] + \left[[\mathcal{H}_0, g^2\mathcal{H}_{2Pt}], g\mathcal{H}_{1t}\right] + \left[[\mathcal{H}_0, g\mathcal{H}_{1Pt}], g^2\mathcal{H}_{2t}\right], \tag{40}$$
$$g^4\mathcal{H}_{t\,4}' = \left[[\mathcal{H}_0, g^4\mathcal{H}_{4Pt}], \mathcal{H}_0\right] + \left[[\mathcal{H}_0, g^3\mathcal{H}_{3Pt}], g\mathcal{H}_{1t}\right] + \left[[\mathcal{H}_0, g^2\mathcal{H}_{2Pt}], g^2\mathcal{H}_{2t}\right] + \left[[\mathcal{H}_0, g\mathcal{H}_{1Pt}], g^3\mathcal{H}_{3t}\right], \tag{41}$$

The solutions to these equations in terms of matrix elements $\mathcal{H}_{t\,ab} = \langle a|\mathcal{H}|b\rangle$ are:

$$\mathcal{H}_{t1\,ab} = f_{t\,ab}\,\mathcal{H}_{01\,ab}, \tag{42}$$
$$\mathcal{H}_{t2\,ab} = f_{t\,ab} \sum_x \mathcal{H}_{01\,ax}\mathcal{H}_{01\,xb}\,\mathcal{B}_{t\,axb} + f_{t\,ab}\,\mathcal{G}_{02\,ab}, \tag{43}$$
$$\mathcal{H}_{t3\,ab} = f_{t\,ab}\,\mathcal{G}_{03\,ab} + f_{t\,ab} \sum_{xy} \mathcal{H}_{01\,ax}\,\mathcal{H}_{01\,xy}\,\mathcal{H}_{01\,yb}\,\mathcal{C}_{t\,axyb} + f_{t\,ab} \sum_x (\mathcal{H}_{01\,ax}\,\mathcal{G}_{02\,xb} + \mathcal{G}_{02\,ax}\,\mathcal{H}_{01\,xb})\,\mathcal{B}_{t\,axb}, \tag{44}$$
$$\mathcal{H}_{t4\,ab} = f_{t\,ab} \sum_{xyz} \mathcal{H}_{01\,ax}\,\mathcal{H}_{01\,xy}\,\mathcal{H}_{01\,yz}\,\mathcal{H}_{01\,zb}\,\mathcal{D}_{t\,axyzb} + f_{\tau\,ab} \sum_{xy} \left(\mathcal{H}_{01\,ax}\,\mathcal{H}_{01\,xy}\,\mathcal{G}_{02\,yb} + \mathcal{H}_{01\,ax}\,\mathcal{G}_{02\,xy}\,\mathcal{H}_{01\,yb} + \mathcal{G}_{02\,ax}\,\mathcal{H}_{01\,xy}\mathcal{H}_{01\,yb}\right)\mathcal{C}_{t\,axyb} + f_{\tau\,ab} \sum_x (\mathcal{G}_{02\,ax}\,\mathcal{G}_{02\,xb} + \mathcal{H}_{01\,ax}\,\mathcal{G}_{03\,xb} + \mathcal{G}_{03\,ax}\,\mathcal{H}_{01\,xb})\,\mathcal{B}_{t\,axb} + f_{\tau\,ab}\,\mathcal{G}_{04\,ab}. \tag{45}$$

The RGPEP factors $\mathcal{A}_{t\,axb}, \mathcal{B}_{t\,axb}, \mathcal{C}_{t\,axyb}$ and $\mathcal{D}_{t\,axyzb}$ result from integration and are given below. They contain in turn form factors $f_{t\,ab} := \exp\left[-ab^2\,t\right]$, with the symbol ab denoting differences of invariant masses squared, $ab := \mathcal{M}_{ab}^2 - \mathcal{M}_{ba}^2$. Every $\mathcal{G}_{t\,ab}$ is defined through $\mathcal{H}_{t\,ab} = f_{t\,ab}\mathcal{G}_{t\,ab}$.

$$\mathcal{A}_{t\,axb} = [ax\,p_{ax} + bx\,p_{bx}]\,f_{t\,ab}^{-1} f_{t\,ax}\,f_{t\,bx}, \tag{46}$$
$$\mathcal{B}_{t\,axb} = \int_0^t \mathcal{A}_{\tau\,axb}\,d\tau, \tag{47}$$
$$\mathcal{C}_{t\,axyb} = \int_0^t \left[\mathcal{A}_{\tau\,axb}\,\mathcal{B}_{\tau\,xyb} + \mathcal{A}_{\tau\,ayb}\,\mathcal{B}_{\tau\,axy}\right] d\tau, \tag{48}$$
$$\mathcal{D}_{t\,axyzb}, = \int_0^t \left[\mathcal{A}_{\tau\,axb}\,\mathcal{C}_{\tau\,xyzb} + \mathcal{A}_{\tau\,azb}\,\mathcal{C}_{\tau\,axyz} + \mathcal{A}_{\tau\,ayb}\,\mathcal{B}_{\tau\,axy}\,\mathcal{B}_{\tau\,yzb}\right]. \tag{49}$$

The Latin indices denote configurations of particles, i.e. a collection of quantum numbers that label creation and annihilation operators. For more details about the notation see Ref. [2].

References

1. S.D. Głazek, Reinterpretation of gluon condensate in dynamics of hadronic constituents. Acta Phys. Polon. B **42**, 1933 (2011)
2. S.D. Głazek, Perturbative formulae for relativistic interactions of effective particles. Acta Phys. Polon. B **43**, 1843 (2012)
3. A.P. Trawiński, S.D. Głazek, S.J. Brodsky, G.F. de Téramond, H.G. Dosch, Effective confining potentials for QCD. Phys. Rev. D **90**(7), 074017 (2014)
4. S.D. Głazek, Calculation of size for bound-state constituents. Phys. Rev. D **90**(4), 045020 (2014)
5. M. Gómez-Rocha, S.D. Głazek, Asymptotic freedom in the front-form Hamiltonian for quantum chromodynamics of gluons. Phys. Rev. D **92**(6), 065005 (2015)
6. S.D. Głazek, K.G. Wilson, Renormalization of hamiltonians. Phys. Rev. D **48**, 5863 (1993)

7. K.G. Wilson, T.S. Walhout, A. Harindranath, W.M. Zhang, R.J. Perry, S.D. Głazek, Nonperturbative QCD: a weak-coupling treatment on the light front. Phys. Rev. D **49**, 6720 (1994)
8. S.D. Głazek, A.P. Trawiński, Effective particles in quantam field theory, *in Proceedings of Light Cone* (2016), arXiv:1612.06211
9. P.A.M. Dirac, Forms of relativistic dynamics. Rev. Mod. Phys. **21**, 392 (1949)
10. N. Brambilla et al., Heavy quarkonium: progress, puzzles, and opportunities. Eur. Phys. J. C **71**, 1534 (2011)
11. A. Pineda, Review of heavy quarkonium at weak coupling. Prog. Part. Nucl. Phys. **67**, 735 (2012)
12. M. Blank, A. Krassnigg, Bottomonium in a Bethe-Salpeter-equation study. Phys. Rev. D **84**, 096014 (2011)
13. T. Hilger, C. Popovici, M. Gomez-Rocha, A. Krassnigg, Spectra of heavy quarkonia in a Bethe-Salpeter-equation approach. Phys. Rev. D **91**(3), 034013 (2015)
14. T. Hilger, M. Gomez-Rocha, A. Krassnigg, Masses of $J^{PC} = 1^{-+}$ exotic quarkonia in a Bethe-Salpeter-equation approach. Phys. Rev. D **91**(11), 114004 (2015)
15. M. Gomez-Rocha, T. Hilger, A. Krassnigg, Effects of a dressed quark-gluon vertex in pseudoscalar heavy-light mesons. Phys. Rev. D **92**(5), 054030 (2015)
16. M. Gómez-Rocha, T. Hilger, A. Krassnigg, Effects of a dressed quark-gluon vertex in vector heavy-light mesons and theory average of the B_c^* meson mass. Phys. Rev. D **93**(7), 074010 (2016)
17. C.S. Fischer, S. Kubrak, R. Williams, Spectra of heavy mesons in the Bethe-Salpeter approach. Eur. Phys. J. A **51**, 10 (2015)
18. S.D. Glazek, J. Mlynik, Boost-invariant Hamiltonian approach to heavy quarkonia. Phys. Rev. D **74**, 105015 (2006)
19. Y. Li, P. Maris, X. Zhao, J.P. Vary, Heavy quarkonium in a holographic basis. Phys. Lett. B **758**, 118 (2016)
20. C. Popovici, P. Watson, H. Reinhardt, Quarks in Coulomb gauge perturbation theory. Phys. Rev. D **79**, 045006 (2009)
21. J. Segovia, D.R. Entem, F. Fernandez, E. Ruiz Arriola, Renormalized quarkonium. Phys. Rev. D **86**, 094027 (2012)
22. J. Greensite, A.P. Szczepaniak, Constituent gluons and the static quark potential. Phys. Rev. D **93**(7), 074506 (2016)
23. P. Guo, A.P. Szczepaniak, G. Galata, A. Vassallo, E. Santopinto, Heavy quarkonium hybrids from Coulomb gauge QCD. Phys. Rev. D **78**, 056003 (2008)
24. S.J. Brodsky, H.C. Pauli, S.S. Pinsky, Quantum chromodynamics and other field theories on the light cone. Phys. Rept. **301**, 299 (1998)
25. S.D. Głazek, Dynamics of effective gluons. Phys. Rev. D **63**, 116006 (2001)
26. D.J. Gross, F. Wilczek, Ultraviolet behavior of non-abelian gauge theories. Phys. Rev. Lett. **30**, 1343 (1973)
27. H.D. Politzer, Reliable perturbative results for strong interactions? Phys. Rev. Lett. **30**, 1346 (1973)
28. K.G. Wilson, Model of coupling-constant renormalization. Phys. Rev. D **2**, 1438 (1970)
29. S.D. Glazek, Harmonic oscillator force between heavy quarks. Phys. Rev. D **69**, 065002 (2004)
30. S.D. Glazek, A.P. Szczepaniak, Special relativity constraints on an effective constituent theory of hybrids. Phys. Rev. D **67**, 034019 (2003)
31. S.D. Glazek, J. Narebski, Special relativity in decays of hybrids. Acta Phys. Polon. B **37**, 389 (2006)

Few-Body Syst (2017) 58:133
DOI 10.1007/s00601-017-1292-4

Stanley J. Brodsky

Advances in Light-Front QCD: Supersymmetric Properties of Hadron Physics from Light-Front Holography and Superconformal Algebra

Received: 20 November 2016 / Accepted: 24 March 2017 / Published online: 19 April 2017

Abstract A remarkable feature of QCD is that the mass scale κ which controls color confinement and light-quark hadron mass scales does not appear explicitly in the QCD Lagrangian. However, de Alfaro, Fubini, and Furlan have shown that a mass scale can appear in the equations of motion without affecting the conformal invariance of the action if one adds a term to the Hamiltonian proportional to the dilatation operator or the special conformal operator. If one applies the same procedure to the light-front Hamiltonian, it leads uniquely to a confinement potential $\kappa^4\zeta^2$ for mesons, where ζ^2 is the LF radial variable conjugate to the $q\bar{q}$ invariant mass. The same result, including spin terms, is obtained using light-front holography—the duality between the front form and AdS_5, the space of isometries of the conformal group—if one modifies the action of AdS_5 by the dilaton $e^{\kappa^2 z^2}$ in the fifth dimension z. When one generalizes this procedure using superconformal algebra, the resulting light-front eigensolutions predict a unified Regge spectroscopy of meson, baryon, and tetraquarks, including remarkable supersymmetric relations between the masses of mesons and baryons of the same parity. One also predicts observables such as hadron structure functions, transverse momentum distributions, and the distribution amplitudes defined from the hadronic light-front wavefunctions. The mass scale κ underlying confinement and hadron masses can be connected to the parameter $\Lambda_{\overline{MS}}$ in the QCD running coupling by matching the nonperturbative dynamics to the perturbative QCD regime. The result is an effective coupling $\alpha_s(Q^2)$ defined at all momenta. The matching of the high and low momentum transfer regimes determines a scale Q_0 which sets the interface between perturbative and nonperturbative hadron dynamics. The use of Q_0 to resolve the factorization scale uncertainty for structure functions and distribution amplitudes, in combination with the principle of maximal conformality for setting the renormalization scales, can greatly improve the precision of perturbative QCD predictions for collider phenomenology. The absence of vacuum excitations of the causal, frame-independent front-form vacuum has important consequences for the cosmological constant. I also discuss evidence that the antishadowing of nuclear structure functions is non-universal; i.e., flavor dependent, and why shadowing and antishadowing phenomena may be incompatible with sum rules for nuclear parton distribution functions.

Contents

1 Introduction 2
2 Color Confinement and Supersymmetry in Hadron Physics from LF Holography 3
3 Light-Front Holography 3
4 Supersymmetric Features of Hadron Physics 6

This article belongs to the Topical Collection "Light Cone 2016".

Invited talk presented at Light Cone 2016, September 5–8, 2016.

S. J. Brodsky (✉)
SLAC National Accelerator Laboratory, Stanford University, Stanford, CA, USA
E-mail: sjbth@slac.stanford.edu

5 Ridge Formation from Flux-Tube Collisions 9
6 Calculations Using LF-Time-Ordered Perturbation Theory and Hadronization at the Amplitude Level 10
7 The Light-Front Vacuum 10
8 The QCD Coupling at all Scales 11
9 Is the Momentum Sum Rule Valid for Nuclear Structure Functions? 11
10 Summary 13
References 14

1 Introduction

Light-front wavefunctions provide a direct link between the QCD Lagrangian and hadron structure. The LFWFs are the eigenstates of the Front-Form Hamiltonian $P^- = i\frac{d}{d\tau}$, the LF time $\tau = x^+ = t+z/c$ evolution operator. When one makes a measurement of a hadron, such as in deep inelastic lepton-proton scattering $\ell p \to \ell' X$, the hadron is observed along a light-front (LF)—in analogy to a flash photograph—not at a fixed time t. This is the underlying principle of the "front form" discussed by Dirac [1].

In the case of QCD, the eigenvalues of the LF invariant Hamiltonian $H_{LF} = P^+P^- - \vec{P}^2_\perp$, where $P^+ = P^0 + P^z$ and $\vec{P}_\perp$ are kinematical, are the squares of the hadron masses M^2_H: $H_{LF}|\Psi_H >= M^2_H|\Psi_H >$ [2]. The eigensolutions of H_{LF} provide the n-particle hadronic LF Fock state wavefunctions (LFWFs) $\psi^H_n(x_i, \vec{k}_{\perp i}, \lambda_i) = \langle n|\Psi_H\rangle$, the projection on the free Fock basis. The LF Hamiltonian, can be derived directly from the QCD Lagrangian. The constituents' physical momenta are $p^+_i = x_i P^+$, and $\vec{p}_{\perp i} = x_i \vec{P}_\perp + \vec{k}_{\perp i}$, and the λ_i label the spin projections S^z_i. The LFWFs are Poincare' invariant: they are independent of P^+ and $P_\perp$ and are thus independent of the motion of the observer. The elastic and transition form factors of hadrons, weak-decay amplitudes and distribution amplitudes are overlaps of LFWFs; structure functions, transverse momentum distributions and other inclusive observables are constructed from the squares of the LFWFs. The calculation of deeply virtual Compton scattering is given in Ref. [3]. Since the LFWFs are independent of the hadron's momentum, there is no length contraction. The absence of length contraction of a photographed object was first noted by Terrell [4] and Penrose [5]. One measures the same structure function in an electron-ion collider as in an electron-scattering experiment where the target hadron is at rest.

Light-front quantization thus provides a physical, frame-independent formalism for hadron dynamics and structure. The LFWFs play the same role as the Schrodinger wavefunctions which encode the structure of atoms in QED. One cannot compute form factors of hadrons or other current matrixelements of hadrons from overlap of the usual "instant" form wavefunctions since one must also include contributions where the photon interacts with connected but acausal vacuum-induced currents. One can show that the anomalous gravitomagnetic moment $B(q^2 = 0)$ vanishes identically for any LF Fock state [6], in agreement with a theorem [7,8] which follows from the equivalence theorem of gravity. One can derive H_{LF} directly from the QCD Lagrangian and avoid ghosts and longitudinal gluonic degrees of freedom by choosing to work in the light-cone gauge $A^+ = 0$. The quark masses appear in the LF kinetic energy as $\sum_i \frac{m^2}{x_i}$. This can be derived from the Higgs theory quantized using LF dynamics. The confined quark field ψ_q couples to the background Higgs field $g_{\overline{\Psi}_q}\langle H\rangle\Psi_q$ via its Yukawa scalar matrix element coupling $g_q\langle H\rangle\bar{u}(p)1u(p) = m_q \times \frac{m_q}{x} = \frac{m^2}{x}$.

PQCD factorization theorems and the DGLAP [9–11] and ERBL [12–15] evolution equations can also be derived using the light-front Hamiltonian formalism [13]. In the case of an electron-ion collider, one can represent the cross section for $e - p$ colisions as a convolution of the hadron and virtual photon structure functions times the subprocess cross-section in analogy to hadron-hadron colisions. This nonstandard description of $\gamma^* p \to X$ reactions gives new insights into electroproduction physics—physics not apparent in the usual infinite-momentum frame description, such as the dynamics of heavy quark-pair production. Intrinsic heavy quarks at high x also play an important role [16].

The LF Heisenberg equation can in principle be solved numerically by matrix diagonalization using the "Discretized Light-Cone Quantization" (DLCQ) [17] method. Anti-periodic boundary conditions in x^- render the k^+ momenta discrete as well as limiting the size of the Fock basis. In fact, one can easily solve $1 + 1$ quantum field theories such as QCD(1+1) [18] for any number of colors, flavors and quark masses using DLCQ. Unlike lattice gauge theory, the nonpertubative DLCQ analysis is in Minkowski space, is frame-independent, and is free of fermion-doubling problems. AdS/QCD, based on the AdS_5 representation of the conformal group in five dimensions, maps to physical 3+1 space-time at fixed LF time; this correspondence, "light-front holography" [19], is now providing a color-confining approximation to H^{QCD}_{LF} for QCD$(3 + 1)$. This method gives a remarkable first approximation to hadron spectroscopy and hadronic LFWFs. A new method for solving nonperturbative QCD "Basis Light-Front Quantization" (BLFQ) [20], uses the eigensolutions of a color-confining approximation to QCD (such as LF holography) as the basis functions, rather than the plane-wave basis used in DLCQ, thus incorporating the full dynamics of QCD. LFWFs can also be determined from the covariant Bethe-Salpeter wavefunction by integrating over k^- [21]. A review of the light-front formalism is given in Ref. [2].

2 Color Confinement and Supersymmetry in Hadron Physics from LF Holography

A fundamental problem in hadron physics is to obtain a color-confining first approximation to QCD which can predict both the hadron spectrum and the LFWFs underlying hadron phenomenology. The QCD Lagrangian with zero quark mass has no explicit mass scale; the classical theory is conformally invariant. A profound question is then to understand how the proton mass and other hadronic mass scales—the mass gap—can arise even when $m_q = 0$. In fact, chiral QCD has no knowledge of units such as MeV. However, a remarkable principle, first demonstrated by de Alfaro, Fubini and Furlan (dAFF) [22] in $1+1$ quantum mechanics, is that a mass scale can appear in a Hamiltonian without affecting the conformal invariance of the action. The essential step is to add to the conformal Hamiltonian H_0 terms proportional to the dilation operator D and the special conformal operator K. The coefficients introduce the mass scale κ, and the result is $H = H_0 + V$, where V a harmonic oscillator potential $V(x) = \kappa^2 x^2$. The action remains conformal when one changes to a new time variable.

De Téramond et al. [23] have shown that a mass gap and a fundamental color confinement scale appear when one extends the dAFF procedure to light-front Hamiltonian theory. Remarkably, the resulting light-front potential has a unique form of a harmonic oscillator $\kappa^4\zeta^2$ in the light-front invariant impact variable ζ where $\zeta^2 = b_\perp^2 x(1-x)$. The result is a single-variable frame-independent relativistic equation of motion for $q\bar{q}$ bound states, a "Light-Front Schrödinger Equation" [19], analogous to the nonrelativistic radial Schrödinger equation in quantum mechanics. The same result, including spin terms, is obtained using light-front holography—the duality between the front form and AdS_5, the space of isometries of the conformal group—if one modifies the action of AdS_5 by the dilaton $e^{\kappa^2 z^2}$ in the fifth dimension z. The reduction of the full light-front Hamiltonian in QCD to an effective $q\bar{q}$ LF Schrödinger Equation for mesons is illustrated in Fig. 1.

The Light-Front Schrödinger Equation incorporates color confinement and other essential spectroscopic and dynamical features of hadron physics, including a massless pion for zero quark mass and linear Regge trajectories with the same slope in the radial quantum number n and internal orbital angular momentum L. When one generalizes this procedure using superconformal algebra, the resulting light-front eigensolutions predict a unified Regge spectroscopy of meson, baryon, and tetraquarks, including remarkable supersymmetric relations between the masses of mesons and baryons of the same parity.

An essential point is that the mass scale κ is not determined absolutely by QCD. Only ratios of masses are determined, and the theory has dilation invariance under $\kappa \to C\kappa$, In a sense, chiral QCD has an "extended conformal invariance." The resulting new time variable which retains the conformal invariance of the action, has finite support, conforming to the fact that the LF time between the interactions with the confined constituents is finite. The finite time difference $\Delta\tau$ between the LF times τ of the quark constituents of the proton could be measured using positronium proton scattering $[e^+e^-]p \to e^+e^-p'$. This process, which measures double diffractive deeply virtual Compton scattering for two spacelike photons, is illustrated in Fig. 2 One can also study the dissociation of relativistic positronium atoms to an electron and positron with light front momentum fractions x and $1-x$ and opposite transverse momenta in analogy to the E791 measurements of the diffractive dissociation of the pion to two jets [24]. The LFWF of positronium in the relativistic domain is the central input. One can produce a relativistic positronium beam using the collisions of laser photons with high energy photons or by using Bethe-Heitler pair production below the e^+e^- threshold. The production of parapositronium via the collision of photons is analogous to pion production in two-photon collisions and Higgs production via gluon-gluon fusion.

3 Light-Front Holography

Five-dimensional AdS_5 space provides a geometrical representation of the conformal group. The identical color-confining light-front equation for mesons of arbitrary spin J can be derived [27] from the holographic

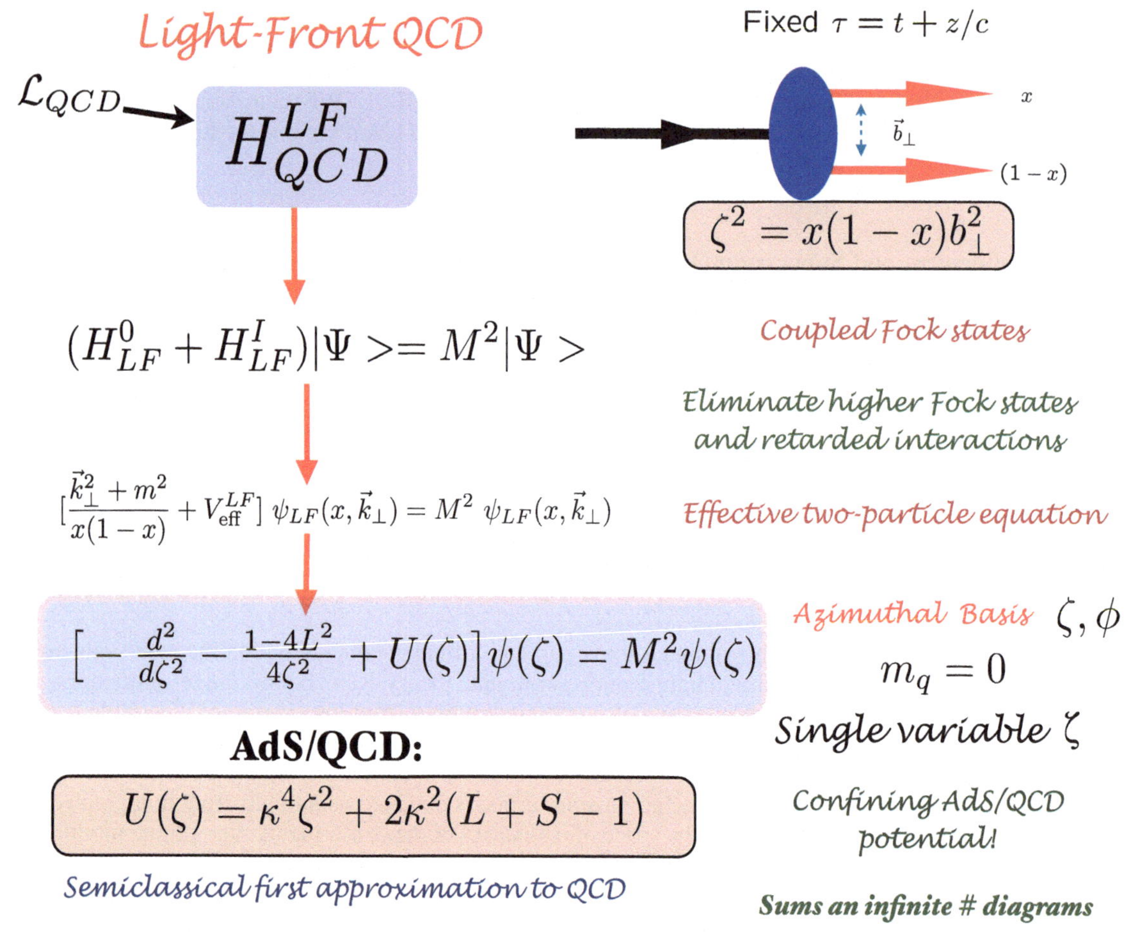

Fig. 1 Reduction of the QCD LF Hamiltonian to an effective $q\bar{q}$ LF Schrödinger equation for mesons. The light-front potential has the unique form of a harmonic oscillator $\kappa^4\zeta^2$ if one applies the DAFF principle or the corresponding soft-wall dilaton in AdS_5

mapping of the "soft-wall model" modification of AdS_5 space for the specific dilaton profile $e^{+\kappa^2 z^2}$, where one identifies the fifth dimension coordinate z with the light-front coordinate ζ [28]. Remarkably, AdS_5 is holographically dual to $3+1$ spacetime at fixed light-front time. The holographic dictionary is summarized in Fig. 3.

The holographic identification of the fifth dimension coordinate z with the LF radial variable ζ is confirmed by the fact that the equations of motion derived from AdS_5 and the QCD LF Hamiltonian in the valence sector are identical; in addition, the Polchinski-Strassler formula [29] for hadronic form factors in AdS_5 is the same as the Drell-Yan-West formula in LF theory [30]. This is true for both electroweak and gravitational currents [27,31].

It should be emphasized that the Gaussian potential is a consequence of the principle of de Alfaro, Fubini, and Furlan [22]: a mass scale can appear in the equations of motion without affecting the conformal invariance of the action if one adds a term to the Hamiltonian proportional to the dilatation operator or the special conformal operator. Applying the same procedure to the light-front Hamiltonian leads to the unique confinement potential $\kappa^4\zeta^2$ for mesons. The same result, including spin terms, is obtained using light-front holography, the duality between the front form and five-dimensional anti-de Sitter space, if one modifies the action by the dilaton $e^{+\kappa^2 z^2}$ in the fifth dimension z of AdS_5.

The combination of light-front dynamics, its holographic mapping to AdS_5 space, and the dAFF procedure provides new insight into the physics underlying color confinement, the nonperturbative QCD coupling, and the QCD mass scale. A comprehensive review is given in Ref. [32]. The $q\bar{q}$ mesons and their valence LF wavefunctions are the eigensolutions of the frame-independent relativistic bound state LF Schrödinger equa-

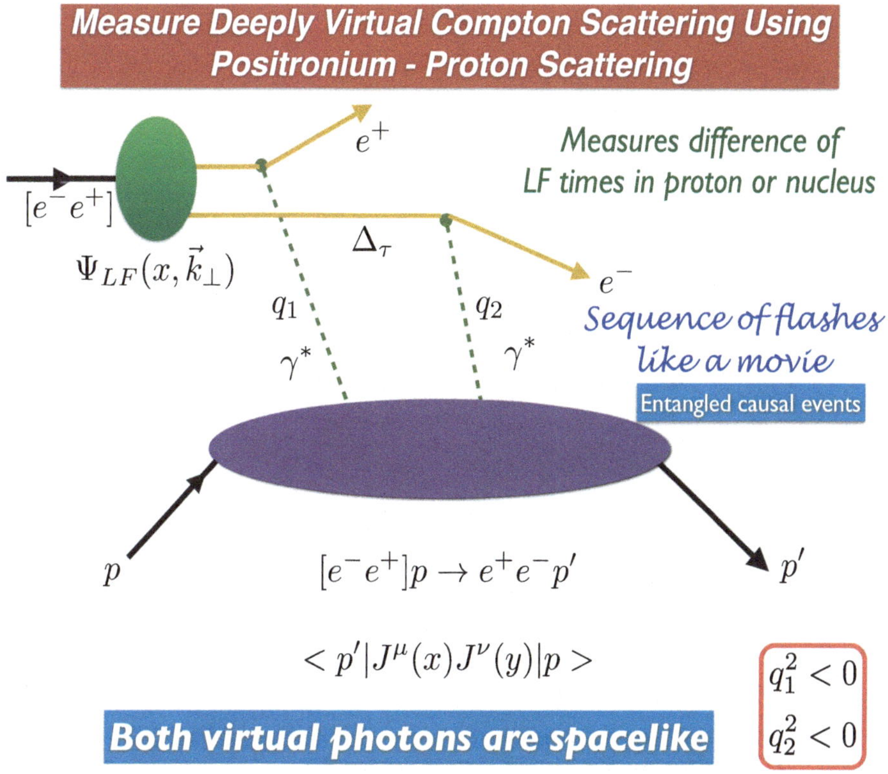

Fig. 2 Doubly Virtual Compton scattering on a proton (or nucleus) can be measured for two *spacelike* photons $q_1^2, q_2^2 < 0$ with minimal, tunable, skewness ξ using positronium-proton scattering $[e^+e^-]p \to e^+e^-p'$. One can also measure double deep inelastic scattering and elastic positronium-proton scattering. An analogous process will create the "true muonium" atom $[\mu^-\mu^-]$ [25,26]

tion. If one include the quark masses, the Gaussian form of the LFWF $\frac{\exp -\mathcal{M}^2}{\kappa^2}$ then vanishes exponentially at $x = 0, 1$. The Gaussian fall-off of the LFWF in $\exp -\frac{\kappa^2}{\zeta^2}$ at large ζ^2 is thus characteristic of confinement. Similarly, the AdS$_5$ solutions $\exp -\kappa^2 z^2$ display IR confinement at large z—this is the central premise of the softwall approach to color confinement.

The mesonic $q\bar{q}$ bound-state eigenvalues for massless quarks are $M^2(n, L, S) = 4\kappa^2(n + L + S/2)$. The equation predicts that the pion eigenstate $n = L = S = 0$ is massless at zero quark mass. The Regge spectra of the pseudoscalar $S = 0$ and vector $S = 1$ mesons are predicted correctly, with equal slope in the principal quantum number n and the internal orbital angular momentum L. A comparison with experiment is shown in Fig. 4.

In the LF formalism, the eigenstates are classified by the internal kinematical orbital angular momentum L^z which specifies the ϕ dependence of the eigenstates in the $\hat{x} - \hat{y}$ plane.. There are 2L+1 degenerate states; thus $L = max|L^z|$ characterizes the internal orbital angular momentum of the quarks in the LF eigenfunction. The LF spin sum rule of Jaffe and Manohar [33] is based on these LF principles. Angular momentum in LF Hamiltonian theory is illustrated for simple LF models in Ref. [6]. A general discussion of LF angular momentum and a derivation of new selection rules is given in [34].

The hadronic LFWFs predicted by LF holography are Gaussian functions in momentum space of the of the invariant mass squared, which is also the LF kinetic energy: $\frac{\mathcal{M}^2 = k_\perp^2}{x(1-x)}$ and the conjugate of the LF radial variable $\zeta^2 = b_\perp^2 x(1-x)$—times a function of $x(1-x)$; they do not factorize as a function of $\vec{k}_\perp^2$ times a function of x. The LF wavefunctions are the exact eigensolutions of the LF Hamiltonian in the $q\bar{q}$ valence sector for massless quarks. The resulting nonperturbative pion distribution amplitude $\phi_\pi(x) = \int d^2\vec{k}_\perp \psi_\pi(x, \vec{k}_\perp) = (4/\sqrt{3}\pi) f_\pi \sqrt{x(1-x)}$, see Fig. 5, which controls hard exclusive process, is consistent with the Belle data for the photon-to-pion transition form factor [35]. The AdS/QCD light-front holographic eigenfunction for the ρ

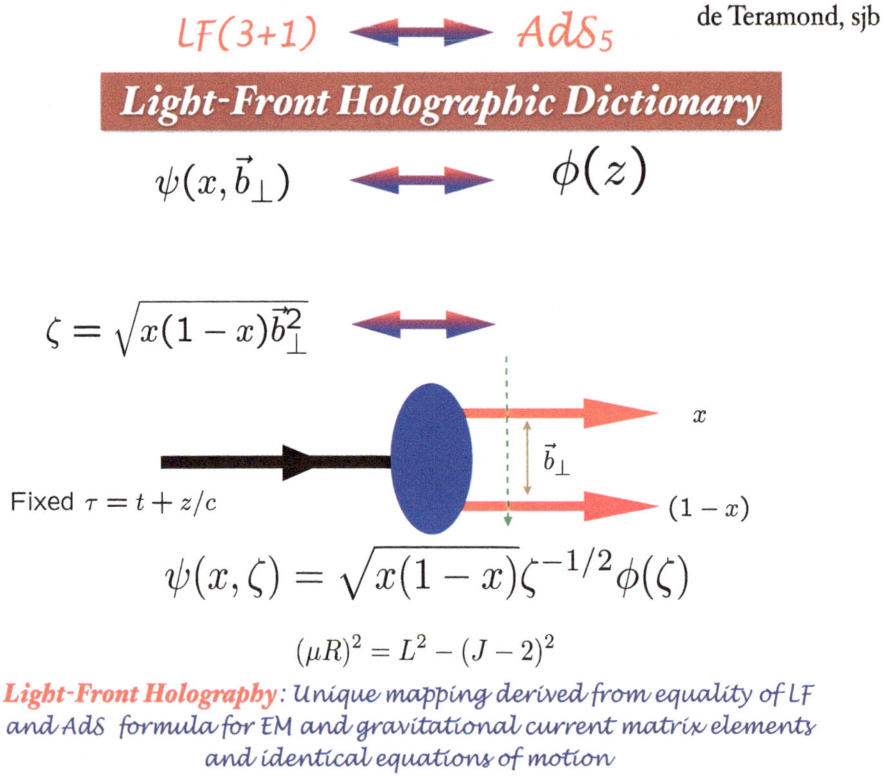

Fig. 3 The holographic dictionary which maps the fifth dimension variable z of the five-dimensional AdS$_5$ space to the LF radial variable ζ where $\zeta^2 = b_\perp^2(1-x)$. The same physics transformation maps the AdS$_5$ and $(3+1)$ LF expressions for electromagnetic and gravitational form factors to each other. From Ref. [27]

meson LFWF $\psi_\rho(x, \vec{k}_\perp)$ gives excellent predictions for the observed features of diffractive ρ electroproduction $\gamma^* p \to \rho p'$, as shown by Forshaw and Sandapen [36].

It is interesting to note that the contribution of the 'H' diagram to $Q\bar{Q}$ scattering is IR divergent as the transverse separation between the Q and the $\bar{Q}$ increases [37]. This is a signal that pQCD is inconsistent without color confinement. The sum of such diagrams could sum to the confinement potential $\kappa^4\zeta^2$ dictated by the dAFF principle that the action remains conformally invariant despite the appearance of the mass scale κ in the Hamiltonian. The $\kappa^4\zeta^2$ confinement interaction between a q and $\bar{q}$ will induce a κ^4/s^2 correction to $R_{e^+e^-}$, replacing the $1/s^2$ signal usually attributed to a vacuum gluon condensate.

4 Supersymmetric Features of Hadron Physics

The conformal group has an elegant 2×2 Pauli matrix representation called "superconformal algebra", originally discovered by Haag et al. [38]. The conformal Hamiltonian operator and the special conformal operators can be represented as anticommutators of Pauli matrices $H = 1/2[Q, Q^\dagger]$ and $K = 1/2[S, S^\dagger]$. As shown by Fubini and Rabinovici [39], a nonconformal Hamiltonian with a mass scale and universal confinement can then be obtained by shifting $Q \to Q + \omega K$, the analog of the dAFF procedure. In effect one has generalized supercharges of the superconformal algebra [39]. The resulting superconformal algebra leads to effective QCD light-front bound-state equations for both mesons and baryons [40–42]. This ansatz provides an elegant representation of the eigensolutions of the LF Hamiltonian, corresponding to meson, baryon and tetraquark eigenstates. The superconformal algebra approach should be distinguished from superconformal QCD—the theory of squarks and gluinos.

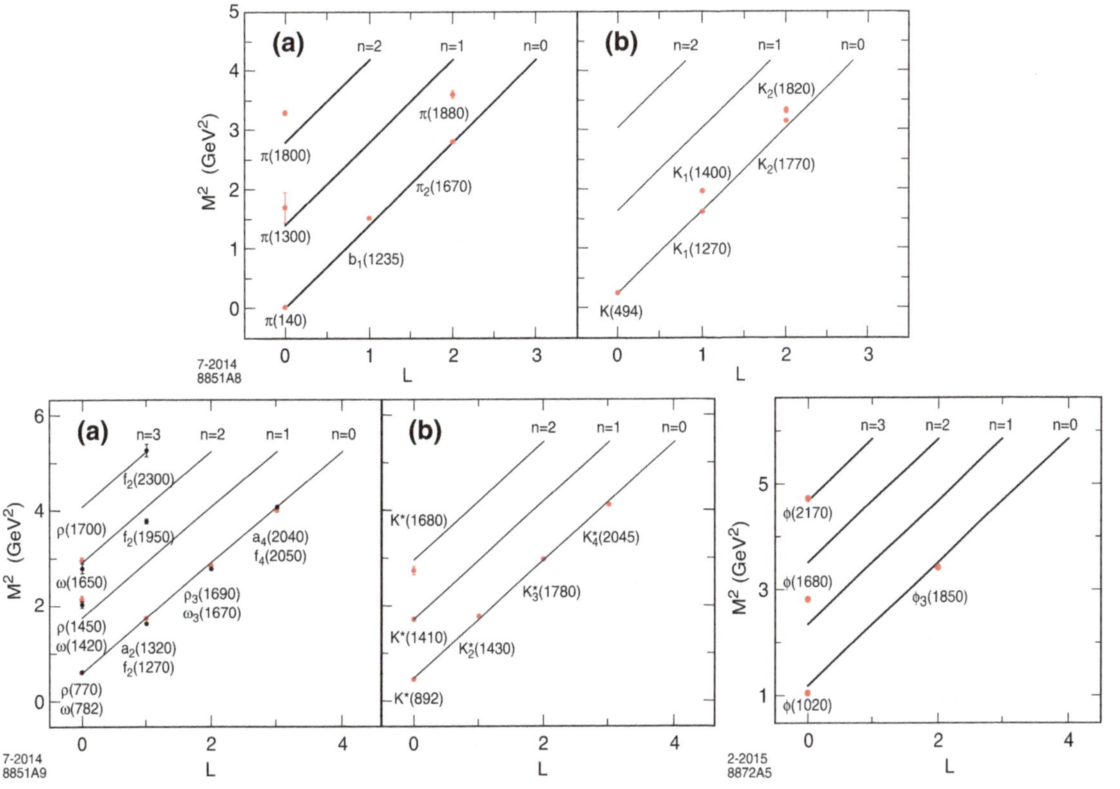

Fig. 4 Comparison of the AdS/QCD prediction $M^2(n, L, S) = 4\kappa^2(n + L + S/2)$ for the orbital L and radial n excitations of the meson spectrum with experiment. The pion is predicted to be massless for zero quark mass. The u, d, s quark masses can be taken into account by perturbing in $\langle m_q^2/x\rangle$. The fitted value of $\kappa = 0.59$ MeV for pseudoscalar mesons, and $\kappa = 0.54$ MeV for vector mesons

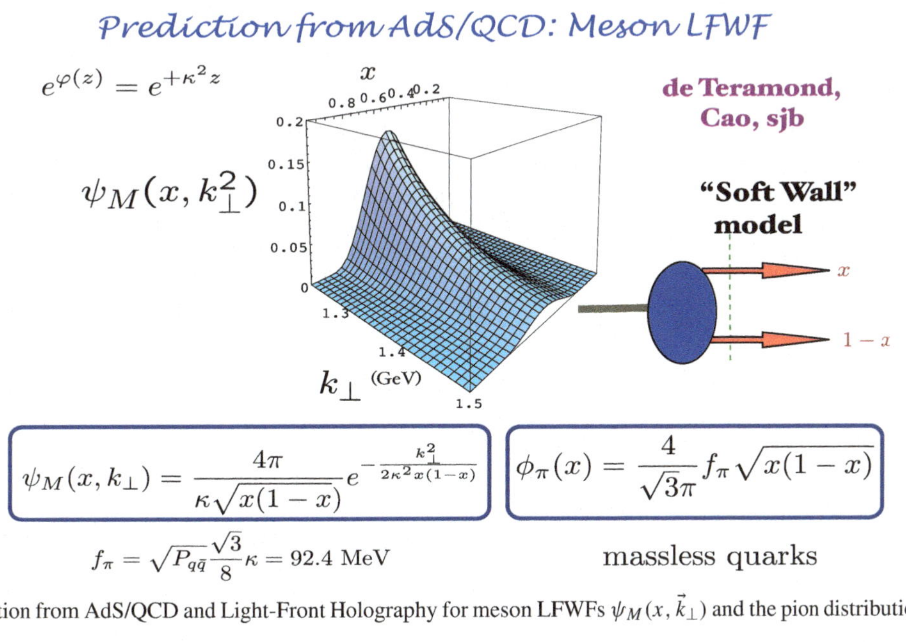

Fig. 5 Prediction from AdS/QCD and Light-Front Holography for meson LFWFs $\psi_M(x, \vec{k}_\perp)$ and the pion distribution amplitude

The supercharges connect the baryon and meson spectra and their Regge trajectories to each other in a remarkable manner: each meson has internal angular momentum one unit higher than its superpartner baryon $L_M = L_B + 1$. The resulting set of LF equations for confined quarks are shown in Fig. 6a. In effect the baryons

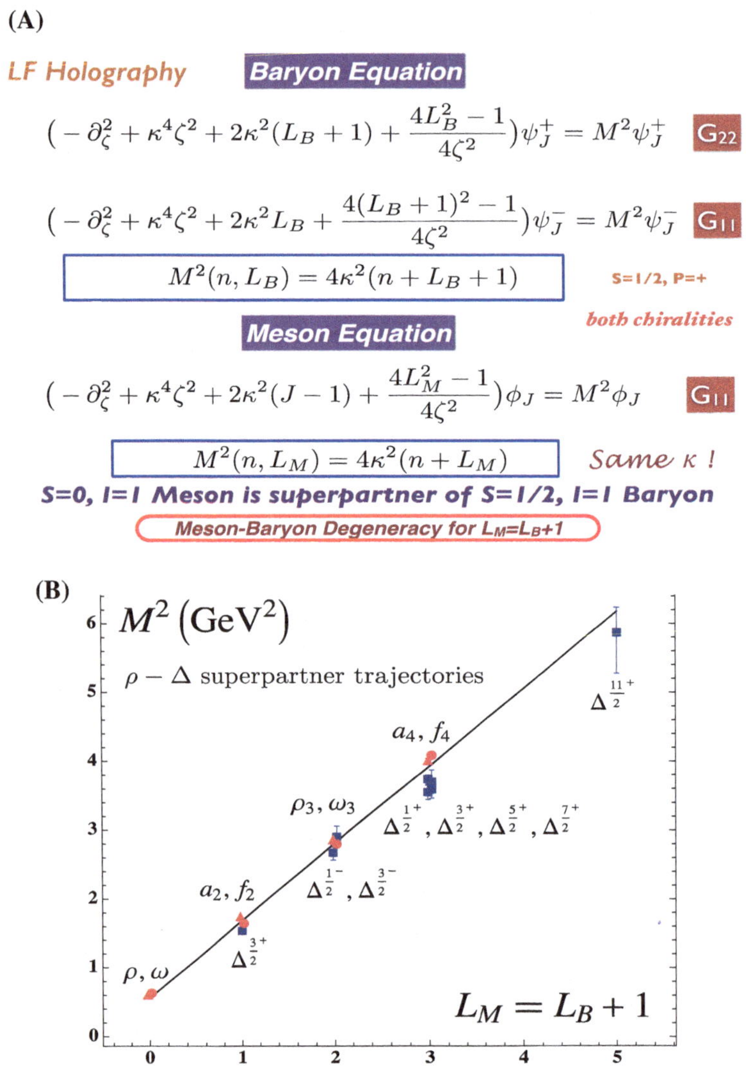

Fig. 6 a The LF Schrödinger equations for baryons and mesons for zero quark mass derived from the Pauli 2×2 matrix representation of superconformal algebra. The $\psi^{\pm}$ are the baryon quark-diquark LFWFs where the quark spin $S_q^z = \pm 1/2$ is parallel or antiparallel to the baryon spin $J^z = \pm 1/2$. The meson and baryon equations are identical if one identifies a meson with internal orbital angular momentum L_M with its superpartner baryon with $L_B = L_M - 1$. See Refs. [40–42]. **b** Comparison of the ρ/ω meson Regge trajectory with the $J = 3/2$ Δ baryon trajectory. Superconformal algebra predicts the degeneracy of the meson and baryon trajectories if one identifies a meson with internal orbital angular momentum L_M with its superpartner baryon with $L_M = L_B + 1$. See Refs. [40,41]

are color-singlet bound-states of color-triplet quarks and $\bar{3}_C$ $[qq]$ diquarks which are themselves $3_C \times 3_C$ clusters. Note that the same slope controls the Regge trajectories of both mesons and baryons in both the orbital angular momentum L and the principle quantum number n. Only one mass parameter $\kappa = \omega^2$ appears; it sets the confinement and the hadron mass scale in the chiral limit, as well as the length scale which underlies hadron structure. "Light-Front Holography" not only predicts meson and baryon spectroscopy successfully, but also hadron dynamics, including vector meson electroproduction, hadronic light-front wavefunctions, distribution amplitudes, form factors, and valence structure functions.

The LF Schrödinger Equations for baryons and mesons derived from superconformal algebra are shown in Fig. 6. In effect the baryons on the proton (Delta) trajectory are bound states of a quark with color 3_C and scalar (vector) diquark with color $\bar{3}_C$ The proton eigenstate labeled ψ^+ (parallel quark and baryon spins) and ψ^- (anti parallel quark and baryon spins) have equal Fock state probability—a feature of "quark chirality invariance". Predictions for the static properties of the nucleons are discussed in Ref. [43]

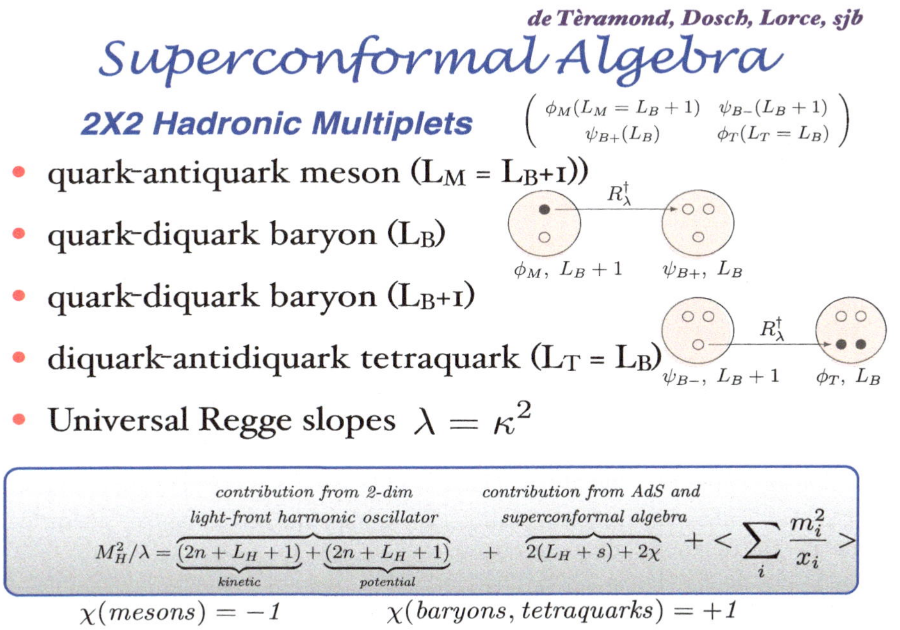

Fig. 7 The eigenstates of superconformal algebra have a 2×2 representation of mass degenerate bosons and fermions: a meson with $L_M = L_B + 1$, a baryon doublet with $L_B, L_B + 1$ components and a tetraquark with $L_T = L_B$. The breakdown of LF kinetic, potential, spin, and quark mass contributions to each hadron is also shown. The virial theorem predicts the equality of the LF kinetic and potential contributions

The comparison between the meson and baryon masses of the ρ/ω Regge trajectory with the spin-3/2 Δ trajectory is shown in Fig. 6b. Superconformal algebra predicts that the bosonic meson and fermionic baryon masses are equal if one identifies each meson with internal orbital angular momentum L_M with its superpartner baryon with $L_B = L_M - 1$; the meson and baryon superpartners then have the same parity, Since $2 + L_M = 3 + L_B$, the meson and baryon superpartners are also the same. Superconformal algebra also predicts that the LFWFs of the superpartners are identical, and thus the corresponding elastic and transition form factors are equal.

The striking phenomenological similarities of mesonic and baryonic Regge trajectories has been noted by Klempt and Mensch [44]. Their paper has many references to the historical literature.

As illustrated in Fig. 7, the hadronic eigensolutions of the superconformal algebra are themselves 2×2 matrices, connected internally by the supersymmetric algebra operators. In addition to the meson and baryon eigenstates, one also predicts color singlet *tetraquark* diquark-antidiquark bound states with the same mass as the baryon. One can also generalize these results to heavy-light $[qQ]$ mesons and $[[qq]Q]$ baryons [42,45]. The mass parameter controlling the Regge slopes will be increased for heavy m_Q; however, the supersymmetric connections between the heavy-light hadrons is predicted to be maintained. The predicted identity of meson and baryon timelike form factors can be tested in $e^+e^- \to H\bar{H}'$ reactions.

5 Ridge Formation from Flux-Tube Collisions

In the case of $ep \to e'X$, one can consider the collisions of the confining QCD flux tube appearing between the q and $\bar{q}$ of the virtual photon with the flux tube between the quark and diquark of the proton. Since the $q\bar{q}$ plane is aligned with the scattered electron's plane, the resulting "ridge" of hadronic multiplicity produced from the $\gamma^* p$ collision will also tend to be aligned with the scattering plane of the scattered electron. The virtual photon's flux tube will also depend on the photon virtuality Q^2, as well as the flavor of the produced pair arising from $\gamma^* \to q\bar{q}$. In the case of high energy $\gamma^*\gamma^*$ collisions, one can control the produced hadron multiplicity and ridge geometry using the scattered electrons' planes or the scattered proton planes in ultra-peripheral collisions at the LHC. The resulting dynamics [46] is a natural extension of the flux-tube collision description of the ridges produced in $p - p$ collisions [47].

6 Calculations Using LF-Time-Ordered Perturbation Theory and Hadronization at the Amplitude Level

LF-time-ordered perturbation theory can be advantageous for perturbative QCD calculations. An excellent example of LF-time-ordered perturbation theory is the computation of multi-gluon scattering amplitudes by Cruz-Santiago and Stasto [48]. In this method, the propagating particles are on their respective mass shells: $k_\mu k^\mu = m^2$, and intermediate states are off-shell in invariant mass; i.e., $P^- \neq \sum k_i^-$. Unlike instant form, where one must sum $n!$ frame-dependent amplitudes, only the τ-ordered diagrams where each propagating particle has positive $k^+ = k^0 + k^z$ can contribute. The number of nonzero amplitudes is also greatly reduced by noting that the total angular momentum projection $J^z = \sum_i^{n-1} L_i^z + \sum_i^n S_i^z$ and the total P^+ are conserved at each vertex. In a renormalizable theory, the change in orbital angular momentum is limited to $\Delta L^z = 0, \pm 1$ at each vertex [34].

A remarkable advantage of LF time-ordered perturbation theory (LFPth) is that the calculation of a subgraph of any order in pQCD only needs to be done once; the result can be stored in a "history" file. This is due to the fact that in LFPth the numerator algebra is independent of the process; the denominator changes, but only by a simple shift of the initial P^-. Another simplification is that loop integrations are three dimensional: $\int d^2\vec{k}_\perp \int_0^1 dx$. Unitarity and explicit renormalization can be implemented using the "alternate denominator" method which defines the required subtraction counterterms [49].

The new insights into color confinement given by AdS/QCD suggest that one could compute hadronization at amplitude level [50] using the confinement interaction and the LFWFs predicted by AdS/QCD and Light-Front Holography. One can postulate that the invariant mass of a color-singlet cluster $\mathcal{M}$ is the variable which separates perturbative and nonperturbative dynamics. For example, consider e^+e^- annihilation using LF τ - ordered perturbation theory. At an early stage in LF time the annihilation will produce jets of quarks and gluons in an intermediate state that are off the P^- energy shell. If a color-singlet cluster of partons in a jet satisfies $\mathcal{M}^2 < \kappa^2$, the cluster constituents will be ruled by the $\kappa^4\zeta^2$ color-confinement potential. At this stage, the LFWF ψ_H converts the off-shell partons to the on-shell hadron. Quarks and gluons only appear in intermediate states, but only hadrons can be produced. Thus the AdS/QCD Light-Front Holographic model suggests how one can implement the transition between perturbative and nonperturbative QCD. For a QED analog, see Refs. [25,26].

7 The Light-Front Vacuum

It is important to distinguish the LF vacuum from the conventional instant-form vacuum. The eigenstates of the instant-form Hamiltonian describe a state defined at a single instant of time t over all space, and they are thus acausal as well as frame-dependent. The instant-form vacuum is defined as the lowest energy eigenstate of the instant-form Hamiltonian. As discussed by Zee [51], the cosmological constant is of order 10^{120} times larger than what is observed if one computes the effects of quantum loops from QED. Similarly, QCD instantons and condensates in the instant-form vacuum give a contribution of order 10^{42}. The contribution of the Higgs VEV computed in the instant form vacuum is 10^{54} times too large.

In contrast, the vacuum in LF Hamlitonian theory is defined as the eigenstate of H_{LF} with lowest invariant mass. It is defined at fixed LF time τ within the causal horizon, and it is frame-independent; i.e., it is independent of the observer's motion. Vacuum loop diagrams from quantum field theory do not appear in the front-form vacuum since the $+$ momenta are positive: $k_i^+ = k_i^0 + k_i^z \geq 0$, and the sum of $+$ momenta is conserved at every vertex. The creation of particles cannot arise from the LF vacuum since $\sum_i k^{+i} \neq P^+_{vacuum} = 0$. Since propagation with negative k^+ does not appear, the LF vacuum is trivial up to possible $k^+ = 0$ "zero" modes. The physics associated with quark and gluon QCD vacuum condensates of the instant form are replaced by physical effects contained within the hadronic LFWFs in the hadronic domain. This is referred to as "in-hadron" condensates [52–54]. In the case of the Higgs theory, the traditional Higgs vacuum expectation value (VEV) is replaced by a "zero mode", analogous to a classical Stark or Zeeman field [55]. The Higgs LF zero mode [55] has no energy-momentum density, so it also gives zero contribution to the cosmological constant.

The universe is observed within the causal horizon, not at a single instant of time. The causal, frame-independent light-front vacuum can thus provide a viable match to the empty visible universe [54]. The huge contributions to the cosmological constant thus do not appear if one notes that the causal, frame-independent light-front vacuum has no quantum fluctuations—in dramatic contrast to to the acausal, frame-dependent instant-form vacuum; the cosmological constant arising from quantum field theory thus vanishes if one uses the front form. The observed nonzero value could could be a property of gravity itself, such as the "emergent gravity" postulated by E. Verlinde [56]. It is also possible that if one solves electroweak theory in a curved universe, the Higgs LF zero mode would be replaced with a field of nonzero curvature which could give a nonzero contribution.

8 The QCD Coupling at all Scales

The QCD running coupling $\alpha_s(Q^2)$ sets the strength of the interactions of quarks and gluons as a function of the momentum transfer Q. The dependence of the coupling Q^2 is needed to describe hadronic interactions at both long and short distances. The QCD running coupling can be defined [57] at all momentum scales from a perturbatively calculable observable, such as the coupling $\alpha^s_{g_1}(Q^2)$, which is defined from measurements of the Bjorken sum rule. At high momentum transfer, such "effective charges" satisfy asymptotic freedom, obey the usual pQCD renormalization group equations, and can be related to each other without scale ambiguity by commensurate scale relations [58].

The dilaton $e^{+\kappa^2 z^2}$ soft-wall modification of the AdS_5 metric, together with LF holography, predicts the functional behavior of the running coupling in the small Q^2 domain [59]: $\alpha^s_{g_1}(Q^2) = \pi e^{-Q^2/4\kappa^2}$. Measurements of $\alpha^s_{g_1}(Q^2)$ are remarkably consistent [60] with this predicted Gaussian form; the best fit gives $\kappa = 0.513 \pm 0.007\ GeV$. See Fig. 8 Deur et al. [59,61,62] have also shown how the parameter κ, which determines the mass scale of hadrons and Regge slopes in the zero quark mass limit, can be connected to the mass scale Λ_s controlling the evolution of the perturbative QCD coupling. The high momentum transfer dependence of the coupling $\alpha_{g1}(Q^2)$ is predicted by pQCD. The matching of the high and low momentum transfer regimes of $\alpha_{g1}(Q^2)$—both its value and its slope—then determines a scale $Q_0 = 0.87 \pm 0.08$ GeV which sets the interface between perturbative and nonperturbative hadron dynamics. This connection can be done for any choice of renormalization scheme, such as the $\overline{MS}$ scheme, as seen in Fig. 8. The result of this perturbative/nonperturbative matching is an effective QCD coupling defined at all momenta. The predicted value of $\Lambda_{\overline{MS}} = 0.339 \pm 0.019\ GeV$ from this analysis agrees well the measured value [63] $\Lambda_{\overline{MS}} = 0.332 \pm 0.019\ GeV$. These results, combined with the AdS/QCD superconformal predictions for hadron spectroscopy, allow us to compute hadron masses in terms of $\Lambda_{\overline{MS}}$: $m_p = \sqrt{2}\kappa = 3.21\ \Lambda_{\overline{MS}}$, $m_\rho = \kappa = 2.2\ \Lambda_{\overline{MS}}$, and $m_p = \sqrt{2} m_\rho$, meeting a challenge proposed by Zee [64]. The value of Q_0 can be used to set the factorization scale for DGLAP evolution of hadronic structure functions and the ERBL evolution of distribution amplitudes. Deur, de Téramond, and I have also computed the dependence of Q_0 on the choice of the effective charge used to define the running coupling and the renormalization scheme used to compute its behavior in the perturbative regime. The use of the scale Q_0 to resolve the factorization scale uncertainty in structure functions and fragmentation functions, in combination with the scheme-indepedent *principle of maximum sensitivity* (PMC) [65] for setting renormalization scales, can greatly improve the precision of pQCD predictions for collider phenomenology.

9 Is the Momentum Sum Rule Valid for Nuclear Structure Functions?

Sum rules for deep inelastic scattering are usually analyzed using the operator product expansion of the forward virtual Compton amplitude, assuming it depends in the limit $Q^2 \to \infty$ on matrix elements of local operators such as the energy-momentum tensor. The moments of structure functions and other distributions can then be evaluated as overlaps of the target hadron's light-front wavefunction, as in the Drell-Yan-West formulae for hadronic form factors [30,66–68]. The real phase of the resulting DIS amplitude and its OPE matrix elements reflects the real phase of the stable target hadron's wavefunction.

The "handbag" approximation to deeply virtual Compton scattering also defines the "static" contribution [69,70] to the measured parton distribution functions (PDF), transverse momentum distributions, etc. The resulting momentum, spin and other sum rules reflect the properties of the hadron's light-front wavefunction. However, final-state interactions which occur *after* the lepton scatters on the quark, can give non-trivial contributions to deep inelastic scattering processes at leading twist and thus survive at high Q^2 and high $W^2 = (q + p)^2$. For example, the pseudo-T-odd Sivers effect [71] is directly sensitive to the rescattering of the struck quark. Similarly, diffractive deep inelastic scattering (DDIS) involves the exchange of a gluon after the quark has been struck by the lepton [72]. In each case the corresponding DVCS amplitude is not given by the handbag diagram since interactions between the two currents are essential. These "lensing" corrections

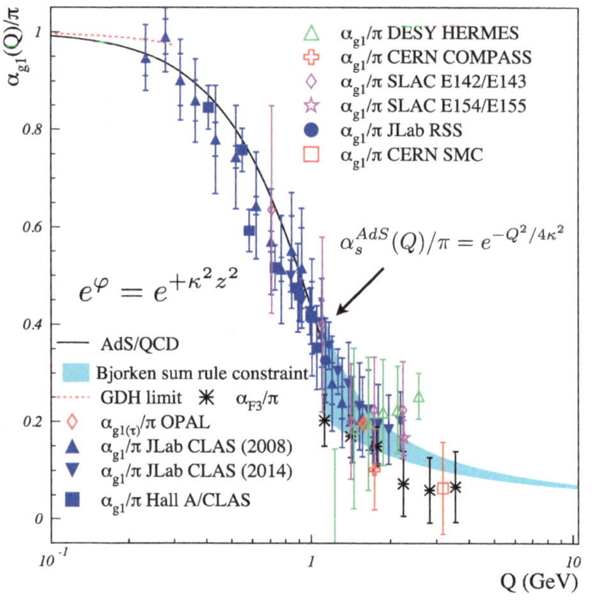

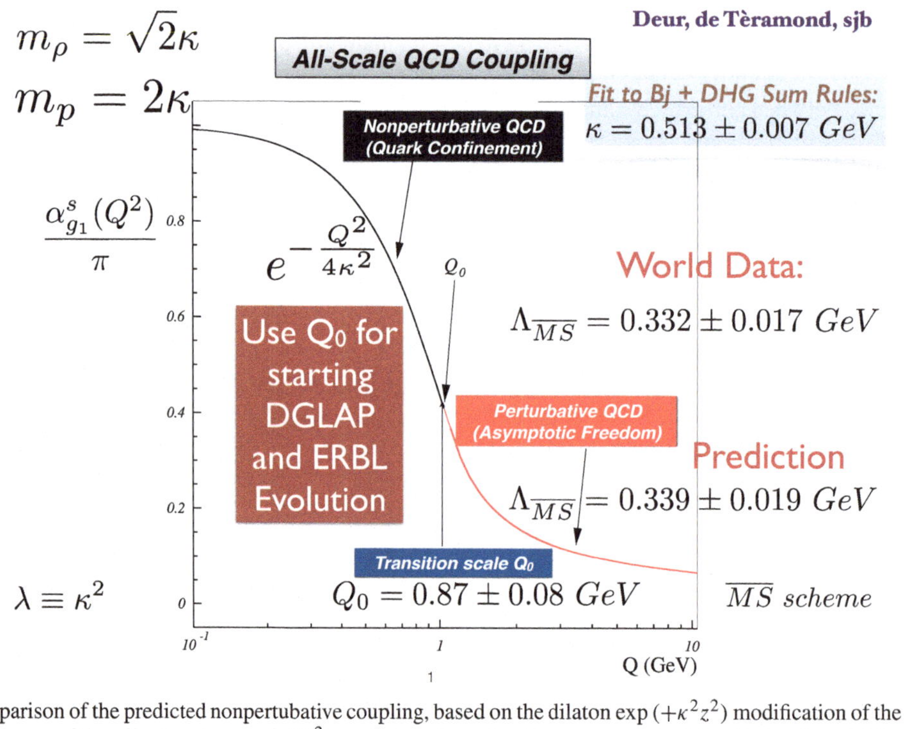

Fig. 8 **a** Comparison of the predicted nonpertubative coupling, based on the dilaton $\exp(+\kappa^2 z^2)$ modification of the AdS$_5$ metric, with measurements of the effective charge $\alpha^s_{g_1}(Q^2)$, as defined from the Bjorken sum rule. **b** Prediction from LF Holography and pQCD for the QCD running coupling $\alpha^s_{g_1}(Q^2)$ at all scales. The magnitude and derivative of the perturbative and nonperturbative coupling are matched at the scale Q_0. This matching connects the perturbative scale $\Lambda_{\overline{MS}}$ to the nonpertubative scale κ which underlies the hadron mass scale. See Ref. [62]

survive when both W^2 and Q^2 are large since the vector gluon couplings grow with energy. Part of the final state phase can be associated with a Wilson line as an augmented LFWF [73] which does not affect the moments.

The Glauber propagation of the vector system V produced by the DDIS interaction on the nuclear front face and its subsequent inelastic interaction with the nucleons in the nuclear interior $V + N_b \to X$ occurs after the lepton interacts with the struck quark. The corresponding DVCS amplitude is not given by the handbag diagram since interactions between the two currents are essential. Because of the rescattering dynamics, the

DDIS amplitude acquires a complex phase from Pomeron and Regge exchange; thus final-state rescattering corrections lead to nontrivial "dynamical" contributions to the measured PDFs; i.e., they are a consequence of the scattering process itself [74]. The $I = 1$ Reggeon contribution to DDIS on the front-face nucleon then leads to flavor-dependent antishadowing [75,76]. This could explain why the NuTeV charged current measurement $\mu A \to \nu X$ scattering does not appear to show antishadowing, in contrast to deep inelastic electron-nucleus scattering as discussed in Ref. [77].

Diffractive deep inelastic scattering is leading-twist. and it is an essential component of the two-step amplitude which causes shadowing and antishadowing of the nuclear PDF. It is important to analyze whether the momentum and other sum rules derived from the OPE expansion in terms of local operators remain valid when these dynamical rescattering corrections to the nuclear PDF are included. The OPE is derived assuming that the LF time separation between the virtual photons in the forward virtual Compton amplitude $\gamma^* A \to \gamma^* A$ scales as $1/Q^2$. However, the propagation of the vector system V produced by the DDIS interaction on the front face and its inelastic interaction with the nucleons in the nuclear interior $V + N_b \to X$ are characterized by a non-vanishing LF time interval in the nuclear rest frame. Note also that shadowing in deep inelastic lepton scattering on a nucleus involves nucleons facing the incoming lepton beam. The geometrical orientation of the shadowed nucleons is not a property the frame-independent nuclear LFWFs used to evaluate the matrix elements of local currents. Thus leading-twist shadowing and antishadowing appear to invalidate the sum rules for nuclear PDFs. The same complications occur in the leading-twist analysis of deeply virtual Compton scattering $\gamma^* A \to \gamma^* A$ on a nuclear target. Thus the leading-twist multi-nucleon processes which produce shadowing and antishadowing in a nucleus are not accounted for using the $Q^2 \to \infty$ OPE analysis.

10 Summary

Light-Front Quantization provides a physical, frame-independent formalism for hadron dynamics and structure. Light-Front Holography and Superconformal Algebra leads to relativistic, Poincar? invariant, light-front quantum mechanical, bound-state wave equations which incorporate quark and gluon confinement and successfully predict many observed spectroscopic and dynamical features of hadron physics, such as form factors and linear Regge trajectories with identical slope in both the radial quantum number and the internal orbital angular momentum. Generalizing this procedure using superconformal algebra, leads to a unified Regge spectroscopy of meson, baryon, and tetraquarks, including remarkable supersymmetric relations between the masses of mesons and baryons of the same parity. The predicted pion mass is massless for zero quark mass, thus satisfying a fundamental principle. One obtains new insights into the hadronic spectrum, light-front wavefunctions, and the $e^{-\frac{Q^2}{4\kappa^2}}$ Gaussian functional form of the QCD running coupling in the nonperturbative domain using light-front holography—the duality between the front form and AdS_5, the space of isometries of the conformal group. Observables such as structure functions, transverse momentum distributions, and distribution amplitudes are defined from the predicted hadronic light-front wavefunctions.

The analytic behavior of the QCD coupling controlling quark and gluon interactions at large and small distances is also determined. The mass scale κ underlying confinement and hadron masses can be connected to the parameter $\Lambda_{\overline{MS}}$ in the QCD running coupling by matching the nonperturbative dynamics, as described by the effective conformal theory mapped to the light-front and its embedding in AdS space, to the perturbative QCD regime. The result is an effective coupling defined at all momenta. This matching of the high and low momentum transfer regimes determines a scale Q_0 which sets the interface between perturbative and nonperturbative hadron dynamics. The result is an effective coupling defined at all momenta and the determination of a momentum scale which sets the interface between perturbative and nonperturbative hadron dynamics. The use of Q_0 to resolve the factorization scale uncertainty for structure functions and distribution amplitudes, in combination with the principle of maximal conformality (PMC) for setting the renormalization scales, can greatly improve the precision of perturbative QCD predictions for collider phenomenology.

The absence of vacuum excitations of the causal, frame-independent front form vacuum has important consequences for the cosmological constant. I have also discussed evidence that the antishadowing of nuclear structure functions is non-universal; i.e., flavor dependent, and why shadowing and antishadowing phenomena may be incompatible with sum rules for nuclear parton distribution functions.

Acknowledgements Presented at Light-Cone 2016, *Challenges in Experimental and Theoretical Physics on the Light Front* September 5–8, 2015, IST, Universidade de Lisboa, Lisbon, Portugal. The results presented here are based on collaborations and discussions with James Bjorken, Kelly Chiu, Alexandre Deur, Guy de Téramond, Guenter Dosch, Susan Gardner, Fred Goldhaber,

Paul Hoyer, Dae Sung Hwang, Rich Lebed, Simonetta Liuti, Cedric Lorce, Matin Mojaza, Michael Peskin, Craig Roberts, Robert Shrock, Ivan Schmidt, Peter Tandy, and Xing-Gang Wu. This research was supported by the Department of Energy, contract DE–AC02–76SF00515. SLAC-PUB-16874.

References

1. P.A.M. Dirac, Rev. Mod. Phys. **21**, 392 (1949)
2. S.J. Brodsky, H.C. Pauli, S.S. Pinsky, Phys. Rep. **301**, 299 (1998). arXiv:hep-th/9705477
3. S.J. Brodsky, M. Diehl, D.S. Hwang, Nucl. Phys. B **596**, 99 (2001). arXiv:hep-th/0009254
4. J. Terrell, Phys. Rev. **116**, 1041 (1959)
5. R. Penrose, Proc. Camb. Philos. Soc. **55**, 137 (1959)
6. S.J. Brodsky, D.S. Hwang, B.Q. Ma, I. Schmidt, Nucl. Phys. B **593**, 311 (2001). arXiv:hep-th/0003082
7. I.Y. Kobzarev, L.B. Okun, Zh. Eksp. Teor. Fiz. **43**, 1904 (1962) [Sov. Phys. JETP **16**, 1343 (1963)]
8. O.V. Teryaev, arXiv:hep-th/9904376
9. V.N. Gribov, L.N. Lipatov, Sov. J. Nucl. Phys. **15**, 438 (1972) [Yad. Fiz. **15**, 781 (1972)]
10. G. Altarelli, G. Parisi, Nucl. Phys. B **126**, 298 (1977)
11. Y.L. Dokshitzer, Sov. Phys. JETP **46**, 641 (1977) [Zh. Eksp. Teor. Fiz. **73**, 1216 (1977)]
12. G.P. Lepage, S.J. Brodsky, Phys. Lett. **87B**, 359 (1979)
13. G.P. Lepage, S.J. Brodsky, Phys. Rev. D **22**, 2157 (1980)
14. A.V. Efremov, A.V. Radyushkin, Phys. Lett. **94B**, 245 (1980)
15. A.V. Efremov, A.V. Radyushkin, Theor. Math. Phys. **42**, 97 (1980) [Teor. Mat. Fiz. **42**, 147 (1980)]
16. S.J. Brodsky, S. Gardner, Phys. Rev. Lett. **116**(1), 019101 (2016). arXiv:1504.00969 [hep-ph]
17. H.C. Pauli, S.J. Brodsky, Phys. Rev. D **32**, 1993 (1985)
18. K. Hornbostel, S.J. Brodsky, H.C. Pauli, Phys. Rev. D **41**, 3814 (1990)
19. G.F. de Teramond, S.J. Brodsky, Phys. Rev. Lett. **102**, 081601 (2009). arXiv:0809.4899 [hep-ph]
20. J.P. Vary, X. Zhao, A. Ilderton, H. Honkanen, P. Maris, S.J. Brodsky, Nucl. Phys. Proc. Suppl. **251–252**, 10 (2014). arXiv:1406.1838 [nucl-th]
21. S.J. Brodsky et al., arXiv:1502.05728 [hep-ph]
22. V. de Alfaro, S. Fubini, G. Furlan, Nuovo Cim. A **34**, 569 (1976)
23. S.J. Brodsky, G.F. De Tramond, H.G. Dosch, Phys. Lett. B **729**, 3 (2014). arXiv:1302.4105 [hep-th]
24. D. Ashery, Nucl. Phys. Proc. Suppl. **90**, 67 (2000) [Nucl. Phys. Proc. Suppl. **108**, 321 (2002)] arXiv:hep-ex/0008036
25. S.J. Brodsky, R.F. Lebed, Phys. Rev. Lett. **102**, 213401 (2009). arXiv:0904.2225 [hep-th]
26. A. Banburski, P. Schuster, Phys. Rev. D **86**, 093007 (2012). arXiv:1206.3961 [hep-th]
27. G.F. de Teramond, H.G. Dosch, S.J. Brodsky, Phys. Rev. D **87**(7), 075005 (2013). arXiv:1301.1651 [hep-th]
28. G.F. de Teramond, S.J. Brodsky, arXiv:1203.4025 [hep-ph]
29. J. Polchinski, M.J. Strassler, Phys. Rev. Lett. **88**, 031601 (2002). arXiv:hep-th/0109174
30. S.J. Brodsky, S.D. Drell, Phys. Rev. D **22**, 2236 (1980)
31. Z. Abidin, C.E. Carlson, Phys. Rev. D **79**, 115003 (2009). arXiv:0903.4818 [hep-ph]
32. S.J. Brodsky, G.F. de Teramond, H.G. Dosch, J. Erlich, Phys. Rep. **584**, 1 (2015). arXiv:1407.8131 [hep-ph]
33. R.L. Jaffe, A. Manohar, Nucl. Phys. B **337**, 509 (1990)
34. K.Y.J. Chiu, S.J. Brodsky, Angular Momentum Conservation Law in Light-Front Quantum Field Theory, Phys. Rev. D **95**(6), 065035 (2017). arXiv:1702.01127 [hep-th]. doi:10.1103/PhysRevD.95.065035
35. S.J. Brodsky, F.G. Cao, G.F. de Teramond, Phys. Rev. D **84**, 075012 (2011). arXiv:1105.3999 [hep-ph]
36. J.R. Forshaw, R. Sandapen, Phys. Rev. Lett. **109**, 081601 (2012). arXiv:1203.6088 [hep-ph]
37. A.V. Smirnov, V.A. Smirnov, M. Steinhauser, Phys. Rev. Lett. **104**, 112002 (2010). arXiv:0911.4742 [hep-ph]
38. R. Haag, J.T. Lopuszanski, M. Sohnius, Nucl. Phys. B **88**, 257 (1975)
39. S. Fubini, E. Rabinovici, Nucl. Phys. B **245**, 17 (1984)
40. G.F. de Teramond, H.G. Dosch, S.J. Brodsky, Phys. Rev. D **91**(4), 045040 (2015). arXiv:1411.5243 [hep-ph]
41. H.G. Dosch, G.F. de Teramond, S.J. Brodsky, Phys. Rev. D **91**(8), 085016 (2015). arXiv:1501.00959 [hep-th]
42. H.G. Dosch, G.F. de Teramond, S.J. Brodsky, Phys. Rev. D **92**(7), 074010 (2015). arXiv:1504.05112 [hep-ph]
43. T. Liu, B.Q. Ma, Phys. Rev. D **92**(9), 096003 (2015). arXiv:1510.07783 [hep-ph]
44. E. Klempt, B.C. Metsch, Eur. Phys. J. A **48**, 127 (2012)
45. H.G. Dosch, G.F. de Teramond, S.J. Brodsky, Phys. Rev. D **95**(3), 034016 (2017). arXiv:1612.02370 [hep-ph]
46. S.J. Brodsky, Nucl. Part. Phys. Proc. **258–259**, 23 (2015). arXiv:1410.0404 [hep-ph]
47. J.D. Bjorken, S.J. Brodsky, A. Scharff Goldhaber, Phys. Lett. B **726**, 344 (2013). arXiv:1308.1435 [hep-ph]
48. C. Cruz-Santiago, P. Kotko, Prog. Part. Nucl. Phys. **85**, 82 (2015)
49. S.J. Brodsky, R. Roskies, R. Suaya, Phys. Rev. D **8**, 4574 (1973)
50. S.J. Brodsky, G.F. de Teramond, arXiv:0901.0770 [hep-ph]
51. A. Zee, Mod. Phys. Lett. A **23**, 1336 (2008)
52. A. Casher, L. Susskind, Phys. Rev. D **9**, 436 (1974)
53. S.J. Brodsky, R. Shrock, Proc. Nat. Acad. Sci. **108**, 45 (2011). arXiv:0905.1151 [hep-th]
54. S.J. Brodsky, C.D. Roberts, R. Shrock, P.C. Tandy, Phys. Rev. C **82**, 022201 (2010). arXiv:1005.4610 [nucl-th]
55. P.P. Srivastava, S.J. Brodsky, Phys. Rev. D **66**, 045019 (2002). arXiv:hep-ph/0202141
56. E.P. Verlinde, arXiv:1611.02269 [hep-th]
57. G. Grunberg, Phys. Lett. **95B**, 70 (1980) Erratum: [Phys. Lett. **110B**, 501 (1982)]
58. S.J. Brodsky, H.J. Lu, Phys. Rev. D **51**, 3652 (1995). arXiv:hep-ph/9405218
59. S.J. Brodsky, G.F. de Teramond, A. Deur, Phys. Rev. D **81**, 096010 (2010). arXiv:1002.3948 [hep-ph]

60. A. Deur, V. Burkert, J.P. Chen, W. Korsch, Phys. Lett. B **650**, 244 (2007). arXiv:hep-ph/0509113
61. A. Deur, S.J. Brodsky, G.F. de Teramond, Phys. Lett. B **750**, 528 (2015). arXiv:1409.5488 [hep-ph]
62. S.J. Brodsky, G.F. de TŽramond, A. Deur, H.G. Dosch, Few Body Syst. **56**(6–9), 621 (2015). arXiv:1410.0425 [hep-ph]
63. K.A. Olive et al., Particle Data Group. Chin. Phys. C **38**, 090001 (2014)
64. A. Zee., Quantum field theory in a nutshell. (Princeton Univ. Pr, Princeton, 2010), 576 p
65. M. Mojaza, S.J. Brodsky, X.G. Wu, Phys. Rev. Lett. **110**, 192001 (2013). arXiv:1212.0049 [hep-ph]
66. S. Liuti, A. Rajan, A. Courtoy, G.R. Goldstein, J.O. Gonzalez Hernandez, Int. J. Mod. Phys. Conf. Ser. **25**, 1460009 (2014). arXiv:1309.7029 [hep-ph]
67. C. Mondal, D. Chakrabarti, Eur. Phys. J. C **75**(6), 261 (2015). arXiv:1501.05489 [hep-ph]
68. C. Lorce, B. Pasquini, M. Vanderhaeghen, JHEP **1105**, 041 (2011). arXiv:1102.4704 [hep-ph]
69. S.J. Brodsky, AIP Conf. Proc. **1105**, 315 (2009). arXiv:0811.0875 [hep-ph]
70. S.J. Brodsky, Nucl. Phys. A **827**, 327C (2009). arXiv:0901.0781 [hep-ph]
71. S.J. Brodsky, D.S. Hwang, I. Schmidt, Phys. Lett. B **530**, 99 (2002). arXiv:hep-ph/0201296
72. S.J. Brodsky, P. Hoyer, N. Marchal, S. Peigne, F. Sannino, Phys. Rev. D **65**, 114025 (2002). arXiv:hep-ph/0104291
73. S.J. Brodsky, B. Pasquini, B.W. Xiao, F. Yuan, Phys. Lett. B **687**, 327 (2010). arXiv:1001.1163 [hep-ph]
74. S.J. Brodsky, D.S. Hwang, Y.V. Kovchegov, I. Schmidt, M.D. Sievert, Phys. Rev. D **88**(1), 014032 (2013). arXiv:1304.5237 [hep-ph]
75. S.J. Brodsky, H.J. Lu, Phys. Rev. Lett. **64**, 1342 (1990)
76. S.J. Brodsky, I. Schmidt, J.J. Yang, Phys. Rev. D **70**, 116003 (2004). arXiv:hep-ph/0409279
77. I. Schienbein, J.Y. Yu, C. Keppel, J.G. Morfin, F. Olness, J.F. Owens, Phys. Rev. D **77**, 054013 (2008). arXiv:0710.4897 [hep-ph]

Few-Body Syst (2017) 58:87
DOI 10.1007/s00601-017-1232-3

Alfredo Vega · Ivan Schmidt · Thomas Gutsche · Valery E. Lyubovitskij

Nonperturbative Strange Sea in Proton Using Wave Functions Inspired by Light Front Holography

Received: 15 November 2016 / Accepted: 13 January 2017 / Published online: 16 February 2017

Abstract We use different light-front wave functions (two inspired by the AdS/QCD formalism), together with a model of the nucleon in terms of meson–baryon fluctuations to calculate the nonperturbative (intrinsic) contribution to the $s(x) - \bar{s}(x)$ asymmetry of the proton sea. The holographic wave functions for an arbitrary number of constituents, recently derived by us, give results quite close to known parametrizations that appear in the literature.

1 Introduction

Hadrons are built by valence quarks together with a sea of $q\bar{q}$ pairs and gluons. This sea plays an important role in the understanding of several hadronic properties and provides an explanation of many experimental results in hadronic physics (for overviews and discussions of experimental and theoretical progress in this field see e.g. Refs. [1–30]). In this vein one of the interesting aspects is the study of the strange quark sea and the $s(x) - \bar{s}(x)$ asymmetry, since both experimental and theoretical analyses indicate the existence of an strange-quark asymmetry (SQA) in the nucleon sea. Therefore, this asymmetry is a major hadronic observable, which study can shed light on our understanding of the nucleon structure.

There are two main mechanisms that can generate a SQA in the nucleon sea—nonperturbative (intrinsic) and perturbative (extrinsic) (see e.g. discussion in Refs. [28,30]), but in this talk we focus on the nonperturbative (or intrinsic) contribution to the strange-quark asymmetry. The last one contribution can be considered as nucleons fluctuating into virtual baryon–meson states (ΛK and ΣK), and can be estimated by using nonperturbative

This article belongs to the Topical Collection "Light Cone 2016".

A. Vega (✉)
Instituto de Física y Astronomía, Universidad de Valparaíso, A. Gran Bretaña 1111, Valparaíso, Chile
E-mail: alfredo.vega@uv.cl

I. Schmidt
Departamento de Física y Centro Científico Tecnológico de Valparaíso (CCTVal), Universidad Técnica Federico Santa María, Casilla 110-V, Valparaíso, Chile

T. Gutsche · V. E. Lyubovitskij
Institut für Theoretische Physik, Universität Tübingen, Kepler Center for Astro and Particle Physics, Auf der Morgenstelle 14, 72076 Tübingen, Germany

V. E. Lyubovitskij
Department of Physics, Tomsk State University, Tomsk, Russia 634050

V. E. Lyubovitskij
Mathematical Physics Department, Tomsk Polytechnic University, Lenin Avenue 30, Tomsk, Russia 634050

models for the nucleons [6,7,14,23] and are dominant in the large-x region. As was shown in Ref. [28], the perturbative SQA, which is significant in the small-x region, is produced by perturbative QCD evolution at next-to-next-to-leading order (NNLO) or at three loops. Such phenomenon occurs even if the SQA vanishes at the initial scale due to nonvanishing u and d valence quark densities [28].

As we mentioned in last paragraph we give predictions for the nonperturbative contribution to the $s(x)-\bar{s}(x)$ asymmetry. We follow the light-front approach proposed by Brodsky and Ma [14]. In Ref. [14] both Gaussian and power-law quark LFWFs were considered, leading to similar results. We intend to clarify this issue, using what we consider are more realistic LFWFs. For this purpose we use three different kinds of wave functions and show that the asymmetry is quite sensitive to the LFWFs [15]. Therefore, the study of the $(s-\bar{s})$ asymmetry could serve as a further tool to distinguish between these LFWFs. In particular, we consider the traditional Gaussian wave function [14] and two variants of LFWFs motivated by AdS/QCD models. These last ones are extracted by matching electromagnetic form factors in LF QCD and AdS/QCD for the massless case, and introducing modifications in order to take into account finite quark masses [31–36]. Notice that the same matching procedure allows for an extraction of GPDs [37,38] and, in addition, for a generalization of the LFWFs for hadrons with an arbitrary number of constituents [39]. In particular, we consider an holographic LFWF (variant I) obtained from matching to LF QCD at large x and a holographic LFWF (variant II) obtained from matching to LF QCD at all values of x and for an arbitrary number of constituents [39]. The second type of holographic LFWF is more useful for our purposes, because it can be applied to the description of hadrons with an arbitrary number of constituents.

This contribution is based on [15], where it is possible found more details.

2 $s(x)-\bar{s}(x)$ Asymmetry in a Light Cone Approach

In the light-front formalism the proton state can be expanded in a series of components as

$$|P\rangle = |uud\rangle\psi_{uud/p} + |uudg\rangle\psi_{uudg/p} + \sum_{q\bar{q}} |uudq\bar{q}\rangle\psi_{uudq\bar{q}/p} + \cdots \tag{1}$$

where $|uud\rangle$, $|uudg\rangle$, $|uudq\bar{q}\rangle, \ldots$ are the contributing Fock states and $\psi_{uud/p}$, $\psi_{uudg/p}$, $\psi_{uudq\bar{q}/p}, \ldots$ are the quark/gluon LFWFs corresponding to these states. In Ref. [14] a different light front approach was proposed, in which the nucleon has components arising from meson–baryon fluctuations, while these hadronic components are composite systems of quarks. This approach is similar to expansions used in meson-cloud models [1–5]. In this case the nonperturbative contributions to the $s(x)$ and $\bar{s}(x)$ distributions in the proton can be expressed as convolutions (see e.g. Ref. [23])

$$s(x) = \int_x^1 \frac{dy}{y} f_{\Lambda/K\Lambda}(y) q_{s/\Lambda}\left(\frac{x}{y}\right) \quad \text{and} \quad \bar{s}(x) = \int_x^1 \frac{dy}{y} f_{K/K\Lambda}(y) q_{\bar{s}/K}\left(\frac{x}{y}\right), \tag{2}$$

where $q_{s/\Lambda}$ and $q_{\bar{s}/K}$ are distributions of s quarks and $\bar{s}$ antiquarks in the $\Lambda^0(\Sigma^0)$ and K^+, respectively. The functions $f_{\Lambda/K\Lambda}(y)$ and $f_{K/K\Lambda}(y)$ describe the probability to find a Λ or a K with light-front momentum fraction y in the $K\Lambda$ state.

In the light-front model proposed in [14] the meson–baryon distribution functions are calculated through the relation

$$f_{B/BM}(y) = \int \frac{d^2k}{16\pi^3} |\psi_{BM}(y,k)|^2. \tag{3}$$

where $\psi_{BM}(y,k)$ is the LFWF describing the distribution of the baryon–meson (BM) components. An important property of these functions, which follows from momentum and charge conservation, is that $f_{BK}(y) = f_{KB}(1-y)$ [8,9]. Here we have defined $f_{BK}(y) = f_{B/BK}(y)$ and $f_{KB}(y) = f_{K/BK}(y)$. In this work we consider a fluctuation probability for $N \to \Lambda K$ of 1.27%, as in Ref. [23].

In the same manner, the distribution functions $q_{s/\Lambda}$ and $q_{\bar{s}/K}$ can be determined by

$$q_{s/\Lambda}(x) = \int \frac{d^2k}{16\pi^3} |\psi_{\Lambda}(x,k)|^2 \quad \text{and} \quad q_{\bar{s}/K}(x) = \int \frac{d^2k}{16\pi^3} |\psi_{K}(x,k)|^2. \tag{4}$$

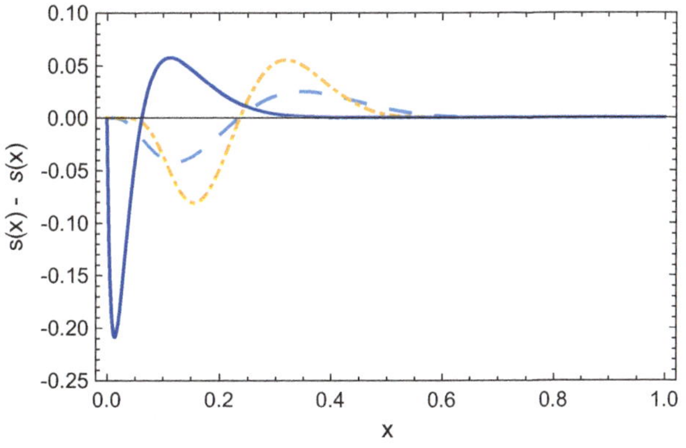

Fig. 1 $s(x) - \bar{s}(x)$ plots for three different types of LFWFs: Gaussian LFWF (*large dashed line*—$\kappa = 330$ MeV), holographic LFWF (variant I), (*dot dashed line*—$\kappa = 350$ MeV) and holographic LFWF (variant II) (*continuous line*—$\kappa = 350$ MeV)

To calculate $f_{\Lambda/K\Lambda}$, $f_{K/K\Lambda}$, $q_{s/\Lambda}$ and $q_{\bar{s}/K}$ it is necessary to know the LFWF for the distribution of Λ and K inside the states ΛK and ΣK, and for the quarks/antiquarks in the Λ and K. As we mentioned before, in this paper we use three different kinds of the LFWFs: (i) a typical Gaussian LFWF [14], (ii) a so-called holographic LFWFs obtained using light-front holography ideas [31–36] at large x and (iii) a further generalization which is extracted from matching at any value of x and that further allows to describe hadrons with an arbitrary number of constituents [39]. In the first two cases we directly follow the ideas of Ref. [14], where two-body wave functions ψ_{BM} are formed by two clusters—baryon (as a quark–diquark bound state) and meson (as the usual quark-antiquark bound state). The third LFWF considers hadrons with an arbitrary number of constituents in the two-body approximation. In particular, the twist-5 wave function corresponds to the LFWF of the baryon–meson bound state ψ_{BM}. For the inclusion of massive constituents (baryon and meson) we follow the procedure proposed and realized in Refs. [31–36]. In the next subsections we give details of all LFWF used in our calculations.

2.1 Gaussian Wave Function

Brodsky and Ma [14] suggested to use two-body Gaussian and power-law wave functions to calculate the $s(x)$ and $\bar{s}(x)$ asymmetry. They got similar results in both cases. Here we first consider the Gaussian wave function specified as

$$\psi(x, k) = A \exp\left[-\frac{1}{8\kappa^2}\left(\frac{k^2}{x(1-x)} + \mu_{12}^2\right)\right], \tag{5}$$

where

$$\mu_{12}^2 = \frac{m_1^2}{x} + \frac{m_2^2}{1-x} \tag{6}$$

and m_1 and m_2 are the masses of the constituents.

In this approach, the functions $f_{\Lambda/K\Lambda}$ and $f_{K/K\Lambda}$ are calculated as

$$f_{\Lambda/K\Lambda}(x) = \frac{\kappa^2 A_{MB}^2}{4\pi^2}\, x(1-x)\, \exp\left[-\frac{\mu_{\Lambda K}^2}{4\kappa^2}\right] \quad \text{and} \quad f_{K/K\Lambda}(x) = \frac{\kappa^2 A_{MB}^2}{4\pi^2}\, x(1-x)\, \exp\left[-\frac{\mu_{K\Lambda}^2}{4\kappa^2}\right]. \tag{7}$$

These LFWFs satisfy the constraint $f_{BK}(x) = f_{KB}(1-x)$. On the other side, the distribution functions $q_{s/\Lambda}$ and $q_{\bar{s}/K}$ are given by

$$q_{s/\Lambda}(x) = \frac{\kappa^2 A_B^2}{4\pi^2}\, x(1-x)\, \exp\left[-\frac{\mu_{sD}^2}{4\alpha^2}\right] \quad \text{and} \quad q_{\bar{s}/K}(x) = \frac{\kappa^2 A_M^2}{4\pi^2}\, x(1-x)\, \exp\left[-\frac{\mu_{\bar{s}K}^2}{4\kappa^2}\right]. \tag{8}$$

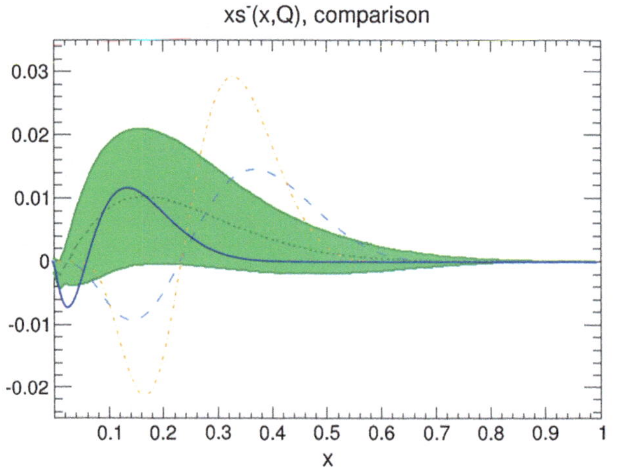

Fig. 2 $xS^- = x(s(x) - \bar{s}(x))$. *Green region* and *small dashed line* correspond to MMHT[27]) that it was generated with APFEL [40]. *Other lines* correspond to same cases in Fig. 1 (color figure online)

Note that all the parameters involved in our calculations—constituent masses of quarks $m_q = 330$ MeV, $m_s = 480$ MeV, of diquark $m_D = 600$ MeV, characteristic internal scale $\kappa = 330$ MeV—have been fixed in Ref. [14] and we use exactly the same set of parameters. The masses of hadrons involved in the calculation are $m_\Lambda = 1115.683$ MeV and $m_{K^+} = 493.677$ MeV. The normalization constants A_{MB}, A_B and A_M are obtained considering that the meson–baryon ($f_{\Lambda/K\Lambda}$, $f_{K/K\Lambda}$) and quark distribution ($q_{s/\Lambda}$, $q_{\bar{s}/K}$) functions are normalized to one:

$$\int_0^1 dx f_{K(\Lambda)/K\Lambda}(x) = \int_0^1 dx q_{s/\Lambda}(x) = \int_0^1 dx q_{\bar{s}/K}(x) = 1. \tag{9}$$

Actually, these normalizations are correct when we do not include the probability for a meson–baryon fluctuation. When this probability is included, which is actually the case for Eqs. (2), then the meson–baryon distribution normalizations should contain this probability. As mentioned before, here we have taken the fluctuation probability for $N \to \Lambda K$ to be 1.27%.

2.2 Holographic Wave Functions

Additionally we use two types of wave functions—variant I (obtained from matching to LF QCD at large x) and variant II (obtained from matching at all values of x and for an arbitrary number of constituents)

$$\psi(x,k) = \frac{A}{\sqrt{x(1-x)}} \exp\left[-\frac{1}{2\kappa^2}\left(\frac{k^2}{x(1-x)} + \mu_{12}^2\right)\right] \tag{10}$$

and

$$\psi_\tau(x,k) = A_\tau f_\tau(x) \exp\left[-\frac{x\log(1/x)}{2\kappa^2(1-x)}\left(\frac{k^2}{x(1-x)} + \mu_{12}^2\right)\right], \tag{11}$$

where

$$f_\tau(x) = \frac{4\pi}{\kappa}\sqrt{\log(1/x)}(1-x)^{\frac{\tau-4}{2}}. \tag{12}$$

Here τ is the twist of the operator that creates these states, which in this case coincides with the number of constituents of the particular state.

The first function satisfies the constraint $f_{BM}(x) = f_{MB}(1-x)$ with masses which are the same as used in the Gaussian case. The second LFWF [39] is also a two-body wave function, describing both clusters as quark–diquark bound states. For both holographic LFWFs we take the same value $\kappa = 350$ MeV fixed previoously from the best description of data on hadron properties [39].

In our specific calculations of the meson–baryon distribution functions we use $\tau = 5$ with masses $m_1 = m_B$ and $m_2 = m_M$. Note that the condition $f_{BM}(x) = f_{MB}(1-x)$ is not satisfied in the case of the second holographic LFWF. To avoid this problem we define the meson–baryon distribution functions as

$$f_{BM}(x) = \int \frac{d^2k}{16\pi^3} |\phi_{BM}(x,k)|^2$$
$$\phi_{BM}(x,k) = \psi_5(x,k) + \psi_5(1-x,k). \tag{13}$$

To calculate $q_{s/\Lambda}$ and $q_{\bar{s}/K}$ we directly use Eq. (11) with $\tau = 3$ and 2 respectively. The parameters used in this case are the same as in the case of the holographic LFWF (variant II), also normalized to one.

3 Results

For each LFWF we have calculated the meson–baryon distribution functions $f_{\Lambda/K\Lambda}$, $f_{K/K\Lambda}$ and the quark distributions $q_{s/\Lambda}$ and $q_{\bar{s}/K}$. Then using these quantities we calculate $(s(x) - \bar{s}(x))$. Figure 1 shows results for all three types of LFWFs using two slightly different values for the scale parameter κ, for reasons explained before. Notice that in both the Gaussian and the first holographic case (variant I) the point where the asymmetry vanishes is at $x \sim 0.7$, whereas in the other holographic case (variant II) it is near $x \sim 0.35$. Figure 2 shows results for $x(s(x) - \bar{s}(x))$ calculated with models discussed here. Considering the value and location of maximum in $x(s(x) - \bar{s}(x))$, and region where start to be zero, we can see that our result for the holographic LFWF (variant II) is consistent with the result of the fit done in MMHT [27].

We consider also results for the second moment $\langle x(s - \bar{s})\rangle$, using the different LFWFs. This quantity is defined as

$$\langle x(s - \bar{s})\rangle = \int_0^1 dx x\Big(s(x) - \bar{s}(x)\Big), \tag{14}$$

where the $s(x)$ and $\bar{s}(x)$ are defined in Eqs. 2. All our results for the second moment are small, positive and consistent with a value $\langle x(s - \bar{s})\rangle \approx 0.005$ mentioned in Ref. [28], which would be required to attribute the NuTeV anomaly [41,42] to the strange asymmetry alone. The value is also in a good agreement with predictions of different models of $|\langle x(s - \bar{s})\rangle| \sim 10^{-4}$ (see discussion in Ref. [28]). For completeness, we also quote some other results for the second moment generated in nonperturbative mechanisms. The result $\langle x(s - \bar{s})\rangle = -0.0027 \pm 0.0013$ [43,44] was deduced from a lowest-order QCD analysis of neutrino data, $-0.001 < \langle x(s - \bar{s})\rangle < 0.004$ from a global QCD fit done in Ref. [24]. Finally we mention the result for the second moment generated in a perturbative mechanism in Ref. [28] of $\langle x(s - \bar{s})\rangle \approx -5 \times 10^{-4}$ at the scale $Q^2 = 20$ GeV^2. Last result does not change too much when evolved to low scales and lies in the band derived in Ref. [24].

4 Conclusions

We calculated the $s(x) - \bar{s}(x)$ asymmetry in a light-front model considering three types of LFWFs. In all of these cases we observe that $s(x) < \bar{s}(x)$ for small values of x and $s(x) > \bar{s}(x)$ in the region of large x. This behavior is exactly opposite to the one obtained in meson-cloud models. For the cases of the Gaussian and the holographic (variant I) LFWFs the asymmetry vanishes at $x \sim 0.6$, while for the second variant of the holographic LFWF this point moves to a lower value of $x \sim 0.35$. The latter case is closer to MMHT [27] (other recent parametrizations are NNPDF [25], MSTW [26], MMHT [27] and nCTEQ [45]). Finally, our predictions for the second moment of the strange quark asymmetry $\langle x(s - \bar{s})\rangle = 0.00102 \pm 0.00055$ are consistent with previous results obtained with nonperturbative mechanisms.

Holographic wave functions has been used to calculate hadron properties, and his parameters was fixed [46,47], therefore, the study of the s-sbar asymmetry serves as independent tool to distinguish these functions. We present the results for fixed value of the κ, although for completeness also allow a variation of this parameter to display the sensitivity of s-sbar asymmetry to a choice of this parameter.

Acknowledgements This work was supported by the German Bundesministerium für Bildung und Forschung (BMBF) under Project 05P2015 - ALICE at High Rate (BMBF-FSP 202): "Jet- and fragmentation processes at ALICE and the parton structure of nuclei and structure of heavy hadrons", Tomsk State University Competitiveness Improvement Program and the Russian Federation program "Nauka" (Contract No. 0.1526.2015, 3854), by FONDECYT (Chile) under Grant No. 1141280 and by CONICYT (Chile) Research Project No. 80140097, and under Grant No. 7912010025. V.E.L. would like to thank Departamento de Física y Centro Científico Tecnológico de Valparaíso (CCTVal), Universidad Técnica Federico Santa María, Valparaíso, Chile and Instituto de Física y Astronomía, Universidad de Valparaíso, Chile for warm hospitality.

References

1. J.D. Sullivan, One pion exchange and deep inelastic electron-nucleon scattering. Phys. Rev. D **5**, 1732 (1972)
2. A.W. Thomas, A limit on the pionic component of the nucleon through SU(3) flavor breaking in the sea. Phys. Lett. B **126**, 97 (1983)
3. M. Burkardt, B. Warr, Chiral symmetry and the charge asymmetry of the s anti-s distribution in the proton. Phys. Rev. D **45**, 958 (1992)
4. C. Boros, J.T. Londergan, A.W. Thomas, Evidence for substantial charge symmetry violation in parton distributions. Phys. Rev. Lett. **81**, 4075 (1998)
5. M. Burkardt, C.A. Miller, W.D. Nowak, Spin-polarized high-energy scattering of charged leptons on nucleons. Rep. Prog. Phys. **73**, 016201 (2010)
6. A.I. Signal, A.W. Thomas, Possible strength of the non-perturbative strange sea of the nucleon. Phys. Lett. B **191**, 205 (1987)
7. H.R. Christiansen, J. Magnin, Strange/anti-strange asymmetry in the nucleon sea. Phys. Lett. B **445**, 8 (1998)
8. H. Holtmann, A. Szczurek, J. Speth, Flavor and spin of the proton and the meson cloud. Nucl. Phys. A **596**, 631 (1996)
9. W. Melnitchouk, J. Speth, A.W. Thomas, Dynamics of light anti-quarks in the proton. Phys. Rev. D **59**, 014033 (1998)
10. S.A. Rabinowitz et al., Measurement of the strange sea distribution using neutrino charm production. Phys. Rev. Lett. **70**, 134 (1993)
11. A.O. Bazarko et al., (CCFR Collaboration). Determination of the strange quark content of the nucleon from a next-to-leading order QCD analysis of neutrino charm production. Z. Phys. C **65**, 189 (1995)
12. M. Arneodo et al., (New Muon Collaboration). Measurement of the proton and deuteron structure functions, F2(p) and F2(d), and of the ratio sigma-L/sigma-T. Nucl. Phys. B **483**, 3 (1997)
13. W.G. Seligman et al., Improved determination of alpha(s) from neutrino nucleon scattering. Phys. Rev. Lett. **79**, 1213 (1997)
14. S.J. Brodsky, B.Q. Ma, The quark/anti-quark asymmetry of the nucleon sea. Phys. Lett. B **381**, 317 (1996)
15. A. Vega, I. Schmidt, T. Gutsche, V.E. Lyubovitskij, Nonperturbative contribution to the strange-antistrange asymmetry of the nucleon sea. Phys. Rev. D **93**(5), 056001 (2016)
16. M. Gluck, E. Reya, M. Stratmann, Mesonic parton densities derived from constituent quark model constraints. Eur. Phys. J. C **2**, 159 (1998)
17. M. Gluck, E. Reya, A. Vogt, Dynamical parton distributions revisited. Eur. Phys. J. C **5**, 461 (1998)
18. M. Gluck, E. Reya, M. Stratmann, W. Vogelsang, Models for the polarized parton distributions of the nucleon. Phys. Rev. D **63**, 094005 (2001)
19. A.D. Martin, R.G. Roberts, W.J. Stirling, R.S. Thorne, Parton distributions and the LHC: W and Z production. Eur. Phys. J. C **14**, 133 (2000)
20. H.L. Lai et al., (CTEQ Collaboration). Global QCD analysis of parton structure of the nucleon: CTEQ5 parton distributions. Eur. Phys. J. C **12**, 375 (2000)
21. S.I. Alekhin, Global fit to the charged leptons DIS data: alpha(s) parton distributions, and high twists. Phys. Rev. D **63**, 094022 (2001)
22. A. Kusina et al., Strange quark PDFs and implications for Drell-Yan Boson Production at the LHC. Phys. Rev. D **85**, 094028 (2012)
23. F.G. Cao, A.I. Signal, On the phenomenological analyses of s-anti-s asymmetry in the nucleon sea. Phys. Rev. D **60**, 074021 (1999)
24. F. Olness et al., Neutrino dimuon production and the strangeness asymmetry of the nucleon. Eur. Phys. J. C **40**, 145 (2005)
25. R.D. Ball et al. (NNPDF Collaboration). Parton distributions for the LHC run II. JHEP **1504**, 040 (2015)
26. A.D. Martin, W.J. Stirling, R.S. Thorne, G. Watt, Parton distributions for the LHC. Eur. Phys. J. C **63**, 189 (2009). doi:10.1140/epjc/s10052-009-1072-5. arXiv:0901.0002 [hep-ph]
27. L.A. Harland-Lang, A.D. Martin, P. Motylinski, R.S. Thorne, Parton distributions in the LHC era: MMHT 2014 PDFs. Eur. Phys. J. C **75**(5), 204 (2015)
28. S. Catani, D. de Florian, G. Rodrigo, W. Vogelsang, Perturbative generation of a strange-quark asymmetry in the nucleon. Phys. Rev. Lett. **93**, 152003 (2004)
29. M. Salajegheh, Intrinsic strange distributions in the nucleon from the light-cone models. Phys. Rev. D **92**(7), 074033 (2015)
30. G.Q. Feng, F.G. Cao, X.H. Guo, A.I. Signal, Perturbative and nonperturbative contributions to the strange quark asymmetry in the nucleon. Eur. Phys. J. C **72**, 2250 (2012)
31. S.J. Brodsky, G.F. de Teramond, Hadronic spectra and light-front wavefunctions in holographic QCD. Phys. Rev. Lett. **96**, 201601 (2006)
32. S.J. Brodsky, G.F. de Teramond, Light-front dynamics and AdS/QCD correspondence: the pion form factor in the space and time-like regions. Phys. Rev. D **77**, 056007 (2008)
33. S.J. Brodsky, G.F. de Teramond, H.G. Dosch, J. Erlich, Light-front holographic QCD and emerging confinement. Phys. Rep. **584**, 1 (2015)
34. A. Vega, I. Schmidt, T. Branz, T. Gutsche, V.E. Lyubovitskij, Meson wave function from holographic models. Phys. Rev. D **80**, 055014 (2009)

35. T. Branz, T. Gutsche, V.E. Lyubovitskij, I. Schmidt, A. Vega, Light and heavy mesons in a soft-wall holographic approach. Phys. Rev. D **82**, 074022 (2010)
36. T. Gutsche, V.E. Lyubovitskij, I. Schmidt, A. Vega, Dilaton in a soft-wall holographic approach to mesons and baryons. Phys. Rev. D **85**, 076003 (2012)
37. A. Vega, I. Schmidt, T. Gutsche, V.E. Lyubovitskij, Generalized parton distributions in AdS/QCD. Phys. Rev. D **83**, 036001 (2011)
38. A. Vega, I. Schmidt, T. Gutsche, V.E. Lyubovitskij, Generalized parton distributions in an AdS/QCD hard-wall model. Phys. Rev. D **85**, 096004 (2012)
39. T. Gutsche, V.E. Lyubovitskij, I. Schmidt, A. Vega, Light-front quark model consistent with Drell-Yan-West duality and quark counting rules. Phys. Rev. D **89**, 054033 (2014)
40. S. Carrazza, A. Ferrara, D. Palazzo, J. Rojo, APFEL Web: a web-based application for the graphical visualization of parton distribution functions. J. Phys. G **42**(5), 057001 (2015)
41. G.P. Zeller et al., (NuTeV Collaboration). A precise determination of electroweak parameters in neutrino nucleon scattering. Phys. Rev. Lett. **88**, 091802 (2002)
42. G.P. Zeller et al., NuTeV Collaboration). A precise determination of electroweak parameters in neutrino nucleon scattering (Erratum. Phys. Rev. Lett. **90**, 239902 (2003)
43. G.P. Zeller et al., (NuTeV Collaboration). On the effect of asymmetric strange seas and isospin violating parton distribution functions on $\sin^2\theta_W$ measured in the NuTeV experiment. Phys. Rev. D **65**, 111103 (2002)
44. G.P. Zeller et al., NuTeV Collaboration). On the effect of asymmetric strange seas and isospin violating parton distribution functions on $\sin^2\theta_W$ measured in the NuTeV experiment (Erratum. Phys. Rev. D **67**, 119902 (2003)
45. A. Kusina et al., nCTEQ15 - Global analysis of nuclear parton distributions with uncertainties. arXiv:1509.01801 [hep-ph]
46. T. Gutsche, V.E. Lyubovitskij, I. Schmidt, A. Vega, Nucleon structure in a light-front quark model consistent with quark counting rules and data. Phys. Rev. D **91**, 054028 (2015)
47. T. Gutsche, V.E. Lyubovitskij, I. Schmidt, A. Vega, Chiral Symmetry Breaking and Meson Wave Functions in Soft-Wall AdS/QCD. Phys. Rev. D **87**, 056001 (2013)

Few-Body Syst (2017) 58:36
DOI 10.1007/s00601-016-1207-9

Jai More · Asmita Mukherjee · Sreeraj Nair

Wigner Distributions of Quarks for Different Polarizations

Received: 19 November 2016 / Accepted: 27 December 2016 / Published online: 19 January 2017

Abstract We calculate quark Wigner distributions using the light-front wave functions in a dressed quark model. In this model, a proton target is replaced by a simplified spin-1/2 state, namely a quark dressed with a gluon. We calculate the Wigner distributions for different polarization configuration of quark and the target state in this model.

1 Introduction

Generalized parton distributions (GPDs) and transverse momentum dependent parton distributions (TMDs) [1] give the three-dimensional picture of the hadronic structure. GPDs [2,3] essentially can provide an understanding of the orbital angular momentum of partons inside the hadrons. GPDs and TMDs are encompassed in a correlation function which describes off-forward scattering amplitudes that were introduced in Ref. [4]. For certain kinematics, if one calculates the matrix element at $z^+ = 0$ or $z^- = 0$, one obtains a reduced matrix element called as generalized transverse momentum dependent parton distributions (GTMDs) [1,4,5]. Wigner distributions are obtained by taking Fourier transform of GTMDs which are also called as mother distribution functions. The main goal is to understand the partonic structure of nucleon that can be achieved by means of position-momentum distributions like Wigner distributions. In a particular limit, these distributions can be related to the spin and angular momentum of quarks and gluons. Wigner distribution is defined as matrix elements between different nucleon states as a function of three position and three momentum of a quark or a gluon inside the nucleon in a particular polarization state.

Quark and gluon Wigner distribution as a function of three position and three momentum variables was first studied by authors in Refs. [6,7] in which relativistic effects were neglected. Wigner distributions [8] do not give probabilistic interpretation as they are not positive definite. Only in the classical limit, they become positive definite. Inspite of this, a phase-space distribution are worth exploring as they can be related to the measurable quantities by integration over transverse position or momentum of the parton. The model dependent

This article belongs to the Topical Collection "Light Cone 2016".
Presented by Jai More at Light Cone 2016, 5–8 September 2016, IST, Universidade de Lisboa, Portugal.

J. More (✉) · A. Mukherjee
Indian Institute of Technology, Bombay, Mumbai 400076, India
E-mail: jai.more@phy.iitb.ac.in

A. Mukherjee
E-mail: asmita@phy.iitb.ac.in

S. Nair
Indian Institute of Science Education and Research, Bhopal, MP 462030, India
E-mail: sreeraj_nair@iiserb.ac.in

calculation of Wigner distributions may provide us information about the correlators of quarks and gluons in a nucleon. Especially, quark orbital angular momentum can be accessed from the Wigner distribution for unpolarized quarks in the longitudinally polarized nucleon (F_{14}) [9–14]. These calculations are also a boon in estimating relations of TMDs and GPDs using GTMDs in a certain limit. Wigner distributions have been studied in various phenomenological models [10,14–20].

In this work, we investigate the quark Wigner distributions in light-front Hamiltonian formalism [21,22]. In this formalism, one gets an intuitive picture of deep inelastic scattering for non-collinear, interacting and massive partons. The target state is a dressed quark state and is expanded in terms of light-front wavefunctions (LFWFs) which are the multiparton wavefunctions. So, in dressed quark model we replace the proton state as a quark dressed with a gluon which incorporates gluonic degrees of freedom in this perturbative model.

The plan of this paper is as follows. In Sect. 2, we calculate the Wigner distribution of quark. For different polarization configuration of quark and target state, one can define 16 Wigner distributions at the leading twist. All these Wigner distributions are calculated in light cone dressed quark model using Hamiltonian perturbation theory. In our model, we obtained 8 independent distributions and we illustrate here only 4 of them, a complete discussion of all the leading twist Wigner distributions can be found in [23]. In Sect. 3 we give numerical results and three-dimensional plots of the independent distributions in momentum space, impact parameter space and mixed space. Finally, we summarize our results in Sect. 4.

2 Quark Wigner Distributions in Dressed Quark Model

The Wigner distribution of quarks is defined as [1,9]

$$\rho^{[\Gamma]}(b_\perp, k_\perp, x, s, s') = \int \frac{d^2\Delta_\perp}{(2\pi)^2} e^{-i\Delta_\perp . b_\perp} W^{[\Gamma]}(\Delta_\perp, k_\perp, x, s, s') \tag{1}$$

where, $\boldsymbol{b}_\perp$ is the impact parameter space conjugate to momentum transfer ($\Delta_\perp$) of a dressed quark in the transverse direction. GTMDs are defined through quark-quark correlator $W^{[\Gamma]}$ at a fixed light-front time as

$$W^{[\Gamma]}(\Delta_\perp, k_\perp, x, s, s') = \int \frac{dz^- d^2 z_\perp}{2(2\pi)^3} e^{ik.z} \Big\langle p^+, \frac{\Delta_\perp}{2}, s \Big| \overline{\psi}(-\frac{z}{2}) \Omega \Gamma \psi(\frac{z}{2}) \Big| p^+, -\frac{\Delta_\perp}{2}, s' \Big\rangle \Big|_{z^+=0} \tag{2}$$

The initial and final dressed quark states are defined in the symmetric frame, with the longitudinal momentum as p^+, the transverse momentum transfer is $\Delta_\perp$ and s, s' are the helicities of initial and final target state. The average four momentum of the quark is k, with $k^+ = xp^+$, here x is the longitudinal momentum fraction of the parton. Ω is the gauge link which is required for color gauge invariance and is chosen to be unity.

The state of a dressed quark with momentum p and fixed helicity s can be written in terms of multiparton wavefunctions (LFWFs) as the expansion of the Fock state [21,22]

$$\Big|p^+, p_\perp, s\Big\rangle = \Phi^s(p) b_s^\dagger(p)|0\rangle + \sum_{s_1 s_2} \int \frac{dp_1^+ d^2 p_1^\perp}{\sqrt{16\pi^3 p_1^+}} \int \frac{dp_2^+ d^2 p_2^\perp}{\sqrt{16\pi^3 p_2^+}} \sqrt{16\pi^3 p^+} \delta^3(p - p_1 - p_2)$$
$$\times \Phi^s_{s_1 s_2}(p; p_1, p_2) b_{s_1}^\dagger(p_1) a_{s_2}^\dagger(p_2)|0\rangle \tag{3}$$

$\Phi^s(p)$ is the single quark state and $\Phi^s_{s_1 s_2}(p; p_1, p_2)$ is the quark-gluon LFWF. s_1 and s_2 are helicities of quark and gluon respectively. $\Phi^s(p)$ gives the wavefunction normalization of the quark. $\Phi^s_{s_1 s_2}(p; p_1, p_2)$ gives the probability amplitude to find a bare quark (gluon) with momentum $p_1 (p_2)$ and helicity $s_1 (s_2)$ inside the dressed quark. Using the Jacobi momenta

$$k_i^+ = x_i P^+ \text{ and } k_i^\perp = q_i^\perp + x_i P^\perp \quad \text{so that } \sum_i x_i = 1, \quad \sum_i q_{i\perp} = 0 \tag{4}$$

the two particle LFWF can be written in terms of boost invariant LFWF as

$$\sqrt{P^+} \Phi(p; p_1, p_2) = \Psi(x_i, q_i^\perp) \tag{5}$$

Using the two particle LFWF [21,22] and two component formalism [24] at leading twist, one obtains only four Dirac operators $\Gamma = \{\gamma^+, \gamma^+\gamma^5, i\sigma^{+1}\gamma^5, i\sigma^{+2}\gamma^5\}$ which corresponds to Wigner distributions for unpolarized, longitudinally polarized and transversely polarized dressed quark. So the quark-quark correlator using two particle LFWFs for different polarization at twist-2 are given by

$$W^{[\gamma^+]}(\Delta_\perp, k_\perp, x, s, s') = \sum_{\lambda_1', \lambda_1, \lambda_2} \Psi^{*s'}_{\lambda_1'\lambda_2}(x, q'^{\perp}) \chi^\dagger_{\lambda_1'} \chi_{\lambda_1} \Psi^{s}_{\lambda_1\lambda_2}(x, q^{\perp}) \tag{6}$$

$$W^{[\gamma^+\gamma^5]}(\Delta_\perp, k_\perp, x, s, s') = \sum_{\lambda_1', \lambda_1, \lambda_2} \Psi^{*s'}_{\lambda_1'\lambda_2}(x, q'^{\perp}) \chi^\dagger_{\lambda_1'} \sigma_3 \chi_{\lambda_1} \Psi^{s}_{\lambda_1\lambda_2}(x, q^{\perp}) \tag{7}$$

$$W^{[i\sigma^{+j}\gamma^5]}(\Delta_\perp, k_\perp, x, s, s') = \sum_{\lambda_1', \lambda_1, \lambda_2} \Psi^{*s'}_{\lambda_1'\lambda_2}(x, q'^{\perp}) \chi^\dagger_{\lambda_1'} \sigma_j \chi_{\lambda_1} \Psi^{s}_{\lambda_1\lambda_2}(x, q^{\perp}) \tag{8}$$

with σ_i are the three Pauli matrices. Equations (6), (7) and (8) gives unpolarized, longitudinally polarized and transversely polarized GTMDs in terms of LFWFs. For various combinations of unpolarized (U), longitudinally polarized (L) and transversely polarized (T) target and quark states, the quark-quark correlators can be parametrized into 16 Wigner distributions [18] at leading twist. We denote Wigner distributions by $\rho_{\lambda,\lambda'}$, where λ and λ' represents polarization of the target state and quark respectively. In this model, we have 10 independent Wigner distributions out of which $\rho^j_{TT}(\boldsymbol{b}_\perp, \boldsymbol{k}_\perp, x)$ is zero with $j = 1, 2$ and $\rho_{LU} = \rho_{UL}$. However, we discuss only four independent Wigner distributions but the details of all the 10 Wigner distributions can be looked in Ref. [23] and in that we have classified Wigner distribution in terms of polarization of the target states.
The unpolarized Wigner distribution

$$\rho_{UU}(\boldsymbol{b}_\perp, \boldsymbol{k}_\perp, x) = \frac{1}{2}\Big[\rho^{[\gamma^+]}(\boldsymbol{b}_\perp, \boldsymbol{k}_\perp, x, \hat{\boldsymbol{e}}_z) + \rho^{[\gamma^+]}(\boldsymbol{b}_\perp, \boldsymbol{k}_\perp, x, -\hat{\boldsymbol{e}}_z)\Big] \tag{9}$$

The unpolarized-longitudinally polarized Wigner distribution

$$\rho_{UL}(\boldsymbol{b}_\perp, \boldsymbol{k}_\perp, x) = \frac{1}{2}\Big[\rho^{[\gamma^+\gamma^5]}(\boldsymbol{b}_\perp, \boldsymbol{k}_\perp, x, \hat{\boldsymbol{e}}_z) + \rho^{[\gamma^+\gamma^5]}(\boldsymbol{b}_\perp, \boldsymbol{k}_\perp, x, -\hat{\boldsymbol{e}}_z)\Big] \tag{10}$$

The longitudinal Wigner distribution

$$\rho_{LL}(\boldsymbol{b}_\perp, \boldsymbol{k}_\perp, x) = \frac{1}{2}\Big[\rho^{[\gamma^+\gamma^5]}(\boldsymbol{b}_\perp, \boldsymbol{k}_\perp, x, \hat{\boldsymbol{e}}_z) - \rho^{[\gamma^+\gamma^5]}(\boldsymbol{b}_\perp, \boldsymbol{k}_\perp, x, -\hat{\boldsymbol{e}}_z)\Big] \tag{11}$$

The transverse Wigner distribution

$$\rho_{TT}(\boldsymbol{b}_\perp, \boldsymbol{k}_\perp, x) = \frac{1}{2}\delta_{ij}\Big[\rho^{[i\sigma^{+j}\gamma^5]}(\boldsymbol{b}_\perp, \boldsymbol{k}_\perp, x, \hat{\boldsymbol{e}}_i) - \rho^{[i\sigma^{+j}\gamma^5]}(\boldsymbol{b}_\perp, \boldsymbol{k}_\perp, x, -\hat{\boldsymbol{e}}_i)\Big] \tag{12}$$

The pretzelous Wigner distribution

$$\rho^{\perp}_{TT}(\boldsymbol{b}_\perp, \boldsymbol{k}_\perp, x) = \frac{1}{2}\epsilon_{ij}\Big[\rho^{[i\sigma^{+j}\gamma^5]}(\boldsymbol{b}_\perp, \boldsymbol{k}_\perp, x, \hat{\boldsymbol{e}}_i) - \rho^{[i\sigma^{+j}\gamma^5]}(\boldsymbol{b}_\perp, \boldsymbol{k}_\perp, x, -\hat{\boldsymbol{e}}_i)\Big] \tag{13}$$

Thus, the analytical expressions for the four Wigner distributions that we study are given by

$$\rho_{UU}(\boldsymbol{b}_\perp, \boldsymbol{k}_\perp, x) = N\int d^2\Delta_\perp \frac{\cos(\Delta_\perp b_\perp)}{D(q_\perp)D(q'_\perp)}\left[\frac{\Big(4k_\perp^2 - \Delta_\perp^2(1-x)^2\Big)(1+x^2)}{x^2(1-x)^3} + \frac{4m^2(1-x)}{x^2}\right] \tag{14}$$

$$\rho_{UL}(\boldsymbol{b}_\perp, \boldsymbol{k}_\perp, x) = -N\int d^2\Delta_\perp \frac{\sin(\Delta_\perp b_\perp)}{D(q_\perp)D(q'_\perp)}\left[\frac{4\Big(k_x\Delta_y - k_y\Delta_x\Big)(1+x)}{x^2(1-x)}\right] \tag{15}$$

$$\rho_{LL}(\boldsymbol{b}_\perp, \boldsymbol{k}_\perp, x) = N\int d^2\Delta_\perp \frac{\cos(\Delta_\perp b_\perp)}{D(q_\perp)D(q'_\perp)}\left[\frac{\Big(4k_\perp^2 - \Delta_\perp^2(1-x)^2\Big)(1+x^2)}{x^2(1-x)^3} - \frac{4m^2(1-x)}{x^2}\right] \tag{16}$$

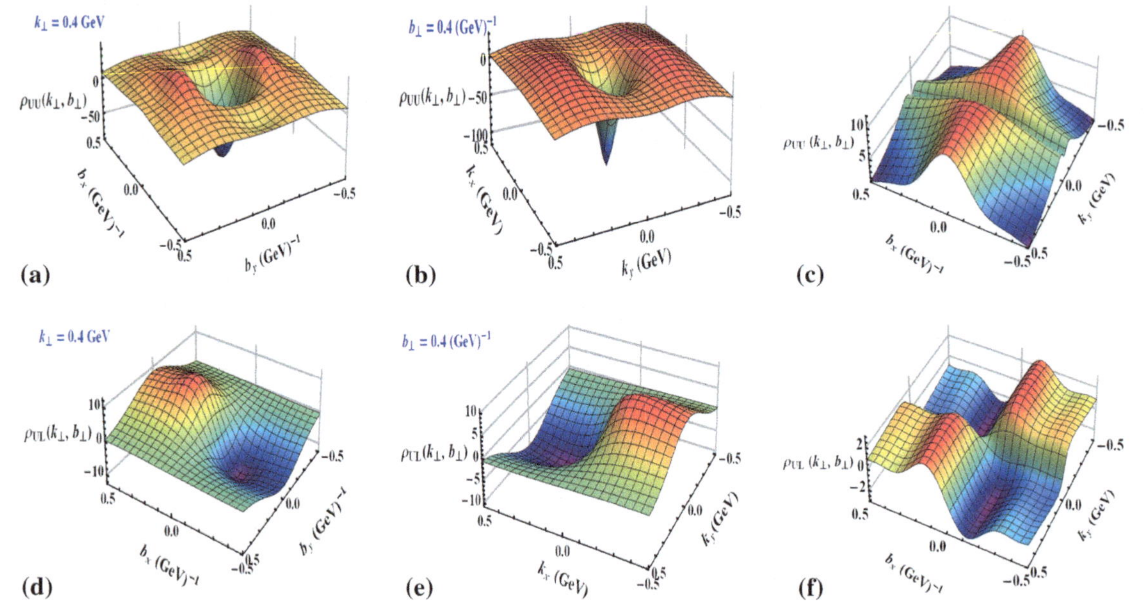

Fig. 1 (color online) 3D plot of Wigner distributions $\rho_{UU}(\mathbf{k}_\perp, \mathbf{b}_\perp)$ and $\rho_{UL}(\mathbf{k}_\perp, \mathbf{k}_\perp)$ in $\boldsymbol{k}_\perp-$space, $\boldsymbol{b}_\perp-$space and $b_x - k_y$ space with $\Delta_{\perp max} = 20$ GeV

$$\rho_{TT}(\boldsymbol{b}_\perp, \boldsymbol{k}_\perp, x) = N \int d^2\Delta_\perp \frac{\cos(\Delta_\perp b_\perp)}{D(q_\perp)D(q'_\perp)} \left[\frac{2\left(4k_\perp^2 - \Delta_\perp^2(1-x)^2\right)}{x(1-x)^3} \right] \tag{17}$$

where,

$$N = \frac{g^2 C_F}{2(2\pi)^5}, \quad C_F \text{ is the color factor}$$

$$D(q_\perp) = \left[m^2 - \frac{m^2 + (k_\perp + \frac{\Delta_\perp(1-x)}{2})^2}{x} - \frac{(k_\perp + \frac{\Delta_\perp(1-x)}{2})^2}{1-x}\right]$$

$$D(q'_\perp) = \left[m^2 - \frac{m^2 + (k_\perp - \frac{\Delta_\perp(1-x)}{2})^2}{x} - \frac{(k_\perp - \frac{\Delta_\perp(1-x)}{2})^2}{1-x}\right] \tag{18}$$

We observed that, $\rho_{TT}^{\perp}(\boldsymbol{b}_\perp, \boldsymbol{k}_\perp, x) = 0$ in this model.

3 Numericals and Plots of Probability Densities in Three Dimension

As mentioned in the previous section, we study 4 independent Wigner distributions given in Eqs. (14)–(17) which are five dimensional function of b_x, b_y, k_x, k_y and x. We integrate over the momentum fraction x which can take values only between (0, 1), in the dressed quark model. We observe the behavior in the transverse momentum and impact parameter space of these 4 distributions. One may also look at the mixed distributions by integrating out transverse momentum and transverse position along the perpendicular transverse directions. In the mixed distribution one expects distribution to be real as the remaining variables are not restricted by Heisenberg's uncertainty principle. In this section, we show the numerical results and the plots of the quark Wigner distributions in transverse momentum space with $b_\perp = 0.4$ GeV^{-1}, transverse position space with $k_\perp = 0.4$ GeV and mixed space. The integration limit of $\Delta_\perp$ should ideally be from zero to infinity.

However, in the numerical integration, we have used a cutoff on $\Delta_\perp$ integration, $\Delta_{\perp max} = 20$ GeV. It is to be noted that in the previous work [14,15], a lower value of $\Delta_{\perp max}$ was used and the results depend on this cutoff. We use Levin's method to perform numerical integration over x and $\Delta_\perp$. It is a very effective method as it gives good convergence and the results do not depend on $\Delta_{\perp max}$.

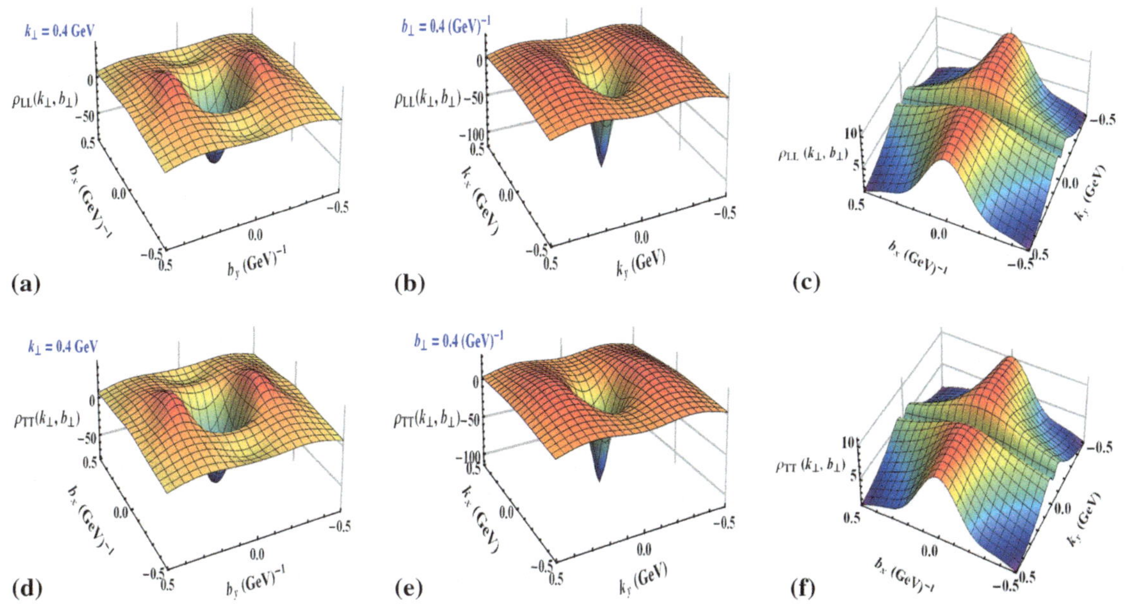

Fig. 2 (color online) 3D plot of Wigner distribution $\rho_{LL}(\mathbf{k}_\perp, \mathbf{b}_\perp)$ and $\rho_{TT}(\mathbf{k}_\perp, \mathbf{b}_\perp)$ in $\boldsymbol{b}_\perp$−space, $\boldsymbol{k}_\perp$−space and $b_x - k_y$ space with $\Delta_{\perp max} = 20$ GeV

For numerical evaluation, in all the plots we have taken $m = 0.33$ GeV and $\Delta_{\perp max} = 20$ GeV. Figure 1a show a three-dimensional plot of Wigner distribution ρ_{UU} in $\boldsymbol{k}_\perp$−space with $\boldsymbol{b}_\perp = 0.4\ \text{GeV}^{-1}\ \hat{\boldsymbol{e}}_y$. ρ_{UU} has a minimum and pick up negative value at $k_x = 0$ and $k_y = 0$. Figure 1b show the Wigner distribution ρ_{UU} in $\boldsymbol{b}_\perp$—space with $\boldsymbol{k}_\perp = 0.4$ GeV $\hat{\boldsymbol{e}}_y$. The behavior of $\boldsymbol{k}_\perp$−space is similar to $\boldsymbol{b}_\perp$—space with a different minima at $b_x = 0$ and $b_y = 0$ but it is still negative. Figure 1c show the Wigner distribution ρ_{UU} in mixed space where we have integrated out k_x and b_y dependence, thus we get the probability densities in the $b_x - k_y$ plane.

Figure 1d shows a three-dimensional plot of Wigner distributions ρ_{UL} in momentum space with $\boldsymbol{b}_\perp = 0.4\ \text{GeV}^{-1}\ \hat{\boldsymbol{e}}_y$. Figure 1e shows a three-dimensional plot of Wigner distributions ρ_{UL} in impact parameter space with $\boldsymbol{k}_\perp = 0.4$ GeV $\hat{\boldsymbol{e}}_y$. Figure 1d and e have similar nature. These two graphs show dipole structure as observed in Refs. [9, 14]. Figure 1f shows a three-dimensional plot of Wigner distributions in mixed space which shows the quadruple structure.

Figure 2a–c shows the three-dimensional plot of Wigner distributions ρ_{LL} in $\boldsymbol{k}_\perp$−space, $\boldsymbol{b}_\perp$−space and the mixed space respectively. Figure 2 for ρ_{LL} shows similar nature as Fig. 1 for ρ_{UU} as expected.

Figure 2d–f shows the three-dimensional plot of Wigner distribution ρ_{TT} in $\boldsymbol{k}_\perp$−space, $\boldsymbol{b}_\perp$−space and mixed space respectively. In this case, both the quark and the target state are transversely polarized in the direction say x-direction. It is important to note that nature of ρ_{TT} is similar to ρ_{UU} and ρ_{LL}.

4 Conclusion

In this work, we include the transverse polarization of the quark and the target state unlike in the previous work [14]. In this model, we study only 4 independent quark Wigner distributions as illustration and the details will be given in Ref. [23]. The unpolarized ρ_{UU}, longitudinally polarized ρ_{LL} and transversity distributions ρ_{TT} show similar nature. ρ_{LU} is equal to ρ_{UL}, which is related to orbital angular momentum of the quark and the pretzelous Wigner distribution $\rho_{TT}^{\perp}$ is zero. We have used an improved method of numerical integration that gives better convergence and the results are independent of $\Delta_{\perp max}$.

Acknowledgements J.M. would like to thank Science and Engineering Research Board (SERB), for providing financial support and Light Cone organizers for their kind hospitality.

References

1. S. Meissner, A. Metz, M. Schlegel, Generalized parton correlation functions for a spin-1/2 hadron. JHEP **0908**, 056 (2009)
2. A.V. Radyushkin, Nonforward parton distributions. Phys. Rev. D **56**, 5524 (1997)
3. X.D. Ji, Gauge-Invariant decomposition of nucleon spin. Phys. Rev. Lett. **78**, 610 (1997)
4. S. Meissner, A. Metz, M. Schlegel, K. Goeke, Generalized parton correlation functions for a spin-0 hadron. JHEP **0808**, 038 (2008)
5. C. Lorcé, B. Pasquini, Structure analysis of the generalized correlator of quark and gluon for a spin-1/2 target. JHEP **1309**, 138 (2013)
6. X.D. Ji, Viewing the proton through 'color' filters. Phys. Rev. Lett. **91**, 062001 (2003)
7. A.V. Belitsky, X.D. Ji, F. Yuan, Quark imaging in the proton via quantum phase space distributions. Phys. Rev. D **69**, 074014 (2004)
8. E.P. Wigner, On the quantum correction for thermodynamic equilibrium. Phys. Rev. **40**, 749 (1932)
9. C. Lorcé, B. Pasquini, Quark Wigner distributions and orbital angular momentum. Phys. Rev. D **84**, 014015 (2011)
10. C. Lorcé, B. Pasquini, X. Xiong, F. Yuan, The quark orbital angular momentum from Wigner distributions and light-cone wave functions. Phys. Rev. D **85**, 114006 (2012)
11. P. Hagler, A. Mukherjee, A. Schafer, Quark orbital angular momentum in the Wandzura-Wilczek approximation. Phys. Lett. B **582**, 55 (2004)
12. K. Kanazawa, C. Lorcé, A. Metz, B. Pasquini, M. Schlegel, Twist-2 generalized transverse-momentum dependent parton distributions and the spin/orbital structure of the nucleon. Phys. Rev. D **90**(1), 014028 (2014)
13. A. Rajan, A. Courtoy, M. Engelhardt, S. Liuti, Parton transverse momentum and orbital angular momentum distributions. Phys. Rev. D **94**(3), 034041 (2016)
14. A. Mukherjee, S. Nair, V.K. Ojha, Quark Wigner distributions and orbital angular momentum in light-front dressed quark model. Phys. Rev. D **90**(1), 014024 (2014)
15. A. Mukherjee, S. Nair, V.K. Ojha, Wigner distributions for gluons in a light-front dressed quark model. Phys. Rev. D **91**(5), 054018 (2015)
16. C. Lorcé, B. Pasquini, M. Vanderhaeghen, Unified framework for generalized and transverse-momentum dependent parton distributions within a 3Q light-cone picture of the nucleon. JHEP **1105**, 041 (2011)
17. T. Liu, Quark orbital motions from Wigner distributions. arXiv:1406.7709 [hep-ph]
18. T. Liu, B.Q. Ma, Quark Wigner distributions in a light-cone spectator model. Phys. Rev. D **91**, 034019 (2015)
19. G.A. Miller, Electron structure: shape, size, and generalized parton distributions in QED. Phys. Rev. D **90**(11), 113001 (2014)
20. C. Lorcé, B. Pasquini, Multipole decomposition of the nucleon transverse phase space. Phys. Rev. D **93**(3), 034040 (2016)
21. A. Harindranath, R. Kundu, W.M. Zhang, Deep inelastic structure functions in light front QCD: radiative corrections. Phys. Rev. D **59**, 094012 (1999)
22. A. Harindranath, R. Kundu, W.M. Zhang, On orbital angular momentum in deep inelastic scattering. Phys. Rev. D **59**, 094013 (1999)
23. J. More, A. Mukherjee, S. Nair, Quark wigner distributions using light-front wave functions. arXiv:1701.00339 [hep-ph]
24. W.M. Zhang, A. Harindranath, Light front QCD. 2: Two component theory. Phys. Rev. D **48**, 4881 (1993)

Few-Body Syst (2017) 58:43
DOI 10.1007/s00601-016-1204-z

J. R. Hiller

Symmetry Breaking in Light-Front ϕ^4 Theory

Received: 15 November 2016 / Accepted: 23 December 2016 / Published online: 19 January 2017

Abstract We consider the symmetric and broken phases of light-front ϕ^4 theory in two dimensions. In both cases the mass of the lowest state is computed and its dependence on the coupling used to infer critical coupling values. The structure of the eigenstate is examined to determine whether it shows the signs of critical behavior, specifically whether the one-body sector becomes improbable relative to the higher Fock sectors. In attempts to establish this behavior, we consider both sector-independent and sector-dependent constituent masses.

1 Introduction

We apply a new computational method to light-front [1–4] ϕ^4 theory in two dimensions, in both the symmetric and broken phases [5–10]. The method is based on an expansion of the Fock-state wave functions in a basis of multivariate symmetric polynomials [11,12]. This allows fine tuning of the resolution, Fock sector by Fock sector, and incorporates small-x behavior that captures an integrable singularity. Both features represent an improvement over the traditional discrete light-cone quantization (DLCQ) [13,14] approach, where the resolution is fixed across all Fock sectors and the integrable singularity at zero momentum fraction is ignored. The presentation here extends earlier work in [15].

The general method and an application to the broken phase are described here; the symmetric phase is discussed specifically by Chabysheva [16]. Section 2 details the structure of the ϕ^4 eigenvalue problem and our method of solution, including the option of a sector-dependent mass. Results are presented in Sect. 3, followed by a brief summary in Sect. 4.

2 Eigenvalue Problem

The Lagrangian for ϕ^4 theory is $\mathscr{L} = \frac{1}{2}(\partial_\mu\phi)^2 - \frac{1}{2}\mu_0^2\phi^2 - \frac{\lambda}{4!}\phi^4$. From it, one obtains the light-front Hamiltonian density $\mathscr{H}^- = \pm\frac{1}{2}\mu^2 :\phi^2: + \frac{\lambda}{4!} :\phi^4:$ and the Hamiltonian $\mathscr{P}^- = \int dx^- \mathscr{H}^-$ that defines the eigenvalue problem $\mathscr{P}^-|\psi\rangle = \frac{M^2}{P^+}|\psi\rangle$. Here, $P^+ = E + p_z$ is the light-front momentum conjugate to the light-front spatial coordinate $x^- \equiv t - z$, and $\mathscr{P}^-$ is the operator that generates translations in light-front time $x^+ \equiv t + z$. The mass of the eigenstate $|\psi\rangle$ is M.

The Hamiltonian is computed with the mode expansion

$$\phi(x^+, x^-) = \int \frac{dp^+}{\sqrt{4\pi p^+}} \left\{ a(p^+)e^{-ipx} + a^\dagger(p^+)e^{ipx} \right\}. \tag{1}$$

This article belongs to the Topical Collection "Light Cone 2016".

J. R. Hiller (✉)
Department of Physics and Astronomy, University of Minnesota-Duluth, Duluth, MN 55812, USA
E-mail: jhiller@d.umn.edu

The creation operators $a^\dagger(p^+)$ satisfy the commutation relation $[a(p^+), a^\dagger(p'^+)] = \delta(p^+ - p'^+)$.

However, before completing the construction of $\mathscr{P}^-$, we consider an asymmetric form obtained by a shift in the field. The generator for the shift is $U = \exp\int dp^+[f(p^+)a^\dagger(p^+) - f^*(p^+)a(p^+)]$. This shifts the creation operator $Ua^\dagger(p^+)U^\dagger = a^\dagger(p^+) + f^*(p^+)$. Following Harindranath and Vary [17], we choose f to correspond to a zero mode $f(p^+) = f^*(p^+) \equiv \sqrt{\pi p^+}\delta(p^+)\phi_s$. The field is then shifted by a constant, $U\phi U^\dagger = \phi + \phi_s$, and, with $\phi_s = \pm\sqrt{6\mu^2/\lambda}$, we obtain

$$U : \mathscr{H}^- : U^\dagger = -\frac{3}{2}\frac{\mu^4}{\lambda} + \frac{1}{2}(2\mu^2) : \phi^2 : + \frac{\lambda\phi_s}{3!} : \phi^3 : + \frac{\lambda}{4!} : \phi^4 : \tag{2}$$

Now substitution of the mode expansion yields $\mathscr{P}^- = \mathscr{P}^-_{11} + \mathscr{P}^-_{22} + \mathscr{P}^-_{13} + \mathscr{P}^-_{31} + \mathscr{P}^-_{12} + \mathscr{P}^-_{21}$, with

$$\mathscr{P}^-_{11} = \int dp \frac{2\mu^2}{p} a^\dagger(p)a(p), \tag{3}$$

$$\mathscr{P}^-_{22} = \frac{\lambda}{4}\int \frac{dp_1 dp_2}{4\pi\sqrt{p_1 p_2}} \int \frac{dp'_1 dp'_2}{\sqrt{p'_1 p'_2}} \delta(p_1 + p_2 - p'_1 - p'_2) a^\dagger(p_1)a^\dagger(p_2)a(p'_1)a(p'_2), \tag{4}$$

$$\mathscr{P}^-_{13} = \frac{\lambda}{6}\int \frac{dp_1 dp_2 dp_3}{4\pi\sqrt{p_1 p_2 p_3 (p_1 + p_2 + p_3)}} a^\dagger(p_1 + p_2 + p_3)a(p_1)a(p_2)a(p_3), \tag{5}$$

$$\mathscr{P}^-_{12} = \mu\sqrt{\frac{3\lambda}{2}}\int \frac{dp_1^+ dp_2^+}{\sqrt{4\pi p_1^+ p_2^+ (p_1^+ + p_2^+)}} a^\dagger(p_1^+ + p_2^+)a(p_1^+)a(p_2^+), \tag{6}$$

$$\mathscr{P}^-_{31} = (\mathscr{P}^-_{13})^\dagger, \quad \mathscr{P}^-_{21} = (\mathscr{P}^-_{12})^\dagger. \tag{7}$$

The eigenstate of $\mathscr{P}^-$, with eigenvalue M^2/P^+, can be expressed as an expansion

$$|\psi(P^+)\rangle = \sum_m P^{+\frac{m-1}{2}} \int \prod_i^m dy_i \delta(1 - \sum_i^m y_i)\psi_m(y_i)|y_i P^+; P^+, m\rangle \tag{8}$$

in terms of Fock states $|y_i P^+; P^+, m\rangle = \frac{1}{\sqrt{m!}}\prod_{i=1}^m a^\dagger(y_i P^+)|0\rangle$ with normalization $1 = \sum_m \int \prod_i^m dy_i \delta(1 - \sum_i^m y_i)|\psi_m(y_i)|^2$. The eigenvalue problem is then reduced to a coupled system of equations

$$\begin{aligned}
&\begin{Bmatrix} +\mu^2 \\ -\mu^2 \\ 2\mu^2 \end{Bmatrix} \sum_i^m \frac{1}{y_i}\psi_m(y_i) + \frac{\lambda}{4\pi}\frac{m(m-1)}{4\sqrt{y_1 y_2}} \int \frac{dx_1 \psi_m(x_1, y_1 + y_2 - x_1, y_3, \ldots, y_m)}{\sqrt{x_1(y_1 + y_2 - x_1)}} \\
&+ \frac{\lambda}{4\pi}\frac{m\sqrt{(m+2)(m+1)}}{6} \int \frac{dx_1 dx_2 \psi_{m+2}(x_1, x_2, y_1 - x_1 - x_2, y_2, \ldots, y_m)}{\sqrt{y_1 x_1 x_2 (y_1 - x_1 - x_2)}} \\
&+ \frac{\lambda}{4\pi}\frac{(m-2)\sqrt{m(m-1)}}{6} \frac{\psi_{m-2}(y_1 + y_2 + y_3, y_4, \ldots, y_m)}{\sqrt{y_1 y_2 y_3 (y_1 + y_2 + y_3)}} \\
&+ \mu\sqrt{\frac{3\lambda}{8\pi}} m\sqrt{m+1} \int \frac{dx_1 \psi_{m+1}(x_1, y_1 - x_1, y_2, \ldots, y_m)}{\sqrt{x_1(y_1 - x_1)y_1}} \\
&+ \mu\sqrt{\frac{3\lambda}{8\pi}} (m-1)\sqrt{m} \frac{\psi_{m-1}(y_1 + y_2, y_3, \ldots, y_m)}{\sqrt{y_1 y_2 (y_1 + y_2)}} = M^2 \psi_m(y_i),
\end{aligned} \tag{9}$$

where the last two terms are kept only for the bottom option of $2\mu^2$, which represents the shifted, asymmetric form of the Hamiltonian. In the symmetric phase, we consider only the odd eigenstate where the number of constituents m is an odd integer.

We solve this system by truncation in Fock space, to $m \leq N_{\max}$, and by expansion of the Fock-state wave functions in terms of multivariate symmetric polynomials $P_{ni}^{(m)}(y_1, \ldots, y_m)$ [11,12]

$$\psi_m(y_1, \ldots, y_m) = \sqrt{y_1 y_2 \cdots y_m} \sum_{ni} c_{ni}^{(m)} P_{ni}^{(m)}(y_1, \ldots, y_m). \tag{10}$$

This converts the system into a generalized matrix eigenvalue problem

$$\sum_{n'i'}\left[\left\{\begin{matrix}+1\\-1\\2\end{matrix}\right\}T^{(m)}_{ni,n'i'}+gV^{(m,m)}_{ni,n'i'}\right]c^{(m)}_{n'i'}+g\sum_{n'i'}V^{(m,m+2)}_{ni,n'i'}c^{(m+2)}_{n'i'}+g\sum_{n'i'}V^{(m,m-2)}_{ni,n'i'}c^{(m-2)}_{n'i'}$$
$$+\sqrt{g}\sum_{n'i'}V^{(m,m+1)}_{ni,n'i'}c^{(m+1)}_{n'i'}+\sqrt{g}\sum_{n'i'}V^{(m,m-1)}_{ni,n'i'}c^{(m-1)}_{n'i'}=\frac{M^2}{\mu^2}\sum_{n'i'}B^{(m)}_{ni,n'i'}c^{(m)}_{n'i'}, \tag{11}$$

with $g=\lambda/(4\pi\mu^2)$. The matrices T, V and B are defined as

$$T^{(m)}_{ni,n'i'}=m\int\left(\prod_j dy_j\right)\delta(1-\sum_j y_j)\left(\prod_{j=2}^m y_j\right)P^{(m)}_{ni}(y_j)P^{(m)}_{n'i'}(y_j), \tag{12}$$

$$V^{(m,m)}_{ni,n'i'}=\frac{1}{4}m(m-1)\int\left(\prod_j dy_j\right)\delta(1-\sum_j y_j)$$
$$\times\int dx_1dx_2\delta(y_1+y_2-x_1-x_2)\left(\prod_{j=3}^m y_j\right)P^{(m)}_{ni}(y_j)P^{(m)}_{n'i'}(x_1,x_2,y_3,\dots,y_m), \tag{13}$$

$$V^{(m,m+2)}_{ni,n'i'}=\frac{1}{6}m\sqrt{(m+2)(m+1)}\int\left(\prod_j dy_j\right)\delta(1-\sum_j y_j)$$
$$\times\int dx_1dx_2dx_3\delta(y_1-x_1-x_2-x_3)\left(\prod_{j=2}^m y_j\right)P^{(m)}_{ni}(y_j)P^{(m+2)}_{n'i'}(x_1,x_2,x_3,y_2,\dots,y_m), \tag{14}$$

$$V^{(m,m+1)}_{ni,n'i'}=\sqrt{\frac{3}{2}}m\sqrt{m+1}\int\left(\prod_j dy_j\right)\delta(1-\sum_j y_j)$$
$$\times\int dx_1dx_2\delta(y_1-x_1-x_2)\left(\prod_{j=2}^m y_j\right)P^{(m)}_{ni}(y_j)P^{(m+1)}_{n'i'}(x_1,x_2,y_2,\dots,y_m), \tag{15}$$

and

$$B^{(m)}_{ni,n'i'}=\int\left(\prod_j dy_j\right)\delta(1-\sum_j y_j)\left(\prod_j^m y_j\right)P^{(m)}_{ni}(y_j)P^{(m)}_{n'i'}(y_j), \tag{16}$$

with $V^{(m,m-2)}_{ni,n'i'}$ and $V^{(m,m-1)}_{ni,n'i'}$ obtained as the adjoints of $V^{(m-2,m)}_{ni,n'i'}$ and $V^{(m-1,m)}_{ni,n'i'}$, respectively.

We convert the matrix problem to an ordinary eigenvalue problem by a singular value decomposition (SVD) [18] of the overlap matrix $B^{(m)}$. This is an implicit orthogonalization of the basis. The SVD is $B^{(m)}=U^{(m)}D^{(m)}U^{(m)T}$, where the columns of $U^{(m)}$ are the eigenvectors of $B^{(m)}$ and $D^{(m)}$ is a diagonal matrix of the eigenvalues. We keep in $U^{(m)}$ only those columns associated with eigenvalues of $B^{(m)}$ that are above some positive threshold, in order to eliminate from the basis those combinations of polynomials that are nearly linearly dependent [19]. We then define new vectors of coefficients $\mathbf{c}^{(m)\prime}=D^{1/2}U^T\mathbf{c}^{(m)}$ and new matrices, such as $T^{(m)\prime}=D^{-1/2}U^TT^{(m)}UD^{-1/2}$. In terms of these, the matrix eigenvalue problem is an ordinary one, which we then diagonalize by standard means.

3 Results

Figure 1 shows the results for the mass squared of the lowest eigenstate as a function of the dimensionless coupling g and for both the symmetric and broken phases. In the symmetric phase, we find a critical coupling of [15] $g=2.1\pm0.05$. In the broken phase, the critical coupling values extrapolate to $g=0.2\pm0.02$.

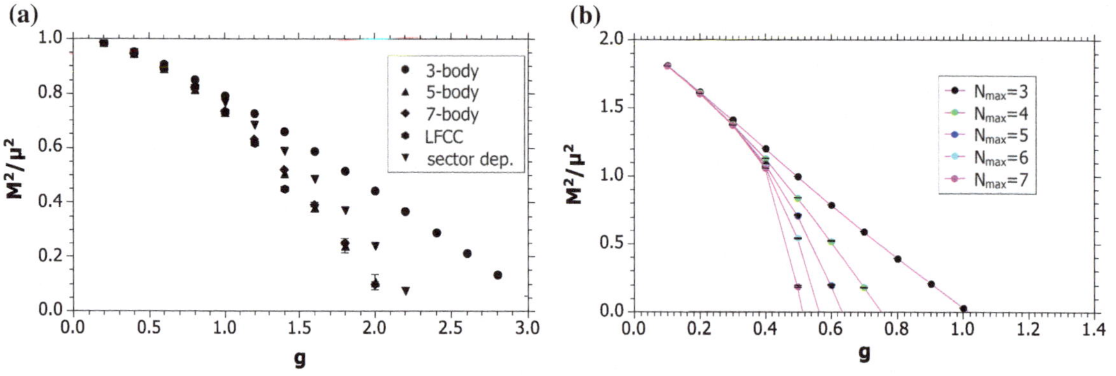

Fig. 1 Mass squared versus coupling strength for the **a** symmetric phase and **b** broken phase. The different Fock-space truncations in (**a**) are the three-body (*triangles*), five-body (*squares*), and seven-body (*diamonds*) Fock sectors. Results for the light-front coupled-cluster method [20] (*circles*) are also included. In (**b**), each set of points corresponds to a different Fock-space truncation to $N_{\max}$ constituents. The different truncations are the four-body (*triangles*), six-body (*squares*), and eight-body (*diamonds*) Fock sectors. *Error bars* are determined by the fits to extrapolation in the polynomial basis size

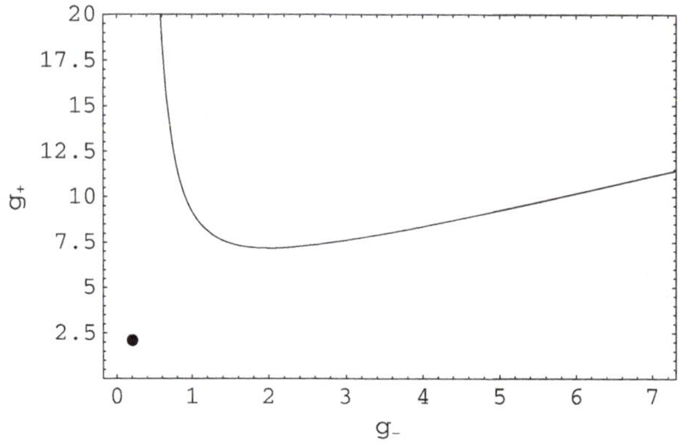

Fig. 2 The duality in couplings between the symmetric and broken phases. The *solid line* is the semi-classical duality of Chang [22]. The point corresponds to our numerical results

These critical coupling values can be compared to Chang's duality [6,17,21,22], which is obtained from the connection between normal orderings with respect to different masses [23]:

$$N_+[\phi^2] = N_-[\phi^2] + \frac{1}{4\pi}\ln\frac{\mu_+^2}{\mu_-^2}, \tag{17}$$

$$N_+[\phi^4] = N_-[\phi^4] + 6\frac{1}{4\pi}\ln\frac{\mu_+^2}{\mu_-^2}N_-[\phi^2] + 3\left(\frac{1}{4\pi}\ln\frac{\mu_+^2}{\mu_-^2}\right)^2. \tag{18}$$

The Hamiltonian density is then written as

$$\mathscr{H}^- = \left(\frac{1}{2}\mu_+^2 + \frac{\lambda}{4}\frac{1}{4\pi}\ln\frac{\mu_+^2}{\mu_-^2}\right)N_-[\phi^2] + \frac{\lambda}{4!}N_-[\phi^4] + \frac{1}{4\pi}\ln\frac{\mu_+^2}{\mu_-^2}\left(2\mu_+^2 + \frac{\lambda}{8}\frac{1}{4\pi}\ln\frac{\mu_+^2}{\mu_-^2}\right). \tag{19}$$

This is equivalent to $\mathscr{H}^-$ with negative mass squared if $\frac{1}{2}\mu_+^2 + \frac{\lambda}{4}\frac{1}{4\pi}\ln\frac{\mu_+^2}{\mu_-^2} = -\frac{1}{2}\mu_-^2$. For the dimensionless couplings $g_\pm \equiv \lambda/4\pi\mu_\pm^2$ this becomes $\frac{1}{g_+} - \frac{1}{2}\ln g_+ = -\frac{1}{2}\ln(g_-) - \frac{1}{g_-}$. The comparison is illustrated in Fig. 2.

The relative probabilities of the higher Fock sectors are readily computed from the Fock-state wave functions obtained in the eigenvector of the Hamiltonian matrix. A plot can be found in [16]. There is no indication of critical behavior at $g = 2.1$, where one would expect that the higher Fock sectors would dominate. A natural

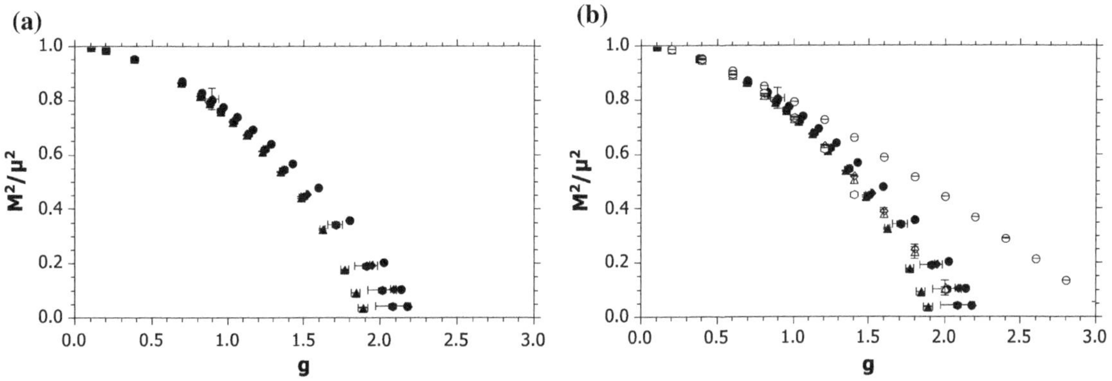

Fig. 3 Mass squared for the lowest eigenstate as a function of the dimensionless coupling g with **a** a sector-dependent constituent mass and **b** with both sector-dependent (*filled symbols*) and independent (*open symbols*). For (**a**), the Fock-space truncations are to a number of constituents $N_{\max} = 3$ (*circles*), 5 (*triangles*), 7 (*diamonds*), and 9 (*hex*); the *error bars* estimate the range of fits for the μ_1 extrapolations used to obtain g and M. In (**b**), the sector-independent, five and seven-constituent results are nearly identical with the nine-constituent sector-dependent results. Plot (**b**) also includes, as open hexagons, the results from a light-front coupled-cluster calculation [20]

assumption for why this might be happening is that the numerical calculation used the same constituent mass μ in every Fock sector, making the invariant mass of the N-constituent Fock sector of order $\sum_i^N \frac{\mu^2}{1/N} = N^2\mu^2$. The higher sectors are then suppressed by this large invariant mass.

This can be avoided by the use of a sector-dependent constituent mass [24–29]; our approach is described in [16]. The results for the lowest odd eigenstate are shown in Fig. 3. They are consistent with the sector-independent approach, and the estimate of the critical coupling remains unchanged at a value of 2.1. As can be seen in [16], the relative probabilities also remain nearly the same.

4 Summary

These calculations have revealed two inconsistencies in the light-front approach to ϕ^4 theory. One is the absence of a vacuum expectation value (VEV) above the critical coupling, where the $\phi \to -\phi$ symmetry is to be broken, and the presumably related behavior in the explicitly broken phase, where the vacuum expectation value remains nonzero above the critical coupling. A Gaussian effective potential analysis [30,31] can determine a VEV that does flip between zero and non-zero, but it does so discontinuously, which is inconsistent with the known second-order nature of the transition.

Another inconsistency is the smooth behavior of the computed Fock sector probabilities as the coupling is increased through the critical value, as presented in more detail in [16]. The relative probabilities for the sector-dependent and independent calculations are essentially the same in the three-body Fock sector. This indicates full convergence with respect to the Fock-space truncation. In Fock sectors with five and seven constituents, the relative probability for the sector-dependent case rises above the probability in the standard case as the critical coupling is approached. This greater probability is expected; however, the full expectation was that these probabilities would increase much more rapidly. The hypothesis, that a sector-dependent mass would reveal the critical behavior, must be incorrect. It seems likely that a coherent-state approach is needed, something that the light-front coupled cluster method can bring [20,32].

Acknowledgements This work was done in collaboration with S.S. Chabysheva and was supported in part by the Minnesota Supercomputing Institute of the University of Minnesota with allocations of computing resources.

References

1. S.J. Brodsky, H.-C. Pauli, S.S. Pinsky, Quantum chromodynamics and other field theories on the light cone. Phys. Rep. **301**, 299–486 (1998)
2. M. Burkardt, Light front quantization. Adv. Nucl. Phys. **23**, 1–74 (2002)

3. B.L.G. Bakker et al., Light-front quantum chromodynamics: a framework for the analysis of hadron physics. Nucl. Phys. Proc. Suppl. **251–252**, 165–174 (2014)
4. J.R. Hiller, Nonperturbative light-front Hamiltonian methods. Prog. Part. Nucl. Phys. **90**, 75–124 (2016)
5. J.S. Rozowsky, C.B. Thorn, Spontaneous symmetry breaking at infinite momentum without P^+ zero modes. Phys. Rev. Lett. **85**, 1614–1617 (2000)
6. V.T. Kim, G.B. Pivovarov, J.P. Vary, Phase transition in light-front ϕ^4_{1+1}. Phys. Rev. D **69**, 085008 (2004)
7. D. Chakrabarti, A. Harindranath, L. Martinovic, J.P. Vary, Kinks in discrete light cone quantization. Phys. Lett. B **582**, 196–202 (2004)
8. D. Chakrabarti, A. Harindranath, L. Martinovic, G.B. Pivovarov, J.P. Vary, Ab initio results for the broken phase of scalar light front field theory. Phys. Lett. B **617**, 92–98 (2005)
9. D. Chakrabarti, A. Harindranath, J.P. Vary, Transition in the spectrum of the topological sector of ϕ^4_2 theory at strong coupling. Phys. Rev. D **71**, 125012 (2005)
10. L. Martinovic, Spontaneous symmetry breaking in light front field theory. Phys. Rev. D **78**, 105009 (2008)
11. S.S. Chabysheva, J.R. Hiller, Basis of symmetric polynomials for many-boson light-front wave functions. Phys. Rev. E **90**, 063310 (2014)
12. S.S. Chabysheva, B. Elliott, J.R. Hiller, Symmetric multivariate polynomials as a basis for three-boson light-front wave functions. Phys. Rev. E **88**, 063307 (2013)
13. H.-C. Pauli, S.J. Brodsky, Solving field theory in one space and one time dimension. Phys. Rev. D **32**, 1993–2000 (1985)
14. H.-C. Pauli, S.J. Brodsky, Discretized light-cone quantization: solution to a field theory in one space and one time dimension. Phys. Rev. D **32**, 2001–2013 (1985)
15. M. Burkardt, S.S. Chabysheva, J.R. Hiller, Two-dimensional light-front ϕ^4 theory in a symmetric polynomial basis. Phys. Rev. D **94**, 065006 (2016)
16. S.S. Chabysheva, Light-front versus equal-time quantization in ϕ^4 theory, in *these proceedings* (2016)
17. A. Harindranath, J.P. Vary, Variational calculation of the spectrum of two-dimensional ϕ^4 theory in light-front field theory. Phys. Rev. D **37**, 3010–3013 (1988)
18. W.H. Press, S.A. Teukoksky, W.T. Vetterling, B.P. Flannery, *Numerical Recipes: The Art of Scientific Computing*, 3rd edn. (Cambridge University Press, Cambridge, 2007)
19. K.G. Wilson, *Ab initio* quantum chemistry: a source of ideas for lattice gauge theorists. Nucl. Proc. B (Proc. Suppl.) **17**, 82–92 (1990)
20. B. Elliott, S.S. Chabysheva, J.R. Hiller, Application of the light-front coupled-cluster method to ϕ^4 theory in two dimensions. Phys. Rev. D **90**, 056003 (2014)
21. S.J. Chang, Quantum fluctuations in a ϕ^4 field theory. 1. The stability of the vacuum. Phys. Rev. D **12**, 1071–1088 (1975)
22. S.J. Chang, The existence of a second order phase transition in the two-dimensional ϕ^4 field theory. Phys. Rev. D **13**, 2778–2788 (1976)
23. S. Coleman, Quantum sine-Gordon equation as the massive Thirring model. Phys. Rev. D **11**, 2088–2097 (1975)
24. R.J. Perry, A. Harindranath, K.G. Wilson, Light front Tamm-Dancoff field theory. Phys. Rev. Lett. **65**, 2959–2962 (1990)
25. R.J. Perry, A. Harindranath, Renormalization in the light front Tamm-Dancoff approach to field theory. Phys. Rev. D **43**, 4051–4073 (1991)
26. J.R. Hiller, S.J. Brodsky, Nonperturbative renormalization and the electron's anomalous moment in large-α QED. Phys. Rev. D **59**, 016006 (1998)
27. V.A. Karmanov, J.-F. Mathiot, A.V. Smirnov, Systematic renormalization scheme in light-front dynamics with Fock space truncation. Phys. Rev. D **77**, 085028 (2008)
28. V.A. Karmanov, J.-F. Mathiot, A.V. Smirnov, Nonperturbative calculation of the anomalous magnetic moment in the Yukawa model within truncated Fock space. Phys. Rev. D **82**, 056010 (2010)
29. S.S. Chabysheva, J.R. Hiller, On the nonperturbative solution of Pauli-Villars regulated light-front QED: a comparison of the sector-dependent and standard parameterizations. Ann. Phys. **325**, 2435–2450 (2010)
30. P.M. Stevenson, Gaussian effective potential II: $\lambda\phi^4$ field theory. Phys. Rev. D **32**, 1389–1408 (1985)
31. G.B. Pivovarov, Gaussian effective potential in light-front ϕ^4_{1+1} (2005). arXiv:hep-th/0501051
32. S.S. Chabysheva, J.R. Hiller, A light-front coupled-cluster method for the nonperturbative solution of quantum field theories. Phys. Lett. B **711**, 417–422 (2012)

Few-Body Syst (2017) 58:120
DOI 10.1007/s00601-017-1277-3

Elżbieta Dzimida-Chmielewska · Jerzy A. Przeszowski

Physical and Nonphysical Modes in the Light Front Formulation for the LC Gauge and the Lorentz Gauge Conditions

Received: 14 November 2016 / Accepted: 7 March 2017 / Published online: 23 March 2017

Abstract The models with free vector fields are analyzed on the light-front hypersurface. The massive vector model—Proca model—can be quantized unambiguously within the novel LF procedure, though it leads to singular terms in x^+ coordinate. Also the modified models, where the Lorentz condition and the LC gauge conditions are explicitly induced by the Lagrange multipliers, are discussed. Possible massless limits are studied for all cases and modes are classified as either physical or nonphysical. A practical definition of physical modes by properties of the 1-particle states is proposed.

1 Introduction

There is a strange idea that for the LF quantization of QED and QCD in the LC gauge $A_- = 0$ one needs to implement additionally the Lorentz condition $\partial^\mu A_\mu = 0$, which means two simultaneous gauge fixing conditions. The outcome of this approach is the Feynman propagator in the momentum representation with three different terms [1]. We would like to use the novel LF quantization procedure, which has been recently proposed for scalar field [2], to check this situation for vector gauge fields. Unfortunately it is quite difficult to study the LC gauge $A_- = 0$ directly, therefore here we make a preliminary analysis with the massive vector field models.

2 Massive Vector Fields: Proca Theory

We begin our analysis with the Proca model [3], where the Lagrangian density is

$$\mathscr{L}_{Proca} = -\frac{1}{4} B_{\mu\nu} B^{\mu\nu} + \frac{m^2}{2} B^2, \quad B_{\mu\nu} = \partial_\mu B_\nu - \partial_\nu B_\mu \tag{1}$$

and m is the mass of all vector field components. This model can be quantized on the equal-time hypersurface and the vacuum expectation value of fields at arbitrary points [4] is

$$\langle 0| \left[B_\mu(x), B_\nu(y) \right] |0\rangle = i \left(g_{\mu\nu} + 1/m^2 \partial_\mu \partial_\nu \right) \Delta(x-y) \tag{2}$$

This article belongs to the Topical Collection "Light Cone 2016".

E. Dzimida-Chmielewska (✉)· J. A. Przeszowski
Faculty of Physics, University of Białystok, ul. Ciołkowskiego 4L, 15-424 Białystok, Poland
E-mail: edzimida@uwb.edu.pl

J. A. Przeszowski
E-mail: j.przeszowski@uwb.edu.pl

with $\Delta(x-y)$ being the Pauli–Villars function for a massive scalar field. We expect that the above commutator function should appear also in the LF procedure. However the terms with $\partial_+\Delta(x-y)$ and $\partial_+^2\Delta(x-y)$ have no limit at the LF hypersurface, as a result of the Lorentz symmetry [5]. Accordingly one cannot define the LF classical Proca field theory at the LF hypersurface. Thus the canonical LF quantization procedure [6,7], which is based on the Dirac procedure for constrained systems [8,9], cannot be applied here [10]. Fortunately the novel LF quantization procedure [2] omits these difficulties and beneath we will implement it without going into many technical details, which will be presented elsewhere.

We start our analysis with the Proca Lagrangian density (1) rewritten explicitly in the LC coordinates and components as

$$\mathcal{L}=\frac{1}{2}\left(\partial_+B_--\partial_-B_+\right)^2-\frac{1}{4}\left(\partial_iB_j-\partial_jB_i\right)^2+\left(\partial_+B_i-\partial_iB_+\right)\left(\partial_-B_i-\partial_iB_-\right)+m^2B_+B_--\frac{m^2}{2}B_i^2. \quad (3)$$

It leads directly to the LF canonical conjugated momenta $\pi^+=0$, $\pi^-=\partial_+B_--\partial_-B_+$, $\pi^i=\partial_-B_i-\partial_iB_-$. Usually one argues that there are two primary constraints $\pi^+\approx 0$ and $\pi^i-B_i+\partial_iB_-\approx 0$, therefore one needs to take the Dirac procedure [8,9]. However in the novel LF formulation one does not introduce such classification of constraints and takes all non-vanishing canonical momenta π^i and π^- for building the component T^{++} of the canonical energy–momentum tensor

$$T^{++}_{can}=\pi^-\partial_-B_-+\pi^i\partial_-B_i=\pi^-\partial_-B_-+\left(\partial_-B_i-\partial_iB_-\right)\partial_-B_i. \quad (4)$$

Before imposing the novel quantization rules, we need to select independent fields in the LF formulation, by inspecting their equations of motion. The Euler–Lagrange equations which follow from the Lagrangian (3) are $(\Box+m^2)B_\mu=\partial_\mu(\partial\cdot B)$, where $\Box=\partial^\mu\partial_\mu=2\partial_+\partial_--\partial_i^2$ and $\partial\cdot B=\partial^\mu B_\mu$. However these equations can be equivalently rewritten as $(\Box+m^2)B_\mu=0$, $\partial^\mu B_\mu=0$, thus all components B_μ satisfy the Klein–Gordon equation of motion, though only three of them are independent modes due to the Lorentz condition. Seemingly, B_- should be taken as one of them, since its conjugated canonical momentum $\pi^-=\partial_+B_--\partial_-B_+$, contains temporal derivative (velocity). However the closer examination of the Euler–Lagrange shows that there are two constraint equations

$$\left(\Delta_\perp-m^2\right)B_-=\partial_-\left(\pi^-+\partial_iB_i\right), \qquad 2\partial_-B_+=\partial_iB_i-\pi^- \quad (5)$$

Since the partial differential operator $(\Delta_\perp-m^2)$ is invertible, then B_- can be unambiguously determined in terms of π^- and B_i fields. Further for avoiding the non-local terms we may introduce new fields Σ and C_i by the differential equations $(\Delta_\perp-m^2)\Sigma=\left(\pi^-+\partial_iB_i\right)$ and $C_i=A_i-\partial_i\Sigma$, respectively. The best choice of three independent fields is Σ and C_i, which satisfy the Klein–Gordon equations $(2\partial_+\partial_--\Delta_\perp+m^2)\Sigma=0$, $(2\partial_+\partial_--\Delta_\perp+m^2)C_i=0$. The constraint equations

$$2\partial_-B_+=2\partial_iC_i+\left(\Delta_\perp+m^2\right)\Sigma, \quad B_i=C_i+\partial_i\Sigma, \quad B_-=\partial_-\Sigma, \quad (6)$$

express B_μ fields in terms of independent fields Σ and C_i, though for B_+ one needs to solve the partial differential equation.[1] Then for new fields we rewrite T^{++}_{can} as $T^{++}_{can}\approx(\partial_-C_i)^2+m^2(\partial_-\Sigma)^2$, where, approximate equality, symbolized by a squiggly equal sign $\approx$, means that terms, which are total partial derivatives, have been omitted. The LF quantization is performed at the arbitrary fixed x^+ coordinate, but for the clarity we just take $x^+=0$ and denote $\bar{x}=(x^-,\boldsymbol{x}_\perp)$.

We postulate as the quantization rule on the LF hypersurface $x^+=0$ the following relations for the independent fields C_i, Σ

$$i\partial_-C_i(\bar{x})=\left[C_i(\bar{x}),P_-\right], \quad i\partial_-\Sigma(\bar{x})=\left[\Sigma(\bar{x}),P_-\right], \quad (7)$$

where the generator of translation in x^- direction is defined as $P_-=\int_{\mathbb{R}^3}d^3\bar{x}\,T^{++}_{can}(\bar{x})$. These relations can be identically satisfied provided the following commutators hold at the LF hypersurface

$$\left[C_i(\bar{x}),\partial_-C_j(\bar{y})\right]=\frac{i}{2}\delta_{ij}\delta^3(\bar{x}-\bar{y}), \quad \left[\Sigma(\bar{x}),\partial_-\Sigma(\bar{y})\right]=\frac{i}{2m^2}\delta^3(\bar{x}-\bar{y}), \quad \left[C_i(\bar{x}),\partial_-\Sigma(\bar{y})\right]=0. \quad (8)$$

[1] We will not introduce the inverse operator $[\partial_-]^{-1}$, which needs the boundary conditions in $\pm\infty$ in x^- direction. Instead later we will solve this equation in the momentum space, which seems to be more physical.

Next we take the vacuum expectation value of these commutators, with the normalized vacuum state $\langle 0|0\rangle = 1$ and further we decompose them into the matrix elements. For achieving this goal we need to introduce smooth fields $C_i[f]$ and $\Sigma[f]$ by smearing with a real-valued test function $f(\bar{x}) \in \mathscr{S}'(\mathbb{R}^3)$, where we write $C_i[f] = \int_{\mathbb{R}^3} d^3\bar{x}\, C_i(\bar{x})\, f(\bar{x})$, $\Sigma[f] = \int_{\mathbb{R}^3} d^3\bar{x}\, \Sigma(\bar{x})\, f(\bar{x})$.[2]

Then we introduce another LF assumption, that the smooth operators $C_i[f]$ and $\Sigma[f]$ when acting on the vacuum state $|0\rangle$ produce 1-particle states,

$$C_i[f]|0\rangle = \int_{\mathbb{R}^3} d\Gamma(\bar{k}) \tilde{f}(\bar{k})\, |\bar{k}, c_i\rangle, \quad \Sigma[f]|0\rangle = \int_{\mathbb{R}^3} d\Gamma(\bar{k}) \tilde{f}(\bar{k})\, |\bar{k}, \sigma\rangle, \quad d\Gamma(\bar{k}) = \frac{d^3\bar{k}}{(2\pi)^3} \frac{\Theta(k^+)}{2k^+}, \tag{9}$$

where $\tilde{f}(\bar{k})$ is the Fourier transform of $f(\bar{x})$, while $|\bar{k}, c_i\rangle$ and $|\bar{k}, \sigma\rangle$ are the 1-particle eigenstates of the kinematical translation generators $\bar{P}_\mu = (P_-, P_i)$.[3] The calculation directly leads to the non-vanishing matrix elements $\langle 0|\Sigma(0)|\bar{k}, \sigma\rangle = 1/m^2$ and $\langle 0|C_i(0)|\bar{k}, c_j\rangle = \delta_{ij}$ which further allows to determine the non-vanishing inner products for 1-particle eigenstates of kinematical momenta $\langle \bar{k}, \sigma|\bar{p}, \sigma\rangle = 1/m^2\, \delta_\Gamma(\bar{k}, \bar{p})$, $\langle \bar{k}, c_i|\bar{p}, c_j\rangle = \delta_{ij}\delta_\Gamma(\bar{k}, \bar{p})$, where $\delta_\Gamma(\bar{k}, \bar{p}) = (2\pi)^3 2k^+ \delta^2(\boldsymbol{k}_\perp - \boldsymbol{p}_\perp)\delta(k^+ - p^+)$.

Then we introduce equations for the quantum field operators in the form identical to the classical Euler–Lagrange equations, since for a free field model there is no need for the renormalization effects and counter-terms. We need to express all partial derivatives in terms of commutators with respective translation generators—for example $i\partial_\mu C_i(x) = \left[C_i(x), P_\mu\right]$. For the second order partial derivatives this leads to double commutators, which will simplify considerably if it acts on the vacuum state, for example we have, $\partial_+\partial_- C_i(x) = -\left[\left[C_i(x), P_-\right], P_+\right]$ and $\partial_+\partial_- C_i(x)|0\rangle = -P_- P_+ C_i(x)|0\rangle$, since $P_\mu|0\rangle = 0$. Accordingly for the independent smooth dynamical fields $\Sigma[f]$ and $C_i[f]$ we obtain the following dynamical LF Schrödinger equations

$$\left(2P_+P_- - P_\perp^2 - m^2\right)\Sigma[f]|0\rangle = 0, \quad \left(2P_+P_- - P_\perp^2 - m^2\right)C_i[f]|0\rangle = 0, \tag{10}$$

which then lead to the eigenvalue equations for the LF Hamiltonian

$$P_+|\bar{k}, \sigma\rangle = \frac{m^2 + k_\perp^2}{2k^+}|\bar{k}, \sigma\rangle \quad P_+|\bar{k}, c_i\rangle = \frac{m^2 + k_\perp^2}{2k^+}|\bar{k}, c_i\rangle, \qquad k^+ > 0. \tag{11}$$

Accordingly the 1-particle states $|\bar{k}, \sigma\rangle$ and $|\bar{k}, c_i\rangle$ are the eigenfunctions (eigenvectors) for all components of momentum operators P_μ. Then we may proceed analogously with other equations and we obtain the non-dynamical LF Schrödinger equations

$$B_-[f]|0\rangle = iP_-\Sigma[f]|0\rangle, \qquad B_j[f]|0\rangle = C_j[f]|0\rangle + iP_j\Sigma[f]|0\rangle, \tag{12a}$$

$$P_-B_+[f]|0\rangle = P_jC_j[f]|0\rangle + i\frac{P_\perp^2 - m^2}{2}\Sigma[f]|0\rangle. \tag{12b}$$

Equation (12b) can be transformed, using the Klein–Gordon equation in the momentum space $(2P_+P_- - P_\perp^2 - m^2)B_+[f] = 0$ into a form, which unambiguously allows for a unique solution

$$(P_\perp^2 + m^2)B_+[f]|0\rangle = 2P_+\left(P_jC_j[f]|0\rangle + i\frac{P_\perp^2 - m^2}{2}\Sigma[f]|0\rangle\right). \tag{12c}$$

All this allows to express $B_\mu[f]|0\rangle$ in its generic form

$$B_\mu[f]|0\rangle = \int_{\mathbb{R}^3} d\Gamma(\bar{k})\tilde{f}(\bar{k}) \sum_{a=1,2,3} \epsilon_\mu^a(\bar{k})|\bar{k}, a\rangle, \tag{13a}$$

[2] These smooth fields are very useful for inferring the commutators (8) from the quantization rules (7).

[3] One may derive this assumption from the nonnegative spectrum of $P_- = P^+$ operator and the Lorentz symmetry of vacuum state $P_\mu|0\rangle = 0$.

where we have introduced new 1-particle eigenfunctions of 4-momenta $|\bar{k},1\rangle = |\bar{k},c_1\rangle$, $|\bar{k},2\rangle = |\bar{k},c_2\rangle$ and $|\bar{k},3\rangle = im|\bar{k},\sigma\rangle$, which have the orthonormal positive inner products $\langle\bar{p},a|\bar{k},b\rangle = \delta_{ab}\,\delta_\Gamma(\bar{p},\bar{k})$, $a,b\in\{1,2,3\}$. The the polarization vectors $\epsilon^a_\mu(\bar{k})$ have the form of one-column matrix[4]

$$\epsilon^1_\mu(\bar{k}) = \frac{1}{k^+}\begin{bmatrix} k_1 \\ 0 \\ k^+ \\ 0 \end{bmatrix}, \quad \epsilon^2_\mu(\bar{k}) = \frac{1}{k^+}\begin{bmatrix} k_2 \\ 0 \\ 0 \\ k^+ \end{bmatrix}, \quad \epsilon^3_\mu(\bar{k}) = \frac{1}{m}\begin{bmatrix} \frac{k_\perp^2-m^2}{2k^+} \\ k^+ \\ k_1 \\ k_2 \end{bmatrix}. \tag{14}$$

Any 1-particle state, which is simultaneously an eigenvector of all Poincare generators P_μ with energy eigenvalue consistent with the Lorentz symmetry and its inner product is positively definite, describes a physical mode. Otherwise it is a nonphysical mode. Our previous analysis evidently shows that there are three physical modes in the Proca model quantized at the LF hypersurface. However the polarization vector ϵ^3_μ contains the overall factor $1/m$, thus it has no massless limit $m\to 0$. Therefore we cannot directly move from the massive Proca fields into the gauge fields, though we have the Lorentz condition $\partial^\mu B_\mu = 0$ already for nonzero mass. For proving the consistency of our result with the usual quantization we may calculate the 2-point Wightman function (the vacuum expectation value of unordered product of two quantum field operators) for arbitrary points

$$\langle 0|B_\mu(x)\,B_\nu(y)|0\rangle = -\left(g_{\mu\nu} + \frac{1}{m^2}\partial_\mu\partial_\nu\right)\int_{\mathbb{R}^3} d\Gamma(\bar{k})e^{-i\bar{k}\cdot(\bar{x}-\bar{y})}e^{-i\frac{m^2+k_\perp^2}{2k^+}(x^+-y^+)}, \tag{15}$$

which agrees with the commutator function (2) since in the LF momentum representation

$$\Delta(x) = i\int_{\mathbb{R}^3} d\Gamma(\bar{k})\left[e^{i\bar{k}(\bar{x}-\bar{y})}e^{i\frac{m^2+k_\perp^2}{2k^+}(x^+-y^+)} - e^{-i\bar{k}(\bar{x}-\bar{y})}e^{-i\frac{m^2+k_\perp^2}{2k^+}(x^+-y^+)}\right]. \tag{16}$$

Thus we have been able to omit all problems which appear in the coordinate space formulation, where terms singular at $x^+=0$ appear.

3 The Modified Proca Model with the Lorentz Condition

Now we introduce a modification to the Proca model by adding the Lorentz condition $\partial\cdot B = 0$, by means of the Lagrange multiplier field Λ, into the Lagrangian density (3). This will slightly change the Euler–Lagrange equations $(\Box + m^2)B_\mu = \partial_\mu\Lambda$, $\partial^\mu B_\mu = 0$ and the canonical momentum $\pi^- = \partial_+B_- - \partial_-B_+ + \Lambda$. Also the equation for a new field is modified $(\Delta_\perp - m^2)\Sigma = \left(\pi^- + \partial_i B_i - \Lambda\right)$ and the energy–momentum tensor component is $T^{++} = (\partial_-C_i)^2 + m^2(\partial_-\Sigma)^2 - 2\partial_-\Sigma\partial_-\Lambda$. Then the independent fields are C_i, Σ, Λ with dynamical equations $(\Box + m^2)C_i = 0$, $(\Box + m^2)\Sigma = \Lambda$, $\Box\Lambda = 0$.

Following steps analogous to those from previous section, we find non-vanishing LF commutators for independent fields

$$\left[C_i(\bar{x}), \partial_-C_j(\bar{y})\right] = \frac{i}{2}\delta_{ij}\,\delta^3(\bar{x}-\bar{y}), \qquad \left[\Lambda(\bar{x}), \partial_-\Lambda(\bar{y})\right] = -\frac{i}{2}m^2\delta^3(\bar{x}-\bar{y}), \tag{17a}$$

$$\left[\Lambda(\bar{x}), \partial_-\Sigma(\bar{y})\right] = -\frac{i}{2}\,\delta^3(\bar{x}-\bar{y}), \qquad \left[\Sigma(\bar{x}), \partial_-\Lambda(\bar{y})\right] = -\frac{i}{2}\,\delta^3(\bar{x}-\bar{y}). \tag{17b}$$

and the 1-particle states with their non-vanishing inner products

$$\langle\bar{k},\lambda|\bar{p},\lambda\rangle = -m^2\delta_\Gamma(\bar{k},\bar{p}), \quad \langle\bar{k},\sigma|\bar{p},\lambda\rangle = -\delta_\Gamma(\bar{k},\bar{p}), \quad \langle\bar{k},c_i|\bar{p},c_j\rangle = \delta_{ij}\delta_\Gamma(\bar{k},\bar{p}), \tag{18}$$

The dynamical equations give rise to the dynamical LF Schrödinger equations

$$2P_+k^+|\bar{k},c_i\rangle = (k_\perp^2+m^2)|\bar{k},c_i\rangle, \quad 2P_+k^+|\bar{k},\lambda\rangle = k_\perp^2\,|\bar{k},\lambda\rangle, \quad 2P_+k^+|\bar{k},\sigma\rangle = (k_\perp^2+m^2)\,|\bar{k},\sigma\rangle - |\bar{k},\lambda\rangle, \tag{19}$$

[4] We denote $\epsilon^a_+(\bar{k})$ in the first row, then $\epsilon^a_-(\bar{k})$ in the second row, then $\epsilon^a_1(\bar{k})$ in the third row and at last $\epsilon^a_2(\bar{k})$ in the fourth row.

which show that the 1-particle state $|\bar{k},\sigma\rangle$ is not an eigenfunction of the LF Hamiltonian P_+. From non-dynamical equations $B_- = \partial_- \Sigma$, $2\partial_- B_+ = 2\partial_i C_i + (\Delta_\perp + m^2)\Sigma - \Lambda$, $B_i = C_i + \partial_i \Sigma$, one finds the non-dynamical Schrödinger equations, which lead to

$$B_-[f]|0\rangle = \int_{\mathbb{R}^3} d\Gamma(\bar{k})\tilde{f}(\bar{k})(ik^+)|\bar{k},\sigma\rangle, \quad B_i[f]|0\rangle = \int_{\mathbb{R}^3} d\Gamma(\bar{k})\tilde{f}(\bar{k})\left(|\bar{k},c_i\rangle + ik_i|\bar{k},\sigma\rangle\right), \tag{20a}$$

$$B_+[f]|0\rangle = \int_{\mathbb{R}^3} d\Gamma(\bar{k})\tilde{f}(\bar{k})\left(\frac{k_i}{k^+}|\bar{k},c_i\rangle + i\frac{k_\perp^2 - m^2}{2k^+}|\bar{k},\sigma\rangle + \frac{i}{2k^+}|\bar{k},\lambda\rangle\right). \tag{20b}$$

We may define new 1-particle states $|\bar{k},0\rangle = \frac{i}{m}|\bar{k},\lambda\rangle$, $|\bar{k},3\rangle = im|\bar{k},\sigma\rangle - \frac{i}{m}|\bar{k},\lambda\rangle$, $|\bar{k},1\rangle = |\bar{k},c_1\rangle$, $|\bar{k},2\rangle = |\bar{k},c_2\rangle$, which are the eigenstates of the Hamiltonian operator P_+, for $k^+ > 0$,

$$P_+|\bar{k},a\rangle = \frac{k_\perp^2 + m^2}{2k^+}|\bar{k},a\rangle, \quad P_+|\bar{k},0\rangle = \frac{k_\perp^2}{2k^+}|\bar{k},0\rangle, \tag{21}$$

where $a \in \{1,2,3\}$. The non-vanishing inner product for these states $\langle\bar{p},a|\bar{k},b\rangle = \delta_{ab}\,\delta_\Gamma(\bar{p},\bar{k})$ and $\langle\bar{p},0|\bar{k},0\rangle = -\delta_\Gamma(\bar{p},\bar{k})$. Therefore there are 3 physical modes with mass m, described by the 1-particle states $|\bar{k},a\rangle$ with $a \in \{1,2,3\}$ and one unphysical massless mode, described by $|\bar{k},0\rangle$.

For the case $m \neq 0$ we find the smooth fields acting on the vacuum state

$$B_\mu[f]|0\rangle = \int_{\mathbb{R}^3} d\Gamma(\bar{k})\tilde{f}(\bar{k})\sum_{a=1,2,3}\epsilon_\mu^a(\bar{k})|\bar{k},a\rangle + \frac{1}{m}P_\mu\int_{\mathbb{R}^3} d\Gamma(\bar{k})\tilde{f}(\bar{k})|\bar{k},0\rangle, \tag{22}$$

where polarization vectors are the same as in the Proca model (14). There is a cancellation of $1/m$ contributions between $|\bar{k},3\rangle$ and $|\bar{k},0\rangle$ states, which allows for the massless limit $m \to 0$. We denote this limit by $B_\mu \to A_\mu$ and we find

$$A_\mu[f]|0\rangle = \int_{\mathbb{R}^3} d\Gamma(\bar{k})\tilde{f}(\bar{k})\sum_{a=1,2,3}\epsilon_\mu^a(\bar{k})|\bar{k},a\rangle + P_\mu\int_{\mathbb{R}^3} d\Gamma(\bar{k})\tilde{f}(\bar{k})\frac{1}{k_\perp}|\bar{k},0\rangle, \tag{23}$$

where now the 1-particle states are slightly redefined $|\bar{k},3\rangle = \frac{i}{k_\perp}|\bar{k},\lambda\rangle$, $|\bar{k},0\rangle = ik_\perp|\bar{k},\sigma\rangle$, with the non-vanishing inner product for them $\langle\bar{k},0|\bar{p},3\rangle = \langle\bar{k},3|\bar{p},0\rangle = -\delta_\Gamma(\bar{k},\bar{p})$. Here we stress that $|\bar{k},3\rangle$ is the eigenfunction of the LF Hamiltonian P_+, while $|\bar{k},0\rangle$ is not

$$P_+|\bar{k},3\rangle = \frac{k_\perp^2}{2k^+}|\bar{k},3\rangle, \quad P_+|\bar{k},0\rangle = \frac{k_\perp^2}{2k^+}|\bar{k},0\rangle - \frac{k_\perp^2}{2k^+}|\bar{k},3\rangle. \tag{24}$$

Therefore for the massless limit there are only 2 physical modes described by the 1-particle states $|\bar{k},a\rangle$ with $a \in \{1,2\}$ and 2 nonphysical modes described by the 1-particle states $|\bar{k},0\rangle$ and $|\bar{k},3\rangle$. In Eq. (23) the polarization vectors are slightly modified, $\epsilon_\mu^{1,2}$ remain unchanged as in Eq. (14), while $\epsilon_+^3(\bar{k}) = \frac{k_\perp}{k^+}$, $\epsilon_-^3(\bar{k}) = 0$, $\epsilon_1^3(\bar{k}) = 0$, $\epsilon_2^3(\bar{k}) = 0$.

If one tries to impose the additional restriction for the gauge field potential $A_- = 0$ one evidently removes the unphysical modes $|\bar{k},0\rangle$ and $|\bar{k},3\rangle$.

4 Modified Proca Model with LC Weyl Condition

It is very difficult to perform the LF quantization for the LC gauge $B_- = 0$. Therefore we may check a simpler possibility by imposing the LC Weyl condition $B_+ = 0$. This gauge has been discussed in [11] within the DLCQ quantization and in [12] within the continuous formulation. Here we impose $B_+ = 0$ by means of the Lagrange multiplier field Λ into the Lagrangian density (3). This leads to the Euler–Lagrange equations which are all dynamical $\partial_+\pi^- = 0$, $\partial_+ B_- = \pi^- + \partial_i B_i$, $(2\partial_+\partial_- - \Delta_\perp)B_i = \partial_i\pi^-$, while the canonical energy–momentum tensor component is $T^{++} = (\partial_- B_i)^2 + \pi^-\partial_- B_-$. Following steps analogous to those in previous sections we find the commutators for independent fields are

$$\left[B_-(\bar{x}),\pi^-(\bar{y})\right] = i\delta^3(\bar{x}-\bar{y}), \qquad 2\left[B_i(\bar{x}),\partial_- B_j(\bar{y})\right] = i\delta_{ij}\delta^3(\bar{x}-\bar{y}), \tag{25}$$

Between the 1-particle states, which appear naturally, the non-vanishing inner products $\langle\bar{k}, a_-|\bar{p}, \pi^-\rangle = 2ik^+\delta_\Gamma(\bar{k}, \bar{p})$ and $\langle\bar{k}, a_i|\bar{p}, a_j\rangle = \delta_{ij}\delta_\Gamma(\bar{k}, \bar{p})$. After the redefinitions of 1-particle states

$$|\bar{k}, 3\rangle = |\bar{k}, a_-\rangle - \frac{2k^+}{k_\perp^2 + m^2} k_j |\bar{k}, a_j\rangle, \quad |\bar{k}, j\rangle = |\bar{k}, a_j\rangle - i\frac{k_j}{k_\perp^2 + m^2}|\bar{k}, \pi\rangle, \quad |\bar{k}, 0\rangle = \frac{1}{2k^+}|\bar{k}, \pi^-\rangle, \tag{26}$$

we find that $|\bar{k}, i\rangle$ and $|\bar{k}, 0\rangle$ are the eigenfunctions of the LF Hamiltonian P_+, while $|\bar{k}, 3\rangle$ is not. Therefore there are 2 physical modes described by the 1-particle states $|\bar{k}, i\rangle$ with $i \in \{1, 2\}$ and 2 nonphysical modes described by the 1-particle states $|\bar{k}, 3\rangle$ and $|\bar{k}, 0\rangle$. Further we find the expression

$$B_\mu[f]|0\rangle = \int_{\mathbb{R}^3} d\Gamma(\bar{k}) \widetilde{f}(\bar{k}) \sum_{a=0,1,2,3} \epsilon_\mu^a(\bar{k})\,|\bar{k}, c_a\rangle, \tag{27}$$

with the polarization vectors given in (28). The massless limit $m \to 0$ can be safely taken and the conclusion that there are 2 physical modes and 2 nonphysical modes remains unchanged.

When one adds the additional condition on the gauge field $\partial^\mu B_\mu = 0$, which in the momentum space takes the form $P_+ B_{[}f]|0\rangle = P_i B_i[f]|0\rangle$, then one removes the nonphysical modes.

5 Conclusions and Prospects for Further Research

In this paper we have considered the gauge field model, where a mass term is added as a regularization of the possible infra-red singularities. These massive vector fields can be quantized on the LF hypersurface within the novel quantization procedure. For the Proca model we have found 3 physical modes, all with mass m. Then we have introduced additional conditions on the vector fields. The Lorentz condition $\partial^\mu B_\mu = 0$ leads to 3 physical modes with mass m and 1 nonphysical massless modes. In the massless limit $m \to 0$, one of physical mode is no longer described by an eigenvector of the LF Hamiltonian thus we conclude that for massless fields there are 2 physical and 2 nonphysical modes, as expected. When we introduce the LF Weyl condition $B_+ = 0$, then the LF quantization leads to 2 physical modes, with mass m and 2 nonphysical modes. This classification survives the massless limit.

If we add additional conditions then we see that some nonphysical modes needs to be discarded. Usually nonphysical modes are needed for reaching a smooth behavior of the Feynman propagator. But for a decisive conclusion one needs to quantize at the LF hypersurface the gauge field system with 2 conditions from the beginning. This will be analyzed in the nearest future.

Appendix: Notation and Formulas

We use the covariant notation, where the LF longitudinal coordinates are defined as $x^\pm = \left(x^0 \pm x^3\right)/\sqrt{2}$ and the partial derivatives are denoted as $\partial_\pm = \partial/\partial x^\pm$. The Minkowski space–time metric tensor has non-vanishing components $\eta_{+-} = \eta_{-+} = 1$, $\eta_{ij} = -\delta_{ij}$, $(i, j) \in \{1, 2\}$. The summation over repeated indexes is always understood.

The massive vector field components, we denote by $B_\mu = (B_+, B_-, B_i)$ with $i \in \{1, 2\}$, for keeping it distinct from the massless vector field, which is commonly described by A_μ.

The polarization vectors for the LF Weyl gauge condition are

$$\epsilon_\mu^0(\bar{k}) = \frac{i\gamma}{k_\perp}\begin{bmatrix} 0 \\ \gamma k_\perp \\ k_1 \\ k_2 \end{bmatrix} \quad \epsilon_\mu^1(\bar{k}) = \frac{1}{k_\perp}\begin{bmatrix} 0 \\ \gamma k_1 \\ 0 \\ k_\perp \\ 0 \end{bmatrix} \quad \epsilon_\mu^2(\bar{k}) = \frac{1}{k_\perp}\begin{bmatrix} 0 \\ \gamma k_2 \\ 0 \\ 0 \\ k_\perp \end{bmatrix} \quad \epsilon_\mu^3(\bar{k}) = \begin{bmatrix} 0 \\ 1 \\ 0 \\ 0 \end{bmatrix} \tag{28}$$

with $\gamma = \frac{2k^+k_\perp}{k_\perp^2+m^2}$, where we denote $\epsilon_+^a(\bar{k})$ in the first row, then $\epsilon_-^a(\bar{k})$ in the second row, then $\epsilon_1^a(\bar{k})$ in the third row and at last $\epsilon_2^a(\bar{k})$ in the fourth row.

References

1. P.P. Srivastava, S.J. Brodsky, Light-front-quantized QCD in light-cone gauge: the doubly transverse gauge propagator. Phys. Rev. D **64**, 045006 (2001)
2. J.A. Przeszowski, Alternate light front quantization procedure for scalar fields. Few-Body Syst. **58**, 102 (2017)
3. A. Proca, Sur la théorie ondulatoire positifs et négativis. J. Phys. Radium. **7**, 347–353 (1936)
4. C. Itzykson, J.-B. Zuber, *Quantum Field Theory* (Dover Publication Inc., Mineola, 1980)
5. J.A. Przeszowski, Lorentz symmetry for the light-front Wightman functions. Acta Phys. Pol. Proc. Suppl. B **6**, 327–333 (2013)
6. M. Burkardt, Light front quantization. Adv. Nucl. Phys. **23**, 1–74 (1996)
7. S.J. Brodsky, H.-C. Pauli, S. Pinsky, Quantum chromodynamics and other field theories on the light cone. Phys. Lett. C (Phys. Rep.) **301**, 299–486 (1998)
8. P.A.M. Dirac, Generalized Hamiltonian dynamics. Can. J. Phys. **2**, 129–148 (1950)
9. P.A.M. Dirac, *Lectures on Quantum Mechanics* (Yeshiva University Press, New York, 1964)
10. D.E. Soper, Massive quantum electrodynamics in the infinite-momentum frame. Phys. Rev. D **4**, 1620–1625 (1971)
11. J. Przeszowski, H.W.L. Naus, A.C. Kalloniatis, Quantum electrodynamics in the light-front Weyl gauge. Phys. Rev. D **54**, 5135–5147 (1996)
12. R. Soldati, M. Morara, Consistent perturbative light-front formulation of quantum electrodynamics. Phys. Rev. D **58**, 105011 (1998)

Few-Body Syst (2017) 58:111
DOI 10.1007/s00601-017-1275-5

S. S. Chabysheva

Light-Front Versus Equal-Time Quantization in ϕ^4 Theory

Received: 3 December 2016 / Accepted: 20 February 2017 / Published online: 15 March 2017

Abstract There is a discrepancy between light-front and equal-time values for the critical coupling of two-dimensional ϕ^4 theory. A proposed resolution is to take into account the difference between mass renormalizations in the two quantizations. This distinction was first discussed by M. Burkardt. It prevents direct comparison of bare parameters; however, a method proposed here allows calculation of the difference and thereby resolves the discrepancy. We also consider the consequences of allowing a sector-dependent constituent mass.

1 Introduction

Previous calculations [1,2] have shown that there is a systematic difference between light-front and equal-time values for the critical coupling of ϕ^4 theory when light-front Hamiltonian methods are used. Here we discuss the resolution of this by considering the difference in mass renormalizations as originally suggested by Burkardt [3]. However, we also show that the expected behavior of the probability for higher Fock sectors, that they should grow dramatically as the critical coupling is approached, is not observed. This interferes with the estimation of the mass renormalization. As a means to resolve this, we consider use of sector-dependent constituent mass in the solution of the Hamiltonian eigenvalue problem. Many of the details of the methods used, particularly for this eigenvalue problem, are presented by Hiller [4].

An alternative approach is the light-front Haag expansion method [5,6], where the interacting ϕ field is represented as an expansion in normal-ordered products of the free field. The coefficient functions of the expansion satisfy equations obtained as projections of the Euler–Lagrange equation for the interacting field. This method retains the zero-mode contributions neglected in the light-front Hamiltonian approach and thereby maintains a direct correspondence with equal-time calculations.

Our light-front estimate of the critical coupling is based on calculations of the lowest massive eigenstates as a function of the coupling for both the odd and even sectors. The values of the eigenmass squared are plotted in Fig. 1. The intersections of the spectra with zero indicate a critical coupling of $g = 2.1 \pm 0.05$. This value is compared with values from other calculations in Table 1.

There is clearly a systematic difference between results for light-front quantization and equal-time quantization. In the following section we describe the resolution of this difference in terms of a shift in the effective constituent mass. We also discuss, in Sect. 3, the implementation of a sector-dependent constituent mass as an attempt to improve the calculation of the shift. A summary is given in Sect. 4.

This article belongs to the Topical Collection "Light Cone 2016".

S. S. Chabysheva (✉)
Department of Physics and Astronomy, University of Minnesota-Duluth, Duluth, MN 55812, USA
E-mail: schabysh@d.umn.edu

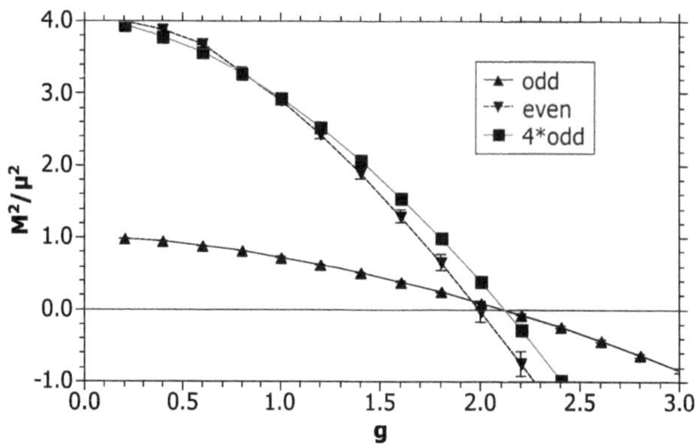

Fig. 1 The lowest masses for the odd and even eigenstates of ϕ^4 theory, from [1]. The threshold for two-particle states, at four times the mass-squared of the odd case, is also shown

Table 1 Critical coupling values, adapted from [7], with use of a slightly different definition of the dimensionless coupling $\bar{g} = \frac{\pi}{6} g = \frac{\lambda}{24\mu^2}$

Method	$\bar{g}_c$	Reported by
Light-front symmetric polynomials	1.1 ± 0.03	This work
DLCQ	1.38	Harindranath and Vary [8]
Light-front Haag expansion	2.50	Salmons et al. [6]
Quasi-sparse eigenvector	2.5	Lee and Salwen [9]
Density matrix renormalization group	2.4954(4)	Sugihara [10]
Lattice Monte Carlo	$2.70\begin{cases}+0.025\\-0.013\end{cases}$	Schaich and Loinaz [11]
	2.79 ± 0.02	Bosetti et al. [12]
Uniform matrix product	2.766(5)	Milsted et al. [13]
Renormalized Hamiltonian truncation	2.97(14)	Rychkov and Vitale [7]

The first two values were computed in light-front quantization with Hamiltonian methods, the third with a light-front Haag expansion, and the remainder in equal-time quantization

2 Mass Renormalization

The bare mass in the Hamiltonian and Lagrangian is renormalized by tadpole contributions in equal-time (ET) quantization but not in light-front (LF) quantization. With the Lagrangian written as $\mathcal{L} = \frac{1}{2}(\partial_\mu \phi)^2 - \frac{1}{2}\mu_0^2\phi^2 - \frac{\lambda}{4!}\phi^4$, this results in a difference that can be expressed as [3]

$$\mu_{\rm LF}^2 = \mu_{\rm ET}^2 + \lambda\left[\left\langle 0\left|\frac{\phi^2}{2}\right|0\right\rangle - \left\langle 0\left|\frac{\phi^2}{2}\right|0\right\rangle_{\rm free}\right]. \tag{1}$$

where the vacuum expectation values (VEV) of ϕ^2 resum the tadpole contributions. Here the subscript *free* indicates the VEV with zero coupling.

To calculate the VEV's, we regulate them with a point splitting by (ϵ^+, ϵ^-) and introduce a decomposition of the identity in terms of the eigenstates $|\psi_n(P)\rangle$ of the light-front Hamiltonian $\mathcal{P}^-$:

$$\left\langle 0\left|\frac{\phi^2}{2}\right|0\right\rangle \to \frac{1}{2}\left\langle 0\left|\phi(\epsilon^+,\epsilon^-)\int_0^\infty dP\sum_n\right|\psi_n(P)\right\rangle\langle\psi_n(P)|\phi(0,0)|0\rangle. \tag{2}$$

The operator $\phi(\epsilon^+,\epsilon^-)$ is obtained by evolving forward in light-front time $x^+ \equiv t+z$ from zero to ϵ^+, so that $\phi(\epsilon^+,\epsilon^-) = e^{i\mathcal{P}^-\epsilon^+/2}\phi(0,\epsilon^-)e^{-i\mathcal{P}^-\epsilon^+/2}$.

The necessary matrix elements for the nth bound state are

$$\langle\psi_n(P)|\phi(0,0)|0\rangle = \left\langle 0\left|\psi_{n1}^* a(P)\int\frac{dp}{\sqrt{4\pi p}}a^\dagger(p)\right|0\right\rangle = \frac{\psi_{n1}^*}{\sqrt{4\pi P}} \tag{3}$$

and

$$\langle 0|\phi(\epsilon^+,\epsilon^-)|\psi_n(P)\rangle = \left\langle 0 \left| \int \frac{dp}{\sqrt{4\pi p}} a(p) e^{-ip\epsilon^-/2} e^{-iM_n^2\epsilon^+/2P} \psi_{n1} a^\dagger(P) \right| 0 \right\rangle = \frac{\psi_{n1}}{\sqrt{4\pi P}} e^{-i(P\epsilon^- + M_n^2\epsilon^+/P)/2}. \tag{4}$$

Similarly, for the one-particle free state, the matrix elements are

$$\langle 0|a(P)\phi(0,0)|0\rangle = \left\langle 0 \left| a(P) \int \frac{dp}{\sqrt{4\pi p}} a^\dagger(p) \right| 0 \right\rangle = \frac{1}{\sqrt{4\pi P}} \tag{5}$$

and

$$\langle 0|\phi(\epsilon^+,\epsilon^-) a^\dagger(P)|0\rangle = \left\langle 0 \left| \int \frac{dp}{\sqrt{4\pi p}} a(p) e^{-ip\epsilon^-/2} e^{-i\mu^2\epsilon^+/2P} a^\dagger(P) \right| 0 \right\rangle = \frac{1}{\sqrt{4\pi P}} e^{-i(P\epsilon^- + \mu^2\epsilon^+/P)/2}. \tag{6}$$

On substitution of these matrix elements, the VEV's can be written as

$$\left\langle 0 \left| \frac{\phi^2}{2} \right| 0 \right\rangle = \frac{1}{2} \sum_n \int_0^\infty dP \frac{|\psi_{n1}|^2}{4\pi P} e^{-i(P\epsilon^- + M_n^2\epsilon^+/P)/2} \quad \text{and} \quad \left\langle 0 \left| \frac{\phi^2}{2} \right| 0 \right\rangle_{\text{free}} = \frac{1}{2} \int_0^\infty dP \frac{1}{4\pi P} e^{-i(P\epsilon^- + \mu^2\epsilon^+/P)/2}. \tag{7}$$

With use of $1 = \sum_n |\psi_{n1}|^2$, the difference can be reduced to

$$\left\langle 0 \left| \frac{\phi^2}{2} \right| 0 \right\rangle - \left\langle 0 \left| \frac{\phi^2}{2} \right| 0 \right\rangle_{\text{free}} = \sum_n \frac{|\psi_{n1}|^2}{8\pi} \int_0^\infty \frac{dP}{P} e^{-iP\epsilon^-/2} \left[e^{-i\frac{M_n^2\epsilon^+}{2P}} - e^{-i\frac{\mu^2\epsilon^+}{2P}} \right]. \tag{8}$$

The mass shift is then proportional to

$$\left\langle 0 \left| \frac{\phi^2}{2} \right| 0 \right\rangle - \left\langle 0 \left| \frac{\phi^2}{2} \right| 0 \right\rangle_{\text{free}} = \sum_n \frac{|\psi_{n1}|^2}{4\pi} \left[K_0(M_n\sqrt{-\epsilon^2 + i\eta}) - K_0(\mu\sqrt{-\epsilon^2 + i\eta}) \right], \tag{9}$$

where η is a convergence factor. From the asymptotic behavior of the Bessel function $K_0(z) \to -\ln(z/2) - \gamma$, we obtain

$$\left\langle 0 \left| \frac{\phi^2}{2} \right| 0 \right\rangle - \left\langle 0 \left| \frac{\phi^2}{2} \right| 0 \right\rangle_{\text{free}} = -\sum_n \frac{|\psi_{n1}|^2}{4\pi} \ln \frac{M_n}{\mu_{\text{LF}}} \equiv -\Delta/4\pi, \tag{10}$$

Finally, the relationship between the masses is

$$\mu_{\text{LF}}^2 = \mu_{\text{ET}}^2 - \frac{\lambda}{4\pi}\Delta \quad \text{or} \quad \frac{\mu_{\text{ET}}^2}{\mu_{\text{LF}}^2} = 1 + g_{\text{LF}}\Delta. \tag{11}$$

This implies coupling constant and mass ratios of

$$g_{\text{ET}} = \frac{g_{\text{LF}}}{\mu_{\text{ET}}^2/\mu_{\text{LF}}^2} = \frac{g_{\text{LF}}}{(1 + g_{\text{LF}}\Delta)} \quad \text{and} \quad \frac{M^2}{\mu_{\text{ET}}^2} = \frac{1}{1 + g_{\text{LF}}\Delta} \frac{M^2}{\mu_{\text{LF}}^2}. \tag{12}$$

The shift Δ is plotted in Fig. 2. As the critical coupling is approached, the shift does not behave as it should. Fits are then done for coupling values much less than the critical value, in order to extrapolate. The extrapolations of the shift then yield $\Delta(g = 2.1) = -0.47 \pm 0.12$. The latest equal-time value for the critical coupling [7], $g_{\text{ETc}} = \frac{6}{\pi}2.97 = 5.67$, implies a shift of $(g_{\text{LFc}}/g_{\text{ETc}} - 1)/g_{\text{LFc}} = -0.30$, which is consistent.

The criterion for the set of coupling values used is determined by examining the behavior of the predicted equal-time values for the mass squared, based on the relationship in (12), as shown in Fig. 3. Above $g = 1$ the equal-time masses begin to increase rather than continue the proper decrease.

The origin of the increase in the mass, and the improper behavior of the shift near the critical coupling, is in the finiteness of the relative probabilities for higher Fock states, displayed in Fig. 4. As the critical coupling is approached there is only a continued gradual increase, implying that the one-body probability $|\psi_{11}|^2$ remains nonzero. This causes the shift Δ to diverge as the eigenmass M_1 goes to zero.

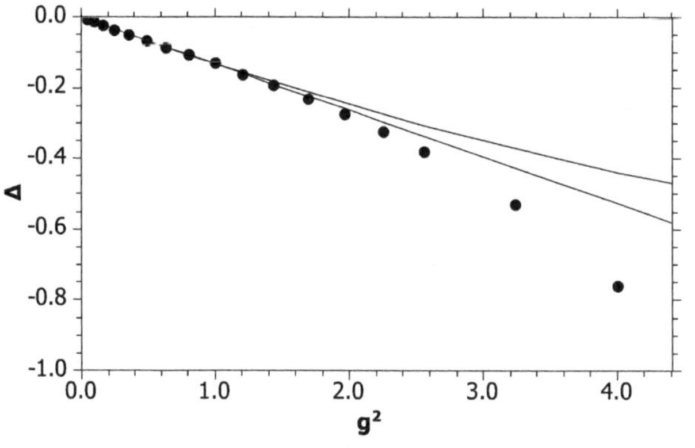

Fig. 2 The renormalization shift Δ as a function of the square of the dimensionless coupling g, as shown in [1]. The *points* displayed are obtained as extrapolations in the polynomial basis size. The *lines* are linear and quadratic fits to shifts below $g = 1$, extrapolated to the region of the critical coupling

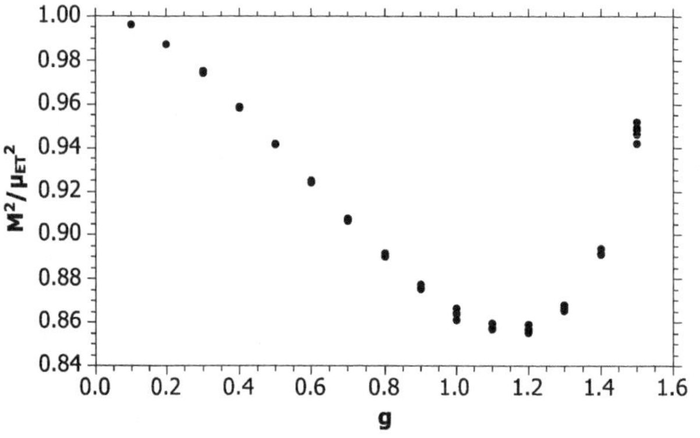

Fig. 3 Lowest equal-time mass eigenvalues for odd numbers of constituents plotted versus the dimensionless light-front coupling g, from [1]. *Different points* at the same g value correspond to different truncations of the polynomial basis size

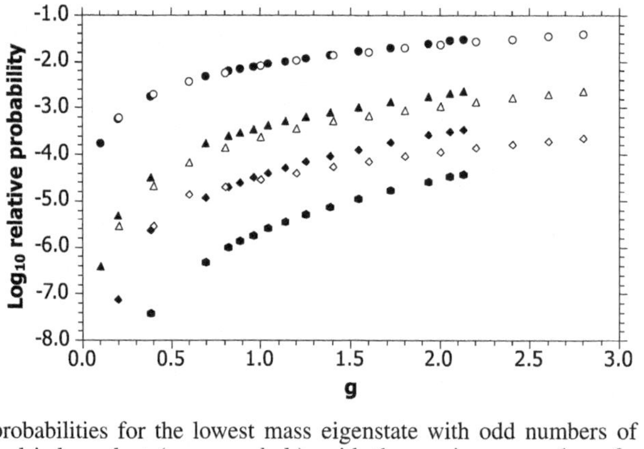

Fig. 4 Relative Fock-sector probabilities for the lowest mass eigenstate with odd numbers of constituents, for both sector-dependent (*closed symbols*) and independent (*open symbols*), with the maximum number of constituents $N_{\max} = 9$ and 7, respectively. The sectors represented include three (*circles*), five (*triangles*), seven (*diamonds*), and nine (*hexagons*) constituents

3 Sector-Dependent Constituent Mass

With a sector-independent constituent mass μ, the invariant mass of a higher Fock state is quite large and such states are then naturally suppressed in any calculation with a Fock-space truncation. An obvious way to avoid this suppression is to use a sector-dependent mass [14–19] μ_m. The matrix eigenvalue problem given in [4] can then be written as

$$\sum_{n'i'}\left[\tilde{\mu}_m^2 T_{ni,n'i'}^{(m)\prime} + V_{ni,n'i'}^{(m,m)\prime}\right]c_{n'i'}^{(m)\prime} + \sum_{n'i'} V_{ni,n'i'}^{(m,m+2)\prime} c_{n'i'}^{(m+2)\prime} + \sum_{n'i'} V_{ni,n'i'}^{(m,m-2)\prime} c_{n'i'}^{(m-2)\prime} = \tilde{M}^2 c_{ni}^{(m)\prime}, \tag{13}$$

with $\tilde{\mu}_m \equiv \mu_m\sqrt{4\pi/\lambda}$ and $\tilde{M} \equiv M\sqrt{4\pi/\lambda}$. The sector-dependent mass then allows for the fact that a Fock-space truncation forces self-energy corrections to be different in each Fock sector; in particular, in the highest sector there is no self-energy correction. When the Fock-space truncation is removed, in the $N_{\max} \to \infty$ limit, the dimensionless sector-dependent mass $\tilde{\mu}_m$ becomes $\tilde{\mu} \equiv \pm 4\pi\mu^2/\lambda$. This convergence to the sector-independent mass happens first in the lowest sector; therefore, the dimensionless coupling $g \equiv \frac{\lambda}{4\pi\mu^2}$ can be extracted as $g \simeq 1/|\tilde{\mu}_1^2|$ and the dimensionless eigenmass as $M^2/\mu^2 = g\tilde{M}^2 = \tilde{M}^2/|\tilde{\mu}_1^2|$.

The sector-dependent case is no longer an explicit eigenvalue problem. The sector-dependent masses $\tilde{\mu}_m$ must be computed recursively for a given value of $\tilde{M}$ and then translated into values for g and M/μ. The recursive nature is that each $\tilde{\mu}_m$ is computed with only the Fock sectors above taken into account. In the top sector, $\tilde{\mu}_{N_{\max}}$ is just $\tilde{M}$. For each sector below the top, the mass $\tilde{\mu}_m$ has been computed by solving the smaller problem where $\tilde{\mu}_m$ is the mass in the one-body sector and the highest sector has $N_{\max} - m + 1$ constituents with mass $\tilde{M}$.

To carry out this calculation, we define a set of matrices $G^{(m)}$, from $m = N_{\max}$ down to 3, as

$$G^{(m)} = \left[\tilde{\mu}_m^2 T^{(m)\prime} + V^{(m,m)\prime} - \tilde{M}^2 I^{(m)} - V^{(m,m+2)\prime} G^{(m+2)} V^{(m+2,m)\prime}\right]^{-1}, \tag{14}$$

$$G^{(N_{\max})} = \left[\tilde{M}^2 T^{(N_{\max})\prime} + V^{(N_{\max},N_{\max})\prime} - \tilde{M}^2 I^{(N_{\max})}\right]^{-1}. \tag{15}$$

The mass in the lowest sector is then simply

$$\tilde{\mu}_1^2 = \frac{1}{T^{(1)}}\left[\tilde{M}^2 - V^{(1,1)\prime} - V^{(1,3)\prime} G^{(3)} V^{(3,1)\prime}\right]. \tag{16}$$

The coefficients for the wave-function expansions are constructed recursively from $m = 3$ up to $N_{\max}$ by

$$\mathbf{c}^{(m)\prime}/c^{(1)} = G^{(m)} V^{(m,m-2)\prime} \mathbf{c}^{(m-2)\prime}/c^{(1)}. \tag{17}$$

The relative probabilities are compared with those of the sector-independent approach in Fig. 4. The sector-dependent results do converge more slowly, with respect to the Fock-space truncation, requiring $N_{\max} = 9$ to reach the convergence found at $N_{\max} = 5$ for the sector-independent calculation. The slower convergence is to be expected, and even desired, given that we expect the higher Fock states to contribute more easily and to be more important as the critical coupling is approached. However, the relative probabilities continue to show no critical behavior.

4 Summary

It is possible to understand the difference between ET and LF values of the critical coupling by taking the different mass renormalizations into account. However, calculation of the mass shift near the critical coupling shows poor behavior of computed eigenstates. The relative probabilities of higher Fock states do not show critical behavior; they should diverge because the one-body probability should go to zero. The use of a sector-dependent constituent mass does not help. The difficulty near the critical coupling is avoided instead by extrapolation from smaller coupling values.

The correct representation of the eigenstates near critical coupling apparently requires a method without Fock-space truncation. In general, this would be a coherent-state basis. A specific implementation might be the light-front coupled cluster method [20] with a nontrivial valence state, such as a linear combination of the one-body and three-body states.

Acknowledgements This work was done in collaboration with M. Burkardt and J.R. Hiller, and was supported in part by the Minnesota Supercomputing Institute of the University of Minnesota with allocations of computing resources.

References

1. M. Burkardt, S.S. Chabysheva, J.R. Hiller, Two-dimensional light-front ϕ^4 theory in a symmetric polynomial basis. Phys. Rev. D **94**, 065006 (2016)
2. S.S. Chabysheva, Light-front ϕ^4_{1+1} theory using a many-boson symmetric-polynomial basis. Few Body Syst. **57**, 675–680 (2016)
3. M. Burkardt, Light-front quantization of the sine-Gordon model. Phys. Rev. D **47**, 4628–4633 (1993)
4. J.R. Hiller, Symmetry breaking in light-front ϕ^4 theory. Few-Body Syst. **58**(2), 1–6 (2017). doi:10.1007/s00601-016-1204-z
5. P. Grangé, P. Ullrich, E. Werner, Continuum version of ϕ^4_{1+1} theory in light-front quantization. Phys. Rev. D **57**, 4981–4989 (1998)
6. S. Salmons, P. Grangé, E. Werner, Critical properties of (1+1)-dimensional ϕ^4 theory in light-cone quantization. Phys. Rev. D **65**, 125014 (2002)
7. S. Rychkov, L.G. Vitale, Hamiltonian truncation study of the ϕ^4 theory in two dimensions. Phys. Rev. D **91**, 085011 (2015)
8. A. Harindranath, J.P. Vary, Solving two-dimensional ϕ^4 theory by discretized light-front quantization. Phys. Rev. D **36**, 1141–1147 (1987)
9. D. Lee, N. Salwen, The diagonalization of quantum field hamiltonians. Phys. Lett. B **503**, 223–235 (2001)
10. T. Sugihara, Density matrix renormalization group in a two-dimensional lambda ϕ^4 Hamiltonian lattice model. J. High Energy Phys. **05**, 007 (2004)
11. D. Schaich, W. Loinaz, An improved lattice measurement of the critical coupling in ϕ^4_2 theory. Phys. Rev. D **79**, 056008 (2009)
12. P. Bosetti, B. De Palma, M. Guagnelli, Monte Carlo determination of the critical coupling in ϕ^4_2 theory. Phys. Rev. D **92**, 034509 (2015)
13. A. Milsted, J. Haegeman, T.J. Osborne, Matrix product states and variational methods applied to critical quantum field theory. Phys. Rev. D **88**, 085030 (2013)
14. R.J. Perry, A. Harindranath, K.G. Wilson, Light front Tamm–Dancoff field theory. Phys. Rev. Lett. **65**, 2959–2962 (1990)
15. R.J. Perry, A. Harindranath, Renormalization in the light front Tamm–Dancoff approach to field theory. Phys. Rev. D **43**, 4051–4073 (1991)
16. J.R. Hiller, S.J. Brodsky, Nonperturbative renormalization and the electron's anomalous moment in large-α QED. Phys. Rev. D **59**, 016006 (1998)
17. V.A. Karmanov, J.-F. Mathiot, A.V. Smirnov, Systematic renormalization scheme in light-front dynamics with Fock space truncation. Phys. Rev. D **77**, 085028 (2008)
18. V.A. Karmanov, J.-F. Mathiot, A.V. Smirnov, Nonperturbative calculation of the anomalous magnetic moment in the Yukawa model within truncated Fock space. Phys. Rev. D **82**, 056010 (2010)
19. S.S. Chabysheva, J.R. Hiller, On the nonperturbative solution of Pauli–Villars regulated light-front QED: a comparison of the sector-dependent and standard parameterizations. Ann. Phys. **325**, 2435–2450 (2010)
20. S.S. Chabysheva, J.R. Hiller, A light-front coupled-cluster method for the nonperturbative solution of quantum field theories. Phys. Lett. B **711**, 417–422 (2012)

Few-Body Syst (2017) 58:102
DOI 10.1007/s00601-017-1260-z

Jerzy A. Przeszowski

Alternate Light Front Quantization Procedure for Scalar Fields

Received: 15 November 2016 / Accepted: 13 February 2017 / Published online: 4 March 2017

Abstract The novel procedure for the light-front (LF) quantization is formulated and applied for models of free scalar fields. The expected well-known results are rediscovered for a single field and new results are obtained for the two fields model. We use fields smeared with a test function on the LF hypersurface as the basic ingredient of our novel quantization procedure.

1 Introduction

The motivation of our present work is to carry out the quantization procedure at the light-front (LF) hypersurface without solving the classical canonical structure in advance. This is an important issue for a system with constraints to decide whether one should begin the classical Dirac procedure [1,2] for finding the Dirac-Poisson brackets and then transform these brackets into respective quantum commutators at the quantization hypersurface or one should start just with a quantum theory. The second possibility is the essence of the novel approach and it can be implemented for systems where there are hints of highly singular behavior of some commutators at the LF. Such information may follow from the equal-time (ET) quantization, which for a free field theory gives the vacuum expectation values (VEV) of the quantum field commutators—the commutator functions—for arbitrary points. Evidently one may expect that these commutator functions should also follow from the appropriate LF formulation. As an example one may consider a scalar massive field $\phi(x)$, which leads to the commutator function $\Delta(x)$, the Pauli-Jordan function, which has singular limits $x^+ \to 0$ for their partial derivatives $\partial_+ \Delta(x)$ and $\partial_+^2 \Delta(x)$ [3]. Thus the commutators $\left[\partial_+ \phi(x), \phi(y)\right]$ and $\left[\partial_+^2 \phi(x), \phi(y)\right]$ do not exist at the LF hypersurface $x^+ \to y^+$. Similar singular commutators will appear also for models with fermion and vector fields. But within the standard canonical Dirac procedure such singular commutators cannot follow from the respective Dirac-Poisson brackets—they need to be regular at the quantization hypersurface [4]. Therefore one needs to have a new procedure for the LF quantization, which will allow for such singular commutators.

The first attempt in this direction has been proposed for the $1+1$ dimensional Yukawa model [5], where the Wightman functions (the vacuum expectation values of unordered product of fields) were determined by using their equations of motion, the Lorentz covariance and the PCT symmetry. However this procedure becomes rather complicated for models in higher dimensions and even inapplicable for models with broken Lorentz symmetry. Below we propose another procedure which is free of such deficiencies. Here we will analyze only two models with scalar field(s) and other models will be discussed elsewhere.

This article belongs to the Topical Collection "Light Cone 2016".

J. A. Przeszowski (✉)
Faculty of Physics, University of Białystok, ul. Ciołkowskiego 4L, 15-424 Białystok, Poland
E-mail: j.przeszowski@uwb.edu.pl

2 Standard LF Procedure for the Massive Scalar Field

The standard LF quantization procedure can be presented in a clear way for the massive scalar field $\phi(x)$ with the Lagrangian density, expressed in the LC coordinates,

$$\mathcal{L} = \partial_+\phi\partial_-\phi - \frac{1}{2}(\partial_i\phi)^2 - \frac{m^2}{2}\phi^2. \tag{1}$$

It leads to the Euler–Lagrange equation of motion $\left(2\partial_+\partial_- - \partial_i^2 + m^2\right)\phi(x) = \left(\Box + m^2\right)\phi(x) = 0$. The canonical momentum forms the primary constraint $\pi - \partial_-\phi \approx 0$, but the Dirac procedure for constrained systems is quite simple and produces Dirac's bracket for the classical fields at the LF hypersurface $2\{\phi(\bar{x}), \partial_-\phi(0)\}_D = \delta^3(\bar{x} - \bar{y})$. Thus the commutator for quantum fields (denoted by hats overhead) at the LF hypersurface is $2\left[\hat{\phi}(\bar{x}), \partial_-\hat{\phi}(0)\right] = i\delta^3(\bar{x} - \bar{y})$. Next one introduces the Fourier representation for the quantum field operator

$$\hat{\phi}(x) = \int_{\mathbb{R}^3} d\Gamma(\bar{k}) \left[e^{-i\bar{k}\cdot\bar{x}} e^{-i\frac{m^2+k_\perp^2}{2k^+}x^+} \hat{a}(\bar{k}) + e^{i\bar{k}\cdot\bar{x}} e^{i\frac{m^2+k_\perp^2}{2k^+}x^+} \hat{a}^\dagger(\bar{k}) \right], \qquad d\Gamma(\bar{k}) = \frac{d^2\boldsymbol{k}_\perp}{(2\pi)^3} \frac{dk^+}{2k^+} \Theta(k^+), \tag{2}$$

which evidently agrees with the Klein–Gordon equation of motion and allows to transform the commutator for fields into to the commutator for the creation and annihilation operators $\left[\hat{a}(\bar{k}), \hat{a}^\dagger(\bar{p})\right] = (2\pi)^3\, 2k^+\, \delta^3(\bar{k} - \bar{p})$. All this leads to the 2-point Wightman function at arbitrary points

$$\langle 0|\hat{\phi}(x)\,\hat{\phi}(0)|0\rangle = \Delta_+(x) = \int_{\mathbb{R}^3} d\Gamma(\bar{k})\, e^{-i\bar{k}\cdot\bar{x}} e^{-i\frac{m^2+k_\perp^2}{2k^+}x^+}, \tag{3}$$

which is singular at the LF hypersurface [3,6]. One obtains the translation generators by inserting (2) into the normal ordered space-time formula $\hat{P}_\mu =: \int_{\mathbb{R}^3} d^3\bar{x}\, T^+{}_\mu(\bar{x})$:, where $T^+{}_\mu$ are the components of the energy-momentum tensor. This sequence of steps is the essence of the standard LF quantization procedure [7] and evidently the classical Dirac bracket is a substantial intermediate step.

3 Novel LF Quantization Procedure

First few preliminary remarks and definitions. For the novel LF quantization procedure one needs to assume the translation symmetry of the model, thus for a scalar field $\phi(x)$[1] we have the infinitesimal translations $i\partial_\mu\phi(x) = \left[\phi(x), P_\mu\right]$ generated by the translation generator operators, which commute $\left[P_\mu, P_\nu\right] = 0$. This allows for the unitary representation

$$\phi(x) = e^{iP\cdot x}\phi(0)e^{-iP\cdot x}, \quad \phi(x)|0\rangle = e^{iP\cdot x}\phi(0)|0\rangle, \tag{4}$$

where the vacuum state is translationally invariant, thus $P_\mu|0\rangle = 0$. This can be extended to higher order partial derivatives of quantum field, thus for the d'Alambert operator $\Box = 2\partial_+\partial_- - \partial_i\partial_i$ we have double commutators

$$\Box\phi(x) = -2e^{iP\cdot x}\left[P_-, \left[P_+, \phi(0)\right]\right]e^{-iP\cdot x} + e^{iP\cdot x}\left[P_i, \left[P_i, \phi(0)\right]\right]e^{-iP\cdot x} \tag{5}$$

which simplify considerably when acting on the vacuum state $\Box\phi(x)|0\rangle = e^{iP\cdot x}\left(-2P_+P_- + P_i^2 + m^2\right)$ $\phi(0)|0\rangle$. At the LF hypersurface $x^+ = 0$, we introduce the smooth fields, by smearing field operators with a test function $f(\bar{x}) \in \mathscr{S}'(\mathbb{R}^3)$ [6]

$$\phi[f] = \int_{\mathbb{R}^3} d^3\bar{x}\, f(\bar{x})\,\phi(\bar{x}), \quad \phi[\partial_{\bar{\mu}}f] = -\int_{\mathbb{R}^3} d^3\bar{x}\, f(\bar{x})\,\partial_{\bar{\mu}}\phi(\bar{x}), \quad \bar{\mu} \in \{-, i\}, \tag{6}$$

where we in integration by parts the boundary terms with test functions vanish at infinities. This smooth fields are crucial ingredients of the novel LF quantization. The commutators with the translation generator operators have the form $\left[\phi[f], P_{\bar{\mu}}\right] = -i\phi[\partial_{\bar{\mu}}f]$ and can be taken as the LF quantization rule. It is sufficient to consider only the case with P_- which at the LF hypersurface is a kinematical operator, thus has no dependence on

[1] Hereafter we will omit overhead hats for the quantum field operators hoping it will not lead to any misunderstanding.

interaction. One can take the explicit form $P_- = \int_{\mathbb{R}^3} d^3\bar{x}\, T^{++}(\bar{x})$, with T^{++} being the component of the canonical energy-momentum tensor.

Now we may begin the novel LF procedure for a simple case of one massive scalar field, which satisfies Klein–Gordon equation of motion, therefore we take the Lagrangian density (1). This leads to non-vanishing canonical momentum $\pi = \partial_-\phi$, but here we do not treat π as a constrained variable, but we just insert it into the canonical energy-momentum component $T^{++} = \pi\,\partial_-\phi = (\partial_-\phi)^2$. This allows us to impose the novel LF quantization rule[2]

$$-i\phi[\partial_- f] = \left[\phi[f], P_-\right] = 2\int_{\mathbb{R}^3} d^3\bar{x}\; \left[\phi[f], \partial_-\phi(\bar{x})\right]\, \partial_-\phi(\bar{x}). \tag{7}$$

Then in the sense of distribution $\mathscr{S}(\mathbb{R}^3)$ one finds the LF commutator for quantum fields in two steps

$$\frac{i}{2} f(\bar{x}) = \left[\phi[f], \partial_-\phi(\bar{x})\right] = \int_{\mathbb{R}^3} d^3\bar{y}\; \left[\phi(\bar{y}), \partial_-\phi(\bar{x})\right]\, f(\bar{y}) \quad\Rightarrow\quad \left[\phi(\bar{y}), \partial_-\phi(\bar{x})\right] = \frac{i}{2}\delta^3(\bar{x}-\bar{y}). \tag{8}$$

We can take the vacuum expectation value of the last form of commutator, assuming that the vacuum state is normalized $\langle 0|0\rangle = 1$

$$\langle 0|\phi(\bar{x})\,\partial_-\phi(\bar{y})|0\rangle - \langle 0|\partial_-\phi(\bar{y})\,\phi(\bar{x})|0\rangle = \frac{i}{2}\delta^3(\bar{x}-\bar{y}) \tag{9}$$

and our next task is to decompose this equation into the matrix elements. However introducing the unitary representation for fields (4) we may rewrite the above equation as

$$\langle 0|\phi(0)P_-\,\phi(\bar{y}-\bar{x})|0\rangle + \langle 0|\phi(0)P_-\,\phi(\bar{x}-\bar{y})|0\rangle = \frac{1}{2}\delta^3(\bar{x}-\bar{y}), \tag{10}$$

which manifestly depends only on the difference of coordinates $\bar{x}-\bar{y}$ as a consequence of the translational symmetry. We may integrate it with a test function $f(\bar{x}-\bar{y}) \in \mathscr{S}'(\mathbb{R}^3)$ where we denote $f_-(\bar{x}) = f(-\bar{x})$,

$$\langle 0|\phi(0)P_-\,\phi[f_-]|0\rangle + \langle 0|\phi(0)P_-\,\phi[f]|0\rangle = \frac{1}{2}f(0). \tag{11}$$

One may expect that at the quantization hypersurface a quantum field operator when acting on the vacuum state produces a 1-particle state. Thus we may make the following LF assumption

$$\phi[f]|0\rangle = \int_{\mathbb{R}^3} d\Gamma(\bar{k})\tilde{f}(\bar{k})|\bar{k}\rangle, \qquad \phi[f_-]|0\rangle = \int_{\mathbb{R}^3} d\Gamma(\bar{k})[\tilde{f}_-](\bar{k})|\bar{k}\rangle = \int_{\mathbb{R}^3} d\Gamma(\bar{k})\tilde{f}(-\bar{k})|\bar{k}\rangle, \tag{12}$$

where $\tilde{f}(\bar{k}) = \int_{\mathbb{R}^3} d^3\bar{x}\, f(\bar{x})\, e^{i\bar{k}\cdot\bar{x}}$ is the Fourier transform of a test function, which also is a test function $\tilde{f}(\bar{k}) \in \mathscr{S}'(\mathbb{R}^3)$. This equation can be interpreted as an indirect definition of 1-particle states $|\bar{k}\rangle = a^\dagger(\bar{k})|0\rangle$ labeled by the kinematical momentum eigenvalues $\bar{k}$.[3] Then using the inverse Fourier transformation (34) for $f(0)$ we express (11) as the distributional equation

$$\int_{\mathbb{R}^3} d\Gamma(\bar{k})k^+\left[\tilde{f}(\bar{k}) + \tilde{f}(-\bar{k})\right]\langle 0|\phi(0)|\bar{k}\rangle = \frac{1}{2}f(0) = \int_{\mathbb{R}^3} d\Gamma(\bar{k})k^+\left[\tilde{f}(\bar{k}) + \tilde{f}(-\bar{k})\right], \tag{13}$$

which immediately leads in the sense distribution $\mathscr{S}(\mathbb{R}^3)$ to the matrix element and the internal product of 1-particle states (for details see the "Appendix")

$$\langle 0|\phi(0)|\bar{k}\rangle = 1, \qquad \langle \bar{k}|\bar{p}\rangle = \delta_\Gamma(\bar{k},\bar{p}), \qquad \delta_\Gamma(\bar{k},\bar{p}) = (2\pi)^3\, 2k^+\, \delta^3(\bar{k}-\bar{p}) \tag{14}$$

The Klein–Gordon equation $(\Box + m^2)\phi(x) = 0$ directly transforms into the equation for the smooth field

$$0 = \left(-2P_+P_- + P_\perp^2 + m^2\right)\phi[f]|0\rangle = \int_{\mathbb{R}^3} d\Gamma(\bar{k})\tilde{f}(\bar{k})\left[-2k^+P_+ + k_\perp^2 + m^2\right]|\bar{k}\rangle. \tag{15}$$

[2] We assume that the commutator of two fields at the quantization hypersurface is proportional to the identity operator.
[3] In the "Appendix" we present another definition of $|\bar{k}\rangle$ and then prove that $P_-|\bar{k}\rangle = k^+|\bar{k}\rangle$ and $P_i|\bar{k}\rangle = k_i|\bar{k}\rangle$.

Thus $|\bar{k}\rangle$ is also an eigenfunction of the LF Hamiltonian operator P_+ since in the sense of distribution $\mathscr{S}(\mathbb{R}^3)$ we find $P_+|\bar{k}\rangle = \frac{k_\perp^2+m^2}{2k^+}|\bar{k}\rangle$. This enables us to evaluate the temporal evolution in x^+ variable for the smooth field acting on the vacuum state

$$e^{iP_+x^+}\phi[f]|0\rangle = \int_{\mathbb{R}^3} d\Gamma(\bar{k})\tilde{f}(\bar{k})e^{iP_+x^+}|\bar{k}\rangle = \int_{\mathbb{R}^3} d\Gamma(\bar{k})\tilde{f}(\bar{k})e^{i\frac{k_\perp^2+m^2}{2k^+}x^+}|\bar{k}\rangle, \tag{16}$$

which further leads to the Wightman function for arbitrary LF time x^+

$$\langle 0|\phi(0)e^{iP_+x^+}\phi[f]|0\rangle = \int_{\mathbb{R}^3} d\Gamma(\bar{k})\tilde{f}(\bar{k})e^{i\frac{k_\perp^2+m^2}{2k^+}x^+}\langle 0|\phi(0)|\bar{k}\rangle = \int_{\mathbb{R}^3} d\Gamma(\bar{k})\tilde{f}(\bar{k})e^{i\frac{k_\perp^2+m^2}{2k^+}x^+}, \tag{17}$$

where the matrix element $\langle 0|\phi(0)|\bar{k}\rangle = 1$ is taken from Eq. (14). Both sides of this equation can be expressed as integrals with the test function $f(\bar{x})$

$$\int_{\mathbb{R}^3} d^3\bar{x}\; f(\bar{x})\,\langle 0|\phi(0)e^{iP_+x^+}\phi(\bar{x})|0\rangle = \int_{\mathbb{R}^3} d^3\bar{x}\; f(\bar{x}) \int_{\mathbb{R}^3} d\Gamma(\bar{k})e^{i\bar{k}\cdot\bar{x}}e^{i\frac{k_\perp^2+m^2}{2k^+}x^+} \tag{18}$$

so in the sense of distributions $\mathscr{S}(\mathbb{R}^3)$ we find $\langle 0|\phi(0)\phi(x)|0\rangle = \Delta_+(-x)$ in consistency with Eq. (3)—due to translational symmetry of the Wightman function, which is explicit in Eq. (10).

Then we may turn to the creation and annihilation operators, which build the bra and ket 1-particle states out of the vacuum state $|\bar{k}\rangle = a^\dagger(\bar{k})|0\rangle$ and $\langle\bar{k}| = \langle 0|a(\bar{k})$, when we impose the annihilation conditions for them $\langle 0|a^\dagger(\bar{k}) = 0 = a(\bar{k})|0\rangle$. Thus from Eq. (14) we find

$$\langle\bar{k}|\bar{p}\rangle = \langle 0|a(\bar{k})a^\dagger(\bar{p})|0\rangle = \delta_\Gamma(\bar{k},\bar{p}), \qquad \left[a(\bar{k}), a^\dagger(\bar{p})\right] = \delta_\Gamma(\bar{k},\bar{p}). \tag{19}$$

This simple algebra of the creation and annihilation operators allows to extract the commutators of $a^\dagger(\bar{p})$ with all translation generators from their eigenvalue equations

$$\left[P_-, a^\dagger(\bar{k})\right] = k^+a^\dagger(\bar{k}), \qquad \left[P_i, a^\dagger(\bar{k})\right] = k_i a^\dagger(\bar{k}), \qquad \left[P_+, a^\dagger(\bar{k})\right] = \frac{k_\perp^2+m^2}{2k^+}a^\dagger(\bar{k}). \tag{20}$$

Now the momentum representation of the translation generators follow immediately

$$P_+ = \int_{\mathbb{R}^3} d\Gamma(\bar{k})\,\frac{m^2+k_\perp^2}{2k^+}\,a^\dagger(\bar{k})a(\bar{k}), \quad P_- = \int_{\mathbb{R}^3} d\Gamma(\bar{k})\,k^+\,a^\dagger(\bar{k})a(\bar{k}), \quad P_i = \int_{\mathbb{R}^3} d\Gamma(\bar{k})\,k_i\,a^\dagger(\bar{k})a(\bar{k}), \tag{21}$$

provided the normal ordering is taken for guaranteeing $P_\mu|0\rangle = 0$, which follows from the translation symmetry of the vacuum state. Finally one can find the Fourier representation of the scalar field at arbitrary points $\phi(x) = e^{iP_+x^+}\phi(\bar{x})e^{-iP_+x^+}$, using its hermiticity, the algebra of creation and annihilation operators (19) and the smooth Wightman function (18). All this gives (2) expected from the standard LF quantization procedure. Thus for a free massive scalar field our novel LF procedure and the standard approach give the same results, but evidently on different ways of logical reasoning.

4 Dipole Scalar Field

Here we will apply the novel LF quantization procedure for the model of two scalar fields ϕ and χ with the Lagrangian density

$$\mathcal{L} = \partial^\mu\phi\partial_\mu\chi + \frac{m^2}{2}\chi^2 - m^2\chi\phi = \partial_+\phi\partial_-\chi + \partial_-\phi\partial_+\chi - \partial_i\phi\partial_i\chi + \frac{m^2}{2}\chi^2 - m^2\chi\phi, \tag{22}$$

We will take steps analogous to those the previous section, but omitting many details we will stress only upon differences, which appear for the present model. Here there are two Euler–Lagrange equations $(\Box + m^2)\phi = m^2\chi$, $(\Box + m^2)\chi = 0$, which effectively give the second order Klein–Gordon equation for ϕ field $(\Box + m^2)^2\phi = 0$. We take two nonzero canonical momenta $\pi^\phi = \partial_-\chi$, $\pi^\chi = \partial_-\phi$ for building $T^{++} = \pi^\phi\partial_-\phi + \pi^\chi\partial_-\chi = 2\partial_-\phi\partial_-\chi$ and $P_- = \int_{\mathbb{R}^3} d^3\bar{x}\,T^{++}(\bar{x})$. At the LF hypersurface $x^+ = 0$, we define two smooth fields by smearing with a test function $f(\bar{x}) \in \mathscr{S}'(\mathbb{R}^3)$ and impose the LF quantization rules for them

$$\phi[f] = \int_{\mathbb{R}^3} d^3\bar{x}\, f(\bar{x})\,\phi(\bar{x}), \quad \chi[f] = \int_{\mathbb{R}^3} d^3\bar{x}\, f(\bar{x})\,\chi(\bar{x}), \quad -i\phi[\partial_- f] = \left[\phi[f], P_-\right], \quad -i\chi[\partial_- f] = \left[\chi[f], P_-\right]. \tag{23}$$

Assuming that ϕ and χ are independent fields we find the LF commutators

$$2\left[\phi(\bar{x}), \partial_-\chi(\bar{y})\right] = i\delta^3(\bar{x}-\bar{y}), \quad \left[\phi(\bar{x}), \partial_-\phi(\bar{y})\right] = \left[\chi(\bar{x}), \partial_-\chi(\bar{y})\right] = 0. \tag{24}$$

Then we decompose the VEV of these commutators, by assuming the action of the smooth operators on the vacuum state

$$\phi[f]|0\rangle = \int_{\mathbb{R}^3} d\Gamma(\bar{k})\tilde{f}(\bar{k})|\bar{k}, a\rangle, \quad \chi[f]|0\rangle = \int_{\mathbb{R}^3} d\Gamma(\bar{k})\tilde{f}(\bar{k})|\bar{k}, b\rangle, \tag{25}$$

with two independent 1-particle states $|\bar{k}, a\rangle$ and $|\bar{k}, b\rangle$. This leads to the matrix elements

$$\langle 0|\phi(0)|\bar{k}, a\rangle = 0, \quad \langle 0|\chi(0)|\bar{k}, a\rangle = 1, \quad \langle 0|\phi(0)|\bar{k}, b\rangle = 1, \quad \langle 0|\chi(0)|\bar{k}, b\rangle = 0. \tag{26}$$

From the equations of motion we may infer the LF Schrödinger equations

$$0 = \left(-2P_+P_- + P_\perp^2 + m^2\right)\chi[f]|0\rangle, \quad \left(-2P_+P_- + P_\perp^2 + m^2\right)\phi[f]|0\rangle = m^2\chi[f]|0\rangle, \tag{27}$$

we find that $|\bar{k}, b\rangle$ is the eigenfunction of the LF Hamiltonian P_+, while $|\bar{k}, a\rangle$ is not

$$P_+|\bar{k}, b\rangle = \frac{m^2 + k_\perp^2}{2k^+}|\bar{k}, b\rangle, \quad P_+|\bar{k}, a\rangle = \frac{m^2 + k_\perp^2}{2k^+}|\bar{k}, a\rangle - \frac{m^2}{2k^+}|\bar{k}, b\rangle. \tag{28}$$

Then from the algebra of the creation and annihilation operators we establish the momentum representation for the translation generator operators P_μ

$$P_+ = \int_{\mathbb{R}^3} d\Gamma(\bar{k})\left[-\frac{m^2}{2k^+}b^\dagger(\bar{k})b(\bar{p}) + \frac{m^2 + k_\perp^2}{2k^+}\left(b^\dagger(\bar{k})a(\bar{k}) + a^\dagger(\bar{k})b(\bar{k})\right)\right], \tag{29a}$$

$$P_- = \int_{\mathbb{R}^3} d\Gamma(\bar{k})\, k^+\left[b^\dagger(\bar{k})a(\bar{k}) + a^\dagger(\bar{k})b(\bar{k})\right], \quad P_i = \int_{\mathbb{R}^3} d\Gamma(\bar{k})\, k_i\left[b^\dagger(\bar{k})a(\bar{k}) + a^\dagger(\bar{k})b(\bar{k})\right]. \tag{29b}$$

Finally we may obtain the momentum representation for scalar fields

$$\chi(x) = \int_{\mathbb{R}^3} d\Gamma(\bar{k})\left[e^{-i\bar{k}\cdot\bar{x}}e^{-i(m^2+k_\perp^2)/(2k^+)x^+}b(\bar{k}) + h.c.\right], \tag{30a}$$

$$\phi(x) = \int_{\mathbb{R}^3} d\Gamma(\bar{k})\left[e^{-i\bar{k}\cdot\bar{x}}e^{-i(m^2+k_\perp^2)/(2k^+)x^+}\left(a(\bar{k}) + i\frac{m^2x^+}{2k^+}b(\bar{k})\right) + h.c.\right], \tag{30b}$$

which allows to calculate the Wightman functions for all scalar field, and for example we find

$$\begin{aligned}\langle 0|\phi(x)\phi(0)|0\rangle &= \int_{\mathbb{R}^3} d\Gamma(\bar{k})\, e^{-i\bar{k}\cdot\bar{x}}e^{-i(m^2+k_\perp^2)/(2k^+)x^+}\, i\frac{m^2x^+}{2k^+}\langle\bar{k}, b|\phi(0)|0\rangle \\ &= im^2x^+\int_{\mathbb{R}^3} d\Gamma(\bar{k})\, e^{-i\bar{k}\cdot\bar{x}}e^{-i(m^2+k_\perp^2)/(2k^+)x^+}\frac{1}{2k^+} = -m^2\frac{\partial}{\partial m^2}\Delta_+(x).\end{aligned} \tag{31}$$

The last expression manifestly agrees with the Lorentz symmetry. It allows for the calculation of the LF Feynman propagator defined as the VEV of the chronologically ordered product in x^+ variable

$$\langle 0|T^+\phi(x)\phi(0)|0\rangle = \theta(x^+)\langle 0|\phi(x)\phi(0)|0\rangle + \theta(-x^+)\langle 0|\phi(0)\phi(x)|0\rangle \tag{32}$$

$$= \frac{im^2}{(2\pi)^3}\int_{\mathbb{R}^4} d^4k\, e^{-ik\cdot x}\int_0^\infty \frac{d\lambda}{4\lambda^3}e^{i\frac{k^2-m^2}{2\lambda}} = -\frac{im^2}{(2\pi)^4}\int_{\mathbb{R}^4} d^4k\, e^{-ik\cdot x}\frac{1}{(k^2-m^2+i0^+)^2}, \tag{33}$$

where the λ-representation [8] follows from the change of variables $k^+ = \lambda|x^+|$. For the last step we take (40) from the "Appendix". We see that the double pole appears naturally, which is the expected result within the ET formulation but it seems to be out of reach for the standard LF formulation.

5 Conclusions and Prospects

The novel LF quantization procedure can be implemented for scalar field models with no limitation on number of fields and their masses. Since no classical field theory must be canonically solved in advance, then we need not to solve any partial differential equations and the calculation is basically reduced to an algebra. The field operators and translation generators are explicitly established in the momentum space. We need to have the translation symmetry but Lorentz symmetry is not obligatory. Further research for fields with higher spin as fermions and vector fields will be undertaken. Also this procedure should work for nonlinear thus interacting models, but the problem of renormalization needs to be reformulated as well.

Acknowledgements We would like to thank the organizers of the LC2016 Workshop for all their efforts and hospitality.

A Appendix: Definitions, Useful Formulas and Details of Calculations

The LF longitudinal and transverse coordinates are defined as $x^{\pm} = (x^0 \pm x^3)/\sqrt{2}$ and $\boldsymbol{x}_{\perp} = (x^1, x^2)$, respectively.Their partial derivatives are denoted as $\partial_\mu = \partial/\partial x^\mu$. Analogously we define 4-momentum coordinates. The Minkowski space-time metric tensor has non-vanishing components $\eta_{+-} = \eta_{-+} = 1$, $\eta_{ij} = -\delta_{ij}$. At the LF hypersurface $x^+ = 0$ the light front momenta and coordinates are $\bar{k} = \left(k^+ \in (0, \infty), \boldsymbol{k}_\perp \in (-\infty, \infty)\right)$, $\bar{x} = \left(x^- \in (-\infty, \infty), \boldsymbol{x}_\perp \in (-\infty, \infty)\right)$. For arbitrary 4-vectors a^μ and b^μ, we define the restricted scalar product $\bar{a} \cdot \bar{b} = a^+ b^- - \boldsymbol{a}_\perp \cdot \boldsymbol{b}_\perp$ and the Lorentz scalar product $a \cdot b = \bar{a} \cdot \bar{b} + a_+ b^+$.

For a discussion of the LF assumption in Eq. (12) we start with the inverse Fourier transform

$$f(\bar{x}) = \int_{\mathbb{R}^3} d\Gamma(\bar{k})\, 2k^+ \left[\tilde{f}(\bar{k}) e^{-i\bar{k}\cdot\bar{x}} + \tilde{f}(-\bar{k}) e^{i\bar{k}\cdot\bar{x}}\right] \tag{34}$$

which allows to express the action of the smooth field operator on the vacuum state as

$$\phi[f]|0\rangle = \int_{\mathbb{R}^3} d\Gamma(\bar{k}) \left[\tilde{f}(\bar{k})|\bar{k}, -\rangle + \tilde{f}(-\bar{k})|\bar{k}, +\rangle\right], \tag{35}$$

where we define 1-particle states as

$$|\bar{k}, \pm\rangle = 2k^+ \int_{\mathbb{R}^3} d^3\bar{x}\, e^{\pm i\bar{k}\cdot\bar{x}}\, \phi(\bar{x})|0\rangle, \tag{36}$$

in accordance with the earlier attempts [9, 10]. Thus the commutator of the smooth field with P_- acting on the vacuum state $-i\phi[\partial_- f]|0\rangle = \left[\phi[f], P_-\right]|0\rangle = -P_-\phi[f]|0\rangle$ can be expressed as

$$\begin{aligned} i\phi[\partial_- f]|0\rangle &= i \int_{\mathbb{R}^3} d\Gamma(\bar{k}) \left[\widetilde{[\partial_- f]}(\bar{k})|\bar{k}, -\rangle + \widetilde{[\partial_- f]}(-\bar{k})|\bar{k}, +\rangle\right] \\ &= \int_{\mathbb{R}^3} d\Gamma(\bar{k}) \left[\tilde{f}(\bar{k}) P_-|\bar{k}, -\rangle + \tilde{f}(-\bar{k}) P_-|\bar{k}, +\rangle\right]. \end{aligned} \tag{37}$$

From the Fourier transform $\widetilde{[\partial_- f]}(\bar{k}) = -ik^+\tilde{f}(\bar{k})$ we find the following eigenvalue equations $P_-|\bar{k}, -\rangle = k^+|\bar{k}, -\rangle$, and $P_-|\bar{k}, +\rangle = -k^+|\bar{k}, +\rangle$. Since P_- cannot have a negative eigenvalue, thus we conclude $|\bar{k}, +\rangle = 0$. When we change slightly the notation $|\bar{k}, -\rangle = |\bar{k}\rangle$ and we prove Eq. (12). Accordingly this LF assumption follows from the nonnegative spectrum of P_-.

Analogously we may express the relations $-i\phi[\partial_j f]|0\rangle = \left[\phi[f], P_j\right]|0\rangle = -P_j\phi[f]|0\rangle$ in its integral representation

$$i\phi[\partial_j f]|0\rangle = i \int_{\mathbb{R}^3} d\Gamma(\bar{k})\, \widetilde{[\partial_j f]}(\bar{k})\, |\bar{k}\rangle = \int_{\mathbb{R}^3} d\Gamma(\bar{k})\, \tilde{f}(\bar{k})\, k_j\, |\bar{k}\rangle = \int_{\mathbb{R}^3} d\Gamma(\bar{k})\, \tilde{f}(\bar{k})\, P_j|\bar{k}\rangle \tag{38}$$

thus in the sense of distribution $\mathscr{S}(\mathbb{R}^3)$ we have the eigenvalue equation $P_j|\bar{k}\rangle = k_j|\bar{k}\rangle$.

From the hermiticity of the scalar field we find from Eq. (14) the matrix element $\langle\bar{k}|\phi(0)|0\rangle = \langle 0|\phi(0)|\bar{k}\rangle = 1$, which when multiplied by the phase factor $e^{i\bar{k}\cdot\bar{x}}$ and the integration with a test function $f(\bar{x}) \in \mathscr{S}'(\mathbb{R}^3)$ leads successively to

$$e^{i\bar{k}\cdot\bar{x}}\langle\bar{k}|\phi(0)|0\rangle = \langle\bar{k}|e^{i\bar{P}\cdot\bar{x}}\phi(0)|0\rangle = \langle\bar{k}|\phi(\bar{x})|0\rangle = e^{i\bar{k}\cdot\bar{x}}, \quad \langle\bar{k}|\phi[f]|0\rangle = \tilde{f}(k). \tag{39}$$

Then inserting Eq. (14) into $\phi[f]|0\rangle$ we find the internal product $\langle\bar{k}|\bar{p}\rangle = \delta_\Gamma(\bar{k}, \bar{p})$, in the sense of distributions $\mathscr{S}(\mathbb{R}^3)$.

For calculating the Feynman propagator we may use the generic formula for the case $n = 2$

$$\int_0^\infty \frac{d\lambda}{\lambda^{n+1}} e^{iA/\lambda} = \frac{i^n(n-1)!}{(A+i0)^n}. \tag{40}$$

References

1. P.A.M. Dirac, Generalized Hamiltonian dynamics. Can. J. Phys. **2**, 129–148 (1950)
2. P.A.M. Dirac, *Lectures on Quantum Mechanics* (Yeshiva University Press, New York, 1964)
3. J.A. Przeszowski, Lorentz symmetry for the light-front Wightman functions. Acta Phys. Polon. Proc. Suppl. B **6**, 327–333 (2013)
4. S.J. Brodsky, H.-C. Pauli, S. Pinsky, Quantum chromodynamics and other field theories on the light cone. Phys. Rep. **301**, 299–486 (1998)
5. J.A. Przeszowski, J. Żochowski, Light-front quantization with explicit Lorentz symmetry for Yukawa model. Few-Body Syst. **56**, 579–585 (2015)
6. M. Herrmann, W.N. Polyzou, Light-front vacuum. Phys. Rev. D **91**, 085043 (2015)
7. A. Harindranath, An introduction to light-front dynamics for pedestrians, in *International School on Light-Front Quantization and NonPerturbative QCD*, ed. by J.P. Vary, F. Woelz (1996)
8. J.A. Przeszowski, E. Dzimida-Chmielewska, J. Żochowski, Light-front perturbation without spurious singularities. Few-Body Syst. **57**, 527–532 (2016)
9. H. Leutwyler, J.R. Klauder, L. Streit, Quantum field theory on lightlike slabs. Nouvo Cimento A **66**, 536–554 (1970)
10. N. Nakanishi, K. Yamawaki, A consistent formulation of the null-plane quantum field theory. Nucl. Phys. B **122**, 15–28 (1977)

Few-Body Syst (2017) 58:97
DOI 10.1007/s00601-017-1268-4

Javier Gomez

The Mirror Nuclei ^{3}H and ^{3}He Program at JLab

Received: 14 January 2017 / Accepted: 13 February 2017 / Published online: 28 February 2017

Abstract Using electron beam energies of up to 11 GeV, Jefferson Lab plans to carry out in the near future a group of four experiments involving the mirror nuclei ^{3}H and ^{3}He. The experiments aim to, *(A)* extract the deep inelastic neutron to proton structure function ratio F_2^n/F_2^p (and u/d quark distributions) for $0.2 \leq x \leq 0.9$, *(B)* study the isospin structure of two-nucleon and search for three-nucleon Short Range Correlations (SRC) for $x < 3$, *(C)* measure the proton momentum distribution of both nuclei at $x = 1.2$ to further our understanding of SRCs in the few-body and *(D)* extract the charge radii of both nuclei at $Q^2 \leq 0.1$ GeV^2.

1 Nuclear Effects in Deep Inelastic Scattering (DIS)

The virtual photo-production cross section for unpolarized deep inelastic lepton scattering (DIS) of a nucleus of atomic weight A and charge Z, $\sigma^{\{A,Z\}}$, can be written as

$$\sigma^{\{A,Z\}} = \sigma_M \frac{F_2^{\{A,Z\}}}{\nu} \left[\frac{1 + \varepsilon R^{\{A,Z\}}}{\varepsilon \left(1 + R^{\{A,Z\}}\right)} \right] \tag{1}$$

where σ_M represents the Mott cross section, $F_2^{\{A,Z\}}$ the DIS structure function, $\nu = E - E'$ the virtual photon energy with E (E') the incident (scattered) lepton energy, $R = \sigma_L/\sigma_T$ the ratio of the virtual photon longitudinal (σ_L) to transverse (σ_T) cross sections and ε represents the degree of longitudinal polarization of the virtual photon. It is given by $\varepsilon = \left[1 + 2(1 + \nu^2/Q^2)\tan^2(\theta/2)\right]^{-1}$ with θ the lepton scattering angle in the laboratory and $Q^2 = 4EE' \sin^2(\theta/2)$ the invariant four-momentum transferred by the photon. Note that $\varepsilon = 1(0)$ when $\theta = 0°(180°)$. The undetected final state has an invariant mass $W^2 = M^2 + Q^2\,(x^{-1} - 1)$ where M the nucleon mass and $x = Q^2/(2\,M\,\nu)$. Bjorken x is interpreted as the fraction of the longitudinal nucleon momentum carried by the struck quark in DIS. Note that $x = 1$ corresponds to quasi-elastic scattering off a nucleon in a nucleus (elastic scattering if the nucleus is hydrogen) and, for example, $x = 2$ to scattering off a di-nucleon in a nucleus.

Inelastic scattering off ^{1}H provides the unbound (free) proton structure function F_2^p. Since there are no equivalent neutron targets, obtaining F_2^n requires measuring a target like ^{2}H and subtracting the proton contribution after accounting for nuclear effects. Measurements [1] in the late 1970's yielded $F_2^n/F_2^p \sim 1 - 0.85x$ for the ratio of unbound neutron-to-proton structure functions.

Measuring the Q^2 evolution of the structure functions expected from Quantum Chromodynamics (QCD), the field theory describing the strong interaction between quarks mediated by gluons, necessitated higher beam

This article belongs to the Topical Collection "Light Cone 2016".

J. Gomez (✉)
Thomas Jefferson National Accelerator Facility, Newport News, VA, USA
E-mail: gomez@jlab.org

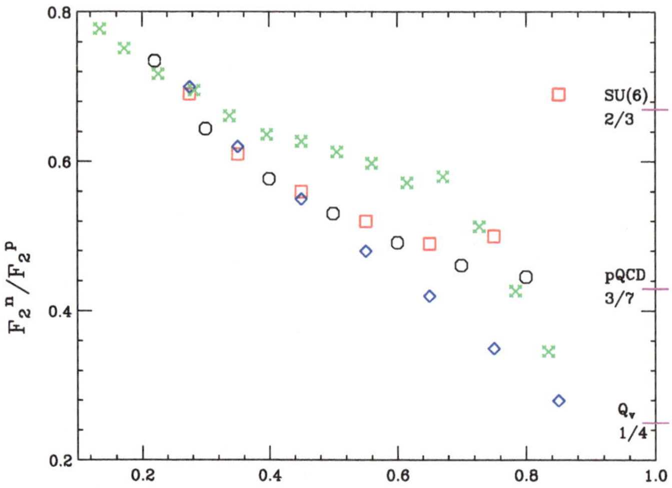

Fig. 1 Results of using different methods to extract F_2^n/F_2^p. *Open diamonds* (*blue*) obtained by [6] using the procedure of [3] with Paris wave-function for ^{2}H. *Open squares* (*red*) obtained by [6] using the method proposed in [5] with a subset of the isoscalar ratios of [7]. *Open circles* (*black*) obtained by [8] using the cross sections of [9] with $Q^2 \geq 5\,\mathrm{GeV}^2$ to perform global fits over nuclei assuming that all protons and neutrons in a given nuclei are affected equally by the EMC effect. Various functional parameterizations of the EMC effect were used (e.g. "average" nuclear density, nucleon overlap) with little change. *Diagonal crosses* (*green*) were extracted by [10] using spectator tagging in ^{2}H, a different technique (and assumptions). Some limiting values for F_2^n/F_2^p as $x \Rightarrow 1$ are also shown. pQCD refers to the work of [11]

energies than those available from fixed target electron machines in the 70's and 80's. Muon and neutrino beams of much higher energy were produced but their low intensity required using very thick targets of high atomic number to reach reasonable luminosities. Targets of carbon, marble, heavy liquids and iron were used. The procedures of [2] and [3] could have been used to extract the free nucleon structure function F_2^N out of these targets. However, these corrections affect mostly data with Bjorken x larger than about 0.6. Since all these early experiments used fixed targets, their higher beam energy extended their reach towards $x = 0$ while their limited luminosity made it difficult for them to get reasonable statistics for x larger than about 0.5. So, there was no need for experiments to even apply the nuclear corrections in most cases. The structure function per nucleon $\mathscr{F}_2^{\{A,Z\}}$ of a nucleus of atomic number A and charge Z for x less than about 0.6 was assumed to be simply $\mathscr{F}_2^{\{A,Z\}} = \left[Z\,F_2^p + (A-Z)F_2^n\right]/A$. Equivalently, one could easily transform the measured per nucleon structure function $\mathscr{F}_2^{\{A,Z\}}$ into a deuterium-like equivalent $\mathscr{F}_2^{\{A,A/2\}}$ using the measured F_2^n/F_2^p ratio. One then expects $\mathscr{F}_2^{\{A,A/2\}} = \mathscr{F}_2^{\{2,1\}}$. The European Muon Collaboration (EMC) performed such comparison [4] and concluded that, at least for iron, such equality did not hold. Reanalysis of existing data and new measurements confirmed the finding and extended it to many new nuclei. The EMC effect was born. The effect is Q^2 independent and the strength of the effect increases from light to heavier nuclei. The exact functional dependence on nuclei is not know. So far, mostly simple parametrizations in terms of nuclear mass or "average nuclear density" derived from low-energy electron scattering (nucleus RMS charge radius) have been used. Many explanations have been put forward but none of them seem to be universally accepted.

Given the existence of the EMC effect, Frankfurt and Strikman [5] brought into question if all nuclear effects in deuterium had been taken into account when extracting the unbound neutron structure function. Figure 1 shows the results of using different methods to extract F_2^n/F_2^p. Basically, the uncertainty in the nuclear corrections to be applied introduce large uncertainties in the extraction of the F_2^n/F_2^p ratio and associated quark momentum distributions.

2 The Ratio F_2^n/F_2^p from ^{3}H and ^{3}He Scattering

A measure of the nuclear effects in ^{3}H and ^{3}He is given by the ratios $Q(^3\mathrm{H}) = F_2^{\{3,1\}}/\left(F_2^p + 2\,F_2^n\right)$ and $Q(^3\mathrm{He}) = F_2^{\{3,2\}}/\left(2\,F_2^p + F_2^n\right)$ where $F_2^{\{3,1\}}$ and $F_2^{\{3,2\}}$ are the experimentally measured deep inelastic structure functions of ^{3}H and ^{3}He respectively. The ratio of ratios $Q_r = Q(^3\mathrm{He})/Q(^3\mathrm{H})$ is then a measure

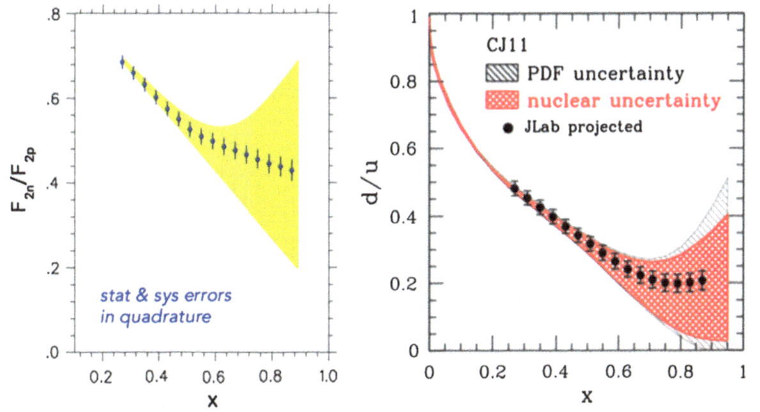

Fig. 2 Expected results for E12-10-103. The *left panel* shows the results expected for F_2^n/F_2^p. Points include statistical and systematic errors added in quadrature. *Yellow band* indicates present uncertainties in the F_2^n/F_2^p ratio due to uncertainties in the nuclear corrections to be applied. The *right panel* shows expected results for the d/u quark ratio extracted from F_2^n/F_2^p. *Gray and red bands* indicate present uncertainties in d/u due to parton density functions (PDF) and nuclear corrections to be used in the extraction

of the difference in nuclear effects between the mirror nuclei ^{3}H and ^{3}He. Studies by several authors [12–15] have concluded that those differences are small and well understood, $Q_r < 2\%$ for $x < 0.9$. We have then a different way to extract F_2^n/F_2^p than those used in the past relying only on the small relative difference of nuclear effects in ^{3}H and ^{3}He. Contrast this with the extraction of F_2^n from ^{2}H via inclusive or semi-inclusive scattering where absolute knowledge of the nuclear effects magnitude is needed and, as discussed in Sect. 1, there are questions if all nuclear effects in ^{2}H have been accounted for. So,

$$\frac{F_2^n}{F_2^p} = \frac{2\,Q_r - \left[F_2^{\{3,2\}}/F_2^{\{3,1\}}\right]}{2\left[F_2^{\{3,2\}}/F_2^{\{3,1\}}\right] - Q_r} \tag{2}$$

where Q_r comes from theory while $F_2^{\{3,1\}}$ and $F_2^{\{3,2\}}$ come from experiment. They are related to the ^{3}H and ^{3}He measured cross sections, $\sigma^{\{3,1\}}$ and $\sigma^{\{3,2\}}$ respectively, by

$$\frac{F_2^{\{3,2\}}}{F_2^{\{3,1\}}} = \frac{\sigma^{\{3,2\}}}{\sigma^{\{3,1\}}}\,\frac{\left(1+\varepsilon\,R^{\{3,1\}}\right)}{\left(1+\varepsilon\,R^{\{3,2\}}\right)}\,\frac{\left(1+R^{\{3,2\}}\right)}{\left(1+R^{\{3,1\}}\right)} \tag{3}$$

Existing data [16] indicates that R is the same for all nuclei within experimental errors—i.e. $R^{\{3,1\}} = R^{\{3,2\}}$. Then, $F_2^{\{3,2\}}/F_2^{\{3,1\}} = \sigma^{\{3,2\}}/\sigma^{\{3,1\}}$ and Eq. (2) becomes,

$$\frac{F_2^n}{F_2^p} = \frac{2\,Q_r - \left[\sigma^{\{3,2\}}/\sigma^{\{3,1\}}\right]}{2\left[\sigma^{\{3,2\}}/\sigma^{\{3,1\}}\right] - Q_r} \tag{4}$$

The objective of Experiment E12-10-103 [17] is to use this technique to,

(a) Extract F_2^n/F_2^p in the range of $[x,\ Q^2(\mathrm{GeV}^2),\ W(\mathrm{GeV}),\ \varepsilon]$ of [.23, 3.0, 3.3 0.62]-to-[0.87, 14.5, 1.8, 0.29]. See Fig. 2 for expected results.
(b) Check A-independence of R at $x = (0.25, 0.35, 0.45, 0.55, 0.65)$ with four measurements in ε at each x: $\sim 0.19 \le \varepsilon \le\sim 0.86$, $W \sim 2\,\mathrm{GeV}$ and $1.3 \le Q^2(\mathrm{GeV}^2) \le 5.7$
(c) Measure the EMC effect in ^{3}H and ^{3}He.

Table 1 RMS charge radius of ^{3}H and ^{3}He

	$\langle r_E^2 \rangle_{^3\mathrm{H}}$	$\langle r_E^2 \rangle_{^3\mathrm{He}}$
GFMC [18]	1.77 (1)	1.97 (1)
χEFT [19]	1.756 (6)	1.962 (4)
SACLAY [20]	1.76 (9)	1.96 (3)
BATES [21]	1.68 (3)	1.97 (3)
Atomic [18]	–	1.959 (4)

First two rows are theoretical predictions while the last three rows are experimental measurements

Table 2 Uncertainties sources and expected error contributions to the ^{3}H and ^{3}He RMS charge radius measurements of E12-14-009 [22]

Sources	Error (%)
Statistics	0.4
Integrated beam charge	<0.5
Relative target thickness	1.5–2
Dead-time, tracking and detector	
Efficiency corrections	<0.5
G_M substraction	0.4
Radiative corrections	0.5
Coulomb and two-photon corrections	0.4
Total	1.8–2.2

A total error of about 2%, will lead to an error in $\delta\langle r_E^2\rangle$ of ± 0.02 fm

3 The Elastic Electric Form Factor of ^{3}H and ^{3}He

In the one-photon approximation, the elastic cross section is given by

$$\frac{d\sigma}{d\Omega} = \frac{\sigma_M}{(1+\tau)}\left[G_E^2 + \frac{\tau}{\varepsilon}\,G_M^2\right] \tag{5}$$

with $\tau = Q^2/4M^2$ and G_E, G_M the elastic electric and magnetic form factors respectively. For low Q^2, $\tau \ll 1$ and $\varepsilon \sim 1$, G_E dominates the cross section and, in the non-relativistic regime, is also related to the charge distribution (ρ) and RMS charge radius ($\langle r_E^2\rangle$),

$$G_E = \int \rho_E(\mathbf{r})\, e^{i\mathbf{q}\cdot\mathbf{r}}\, d^3\mathbf{r} \approx 1 - \frac{1}{6}q^2\left\langle r_E^2\right\rangle + \cdots \tag{6}$$

or

$$\left\langle r_E^2\right\rangle = -6\frac{dG_E}{dQ^2}|_{Q^2=0} \tag{7}$$

Table 1 shows theoretical expectations for the RMS charge radius of ^{3}H and ^{3}He as well as the experimentally obtained values. Basically, the difference (δ) in RMS charge radius obtained by SACLAY is in line with theoretical expectations $\left(\delta\left\langle r_E^2\right\rangle = 0.20(10)\right)$ but the more precise measurements from BATES give a difference of $\delta\left\langle r_E^2\right\rangle = 0.29(04)$, nearly a 50% larger value. Only the RMS charge radius of ^{3}He has been measured by atomic physics experiments and the value measure is in line with expectations so, it can not help solve the discrepancy. The aim of Experiment E12-14-009 [22] is to measure precisely the RMS charge radius of ^{3}H and ^{3}He. Table 2 summarizes the expected uncertainty sources and their error contribution to the measurements.

4 Short Range Correlations (SRCs)

Frankfurt et al. [23] pointed out that:

- Access to short-range correlations in nuclei—i.e. to nucleon pairs in the nuclear ground state with large relative momentum and small center-of-mass momentum, could be enhanced by looking at inclusive scattering of a nucleus A with $Q^2 > 1$ GeV, $x > 1$ and $\nu > k$ where k is the Fermi momentum ($\sim 0.3 - 0.4$ GeV).

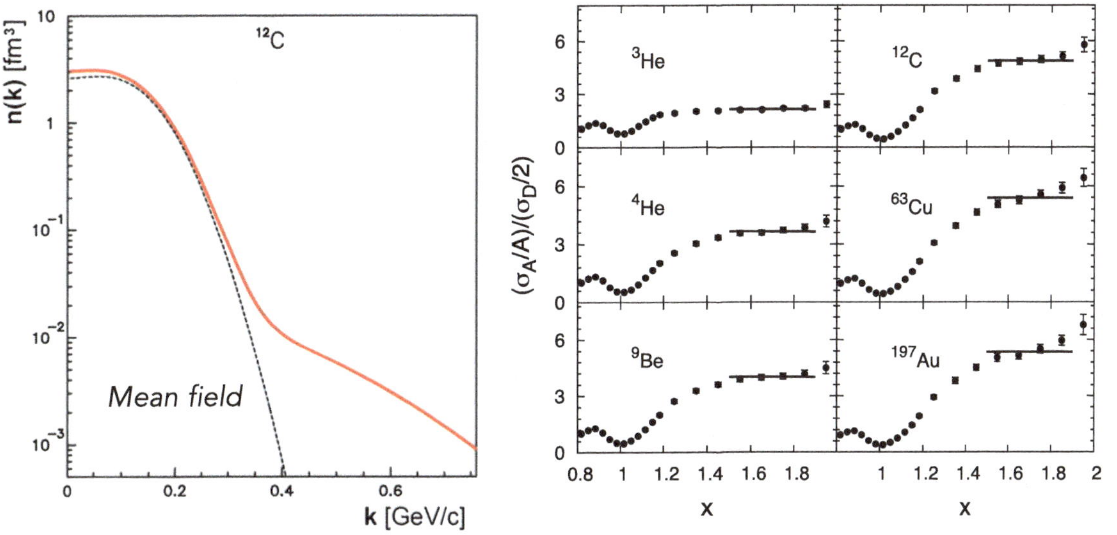

Fig. 3 The *left panel* shows nucleon momentum probability in ^{12}C without correlations (Mean field, *dotted curve*) and with (*red curve*). The *right panel* shows recent data [24] from JLab. Plotted is the per nucleon ratio of heavy nuclei to deuterium as a function of Bjorken x. The *solid lines* indicate the region where quasi-scaling is observed indicating the presence of two-nucleon SRCs

– The presence of two-nucleon SRCs should lead to scaling when comparing heavy nuclei to deuterium.

The authors went on to analyze existing data to show the presence of the predicted effect. Figure 3 shows results from a recent JLab experiment.

More recently, the relative abundance of p-n, p-p and n-n SRC pairs in ^{12}C was stablished by looking at the reactions ^{12}C(*e,e'p*), ^{12}C(*e,e'pn*) and ^{12}C(*e,e'pp*) [25]. Analysis of the data indicates that 80% of the nucleons in the ^{12}C nucleus acted independently, 18% were n-p SRC pairs with the remaining 2% distributed equally between $p-p$ and $n-n$ pairs. Also, the relative momentum of the $n-p$ pair is larger than for the $p-p$ and $n-n$ pairs, as expected from theory (see, for example, [26]).

The goal of Experiment E12-14-011 [27] is to confirm that, like in the case of ^{12}C, short-range correlations in ^{3}H and ^{3}He are also dominated by n-p pairs with a large relative momentum. If this is case, one expects an enhancement in the ratio of the number of protons with low-momentum k knocked-out from ^{3}He compared to ^{3}H $[n_{^3\mathrm{He}}(k)/n_{^3\mathrm{H}}(k)]$ in a A$(e, e'p)$ type of reaction. This is basically due to a large fraction of the lonely protons in ^{3}H being shifted to higher momentum by the SRCs. The ratio should approach unity when probing higher momentum protons simply because the SRCs in ^{3}He and ^{3}H are expected to be similar. This behavior of the ratio is shown in the left panel of Fig. 4.

Experiment E12-11-112 [28] will use inclusive scattering to search for scaling in the per nucleon cross section ratios $\sigma^N(^3\mathrm{He})/\sigma^N(^2\mathrm{H})$ and $\sigma^N(^3\mathrm{H})/\sigma^N(^2\mathrm{H})$ looking for two-nucleon and three-nucleon SRCs. The experiment will also measure the quasi-elastic ^{3}H and ^{3}He cross section in the region $0.7 \leq x \leq 1.2$ and $0.6 \leq Q^2(\mathrm{GeV}^2) \leq 3.2$. The right panel of Fig. 4 shows some of the expected results form this experiment.

5 The JLab Accelerator and Experimental Hall A

Figure 5 shows a Google Maps view of the Thomas Jefferson National Accelerator Facility (JLab) in Newport News, Virgina (USA). The accelerator has a racetrack design about 1.4 Km long with two superconducting linacs operating at 2.08 K (straight sections of the racetrack). The "continuous wave" electron beam structure consists of 1.5×10^9 electron bunches per second, each about 300 femtoseconds long and 100 μm full width at half maximum (FWHM). In the 2011–2012 period, the energy of the machine was upgraded and a new hall (D) added to the existing three halls (A, B and C). Starting at the injector (INJ in Fig. 5), a bunch goes up to five times around the machine before reaching Halls A, B or C. If the set energy gain per pass is δE, the halls can receive beams with energies $\delta E, 2\delta E, \ldots, 5\delta E$. Halls A, B and C can simultaneously receive beam with energy of $5\delta E$ but only one of them can be at each of the lower energy multiples. To reach Hall D, an electron bunch has to go 5.5 times around the machine (beam energy $5.5\delta E$). Hall D is a photon-only hall. A

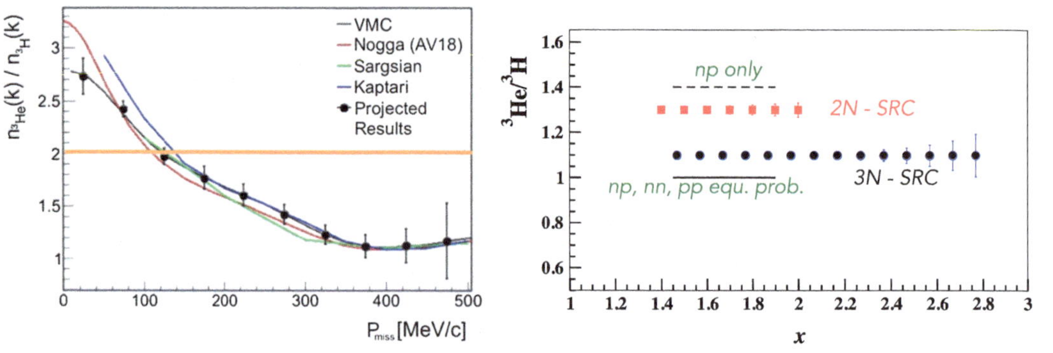

Fig. 4 The *left panel* shows the results expected from Experiment E12-14-011 [27] (see main text for details). The *right panel* shows the results expected from Experiment E12-11-112 [28]. The expected systematic errors will add $\pm 1.2\%$ point-to-point to the error bars shown. An additional, overall, $\pm 1.1\%$ normalization is also expected. In the case of 2 N correlations, the experiment should be able to easily separate the case of SRCs dominated by $n-p$ pairs from the case of equal probability for all nucleon-type pairs (*dashed and solid horizontal lines* respectively)

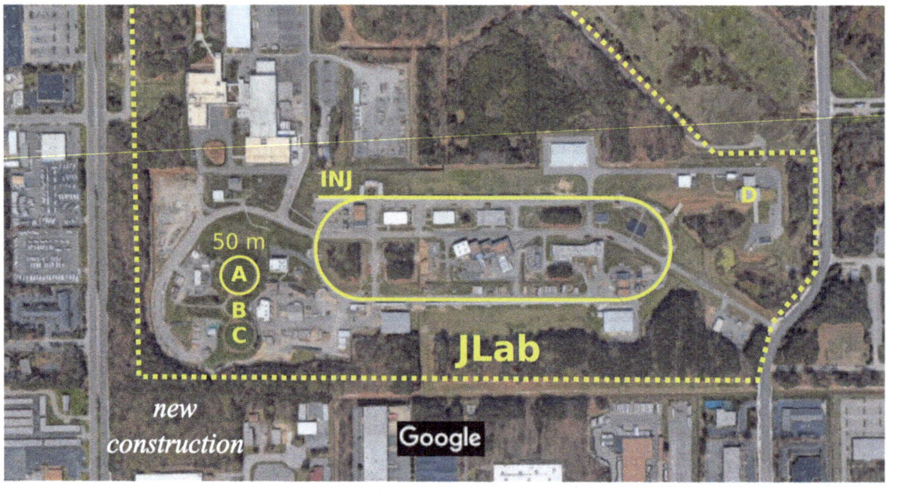

Fig. 5 Satellite view of JLab. Halls A, B and C are located to the left of the picture. For reference, the 50 m diameter of Hall A is highlighted. Hall D is located on the opposite side of the accelerator

thin radiator is used to produce tagged quasi-monochromatic photons, polarized or unpolarized. The photons are directed 96 m downstream to the Hall D target and detector system while the electron beam is disposed in a dump. The maximum δE is about 2.2 GeV/pass or 11 GeV maximum electron energy for Halls A, B and C and 12 GeV for Hall D. The photons sent to Hall D peak at 9 GeV. Hall A and C beam dumps can handle up to 1 MW of beam power, beam current and energy can be traded to stay below that limit.

Experimental Hall A [29] is the largest of the four experimental halls with a diameter of about 50 m. It houses two magnetic High-Resolution Spectrometers (HRSs). Table 1 of [29] summarizes the main characteristics of these spectrometers. These spectrometers were originally designed to study reactions of the type $A(e, e'p)$ requiring high-accuracy and momentum resolution ($\delta p/p$) of 10^{-4} or better. Their maximum momentum however is limited to 4 GeV. This does not represent a problem for inclusive reactions at large Q^2 where only the scattered electron is detected after loosing a large fraction of its incident energy (like on some of the experiments presented here). It does present a problem if one needs to detect particles produced in elastic or semi-inclusive scattering by the higher energy beams now available at JLab. In the future, Hall A is expected to focus on special setups with one-of-a-kind detector systems.

6 Target System

The target assembly consist of four identical cells. Each cell is machined out of a single block of Aluminum 7075 (no seams). It is 250 mm long, 12.7 mm inside diameter with 0.457, 0.254 and 0.279 mm thick side, entrance and exit walls respectively. A 12.7 mm thick collimator with a 10.2 mm diameter hole for the beam to pass through covers the beam entrance side of the cell. Its purpose is to stop the beam from hitting the cell side walls which could cause cell failure. Such situation could arise if the target assembly position encoders fail to operate correctly leading to position a target incorrectly in the beam.

The cells will be filled at 300 K with ^{3}H, ^{1}H, ^{2}H and ^{3}He at 200, 500, 500 and 500 psia respectively and then sealed (no gas circulation). In the case of ^{3}H, we expect a pressure uncertainty of less that 1% and a purity of 99.8% (main contaminants are ^{3}He and ^{2}H). At the operating temperature of 40 K, the expected gas densities are 0.00363 g/cm^3 (^{3}H), 0.00425 g/cm^3 (^{3}He) and 0.00568 g/cm^3 (^{2}H).

The cells are approved to be operated with a minimum beam raster size of $2 \times 2\,\text{mm}^2$ at beam currents of up to 20 μA except the ^{3}He cell which can be operated up to 25 μA. The beam raster frequencies are in the neighborhood of 25 kHz. Beam induced target density changes along the beam path are expected to be of order 6% at 20 μA.

The target assembly also includes an empty cell to measure target wall contributions, alignment targets, carbon optics targets and several solid targets. The target assembly moves vertically to bring the desired target into the beam path. The entire target assembly is contained in an isolated scattering chamber. The accelerator and scattering chamber vacuum spaces are separated by a 0.2 mm thick Be window mounted on a 102 mm long water cooled collimator (Densimet) with a 10.2 mm diameter hole for the beam to go through. The purpose of this collimator is to ensure that a mis-steered beam with a large incident angle, does not hit the target cell side walls. A small gap separates this collimator assembly from those collimators mounted directly on the target cells.

6.1 Special Considerations Needed to Use a ^{3}H Target at JLab

Jefferson Lab, as it can be seen in Fig. 5, is surrounded by many businesses (industrial and shopping) as well as apartment complexes. The target design was directed to (A) meet the needs of the experiments, (B) minimize the likelihood that ^{3}H containment is breached and, (C) take measures to limit human exposure if all the ^{3}H is released unexpectedly. Key points to achieving those goals are:

- Minimize the amount of ^{3}H to be used by the experiments. One kCi was chosen, an amount that can be shipped via commercial carrier. For comparison, some ^{3}H experiments carried out at other labs in the past used up to 180 kCi.
- Use a sealed target cell design, i.e., no manipulation of ^{3}H gas will be done at JLab. The ^{3}H target will be filled and sealed at Savannah River Tritium Enterprise (SRTE), then shipped to JLab in a specialized dual vessel container for protection. The dual vessel allows to determine if the cell has failed without having to open the vessel.
- The ^{3}H target cell will be kept in its transportation/protection container until all other target system components have been tested and the scattering chamber is ready to be closed.
- A hut will enclose the scattering chamber for the duration of these experiments. A large diameter duct will connect the hut to a 20 m high chimney located above Hall A dome. Forced air will be flowed constantly from the hut to the outside. All ^{3}H target cell handling operations will take place inside this hut.
- Access to the experimental hall will be restricted to those with proper training and need to be there. Access to the scattering chamber hut will be restricted to only a couple of people. Work near the hut will be prohibited after the ^{3}H has been taken out of its transportation/protection container.

The ^{3}H target will be used at JLab for about a year and then shipped back to SRTE to avoid having JLab treated as a nuclear facility (i.e. a nuclear power plant). We expect to loose about 5% of the ^{3}H in a year (12.32 years half-life).

6.2 Handling of the ^{3}H Cell and Abnormal Situations

(a) The shipping/protection container is to be opened/closed at two locations only: SRTE and inside the scattering chamber hut in Hall A.
(b) A Residual Gas Analyzer (RGA) will be used to check the integrity of the ^{3}H cell before opening the shipping/protection container—the container will not be opened if ^{3}H gas is detected.

(c) If the ^{3}He cell fails and it is not contained inside the scattering chamber, the ^{3}H gas is automatically vented outdoors through the chimney by the constant air flow of the hut. This failure mode could happen while placing/removing the cell from the target assembly or if the cell fails inside the scattering chamber followed by a failure of the scattering chamber window.
(d) If the ^{3}H cell fails and remains contained within the scattering chamber, an uranium getter gets activated to collect the ^{3}H gas.

Computer simulations of case (c) indicate that the largest exposure (about 1 mrem) will occur to a person located at about 200 m from the exhaust chimney above Hall A. This exposure is much less than the 360 mrem/year than the average person receives in the US due to natural (e.g. Radon gas) and man-made sources (e.g. medical X-rays).

A worse case scenario was also considered. For this case, we assume protocol is violated and the shipping container is opened (and the cell fails) outside the hut. Several points around JLab were simulated using weather patterns that enhance the formation of Tritium Oxide. The worst case has an exposure of 1.5 rem for the individual handling the cell steadily dropping to 10 mrem at 200 m and 1 mrem at 1 Km. For comparison, the Department of Energy (DOE) limit for routine whole-body exposure of radiological workers is 5 rem.

7 Summary

Comparison of the mirror nuclei ^{3}H and ^{3}He are expected to provide a major step forward in our understanding of the valence quark momentum distributions in nucleons near $x \sim 1$, the EMC effect and the nature of short range correlations in light nuclei. The experiments described here are scheduled to begin in Fall 2017.

References

1. A. Bodek et al., Experimental studies of the neutron and proton electromagnetic structure functions. Phys. Rev. D **20**, 1471–1552 (1979)
2. L.L. Frankfurt, M.I. Strikman, On the problem of extracting the neutron structure function from eD scattering. Phys. Lett. B **76**, 333–336 (1978)
3. L.L. Frankfurt, M.I. Strikman, High-energy phenomena, short-range nuclear structure and QCD. Phys. Rep. **76**, 215–347 (1981)
4. J.J. Aubert et al., The ratio of the nucleon structure functions F_2^N for iron and deuterium. Phys. Lett. B **123**, 275–278 (1983)
5. L.L. Frankfurt, M.I. Strikman, Hard nuclear processes and microscopic nuclear structure. Phys. Rep. **160**, 235–427 (1988)
6. L.W. Whitlow et al., Precise measurements of the proton and deuteron structure functions from global analysis of the SLAC deep inelastic electron scattering cross sections. Phys. Lett. B **282**, 475–482 (1992)
7. R.G. Arnold et al., Measurements of the *A* Dependence of Deep-Inelastic Electron Scattering from Nuclei. Phys. Rev. Lett. **52**, 727–730 (1984)
8. J. Gomez, Private communication (1991)
9. J. Gomez et al., Measurement of the *A* dependence of deep-inelastic electron scattering. Phys. Rev. D **49**, 4348–4372 (1994)
10. N. Baillie et al., Measurement of the neutron F_2 structure function via spectator tagging with CLAS. Phys. Rev. Lett. **108**, 142001 (2012)
11. G.R. Farrar, D.R. Jackson, Pion and nucleon structure functions near $x = 1$. Phys. Rev. Lett. **35**, 1416–1419 (1975)
12. I.R. Afnan et al., Neutron structure function and $A = 3$ mirror nuclei. Phys. Lett. B **493**, 36–42 (2000)
13. I.R. Afnan et al., Deep inelastic scattering from $A = 3$ nuclei and the neutron structure function. Phys. Rev. C **68**, 035201 (2003)
14. E. Pace et al., Neutron structure function $F_2^n(x)$ from deep inelastic electron scattering off few-nucleon systems. Phys. Rev. C **64**, 055203 (2001)
15. M.M. Sargsian et al., Neutron structure function and inclusive deep inelastic scattering from ^{3}H and ^{3}He at large Bjorken x. Phys. Rev. C **66**, 024001 (2002)
16. L.H. Tao et al., Precision measurement of $R = \sigma_L/\sigma_T$ on hydrogen, deuterium, and beryllium targets in deep inelastic electron scattering. Z. Phys. C **70**, 387–390 (1996)
17. E12-10-103 Measurement of the F_2^n/F_2^p, d/u Ratios and $A = 3$ EMC Effect in Deep Inelastic Electron Scattering off the Tritium and Helium Mirror Nuclei. https://www.jlab.org/exp_prog/proposals/10/PR12-10-103
18. L.-B. Wang, *Ph.D. Thesis* (University of Illinois, Urbana, Champaing, 2004)
19. M. Piarulli et al., Electromagnetic structure of $A = 2$ and 3 nuclei in chiral effective field theory. Phys. Rev. C **87**, 014006 (2013)
20. A. Amroun et al., ^{3}H and ^{3}He electromagnetic form factors. Nucl. Phys. A **579**, 596–626 (1994)
21. D.H. Beck et al., Tritium form factors at low q. Phys. Rev. C **30**, 1403–1408 (1984)
22. E12-14-009 Ratio of the electric form factor in the mirror nuclei ^{3}He and ^{3}H. https://www.jlab.org/exp_prog/proposals/14/PR12-14-009
23. L.L. Frankfurt et al., Evidence for short-range correlations from high $Q^2(e, e')$ reactions. Phys. Rev. C **48**, 2451–2461 (1993)

24. N. Fomin et al., New measurements of high-momentum nucleons and short-range structures in nuclei. Phys. Rev. Lett. **108**, 092502 (2012)
25. R. Subedi et al., Probing cold dense nuclear matter. Science **320**, 1476–1478 (2008)
26. R. Schiavilla et al., Tensor forces and the ground-state structure of nuclei. Phys. Rev. Lett. **98**, 132501 (2007)
27. E12-14-011 Proton and Neutron Momentum Distributions in $A = 3$ Asymmetric Nuclei. https://www.jlab.org/exp_prog/proposals/14/PR12-14-011
28. E12-11-112 Precision measurement of the isospin dependence in the 2N and 3N short range correlation region. https://www.jlab.org/exp_prog/proposals/11/PR12-11-112
29. J. Alcorn et al., Basic instrumentation for Hall A at Jefferson Lab. Nucl. Instrum. Methods Phys. Res. A **522**, 294–346 (2004)

Few-Body Syst (2017) 58:55
DOI 10.1007/s00601-016-1197-7

Stefan Kofler · Wolfgang Schweiger

A Perturbative QCD Approach to $\pi^- p \rightarrow D^- \Lambda_c^+$

Received: 15 November 2016 / Accepted: 16 December 2016 / Published online: 23 January 2017

Abstract We employ the generalized parton picture to analyze the reaction $\pi^- p \rightarrow D^- \Lambda_c^+$. Thereby it is assumed that the process amplitude factorizes into one for the perturbatively calculable subprocess $\bar{u}\, u \rightarrow \bar{c}\, c$ and hadronic matrix elements that can be parameterized in terms of generalized parton distributions for the $\pi^- \rightarrow D^-$ and $p \rightarrow \Lambda_c^+$ transitions, respectively. Representing these parton distributions in terms of valence-quark light-cone wave functions for π, D, p and Λ_c allows us to make numerical predictions for unpolarized differential and integrated cross sections as well as spin observables. In the kinematical region where this approach is supposed to work, i.e. $s \gtrsim 20\,\mathrm{GeV}^2$ and in the forward hemisphere, the resulting cross sections are of the order of nb. This is a finding that could be of interest in view of plans to measure $\pi^- p \rightarrow D^- \Lambda_c^+$, e.g., at J-PARC or COMPASS.

1 Motivation

Exclusive production of charmed hadrons is still a very controversial topic. Experimental data are very scarce and theoretical predictions differ by orders of magnitude, depending on the approach used. From general scaling considerations [1] one expects, e.g., that the $\bar{p}p \rightarrow \bar{\Lambda}_c^- \Lambda_c^+$ cross section is suppressed by at least two to three orders of magnitude as compared to the $\bar{p}\, p \rightarrow \bar{\Lambda}\, \Lambda$ cross section. This means that one probably has to deal with cross sections of the order of nb, a challenge which nevertheless seems to be experimentally treatable, as the measurement of $e^+e^- \rightarrow \Lambda_c^+ \bar{\Lambda}_c^-$ cross sections has shown [2]. A considerable improvement of the experimental situation on pair production of charmed hadrons is to be expected from the $\bar{\mathrm{P}}$ANDA detector at FAIR [3]. Another class of reactions for which experimental data may become available even in the near future is the pion-induced exclusive production of charmed hadrons as planned, e.g., at J-PARC [4].

In the present contribution we are going to present a theoretical analysis of $\pi^- p \rightarrow D^- \Lambda_c^+$ based on the generalized parton picture. Assuming the intrinsic charm of the p (and the π) to be negligible, the charmed hadrons in the final state are produced via a handbag-type mechanism (see Fig. 1). The blobs in Fig. 1 indicate soft hadronic matrix elements that are parameterized in terms of generalized parton distributions (GPDs), the $\bar{c}\, c$ pair is produced perturbatively with the c-quark mass m_c acting as a hard scale. A model for the $p \rightarrow \Lambda_c^+$ GPDs is available from foregoing work on $\bar{p}\, p \rightarrow \bar{\Lambda}_c^- \Lambda_c^+$ [5]. The new ingredients are the $\pi^- \rightarrow D^-$ transition GPDs which we model in analogy to Ref. [5] as overlap of valence-quark light-cone wave functions

This article belongs to the Topical Collection "Light Cone 2016".

S. Kofler · W. Schweiger (✉)
Institut für Physik, FB Theoretische Physik, Universität Graz, Graz, Austria
E-mail: wolfgang.schweiger@uni-graz.at

S. Kofler
E-mail: stefan.kofler@uni-graz.at

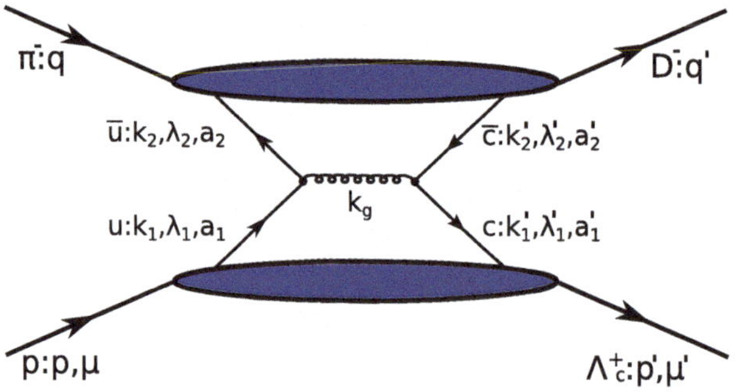

Fig. 1 The double-handbag contribution to $\pi^- \, p \to D^- \, \Lambda_c^+$ (with momenta, LC-helicities and color of the individual particles indicated)

for π^- and D^-. With these models for the $p \to \Lambda_c^+$ and $\pi^- \to D^-$ GPDs we will estimate the contribution of our handbag-type mechanism to the $\pi^- \, p \to D^- \, \Lambda_c^+$ cross section. In Sect. 2 we will sketch the steps and assumptions which finally give us a factorized form of the hadronic $\pi^- \, p \to D^- \, \Lambda_c^+$ scattering amplitude. Here also the model wave functions leading to the GPDs used for the numerical calculations are presented. Section 3 contains the numerical predictions and a short discussion of competing production mechanisms based on hadrondynamics and our conclusions.

2 Double-Handbag Amplitude, GPDs and Transition Form Factors

We consider $\pi^- \, p \to D^- \, \Lambda_c^+$ in a symmetric CM frame. This means that the transverse component of the momentum transfer $\Delta = (p' - p) = (q - q')$ is symmetrically shared between the particles and the 3-vector part of the average momentum $\bar{p} = (p + p')/2$ is aligned along the z-axis (for assignments of momenta see Fig. 1). Expressed in light-cone coordinates the particle four-momenta can then be written as

$$\begin{aligned} p &= \left[(1+\xi)\bar{p}^+, \frac{m_p^2 + \mathbf{\Delta}_\perp^2/4}{2(1+\xi)\bar{p}^+}, -\frac{\mathbf{\Delta}_\perp}{2}\right], \quad q = \left[\frac{m_\pi^2 + \mathbf{\Delta}_\perp^2/4}{2(1+\eta)\bar{q}^-}, (1+\eta)\bar{q}^-, \frac{\mathbf{\Delta}_\perp}{2}\right], \\ p' &= \left[(1-\xi)\bar{p}^+, \frac{M_{\Lambda_c}^2 + \mathbf{\Delta}_\perp^2/4}{2(1-\xi)\bar{p}^+}, \frac{\mathbf{\Delta}_\perp}{2}\right], \quad q' = \left[\frac{M_D^2 + \mathbf{\Delta}_\perp^2/4}{2(1-\eta)\bar{q}^-}, (1-\eta)\bar{q}^-, -\frac{\mathbf{\Delta}_\perp}{2}\right], \end{aligned} \tag{1}$$

where we have introduced the skewness parameter $\xi = -\Delta^+/2\bar{p}^+$. The minus component of the average momentum $\bar{q}^- = (q^- + q'^-)/2$ and the skewness parameter $\eta = \Delta^-/2\bar{q}^-$ have been introduced for convenience, but are determined by ξ, $\bar{p}^+$ and $\mathbf{\Delta}_\perp^2$.

The hadronic amplitude as depicted in Fig. 1 can then be written in the form

$$\begin{aligned} \mathcal{M} = &\int d^4k_1^{\rm av}\theta\left(k_1^{\rm av+}\right)\int\frac{d^4z_1}{(2\pi)^4}e^{ik_1^{\rm av}z_1}\int d^4k_2^{\rm av}\theta\left(k_2^{\rm av-}\right)\int\frac{d^4z_2}{(2\pi)^4}e^{ik_2^{\rm av}z_2} \\ &\times\langle\Lambda_c^+ : p'\mu'|T\,\bar{\Psi}^c(-z_1/2)\Psi^u(z_1/2)|p: p\mu\rangle\,\tilde{H}\left(k_1', k_2'; k_1, k_2\right)\,\langle D^- : q'\nu'|T\,\bar{\Psi}^u(z_2/2)\Psi^c(-z_2/2)|\pi^- : q\nu\rangle\,, \end{aligned} \tag{2}$$

with $\tilde{H}$ denoting the perturbatively calculable kernel that represents the partonic subprocess $\bar{u}\,u \to g \to \bar{c}\,c$. Here two of the four integrations over the quark 4-momenta (and the corresponding Fourier transforms) have already been eliminated by introducing average quark momenta $k_i^{\rm av} = (k_i + k_i')/2$ and the fact that the momentum transfer on hadron and parton level should be the same, i.e. $(k_1 - k_1') = (p - p')$ and $k_2 - k_2' = (q - q')$ (a consequence of translational invariance). The further analysis of $\mathcal{M}$ makes use of the collinear approximation for the active partons. This means that their momenta are replaced by

$$
\begin{aligned}
k_1 &\approx \left[k_1^+, \frac{x_1^2 \mathbf{\Delta}_\perp^2}{8k_1^+}, -\frac{1}{2}x_1 \mathbf{\Delta}_\perp \right], \quad k_1' \approx \left[k_1'^+, \frac{m_c^2 + x_1'^2 \mathbf{\Delta}_\perp^2/4}{2k_1'^+}, \frac{1}{2}x_1' \mathbf{\Delta}_\perp \right], \\
k_2 &\approx \left[\frac{x_2^2 \mathbf{\Delta}_\perp^2}{8k_2^-}, k_2^-, \frac{1}{2}x_2 \mathbf{\Delta}_\perp \right], \quad k_2' \approx \left[\frac{m_c^2 + x_2'^2 \mathbf{\Delta}_\perp^2/4}{2k_2'^-}, k_2'^-, -\frac{1}{2}x_2' \mathbf{\Delta}_\perp \right],
\end{aligned} \tag{3}
$$

with the momentum fractions $x_i^{(\prime)}$ defined by $k_1^{(\prime)+} = x_1^{(\prime)} p^{(\prime)+}$ and $k_2^{(\prime)-} = x_2^{(\prime)} q^{(\prime)-}$. For later purposes it is also convenient to introduce average momentum fractions $\bar{x}_i$ via the relations $x_1^{(\prime)} = (\bar{x}_1 \pm \xi)/(1 \pm \xi)$ and $x_2^{(\prime)} = (\bar{x}_2 \pm \eta)/(1 \pm \eta)$. The justification of the collinear approximation rests on the physically plausible assumptions that the parton virtualities and (intrinsic) transverse momenta are restricted by a typical hadronic scale of the order of 1 GeV and that the GPDs exhibit a pronounced peak at large values of $\bar{x}_1$ ($\bar{x}_2$) close to the ratio $\bar{x}_{10} = m_c/M_{\Lambda_c}$ ($\bar{x}_{20} = m_c/M_D$) (for details, see Ref. [5]). As a consequence of the collinear approximation $\tilde{H}$ does not depend on $k_1^{\text{av}-}$, $\mathbf{k}_{1\perp}^{\text{av}}$, $k_2^{\text{av}+}$, $\mathbf{k}_{2\perp}^{\text{av}}$ and the corresponding integrations can be carried out leading to delta functions in the associated z-variables. These delta functions force the products of the quark-field operators onto the light cone ($z_1 \to z_1^-$ and $z_2 \to z_2^+$) and the time ordering can be dropped [6].

To proceed further one picks out the "leading twist" contributions from the bilocal quark-field operator products:

$$
\langle \Lambda_c^+ | \overline{\Psi}^c(-z_1^-/2) \Psi^u(z_1^-/2) \, | p \rangle : \langle \Lambda_c^+ | \overline{\Psi}^c(-z_1^-/2) \{ \gamma^+, \gamma^+ \gamma_5, i\sigma^{+j} \} \Psi^u(z_1^-/2) \, | p \rangle , \tag{4}
$$

$$
\langle D^- | \overline{\Psi}^u(z_2^+/2) \Psi^c(-z_2^+/2) \, | \pi^- \rangle : \langle D^- | \overline{\Psi}^u(z_2^+/2) \{ \gamma^-, \gamma^- \gamma_5, i\sigma^{-j} \} \Psi^c(-z_2^+/2) \, | \pi^- \rangle . \tag{5}
$$

The three Dirac structures showing up in Eqs. (4) and (5) can be considered as + or − components of (bilocal) vector, pseudovector and tensor currents, respectively. These currents are then Fourier transformed (with respect to z_1^- or z_2^+) and decomposed into appropriate hadronic covariants. The coefficients in front of these covariants are the quantities which are usually understood as GPDs. Due to parity invariance the matrix elements $\langle D^- | \overline{\Psi}^u \gamma^- \gamma_5 \Psi^c \, | \pi^- \rangle$ vanish and the covariant decomposition of the remaining vector and tensor currents gives rise to two $\pi^- \to D^-$ transition GPDs, $H_{\pi D}^{\overline{c}u}$ and $E_{T\pi D}^{\overline{c}u}$, which are defined by:

$$
\begin{aligned}
&\bar{q}^- \int \frac{dz_2^+}{2\pi} e^{i\bar{x}_2 \bar{q}^- z_2^+} \langle D^- : q' | \overline{\Psi}^u \left(z_2^+/2 \right) \left\{ \gamma^-, i\sigma^{-j} \right\} \Psi^c \left(-z_2^+/2 \right) | \pi^- : q \rangle \\
&\quad = \left\{ 2\bar{q}^- \, H_{\pi D}^{\overline{c}u}(\bar{x}_2, \eta, t), \; \frac{\bar{q}^- \Delta^j - \Delta^- \bar{q}^j}{m_\pi + M_D} E_{T\pi D}^{\overline{c}u}(\bar{x}_2, \eta, t) \right\} .
\end{aligned} \tag{6}
$$

An analogous analysis leads to eight GPDs for the $p \to \Lambda_c^+$ transition [5]. These are functions of $\bar{x}_1$, ξ and $t = \Delta^2$.

Having expressed the soft hadronic matrix elements in terms of generalized parton distributions one ends up with an integral in which these parton distributions, multiplied with the hard partonic scattering amplitude $H_{\lambda_1' \lambda_2', \lambda_1 \lambda_2} \left(\bar{x}_1 \bar{p}^+, \bar{x}_2 \bar{q}^- \right)$, are integrated over $\bar{x}_1$ and $\bar{x}_2$. The fact that a heavy $\bar{c}\,c$ pair has to be produced (since we neglect non-perturbative intrinsic charm in the light hadrons) means that the virtuality of the intermediate gluon should be larger than $4m_c^2 \approx 6.3\,\text{GeV}^2$. This justifies the perturbative treatment of $\bar{u}\,u \to \bar{c}\,c$ and puts (for fixed $s > (M_{\Lambda_c} + M_D)^2 \approx 17.27\,\text{GeV}^2$) kinematical constraints on $\bar{x}_1$ and $\bar{x}_2$. For s well above the production threshold ($s \gtrsim 20\,\text{GeV}^2$) and in the forward-scattering hemisphere it can be checked numerically that these constraints imply $\bar{x}_1 > \xi$ and $\bar{x}_2 > \eta$. This means that only the DGLAP region of the GPDs ($|\bar{x}_1| > \xi$, $|\bar{x}_2| > \eta$) can contribute to our handbag-type mechanism, an important observation which simplifies the modeling of the GPDs.

The supposition that the $p \to \Lambda_c^+$ and $D^- \to \pi^-$ GPDs are strongly peaked at $\bar{x}_{10}$ and $\bar{x}_{20}$, respectively, leads to a further simplification of the $\pi^- p \to D^- \Lambda_c^+$ amplitude. The major contributions to the $\bar{x}_1$ and $\bar{x}_2$ integrals will then come from $\bar{x}_1 \approx \bar{x}_{10}$ and $\bar{x}_2 \approx \bar{x}_{20}$. One can thus replace the hard partonic scattering amplitude by its value at the peak position, $H_{\lambda_1' \lambda_2', \lambda_1 \lambda_2} \left(\bar{x}_{10} \bar{p}^+, \bar{x}_{20} \bar{q}^- \right)$ and take it out of the integral. What one is left with are separate integrals over the GPDs which may be interpreted as generalized $p \to \Lambda_c^+$ and $D^- \to \pi^-$ transition form factors. With this "peaking approximation" our final expressions for the $\pi^- p \to D^- \Lambda_c^+$ amplitudes become:

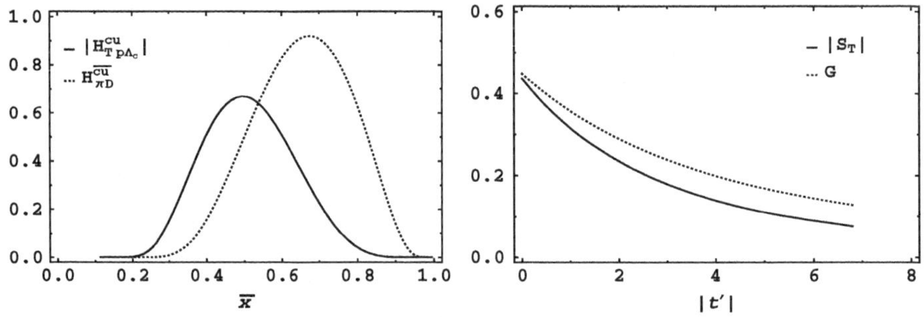

Fig. 2 *Left panel* The $p \to \Lambda_c^+$ (*solid line*) and $\pi^- \to D^-$ (*dotted line*) transition GPDs $H^{cu}_{T\,p\Lambda_c}$ and $H^{\overline{cu}}_{\pi D}$ obtained with the wave functions (9) and (10) and the KK mass exponential for $s = 25\,\text{GeV}^2$ and $\mathbf{\Delta}_\perp^2 = 0$. *Right panel* The corresponding transition form factors $|S_T|$ and G [see Eq. (8)] as functions of $|t'| = |t - t_0|$, where t_0 is the t value for forward scattering

$$\mathcal{M}_{+,+} = \mathcal{M}_{-,-} = \frac{1}{4}\sqrt{1-\xi^2}\, H_{+-,+-}\, R_V\, G,$$
$$\mathcal{M}_{+,-} = -\mathcal{M}_{-,+} = \frac{1}{4}\sqrt{1-\xi^2}\, H_{++,-+}\, S_T\, G, \tag{7}$$

with the $D^- \to \pi^-$ transition form factor

$$G(\eta, t) = \int_\eta^1 \frac{d\bar{x}_2}{\sqrt{\bar{x}_2^2 - \eta^2}} H^{\overline{cu}}_{\pi D}(\bar{x}_2, \eta, t). \tag{8}$$

In Eq. (7) we have restricted ourselves to the two most important $p \to \Lambda_c$ GPDs, $H^{cu}_{p\Lambda_c}$ and $H^{cu}_{T\,p\Lambda_c}$ (these are associated with γ^+ and σ^{+j}). The respective form factors are R_V and S_T, defined analogously to Eq. (8). The underlying assumption is that those GPDs (and corresponding form factors) which involve non-zero orbital angular momentum of the (anti)quarks that make up the hadrons are suppressed. It leads also to omission of $E^{\overline{cu}}_{T\pi D}$.

This concludes the general analysis of our process. The next step is the modeling of the GPDs. As mentioned already above, we have to consider only the DGLAP region There it is possible to model the GPDs as overlaps of light-cone wave functions for the valence Fock states of the respective hadrons [8]. For the pion and the proton we take the parameterizations of the light-cone wave functions proposed in Refs. [9] and [10]

$$\psi_\pi(x, \mathbf{k}_\perp) = N_\pi \exp\left[\frac{-a_\pi^2\,\mathbf{k}_\perp^2}{x(1-x)}\right], \quad \psi_p(x_i, \mathbf{k}_{\perp i}) = N_p(1+3x_1)\exp\left[-a_p^2 \sum \frac{\mathbf{k}_{\perp i}^2}{x_i}\right]. \tag{9}$$

These forms are supported by several phenomenological applications. Similar wave functions are taken for the D and the Λ_c:

$$\psi_D(x, \hat{\mathbf{k}}_\perp) = N_D \exp[-f(x)] \exp\left[\frac{-a_D^2\,\mathbf{k}_\perp^2}{x(1-x)}\right], \quad \psi_{\Lambda_c}(x_i, \mathbf{k}_{\perp i}) = N_{\Lambda_c}\exp[-f(x_1)]\exp\left[-a_{\Lambda_c}^2 \sum \frac{\mathbf{k}_{\perp i}^2}{x_i}\right]. \tag{10}$$

The mass exponential generates the expected peak at $x_{(1)} \approx \bar{x}_{i0}$ (with $x_{(1)}$ being the momentum fraction of the heavy quark). The parameters N_D and a_D are chosen such that the experimental value of the D-meson decay constant $f_D = 0.207\,\text{GeV}$ is reproduced and the valence-Fock-state probability becomes 0.9 [7]. An appropriate choice of parameters for the Λ_c wave function can be found in Ref. [5], where an overlap representation of $p \to \Lambda_c$ GPDs has been derived and applied to $\bar{p}p \to \bar{\Lambda}_c^- \Lambda_c^+$. We use two types of mass exponentials that have been suggested in the literature [11,12]:

$$f_{KK}(x) = \frac{a^2_{\Lambda_c(D)} M^2_{\Lambda_c(D)} (x - \bar{x}_{i0})^2}{x(1-x)}, \quad f_{BB}(x) = a^2_{\Lambda_c(D)} M_{\Lambda_c(D)}(1-x). \tag{11}$$

For a more detailed account of the formalism and the modeling of the GPDs we refer to Refs. [5,7], where also analytical formulae for the model GPDs, hard scattering amplitudes, etc. can be found. An impression

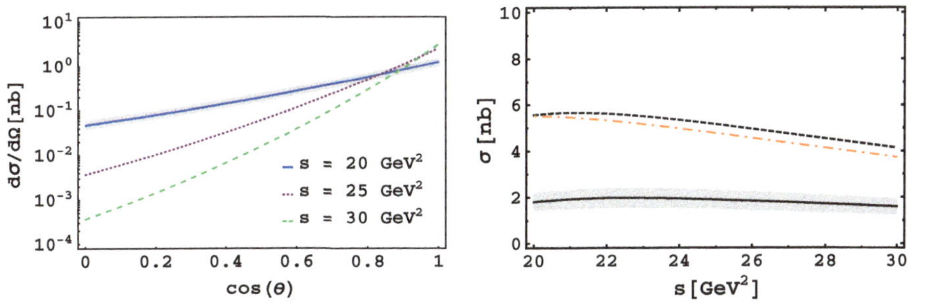

Fig. 3 *Left panel* The differential $\pi^- p \to D^- \Lambda_c^+$ cross section in the CM system versus $\cos\theta$ obtained with the KK mass exponential for $s = 20, 25, 30\,\text{GeV}^2$. Right panel: The integrated $\pi^- p \to D^- \Lambda_c^+$ cross section versus Mandelstam s for the KK (*solid line*) and the BB (*dashed line*) mass exponential in comparison with predictions from a hybrid Regge model [15] (*dash-dotted line*). The effects of uncertainties in the Λ_c and D^- wave-function parameters (in case of the KK mass exponential) are indicated by the shaded band

how our model GPDs and corresponding form factors for the $\pi^- \to D^-$ and $p \to \Lambda_c^+$ transitions look like can be gained from Fig. 2. The general behavior of the GPDs is that with increasing s the maximum moves to larger values of $\bar{x}$ and becomes smaller. It should also be mentioned that the two $p \to \Lambda_c^+$ form factors (and corresponding GPDs) that show up in Eq. (7) are approximately the same for reasonably small probability ($\lesssim$10%) to find a c-quark with helicity opposite to the Λ_c helicity in the Λ_c [5].

3 Results and Discussion

With these model GPDs (and corresponding form factors) we are now able to calculate the hadronic scattering amplitudes [see Eq. (7)] and cross sections. Predictions for unpolarized differential and integrated cross sections obtained with the KK mass exponential are displayed in Fig. 3. The shaded band exhibits the variation of the cross section, if the wave-function parameters are varied within a reasonable range. Integrated cross-section results are also shown for the KK mass exponential. Obviously the differences between predictions obtained with different analytical forms of the wave functions are larger than the variations coming from parametric errors in the wave functions. The integrated cross sections are of the order of nb with the BB mass exponential giving the larger results. This is the order of magnitude that has also been found for $\bar{p}p \to \bar{\Lambda}_c^- \Lambda_c^+$ [5] and $\bar{p}p \to \bar{D}^0 D^0$ [13], when treated within the generalized parton framework. It seems to be in accordance with old AGS experiments at $s \approx 25$ GeV^2 which found upper bounds of 7 nb for $\pi^- p \to D^{*-} \Lambda_c^+$ and ≈ 15 nb for $\pi^- p \to D^- \Lambda_c^+$ [14].

Interestingly, an integrated cross section comparable to ours has been found by the authors of Ref. [15] (dash-dotted line in Fig. 3), who used a hybridized Regge model, i.e. a production mechanism mainly based on hadron dynamics. This is in contrast to the findings for $\bar{p}p \to \bar{\Lambda}_c^- \Lambda_c^+$ and $\bar{p}p \to \bar{D}^0 D^0$, where models using simple [16,17] or Reggeized hadron exchange [18,19] find cross sections which are two to three orders of magnitude larger than those from the handbag-type mechanism. The crucial point in these models seems to be the strength of the $D^{(*)} p Y_c$ coupling which is either fixed by $SU(4)$-flavor symmetry [16–18] or by QCD sum rules [19]. In our case this coupling corresponds to the form factors G and S_T. These are the quantities which make the cross section that small. The flavor dependence of the wave functions leads to a strong mismatch of light and heavy hadron wave functions in the overlap integral. If this mismatch would not exist, our cross section would be about two orders of magnitude larger, as we have seen in a similar calculation of $\bar{p}p \to \bar{\Lambda}_c^- \Lambda_c^+$ performed within a quark-diquark model [20].

Cross sections as large as predicted by the hadronic models would also indicate that, in contrast to our assumption, charm is produced non-perturbatively which means that (non-perturbative) intrinsic charm of the proton must be taken into account. This could, in principle, be done within our approach. Then the charmed hadrons in the final state would, in addition to the handbag mechanism, be produced via mechanisms which are fed by the ERBL region of the GPDs. It is, however, hardly conceivable that the small amount of intrinsic charm in the proton that is compatible with inclusive data [21] could increase the cross section for the exclusive production of charmed hadrons by two or three orders of magnitude. This holds in particular for the kinematic situations we are considering, where the skewness parameter and thus also the ERBL region becomes small. Experimental data for processes like $\pi^- p \to D^- \Lambda_c^+$, $\bar{p}p \to \bar{\Lambda}_c^- \Lambda_c^+$, $\gamma p \to \bar{D}^0 \Lambda_c^+$ and $\bar{p}p \to D^0 \bar{D}^0$

up to several GeV above production threshold would thus be highly desirable to pin down the production mechanism of charmed hadrons and shed some more light on the question of non-perturbative intrisic charm in the proton.

Acknowledgements Open access funding provided by Austrian Science Fund (FWF). S. Kofler acknowledges the support of the Fonds zur Förderung der wissenschaftlichen Forschung in Österreich (Grant No. FWF DK W1203-N16). We are thankful to S.-H. Kim for providing us with the numerical data of his results that are shown in Fig. 3.

References

1. S.J. Brodsky, QCD in $\bar{p}$ reactions, in *Physics at LEAR with low energy antiprotons*, ed. by C. Amsler, et al. (Harwood Academic Publishers, Villars-sur-Ollon, 1987), pp. 317–328
2. G. Pakhlova et al., [Belle Collaboration], Observation of a near-threshold enhancement in the $e^+e^- \to \Lambda_c^+ \Lambda_c^-$ cross section using initial-state radiation. Phys. Rev. Lett. **101**, 172001 (2008)
3. M.F. Lutz, et al., [PANDA Collaboration], Physics performance report for PANDA: strong interaction studies with antiprotons. (2009) arXiv:0903.3905 [hep-th]
4. K. Shirotori et al., Charmed baryon spectroscopy experiment at J-PARC. JPS Conf. Proc. **8**, 022012 (2015)
5. A.T. Goritschnig, P. Kroll, W. Schweiger, Proton-antiproton annihilation into a $\Lambda_c^+ \bar{\Lambda}_c^-$ pair. Eur. Phys. J. A **42**, 43 (2009)
6. M. Diehl, T. Gousset, Time ordering in off diagonal parton distributions. Phys. Lett. B **428**, 359 (1998)
7. S. Kofler, P. Kroll, W. Schweiger, $\pi^- p \to D^- \Lambda_c^+$ within the generalized parton picture. Phys. Rev. D **91**, 054027 (2015)
8. M. Diehl, T. Feldmann, R. Jakob, P. Kroll, The overlap representation of skewed quark and gluon distributions. Nucl. Phys. B **596**, 33 (2001); Erratum. Nucl. Phys. B **605**, 647 (2001)
9. J. Bolz, P. Kroll, Modeling the nucleon wave function from soft and hard processes. Z. Phys. A **356**, 327 (1996)
10. T. Feldmann, P. Kroll, Skewed parton distributions for $B \to \pi$ transitions. Eur. Phys. J. C **12**, 99 (2000)
11. J.G. Körner, P. Kroll, Heavy quark symmetry at large recoil: the case of baryons. Z. Phys. C **57**, 383 (1993)
12. P. Ball, V.M. Braun, E. Gardi, Distribution amplitudes of the Λ_b baryon in QCD. Phys. Lett. B **665**, 197 (2008)
13. A.T. Goritschnig, B. Pire, W. Schweiger, Double handbag description of proton-antiproton annihilation into a heavy meson pair. Phys. Rev. D **87**, 014017 (2013); Erratum. Phys. Rev. D **88**, 079903 (2013)
14. J.H. Christenson, E. Hummel, G.A. Kreiter, J. Sculli, P. Yamin, Limits on charm production in hadronic interactions near threshold. Phys. Rev. Lett. **55**, 154 (1985)
15. S.H. Kim, H.C. Kim, H. Hosaka, $K^0\Lambda$ and $D^-\Lambda_c^+$ production induced by pion beams off the nucleon. Phys. Rev. D **94**, 094025 (2016)
16. J. Haidenbauer, G. Krein, Production of charmed baryons in $\bar{p}\,p$ collisions close to their thresholds (2016) arXiv:1611.02985 [hep-th]
17. J. Haidenbauer, G. Krein, Production of charmed pseudoscalar mesons in antiproton-proton annihilation. Phys. Rev. D **89**, 114003 (2014)
18. A.I. Titov, B. Kämpfer, Exclusive charm production in $\bar{p}\,p$ collisions at $\sqrt{s} \lesssim 15$ GeV. Phys. Rev. C **78**, 025201 (2008)
19. A. Khodjamirian, C. Klein, T. Mannel, Y.M. Wang, How much charm can PANDA produce? Eur. Phys. J. A **48**, 31 (2012)
20. P. Kroll, B. Quadder, W. Schweiger, Exclusive production of heavy flavors in proton-antiproton annihilation. Nucl. Phys. B **316**, 373 (1989)
21. P. Jimenez-Delgado, T.J. Hobbs, J.T. Londergan, W. Melnitchouk, New limits on intrinsic charm in the nucleon from global analysis of parton distributions. Phys. Rev. Lett. **114**, 082002 (2015)

Few-Body Syst (2017) 58:74
DOI 10.1007/s00601-017-1244-z

B. Pire · K. Semenov-Tian-Shansky · L. Szymanowski

Baryon-to-Meson Transition Distribution Amplitudes: Formalism and Models

Received: 14 November 2016 / Accepted: 20 January 2017 / Published online: 3 February 2017

Abstract In specific kinematics, hard exclusive amplitudes may be factorized into a short distance dominated part computable in a perturbative way on the one hand, and universal, confinement related hadronic matrix elements on the other hand. The extension of this description to processes such as backward meson electroproduction and forward meson production in antiproton-nucleon scattering leads to define new hadronic matrix elements of three quark operators on the light cone, the nucleon-to-meson transition distribution amplitudes, which shed a new light on the nucleon structure.

1 Introduction

Accessing transition distribution amplitudes (TDAs) in specific exclusive reactions is an important goal to progress in our understanding of quark and gluon confinement. The baryon-to-meson TDAs are defined through baryon-meson matrix elements of the nonlocal three quark (antiquark) operators on the light cone ($n^2 = 0$):

$$\hat{O}^{\alpha\beta\gamma}_{\rho\tau\chi}(\lambda_1 n, \lambda_2 n, \lambda_3 n) = \varepsilon_{c_1 c_2 c_3} \Psi^{c_1\alpha}_{\rho}(\lambda_1 n) \Psi^{c_2\beta}_{\tau}(\lambda_2 n) \Psi^{c_3\gamma}_{\chi}(\lambda_3 n), \tag{1}$$

where α, β, γ stand for the quark (antiquark) flavor indices and ρ, τ, χ denote the Dirac spinor indices. Antisymmetrization is performed over the color group indices $c_{1,2,3}$. Gauge links in (1) are omitted by adopting the light-like gauge $A \cdot n = 0$. These non-perturbative objects, first studied in [1–3], share common features both with baryon distribution amplitudes (DAs) introduced in [4,5] as baryon-to-vacuum matrix elements of the same operators (1) and with generalized parton distributions (GPDs), since the matrix element in question depends on the longitudinal momentum transfer $\Delta^+ = (p_{\mathcal{M}} - p_N) \cdot n$ between a baryon and a meson characterized by the skewness variable $\xi = -\frac{(p_{\mathcal{M}} - p_N)\cdot n}{(p_{\mathcal{M}} + p_N)\cdot n}$ and by a transverse momentum transfer $\boldsymbol{\Delta}_T$.

For the QCD evolution equations obeyed by baryon-to-meson TDAs one distinguishes the Efremov–Radyushkin–Brodsky–Lepage (ERBL)-like domain in which all three momentum fractions of quarks are

This article belongs to the Topical Collection "Light Cone 2016".

B. Pire
Centre de Physique Théorique, École Polytechnique, CNRS, Université Paris-Saclay, 91128 Palaiseau, France
E-mail: bernard.pire@polytechnique.edu

K. Semenov-Tian-Shansky
Petersburg Nuclear Physics Institute, 188300 Gatchina, Russia
E-mail: kirill.semenov@polytechnique.edu

L. Szymanowski (✉)
National Centre for Nuclear Research (NCBJ), 00-681 Warsaw, Poland
E-mail: Lech.Szymanowski@ncbj.gov.pl

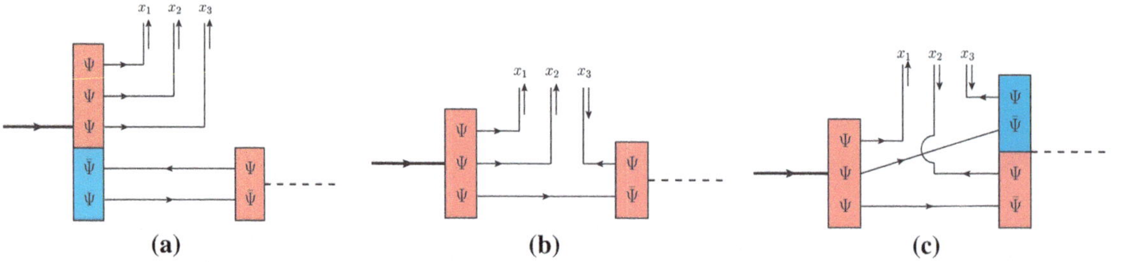

Fig. 1 Interpretation of baryon-to-meson TDAs at low normalization scale. *Small vertical arrows* show the flow of the momentum. **a** Contribution in the ERBL region (all x_i are positive). **b** Contribution in the DGLAP I region (one of x_i is negative). **c** Contribution in the DGLAP II region (two x_i are negative)

positive and two kinds of Dokshitzer–Gribov–Lipatov–Altarelli–Parisi (DGLAP)-like regions in which either one or two momentum fractions of quarks are negative.

The physical picture encoded in baryon-to-meson TDAs is conceptually close to that contained in baryon GPDs and baryon DAs. Baryon-to-meson TDAs characterize partonic correlations inside a baryon and give access to the momentum distribution of the baryonic number inside a baryon. The same operator also defines the nucleon DA, which can be seen as a limiting case of baryon-to-meson TDAs with the meson state replaced by the vacuum. In the language of the Fock state decomposition, baryon-to-meson TDAs are not restricted to the lowest Fock state as DAs. They rather probe the non-minimal Fock components with additional quark-antiquark pair:

$$|\text{Nucleon}\rangle = |\Psi\Psi\Psi\rangle + |\Psi\Psi\Psi;\ \bar{\Psi}\Psi\rangle + \cdots\ ; \qquad |\text{Meson}\rangle = |\bar{\Psi}\Psi\rangle + |\bar{\Psi}\Psi;\ \bar{\Psi}\Psi\rangle + \cdots \tag{2}$$

depending on the particular support region in question (see Fig. 1). Note that this interpretation can be justified only at a very low normalization scale and can be significantly altered at higher scales due to the evolution effects.

Similarly to GPDs, by Fourier transforming baryon-to-meson TDAs to the impact parameter space ($\boldsymbol{\Delta}_T \to \mathbf{b}_T$), one obtains additional insight on the baryon structure in the transverse plane. This allows one to perform the femto-photography of hadrons [6] from a new perspective.

2 Parametrization and Phenomenological Models for Baryon-to-Meson TDAs

Although baryon-to-meson TDAs can be introduced for all types of baryons and mesons, we start our consideration from the simplest case of nucleon-to-pion (πN) TDAs. For a given flavor contents (e.g. uud proton-to-π^0 TDA) the parametrization of the leading twist-3 πN TDA involves 8 invariant functions, each depending on the 3 longitudinal momentum fractions x_i, the skewness parameter ξ, momentum transfer squared Δ^2 as well as on the factorization scale μ^2:

$$\begin{aligned}
&4(p\cdot n)^3 \int \left[\prod_{j=1}^{3} \frac{d\lambda_j}{2\pi}\right] e^{i\sum_{k=1}^{3} x_k \lambda_k (p\cdot n)} \langle \pi^0(p_\pi)|\, \varepsilon_{c_1c_2c_3} u_\rho^{c_1}(\lambda_1 n) u_\tau^{c_2}(\lambda_2 n) d_\chi^{c_3}(\lambda_3 n)\, |N^p(p_N, s_N)\rangle \\
&= \delta(x_1+x_2+x_3-2\xi) i \frac{f_N}{f_\pi}\Big[V_1^{(p\pi^0)}(x_{1,2,3},\xi,\Delta^2)(\hat{p}C)_{\rho\tau}(U^+)_\chi \\
&+ A_1^{(p\pi^0)}(x_{1,2,3},\xi,\Delta^2)(\hat{p}\gamma^5 C)_{\rho\tau}(\gamma^5 U^+)_\chi + T_1^{(p\pi^0)}(x_{1,2,3},\xi,\Delta^2)(\sigma_{p\mu}C)_{\rho\tau}(\gamma^\mu U^+)_\chi \\
&+ m_N^{-1} V_2^{(p\pi^0)}(x_{1,2,3},\xi,\Delta^2)(\hat{p}C)_{\rho\tau}(\hat{\Delta}_T U^+)_\chi + m_N^{-1} A_2^{(p\pi^0)}(x_{1,2,3},\xi,\Delta^2)(\hat{p}\gamma^5 C)_{\rho\tau}(\gamma^5 \hat{\Delta}_T U^+)_\chi \\
&+ m_N^{-1} T_2^{(p\pi^0)}(x_{1,2,3},\xi,\Delta^2)(\sigma_{p\Delta_T}C)_{\rho\tau}(U^+)_\chi + m_N^{-1} T_3^{(p\pi^0)}(x_{1,2,3},\xi,\Delta^2)(\sigma_{p\mu}C)_{\rho\tau}(\sigma^{\mu\Delta_T} U^+)_\chi \\
&+ m_N^{-2} T_4^{(p\pi^0)}(x_{1,2,3},\xi,\Delta^2)(\sigma_{p\Delta_T}C)_{\rho\tau}(\hat{\Delta}_T U^+)_\chi \Big].
\end{aligned} \tag{3}$$

Here $f_\pi = 93\,\text{MeV}$ is the pion weak decay constant and f_N determines the value of the nucleon wave function at the origin. Throughout this paper we adopt Dirac's "hat" notation $\hat{v} \equiv v_\mu \gamma^\mu$; $\sigma^{\mu\nu} = \frac{1}{2}[\gamma^\mu, \gamma^\nu]$;

$\sigma^{v\mu} \equiv v_\lambda \sigma^{\lambda\mu}$; C is the charge conjugation matrix and $U^+ = \hat{p}\hat{n}\, U(p_N, s_N)$ is the large component of the nucleon spinor. The parametrization (3), originally introduced in Ref. [7], is extremely convenient for the phenomenological applications since in the limit $\Delta_T \to 0$ just three invariant amplitudes $V_1^{(p\pi^0)}$, $A_1^{(p\pi^0)}$ and $T_1^{(p\pi^0)}$ do survive. An alternative parametrization which is better suited to address the symmetry properties of πN TDAs was introduced in [8]. In particular, within this latter parametrization πN TDAs turn to satisfy the polynomiality property of the Mellin moments in x_i in its most simple form. The relation between the two parametrizations in presented in the Appendix C of [8].

Working out the physical normalization of baryon-to-meson TDAs and building up consistent phenomenological models for them represents a considerable task. Below we present a short overview of the approaches available at the present moment.

The first attempt to provide a model for πN TDAs relying on chiral dynamics was performed in [7] (see Ref. [8] for the detailed derivation). The soft-pion theorem [9,10] allows to express πN TDAs $V_1^{(p\pi)}$, $A_1^{(p\pi)}$ and $T_1^{(p\pi)}$ in the soft pion limit ($\xi = 1$; $\Delta^2 = m_N^2$) in terms of the 3 leading twist nucleon DAs $\{V^p, A^p, T^p\}(y_1, y_2, y_3) \equiv \{V^p, A^p, T^p\}(y_i)$. This results in a simple model for πN TDAs summarized in the Erratum to [7][1]:

$$\{V_1^{(p\pi^0)},\, A_1^{(p\pi^0)}\}(x_i, \xi) = -\frac{1}{2} \times \frac{1}{4\xi}\{V^p,\, A^p\}\left(\frac{x_i}{2\xi}\right); \quad T_1^{(p\pi^0)}(x_i, \xi) = -\frac{1}{2} \times \frac{3}{4\xi} T^p\left(\frac{x_i}{2\xi}\right); \tag{4}$$

Despite its obvious drawbacks (like the very narrow validity range limited to the close vicinity of the threshold and lack of an intrinsic Δ^2-dependence) this simple model for the first time provided a quantitative estimate of the physical normalization for πN TDAs. In particular, the predictions of the revised soft-pion-limit model (4) were recently employed in the first feasibility studies [11] for accessing πN TDAs with $\bar{\text{P}}$ANDA through $\bar{p}p \to \gamma^* \pi^0$.

Another simple model for πN TDAs suggested in Ref. [8] accounts for the contribution of the cross-channel nucleon exchange. As one can see from Fig. 2, this model is conceptually similar to the pion exchange model for the polarized nucleon GPD $\tilde{E}$ suggested in Ref. [12]. With the use of the πN TDA parametrization (3) the nucleon pole model reads:

$$\begin{aligned}
&\left\{V_1,\, A_1,\, T_1\right\}^{(p\pi^0)}(x_i, \xi, \Delta^2)\Big|_{N(940)} = \Theta_{\text{ERBL}}(x_k) \times \frac{g_{\pi NN}\, m_N f_\pi}{\Delta^2 - m_N^2} \frac{1}{(2\xi)} \frac{1-\xi}{1+\xi} \{V^p,\, A^p,\, T^p\}\left(\frac{x_i}{2\xi}\right);\\
&\left\{V_2,\, A_2,\, T_2,\, T_3\right\}^{(p\pi^0)}(x_i, \xi, \Delta^2)\Big|_{N(940)} = \Theta_{\text{ERBL}}(x_k) \times \frac{g_{\pi NN}\, m_N f_\pi}{\Delta^2 - m_N^2} \frac{1}{(2\xi)} \{V^p,\, A^p,\, T^p,\, T^p\}\left(\frac{x_i}{2\xi}\right);\\
&T_4^{(p\pi^0)}(x_i, \xi, \Delta^2)\Big|_{N(940)} = 0;\\
&\left\{V_{1,2},\, A_{1,2},\, T_{1,2,3,4}\right\}^{(p\pi^+)}(x_i, \xi, \Delta^2)\Big|_{N(940)} = -\sqrt{2}\left\{V_{1,2},\, A_{1,2},\, T_{1,2,3,4}\right\}^{(p\pi^0)}(x_i, \xi, \Delta^2)\Big|_{N(940)}.
\end{aligned} \tag{5}$$

Here $\Theta_{\text{ERBL}}(x_k) \equiv \prod_{k=1}^{3} \theta(0 \le x_k \le 2\xi)$ ensures the pure ERBL-like support of TDAs and $g_{\pi NN} \approx 13$ is the pion-nucleon phenomenological coupling. This turns to be a consistent model for πN TDAs in the ERBL-like region and satisfies the polynomiality conditions and the appropriate symmetry relations.

By an obvious change of couplings the model (5) can be generalized to the case of other light mesons (η, η', K, ... etc.). Also it is not necessarily limited to the contribution of the cross-channel nucleon exchange. In Ref. [8] the contribution of the cross-channel $\Delta(1232)$ into πN TDAs was worked out explicitly. Finally, very recently in Ref. [13] the cross-channel nucleon exchange model was generalized for the case of nucleon-to-vector meson TDAs.

Still the aforementioned baryon-to-meson TDA model describes TDAs only within the ERBL-like support region. To get a model defined on the complete support domain one may rely on the spectral representation

[1] Note the relative signs and the overall factor $\frac{1}{2}$ missed in the Erratum to [7].

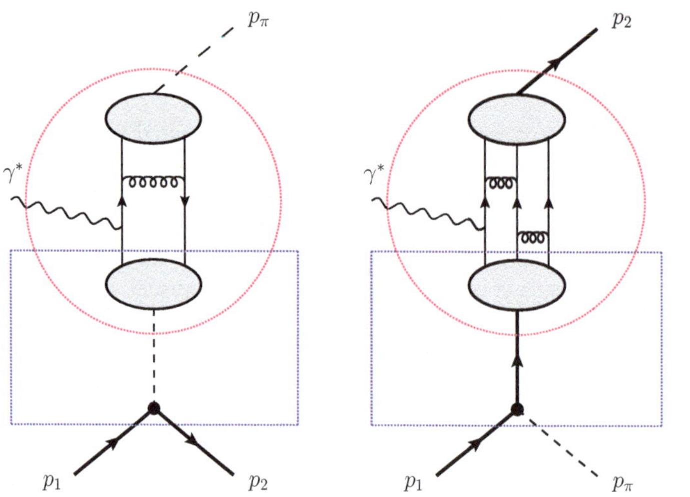

Fig. 2 *Left* Pion pole exchange model for the polarised GPD $\tilde{E}$; *lower* and *upper* blobs depict pion DAs; the *dashed circle* contains a typical LO graph for the pion electromagnetic form factor in perturbative QCD; the *rectangle* contains the pion pole contribution into GPD. *Right* Nucleon pole exchange model for πN TDAs; *dashed circle* contains a typical LO graph for the nucleon electromagnetic form factor in perturbative QCD; the *rectangle* contains the nucleon pole contribution into πN TDAs

for baryon-to-meson TDAs in terms of quadruple distributions suggested in Ref. [14]:

$$H^{(\mathcal{M}N)}(x_1, x_2, x_3 = 2\xi - x_1 - x_2, \xi, t)$$
$$= \left[\prod_{i=1}^{3} \int_{\Omega_i} d\beta_i d\alpha_i \right] \delta(x_1 - \xi - \beta_1 - \alpha_1 \xi)\, \delta(x_2 - \xi - \beta_2 - \alpha_2 \xi)\, \delta(\sum_i \beta_i) \delta(\sum_i \alpha_i + 1) f(\beta_i, \alpha_i, t), \tag{6}$$

where Ω_i denote three copies of the usual domain in the spectral parameter space. The spectral density f is an arbitrary function of six variables, which are subject to two constraints and therefore effectively is a quadruple distribution.

Similarly to the familiar double distribution representation for GPDs the quadruple distribution representation (6) for TDAs turns to be the most general way to implement the support properties of TDAs as well as the polynomiality property for the x_i-Mellin moments, which is a direct consequence of Lorentz invariance (see Ref. [15]).

Contrarily to GPDs, TDAs do not possess the comprehensive forward limit $\xi \to 0$. This complicates the construction of a phenomenological Ansatz for quadruple distributions. However, a partial solution was proposed for the case of πN TDAs. In this case one can rely on the complementary $\xi \to 1$ limit, in which, as already discussed above, πN TDAs $V_1^{\pi N}$, $A_1^{\pi N}$, $T_1^{\pi N}$ are constrained by chiral dynamics (see Eq. (4) for $\xi = 1$). In Ref. [15] we proposed a factorized Ansatz designed as a universal profile function multiplying the nucleon DA combination to which the πN TDA in question reduces in the $\xi = 1$ limit.

In loose words our Ansatz for quadruple distributions is based on "skewing" the $\xi = 1$ limit. At this point we are similar to the famous Radyushkin factorized Ansatz for double distributions [16]. In that case rather the $\xi = 0$ limit, where GPDs reduce to usual parton densities, is "skewed" to provide a non trivial ξ-dependence for GPDs. For practical details we refer the reader to Ref. [15].

Also similarly to the GPD case, in order to satisfy the polynomiality condition in its complete form, the spectral part (6) is to be complemented by a D-term-like contribution defined solely in the ERBL-like region and responsible for the highest possible power of ξ occurring for a given x_i-Mellin moment. The simplest possible model for such a D-term is the contribution of the nucleon exchange into πN TDAs (5). Note that we avoid double counting since the nucleon exchange contribution into πN TDAs $V_1^{(\pi N)}$, $A_1^{(\pi N)}$, $T_1^{(\pi N)}$ dies out in the $\xi \to 1$ limit.

In this way we come to the so-called "two component" model for πN TDAs [15] which contains a spectral part fixed from the $\xi \to 1$ limit and a D-term-like contribution coming from the cross-channel nucleon

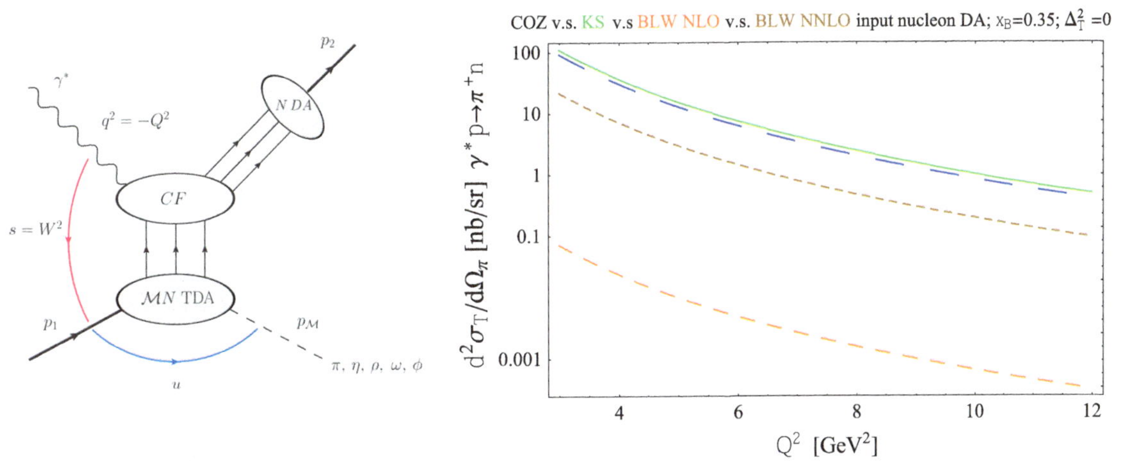

Fig. 3 *Left* Collinear factorization of $\gamma^* N \to N\mathcal{M}$ in the near-backward kinematics regime. $\mathcal{M}N$ TDA stands for the nucleon-to-meson TDA; N DA stands for the nucleon DA; CF denotes hard subprocess amplitudes (coefficient functions). *Right* Unpolarized cross section $\frac{d^2\sigma_T}{d\Omega_\pi}$ (in nb/sr) for strictly backward $\gamma^* p \to \pi^+ n$ as a function of Q^2 for $x_B = 0.35$ with various input phenomenological nucleon DA solutions: COZ (*long blue dashes*); KS (*solid green line*); BLW NLO (*medium orange dashes*) and NNLO modification [21] of BLW (*short brown dashes*) (color figure online)

exchange. This model provides a lively x_i and ξ dependence for πN TDAs. However, integrating a flexible but reasonable t−dependence still remains an open question.

Reference [15] contains the first attempts of phenomenological application of the "two component" πN TDA model for the description of backward pion electroproduction off nucleons for the typical JLab kinematical conditions. πN TDAs within this model do not nullify at the cross-over trajectories separating the ERBL-like and DGLAP-like TDA support regions. This results in the non-zero contribution into the imaginary part of the relevant elementary amplitude. The observable quantity sensitive to this issue is the transverse target single spin asymmetry (STSA) [17]. The "two component" πN TDA model predicts a sizeable value of STSA for x_B typical for JLab@12 GeV conditions.

3 Phenomenology

3.1 Baryon-to-Meson TDAs from Backward Meson Electroproduction at JLab

Within the generalized Bjorken limit, in which $Q^2 = -q^2$ and $s = (q + p_1)^2 = W^2$ are large, $x_B \equiv \frac{Q^2}{2p_1 \cdot q}$ is fixed and the u-channel momentum transfer squared is small compared to Q^2 and s ($|u| \equiv |(p_2 - p_1)^2| \ll Q^2,\ s$) the amplitude of the hard subprocess of the exclusive electroproduction of mesons off nucleons

$$e(k) + N(p_1, s_1) \to \big(\gamma^*(q, \lambda_\gamma) + N(p_1, s_1)\big) + e(k') \to e(k') + N(p_2, s_2) + \mathcal{M}(p_{\mathcal{M}}), \qquad (7)$$

is supposed to admit a collinear factorized description in terms of nucleon-to-meson TDAs and nucleon DAs, as it is shown on Fig. 3. The small u corresponds to the meson being produced in the near-backward direction in the $\gamma^* N$ center-of-mass system (CMS). This regime, referred as the near-backward kinematics, is complementary to the more conventional near-forward kinematical regime ($Q^2 = -q^2$, s – large, x_B – fixed, $|t| \equiv |(p_2 - p_1)^2| \ll Q^2,\ s$) in which the factorized description in terms of GPDs and meson DAs applies to hard meson production subprocess.

The details of the formalism for the case of backward electroproduction of light pseudoscalar mesons (π, η) and rough cross section estimates for the kinematics conditions of JLab can be found in Refs. [7,15]. A generalization for the case of vector mesons (ρ, ω, ϕ) was proposed in [13].

As an example of our predictions, on the right panel of Fig. 3 we present out estimates of the backward $\gamma^* p \to \pi^+ n$ cross section within the factorized description involving πN TDAs using the cross channel nucleon exchange model (5). The cross section turns out to very sensitive to the form of the input phenomenological solution for nucleon DA. We show the results for Chernyak–Ogloblin–Zhitnitsky (COZ) [18] (long blue dashes); King–Sachrajda (KS) [19] (solid green line); Braun–Lenz–Wittmann next-to-leading-order

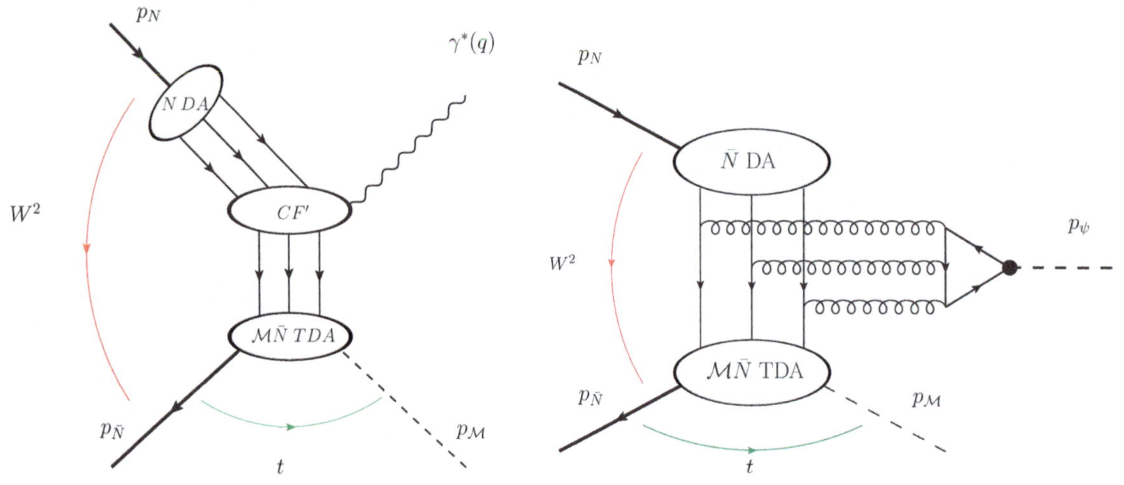

Fig. 4 *Left* Collinear factorization of the annihilation process $N\bar{N} \to \gamma^*(q)\mathcal{M}(p_{\mathcal{M}})$. (near-forward kinematics). *Right* Collinear factorization of the annihilation process $N + \bar{N} \to J/\psi + \mathcal{M}$ (near-forward kinematics). DA stands for the distribution amplitude of a nucleon; $\mathcal{M}\bar{N}$ TDA stands for the TDA from an antinucleon to a meson. By *thick black dot* we denote the light-cone wave function of heavy quarkonium

(BLW NLO) [20] (medium orange dashes) and NNLO modification [21] of BLW (short brown dashes). The solutions close to the asymptotic form of the nucleon DA $\phi(y_i) = 120 y_1 y_2 y_3$ result in a very small cross section, while those significantly different from the asymptotic form like COZ and KS result in larger cross sections. We refer the reader e.g. to the discussion in Ref. [22] on various phenomenological inputs for nucleon DAs.

A first experimental signal may have been detected at JLab [23] in backward kinematics for $e^- N \to e^- \pi N$ and for $e^- N \to e^- \omega N$ [24]. We expect the accessible kinematical domain to be more adequate to a factorized leading twist analysis in higher energy experiments at JLab@12 GeV.

3.2 Study of Baryon-to-Meson TDAs at $\bar{\text{P}}$ANDA

Another tempting possibility to get experimental access to baryon-to-meson TDAs is to consider the time-like counterpart of reaction (7): the nucleon-antinucleon annihilation into a high invariant mass lepton pair in association with a light meson $\mathcal{M}$:

$$\bar{N}(p_{\bar{N}}, s_{\bar{N}}) + N(p_N, s_N) \to \gamma^*(q) + \mathcal{M}(p_{\mathcal{M}}) \to \ell^+(p_{\ell^+}) + \ell^-(p_{\ell^-}) + \mathcal{M}(p_{\mathcal{M}}). \quad (8)$$

The factorization mechanism for (8) suggested in [25,26] applies within two kinematic regimes referred as the near-forward[2] with ($s = (p_N + p_{\bar{N}})^2 \equiv W^2$, Q^2 large with $\xi^t = -\frac{(p_{\mathcal{M}} - p_{\bar{N}})\cdot n^t}{(p_{\mathcal{M}} + p_{\bar{N}})\cdot n^t}$ fixed; and $|t| = |(p_{\mathcal{M}} - p_{\bar{N}})^2| \sim 0$) (see the left panel of Fig. 4) and the near-backward one, which differs by the obvious change of kinematical variables ($p_N \to p_{\bar{N}}, p_{\bar{N}} \to p_N, t \to u$). The charge conjugation invariance results in perfect symmetry between the two kinematical regimes. The suggested reaction mechanism manifests itself through the characteristic forward and backward peaks of the $N\bar{N} \to \gamma^* \mathcal{M}$ cross section. The characteristic features of the TDA-based description of (8) are the scaling behavior in $1/q^2$ of the cross-section and the specific behavior in $\cos\theta_\ell^*$ (by θ_ℓ^* we denote the lepton polar angle defined in the CMS of the lepton pair) resulting from the leading twist dominance of the transverse polarization of the virtual time-like photon. A detailed feasibility study for accessing πN TDAs through $\bar{p}p \to \gamma^* \pi^0 \to e^+e^-\pi^0$ at $\bar{\text{P}}$ANDA was performed in [11]. As a phenomenological input, the predictions of the simple πN TDA model (4) were employed. The results of this analysis are promising concerning the experimental perspectives for accessing πN TDAs within $\bar{\text{P}}$ANDA@GSI-FAIR experiment. On Fig. 5 we present the integrated cross section $d\bar{\sigma}/dQ^2$ with $|\Delta_T^2| \leq 0.2\,\text{GeV}^2$ for $\bar{p}p \to e^+e^-\pi^0$ as a function of Q^2 for $W^2 = 10\,\text{GeV}^2$ and $W^2 = 20\,\text{GeV}^2$ within the collinear factorization approach. We employ the cross channel nucleon exchange model (5) for πN TDAs with various phenomenological DAs used as inputs.

[2] With respect to the positive direction defined along the antiproton beam.

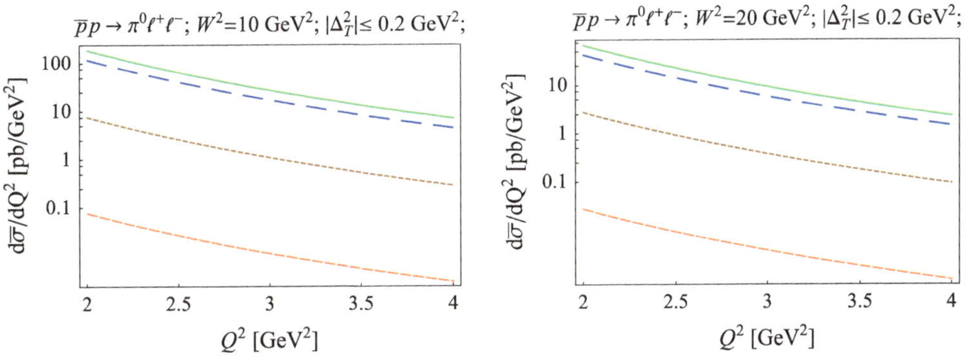

Fig. 5 Integrated cross section $d\bar{\sigma}/dQ^2$ for $\bar{p}p \to \ell^+\ell^-\pi^0$ as a function of Q^2 for $W^2 = 10\,\text{GeV}^2$ and $W^2 = 20\,\text{GeV}^2$ for various phenomenological nucleon DA solutions: COZ (*long blue dashes*); KS (*solid green line*); BLW NLO (*medium orange dashes*) and NNLO modification [21] of BLW (*short brown dashes*) (color figure online)

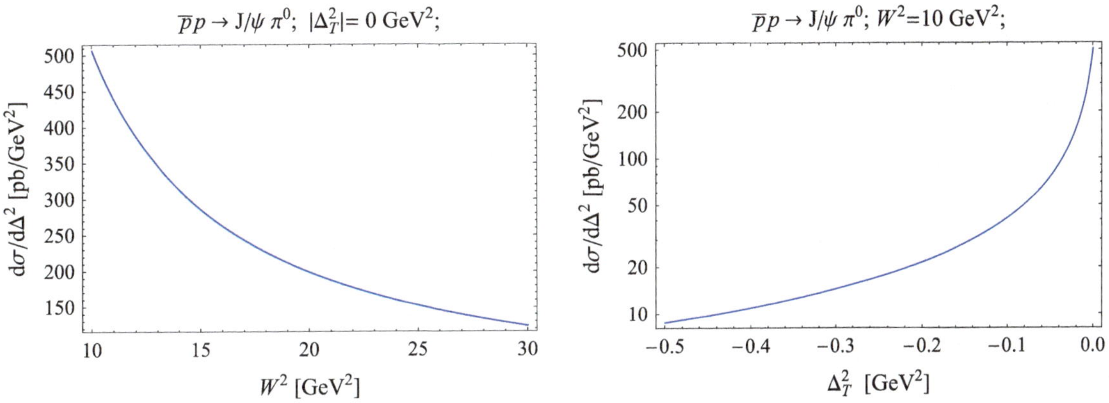

Fig. 6 *Left* Cross section $\frac{d\sigma}{d\Delta^2}$ for $p\bar{p} \to J/\psi\pi^0$ as a function of W^2 for $\Delta_T^2 = 0$; *Right* Same cross section for $W^2 = 10\,\text{GeV}^2$ as a function of the transverse momentum transfer squared Δ_T^2

The mechanism involving nucleon-to-meson TDAs was also proposed in Ref. [27] for the reaction

$$N(p_N) + \bar{N}(p_{\bar{N}}) \to J/\psi(p_\psi) + \mathcal{M}(p_\mathcal{M}), \tag{9}$$

which can also be studied at $\bar{\text{P}}$ANDA alongside with the investigation of the spectrum of charmonium states. Similarly to (8), one can consider the near-forward (see the right panel of Fig. 4) and the near-backward kinematic regimes symmetric due to charge conjugation invariance. On Fig. 6 we show the differential cross section $\frac{d\sigma}{d\Delta^2}$ for $p\bar{p} \to J/\psi\,\pi^0$ as a function of W^2 for exactly forward (backward) pion production ($\Delta_T^2 = 0$) and as a function of Δ_T^2 for a fixed value of W^2. The cross channel nucleon exchange model (5) is employed for the relevant πN TDAs. To cope with the strong $\sim\alpha_s^6$ dependence of the cross section we fix the value of α_s from the requirement that it allowed to reproduce the experimental value of $\Gamma(J/\psi \to p\bar{p})$ within the perturbative QCD description of Ref. [18]. Based on the cross-section estimates presented in Ref. [27] a detailed feasibility study for accessing $p\bar{p} \to J/\psi\pi^0$ at $\bar{\text{P}}$ANDA has been performed [11,28,29].

4 Conclusion

Baryon-to-meson TDAs are new non-perturbative objects which have been designed to help us scrutinize the inner structure of nucleons. Experimentally, one may access TDAs both in the space-like domain with backward electroproduction of mesons at JLab and COMPASS and in the time-like domain in antiproton nucleon annihilation at $\bar{\text{P}}$ANDA. We also expect [30] the time-reversed $\pi \to N$ TDAs to be accessible at J-Parc through the reactions: $\pi^- p \to J/\psi\, n$; $\pi^- p \to \gamma^* n \to \mu^+\mu^- n$. Extracting TDAs from space-like and time-like reactions will be a stringent test of their universality [31], and hence of the factorization property of hard exclusive amplitudes. This hopefully will help us to disentangle details of the complex dynamics of quark and gluon confinement in hadrons.

The special πN TDA case does not exhaust all interesting possibilities, and the vector meson sector should be experimentally accessible as well as the pseudoscalar meson sector [13]. A double handbag description of other processes such as charm meson pair production in proton-antiproton annihilation may also necessitate the introduction of baryon to charmed meson TDAs [32].

Acknowledgements This work is partly supported by Grant No. 2015/17/B/ST2/01838 by the National Science Center in Poland, by the French grant ANR PARTONS (Grant No. ANR-12-MONU-0008-01), by the COPIN-IN2P3 agreement, by the Labex P2IO and by the Polish-French collaboration agreement Polonium. K.S. acknowledges the support from the Russian Science Foundation (Grant No. 14-22-00281).

References

1. L.L. Frankfurt, P.V. Pobylitsa, M.V. Polyakov, M. Strikman, Hard exclusive pseudoscalar meson electroproduction and spin structure of a nucleon. Phys. Rev. D **60**, 014010 (1999)
2. B. Pire, L. Szymanowski, Hadron annihilation into two photons and backward VCS in the scaling regime of QCD. Phys. Rev. D **71**, 111501 (2005)
3. B. Pire, L. Szymanowski, QCD analysis of $\bar{p}N \to \gamma^*\pi$ in the scaling limit. Phys. Lett. B **622**, 83 (2005)
4. G.P. Lepage, S.L. Brodsky, Exclusive processes in perturbative quantum chromodynamics. Phys. Rev. D **22**, 2157 (1980)
5. V.L. Chernyak, A.R. Zhitnitsky, Asymptotic behavior of exclusive processes in QCD. Phys. Rep. **112**, 173 (1984)
6. J.P. Ralston, B. Pire, Femto-photography of protons to nuclei with deeply virtual Compton scattering. Phys. Rev. D **66**, 111501 (2002)
7. J.P. Lansberg, B. Pire, L. Szymanowski, Hard exclusive electroproduction of a pion in the backward region. Phys. Rev. D **75**, 074004 (2007). [Erratum-ibid. D **77**, 019902]
8. B. Pire, K. Semenov-Tian-Shansky, L. Szymanowski, πN transition distribution amplitudes: their symmetries and constraints from chiral dynamics. Phys. Rev. D **84**, 074014 (2011)
9. P.V. Pobylitsa, M.V. Polyakov, M. Strikman, Soft pion theorems for hard near threshold pion production. Phys. Rev. Lett. **87**, 022001 (2001)
10. V.M. Braun, D.Y. Ivanov, A. Lenz, A. Peters, Deep inelastic pion electroproduction at threshold. Phys. Rev. D **75**, 014021 (2007)
11. B.P. Singh, et al. [PANDA Collaboration], Experimental access to transition distribution amplitudes with the PANDA experiment at FAIR. Eur. Phys. J. A **51**(8), 107 (2015)
12. K. Goeke, M.V. Polyakov, M. Vanderhaeghen, Hard exclusive reactions and the structure of hadrons. Prog. Part. Nucl. Phys. **47**, 401 (2001)
13. B. Pire, K. Semenov-Tian-Shansky, L. Szymanowski, QCD description of backward vector meson hard electroproduction. Phys. Rev. D **91**(9), 094006 (2015)
14. B. Pire, K. Semenov-Tian-Shansky, L. Szymanowski et al., A spectral representation for baryon to meson transition distribution amplitudes. Phys. Rev. D **82**, 094030 (2010)
15. J.P. Lansberg, B. Pire, K. Semenov-Tian-Shansky, L. Szymanowski, A consistent model for πN transition distribution amplitudes and backward pion electroproduction. Phys. Rev. D **85**, 054021 (2012)
16. I.V. Musatov, A. Radyushkin, Evolution and models for skewed parton distributions. Phys. Rev. D **61**, 074027 (2000)
17. J.P. Lansberg, B. Pire, L. Szymanowski, Spin observables in transition-distribution-amplitude studies. J. Phys. Conf. Ser. **295**, 012090 (2011)
18. V.L. Chernyak, A.A. Ogloblin, I.R. Zhitnitsky, Calculation of exclusive processes with baryons. Z. Phys. C **42**, 583 (1989). [Yad. Fiz. **48**, 1398] [Sov. J. Nucl. Phys. **48**, 889]
19. I.D. King, C.T. Sachrajda, Nucleon wave functions and QCD sum rules. Nucl. Phys. B **279**, 785 (1987)
20. V.M. Braun, A. Lenz, M. Wittmann, Nucleon form factors in QCD. Phys. Rev. D **73**, 094019 (2006)
21. A. Lenz, M. Gockeler, T. Kaltenbrunner, N. Warkentin, The nucleon distribution amplitudes and their application to nucleon form factors and the $N \to \Delta$ transition at intermediate values of Q^2. Phys. Rev. D **79**, 093007 (2009)
22. N.G. Stefanis, The physics of exclusive reactions in QCD: theory and phenomenology. Eur. Phys. J. Direct C **7**, 1 (1999)
23. A. Kubarovskiy, Electroproduction of π^0 at high momentum transfers in non-resonant region with CLAS. AIP Conf. Proc. **1560**, 576 (2013)
24. Huber G, [*Private Communication*]
25. J.P. Lansberg, B. Pire, L. Szymanowski, Production of a pion in association with a high-Q^2 dilepton pair in antiproton-proton annihilation at GSI-FAIR. Phys. Rev. D **76**, 111502 (2007)
26. J.P. Lansberg, B. Pire, K. Semenov-Tian-Shansky, L. Szymanowski, Accessing baryon to meson transition distribution amplitudes in meson production in association with a high invariant mass lepton pair at GSI-FAIR with PANDA. Phys. Rev. D **86**, 114033 (2012)
27. B. Pire, K. Semenov-Tian-Shansky, L. Szymanowski, QCD description of charmonium plus light meson production in $\bar{p} - N$ annihilation. Phys. Lett. B **724**, 99 (2013) [Erratum-ibid. B764 (2017) 335]

28. B. Ma, B. Pire, K. Semenov-Tian-Shansky, L. Szymanowski, πN TDAs from charmonium production in association with a forward pion at $\overline{\text{P}}$ANDA. EPJ Web Conf. **73**, 05006 (2014)
29. B. Singh, et al. [PANDA Collaboration] Feasibility study for the measurement of πN TDAs at PANDA in $\bar{p}p \to J/\psi\pi^0$. Phys. Rev. D. arXiv:1610.02149 [nucl-ex] (2016) (to appear)
30. B. Pire, K. Semenov-Tian-Shansky, L. Szymanowski, Backward charmonium production in πN collisions. Phys. Rev. D. arXiv:1611.07234 [hep-ph] (2016) (to appear)
31. D. Müller, B. Pire, L. Szymanowski, J. Wagner, On timelike and spacelike hard exclusive reactions. Phys. Rev. D **86**, 031502 (2012)
32. A.T. Goritschnig, B. Pire, W. Schweiger, Double handbag description of proton–antiproton annihilation into a heavy meson pair. Phys. Rev. D **87**, 014017 (2013)

Few-Body Syst (2017) 58:69
DOI 10.1007/s00601-017-1236-z

Liliana Apolinário

In-Medium Parton Branching Beyond Eikonal Approximation

Received: 14 November 2016 / Accepted: 17 January 2017 / Published online: 2 February 2017

Abstract The description of the in-medium modifications of partonic showers has been at the forefront of current theoretical and experimental efforts in heavy-ion collisions. It provides a unique laboratory to extend our knowledge frontier of the theory of the strong interactions, and to assess the properties of the hot and dense medium (QGP) that is produced in ultra-relativistic heavy-ion collisions at RHIC and the LHC. The theory of jet quenching, a commonly used alias for the modifications of the parton branching resulting from the interactions with the QGP, has been significantly developed over the last years. Within a weak coupling approach, several elementary processes that build up the parton shower evolution, such as single gluon emissions, interference effects between successive emissions and corrections to radiative energy loss of massive quarks, have been addressed both at eikonal accuracy and beyond by taking into account the Brownian motion that high-energy particles experience when traversing a hot and dense medium. In this work, by using the setup of single gluon emission from a color correlated quark–antiquark pair in a singlet state ($q\bar{q}$ antenna), we calculate the in-medium gluon radiation spectrum beyond the eikonal approximation. The results show that we are able to factorize broadening effects from the modifications of the radiation process itself. This constitutes the final proof that a probabilistic picture of the parton shower evolution holds even in the presence of a QGP.

1 Introduction

Ultra-relativistic heavy-ion collisions (HIC) are a unique window of opportunity to study, in a controlled environment, a new state of matter called the quark-gluon plasma (QGP). This hot and dense medium can be described using asymptotically free quarks and gluons, the fundamental degrees of freedom of the theory of the strong interactions, the quantum chromodynamics (QCD). Therefore, HIC are a perfect laboratory to assess the QCD fundamental properties, such as asymptotic freedom and the Deconfinement/Confinement transition. The observation of this new state of matter has to made indirectly, through self-generated probes, due to its very short lifetime. Among the several possibilities, hard probes, such as high momentum particles or jets, can provide information on the macroscopic properties of the created medium through the modifications that are observed with respect to extrapolations to proton–proton (pp) collisions—a phenomena that is generically known as *Jet Quenching*. Moreover, the description of leading particle and jet cross-sections has been successfully addressed by perturbative QCD (pQCD) in pp collisions. The new challenge now is to understand at what extent pQCD can be applicable to such dense and hot systems as HIC.

This article belongs to the Topical Collection "Light Cone 2016".

L. Apolinário (✉)
LIP, Av. Elias Garcia 14, 1, 1000-149 Lisbon, Portugal
E-mail: liliana.apolinario@tecnico.ulisboa.pt

Factorization theorem states that:

$$d\sigma_{(vac)}^{AA\to h+rest} = \sum_{ijk} f_{i/A}(x_1, Q^2) \otimes f_{j/A}(x_2, Q^2) \otimes \hat{\sigma}_{ij\to k} \otimes D_{f\to h}^{(vac)}(z, \mu_F), \tag{1}$$

where $f_{i(j)/A}$ describe the probability of finding a parton $i(j)$ in the incoming hadron with a virtuality Q, $\hat{\sigma}_{ij\to j+k}$ the elementary cross-section and $D_{f\to h}(z, \mu_F^2)$ the universal hadronisation functions whose definition depends on some factorisation scale, μ_F.

Probes originated from a hard process, with a large momentum transfer, Q, take place during a time and length scale $\propto Q^{-1}$ that should not be resolved by the medium. Moreover, recent results [3] show that there is no strong dependency on the suppression of different hadronic species at high $p_\perp$, thus supporting a picture in which hadrons are formed already outside the medium. As such, even though the medium scale may be below the Λ_{QCD}, it should be possible to describe observables related to hard processes in heavy-ion collisions within a pQCD approach. Assuming a factorisation scheme in which the medium-modifications are accounted for a modification of the parton branching structure it follows:

$$d\sigma_{med}^{AA\to h+rest} = \sum_{ijk} f_{i/A}(x_1, Q^2) \otimes f_{j/A}(x_2, Q^2) \otimes \hat{\sigma}_{ij\to f+k} \otimes P_f(\Delta E, L, \hat{q}, \ldots) \otimes D_{f\to h}(z, \mu_F^2), \tag{2}$$

where the modifications to the showering of the parton f, P_f, will depend on the energy loss, ΔE, and on the several parameters necessary to describe the medium characteristics (medium length, L, transport coefficient, $\hat{q}$, ...).

This manuscript is organized as follow: Sect. 2, will be dedicated to make a brief overview about the main building blocks that are necessary to build the parton shower evolution in vacuum. This constitutes a well defined baseline to understand the modifications of the parton branching evolution in the presence of a QCD medium. In Sect. 3, it will be discussed the medium modifications to the elementary processes that build up the in-medium parton shower evolution within a pQCD approach. Main focus will be given to the contribution of the work developed by the author in this manuscript. Final conclusions will follow in Sect. 4.

2 Vacuum Parton Shower

The pQCD description of a high virtuality parton, in vacuum, is dominated by gluon bremsstrahlung. The process is illustrated in Fig. 1 (left). The probability of emitting a gluon of energy ω and transverse momentum $k_\perp$ is given by:

$$dP^{q\to qg} \sim \alpha_s C_R \frac{d\omega}{\omega} \frac{dk_\perp^2}{k_\perp^2}, \tag{3}$$

where $\alpha_s = g^2/(4\pi)$, g is the coupling constant and C_R the Casimir color factor ($R = F$ for parent quarks and $R = A$ for parent gluons). An evident feature from Eq. (3) is the double log enhancement for collinear and soft gluon emissions. However, the resummation of such contributions alone fail to describe the jet fragmentation functions by overpopulating the soft modes. To correctly account for multiple gluon emissions, it is necessary to check the interferences between subsequent emissions. It can be shown (see, for instance [18,19] for a complete review) that such interferences actually reduce the available phase space for subsequent radiation, originating the known property of *angular ordering*. This phenomena constitutes the basis of every Monte Carlo event generator. This can be more easily observed when considering the emission of a soft gluon from a

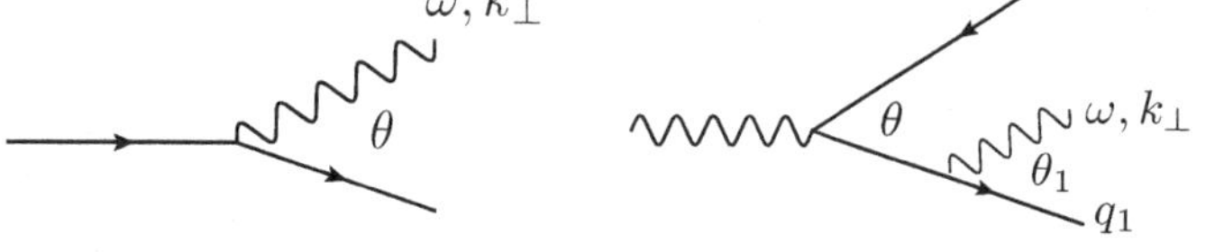

Fig. 1 Diagrams of the processes that constitute the building blocks of a QCD parton shower: *left* gluon bremsstrahlung; *right* gluon emission from a quark–antiquark antenna

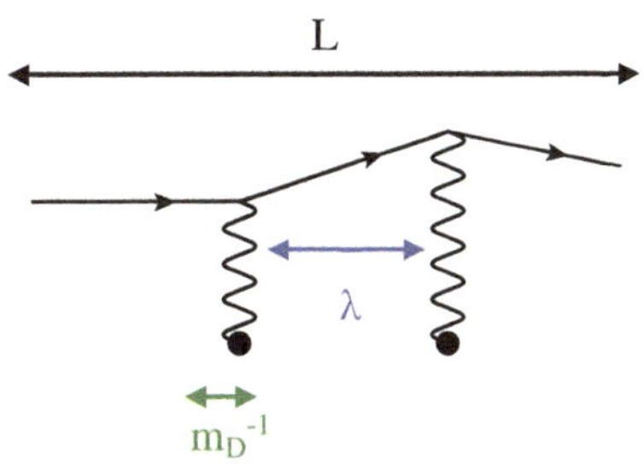

Fig. 2 Illustration of the ordering of the medium scales ($L \gg \lambda \gg m_D^{-1}$) to make a perturbative treatment of the medium dynamics and medium–probe interactions. The scattering centres are represented as *black dots* and the probe as an *arrow*. Each scattering centre induces a transverse momentum *kick* in the transverse direction

quark–antiquark pair that was initiated by a photon[1] (see Fig. 1, right). After integrating over azimuthal angle, the spectrum of radiated gluons of the quark, and similarly for the antiquark, goes as:

$$dN_q^{\omega\to 0} \sim \alpha_s C_F \frac{d\omega}{\omega}\frac{\sin\theta d\theta}{1-\cos\theta}\Theta(\cos\theta_1 - \cos\theta), \tag{4}$$

where θ_1 is the angle between the radiated gluon and the parent parton and $C_F = (N^2-1)/(2N)$ being $N = 3$ the number of colors. The Θ function implies that $\theta > \theta_1 > \theta_2 > \cdots$, thus suppressing large angle emissions. This effect was experimentally observed by the TASSO [14] and OPAL [2] collaborations.

3 Medium Parton Shower

3.1 In-Medium Propagation

The considered medium (QGP) is a strongly coupled fluid and, therefore, its interactions are at a scale that is non perturbative. Nonetheless, the evolution of a hard probe (particle or jet) that is propagating though this matter is, as seen in Sect. 2, described by pQCD since it has a virtuality scale $\sqrt{Q^2} \gg \Lambda_{QCD}, T$, where T is the temperature of the plasma. In this work, we assume that the medium description and its interaction with a hard probe can be described within such perturbative approximation. As such, the medium is seen as a collection of independent static scattering centres such that $L \gg \lambda \gg m_D^{-1}$ where L is the medium length, λ the mean-free path of the particle inside the medium and m_D the Debye screening mass, see Fig. 2. The average squared transverse momentum acquired by the propagating particle, $\langle k_\perp^2\rangle$, per each λ,

$$\hat{q} = \frac{\langle k_\perp^2\rangle}{\lambda}, \tag{5}$$

is called the transport coefficient and, together with the medium length L, are the only parameters necessary to describe the medium characteristics within the employed approximations.

Using the high energy approximation,[2] $k_+ \gg k_\perp$. No energy loss is assumed to occur in the longitudinal direction, and the result of the several interactions with the medium scattering centres is a color phase rotation of the incoming particle, that is described by a Wilson Line:

$$W_{BA}(x_{0+}, L_+; x_\perp) = \mathcal{P}\exp\left\{ig\int_{x_{0+}}^{L_+} dx_+ A_-(x_+, x_\perp)\right\}, \tag{6}$$

[1] When the parent parton of the the quark–antiquark antenna is a gluon, an additional term $\propto C_A\Theta(\cos\theta - \cos\theta_1)$ appears. These emissions, since they are proportional to the color of the gluon, are re-interpreted as emissions from the parent parton before the splitting. As such, the same angular-ordering interpretation holds.

[2] In the remaining of this manuscript, it will be used light cone coordinates, $x = (x_+, x_-, x_\perp)$, whose relation with Minkowski coordinates are as follows: $x_+ = \frac{x_0+x_3}{\sqrt{2}}$, $x_- = \frac{x_0-x_3}{\sqrt{2}}$ and $x_\perp = (x_1, x_2)$.

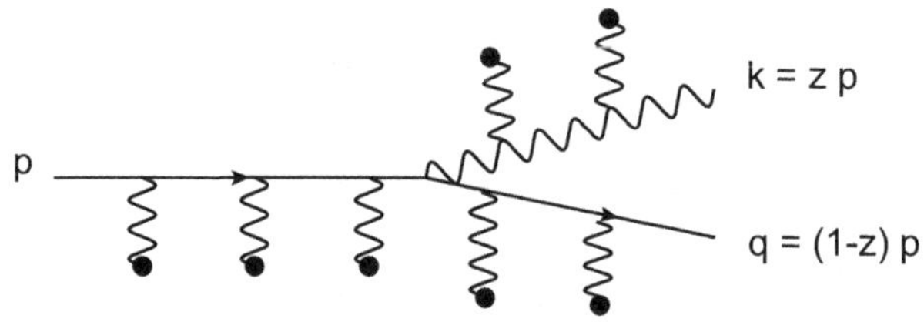

Fig. 3 Illustration of the in-medium gluon bremsstrahlung process. The initial quark has 4-momenta p and emits a gluon with a fraction z of its longitudinal energy

where A and B are the initial and final color states of the incoming particle and A_- the medium fields that are path ordered in the longitudinal direction x_+ and transverse position $x_\perp$. The longitudinal boundaries of the medium are made explicit in the integral. On top of this modification, the particle receives a transverse momentum *kick* from each interaction. Such effect will induce a Brownian motion in the transverse plane, from $x_{0,\perp}$ at the longitudinal position x_{0+} to $x_\perp$ at L_+, that is described by a Green's function:

$$G_{BA}(x_{0+}, x_{0,\perp}, L_+, x_\perp | p_+) = \int_{r_\perp(x_{0+})}^{r_\perp(L_+)} \mathcal{D}r_\perp(\xi) \exp\left\{ \frac{ik_+}{2} \int_{x_{0+}}^{L_+} d\xi \left(\frac{dr_\perp}{d\xi} \right)^2 \right\} \times W_{BA}(x_{0+}, L_+; r_\perp(\xi)). \tag{7}$$

Within the considered limit, it is understood that the dynamics that lead to the modifications of the medium colour structure occur in a timescale that is much larger than the propagation time of the penetrating particle. The computation of the resulting modifications can be done assuming a frozen medium color configuration while an average over the medium configurations ensemble should be carried out at the end. Moreover, one can decouple the color from the kinematic information from Eq. (7) when calculating the medium averages. The former accounts for the computation of N-field correlators (Eq. (6)) that are taken up to second order in the fields, $\langle A(x_\perp)A(y_\perp)\rangle$. The re-exponentiation of the result leads to:

$$\frac{1}{N} Tr \left\langle W(x_\perp) W^\dagger(y_\perp) \right\rangle = \exp\left\{ -\frac{C_F}{2} \int dx_+ \sigma(x_\perp - y_\perp) n(x_+) \right\}, \tag{8}$$

where $n(x_+)$ is the longitudinal density of scattering centres and σ the dipole cross-section. The latter, within a multiple soft scattering approximation, is approximated by its short distance term such that:

$$C_F n(\xi) \sigma(r) \simeq \frac{1}{2} \hat{q} r^2 + \mathcal{O}(r^2 \ln r^2). \tag{9}$$

As for the kinematic information, a semi-classical approximation is usually employed to evaluate the corresponding path-integrals. The dominant contribution is given by the classical trajectory that is found from the classical action, $R_{cl} = \int_{x_+}^{y_+} d\xi \mathcal{L}$, with some additional fluctuations:

$$\begin{aligned} G_0(x_+, x_\perp; y_+, y_\perp) &= \int_{r_\perp(x_+)=x_\perp}^{r_\perp(y_+)=y_\perp} \mathcal{D}r_\perp(\xi) \exp\left\{ \frac{ip_+}{2} \int_{x_+}^{y_+} d\xi \left(\frac{dr_\perp}{d\xi} \right)^2 \right\} \\ &= \frac{1}{(2\pi i)^{D/2}} \left| det\left(-\frac{\partial^2 R_{cl}}{\partial y_i \partial x_i} \right) \right|^{1/2} \exp\{ i R_{cl}(x_+, x_\perp, y_+, y_\perp) \}, \end{aligned} \tag{10}$$

where $i, j \in \{1, \ldots, D\}$ and D is the number of dimensions.

3.2 In-Medium Radiation

Multiple soft scatterings will enhance the resulting gluon radiation due to the accumulation of momenta by each scattering. This process, illustrated in Fig. 3, has been extensively addressed in the last years (see [4,7–9,26–28] for works done in the same path-integral approach as this manuscript and [21,22] for works done

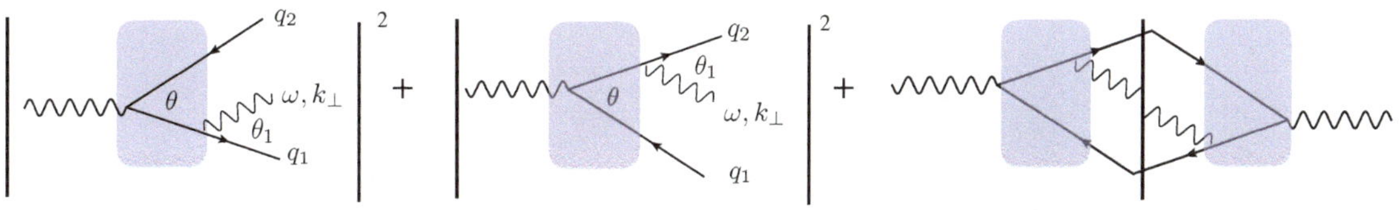

Fig. 4 Diagrams that contribute to the single-gluon emission energy spectrum from a quark–antiquark antenna setup in a finite medium (represented as the *purple region*). The three contributions correspond to the sum of the three terms in Eq. (12) (colour figure online)

using a soft and collinear effective theory). More recent works [5,10,12] were able to include all non-eikonal corrections in the description of the in-medium gluon bremsstrahlung. They have shown that the probability of such process is proportional to the independent broadening of the two outgoing particles with a correcting factor [5]:

$$\Delta_{coh} = 1 + \int_{x_+}^{L_+} d\tau \hat{q}\, (x_q - x_{\bar{q}}) \cdot (x_g - x_{\bar{g}})\big|_\tau \exp\left\{ -\hat{q}_F \int_{x_+}^{\tau} d\xi (x_q - x_{\bar{g}}) \cdot (x_{\bar{q}} - x_g) \right\}, \tag{11}$$

where $x_{q(g)}$ stands for the transverse position of the quark (gluon) and in the amplitude and $x_{\bar{q}(\bar{g})}$ in the complex conjugate amplitude. This additional term states that, on top of a complete factorisation of the showering process, there is an additional correction in which both final particles remain in a coherent state. There were several works that address the phenomenology of the parton cascade when neglecting this correction factor [11,13,20,23]. In this work, we devoted ourselves to study the consequences of such correction for subsequent emissions. For that, we consider the gluon emission off a quark–antiquark antenna that is propagating through a finite medium. The corresponding diagrams that need to be evaluated are schematically represented in Fig. 4.

3.3 In-Medium QCD Antenna

The calculation of the diagrams in Fig. 4 was done in the eikonal approximation in the works by [15,16,24,25]. The result for the spectrum of radiated gluons can be written as:

$$\frac{dI}{d\Omega_k} = R_q + R_{\bar{q}} - 2J(1 - \Delta_{med}), \tag{12}$$

where the gluon phase space is $d\Omega_k = dk_+/2k_+\, d^2k_\perp/(2\pi)^3$. The Dirac contributions corresponding to the diagrams in Fig. 4 are defined as:

$$\begin{aligned} R_q &\sim \alpha_s C_F \frac{q_{1+}}{(k \cdot q_1)}, \quad R_{\bar{q}} \sim \alpha_s C_F \frac{q_{2+}}{(k \cdot q_2)} \\ 2J &\sim \alpha_s C_F \left[\frac{q_{1+}}{(k \cdot q_1)} + \frac{q_{2+}}{(k \cdot q_2)} - \frac{k_+(q_1 \cdot q_2)}{(k \cdot q_1)(k \cdot q_2)} \right], \end{aligned} \tag{13}$$

where:

$$1 - \Delta_{med} = \frac{1}{N^2} Tr\left\langle W_A(x_q) W_A^\dagger(x_{\bar{q}}) \right\rangle, \tag{14}$$

is the non-trivial color structure from the third diagram in Fig. 4. The A subscript refers to the adjoint representation of Eq. (6) and it is made explicit the transverse coordinates for the quark (x_q) and anti-quark ($x_{\bar{q}}$). Finally, it follows that:

$$\frac{dI}{d\Omega_k} = R_{coh} + 2J\Delta_{med}, \tag{15}$$

where

$$R_{coh} \sim \alpha_s C_F \frac{k_+(q_1 \cdot q_2)}{(k \cdot q_1)(k \cdot q_2)}, \tag{16}$$

is the result from the *vacuum* antenna (Eq. (4) after integrating in azimuthal angle).

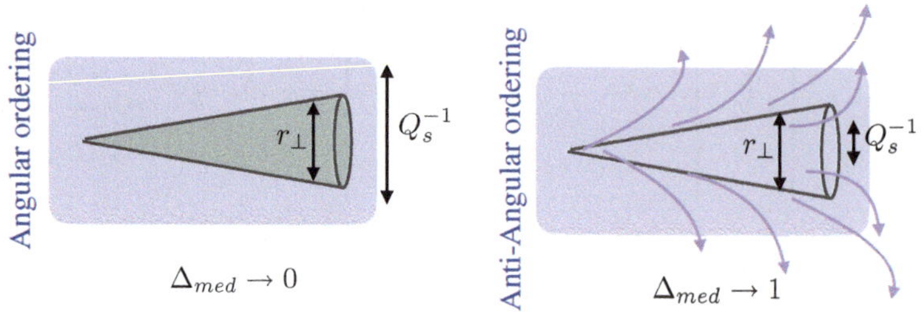

Fig. 5 Schematic interpretation of the two angular ordering regimes of Eq. (17). The *left* diagram corresponds to $\Theta(\cos\theta_1 - \cos\theta)$ function, representing the vacuum angular ordering property. The *right* diagram corresponds to the $\Theta(\cos\theta - \cos\theta_1)$ function and stands for the anti-angular ordering property

After integrating in azimuthal angle, and limiting the present discussion to the soft limit, the spectrum of radiated gluons is proportional to:

$$dN_q^{\omega\to 0} \sim \alpha_s C_F \frac{d\omega}{\omega}\frac{\sin\theta d\theta}{1-\cos\theta}\left[\Theta(\cos\theta_1 - \cos\theta) + \Delta_{med}\Theta(\cos\theta - \cos\theta_1)\right], \tag{17}$$

where (within this limit),

$$\Delta_{med} \approx 1 - \exp\left\{-\frac{1}{12}Q_s^2 r_\perp^2\right\}. \tag{18}$$

The transverse antenna resolution is $r_\perp = \theta L$ and the medium transverse scale $Q_s^{-1} = (\hat{q}L)^{-1/2}$. From there, it is clear the interplay between these two scales: when $r_\perp \ll Q_s^{-1}$, $\Delta_{med} \to 0$ and the vacuum result, Eq. (4), where subsequent emissions follow the angular ordering behaviour, is recovered; in the opposite limit, $r_\perp \gg Q_s^{-1}$, $\Delta_{med} \to 1$ and the available phase space for radiation opens to allow also *anti-angular ordering* emissions, where $\theta < \theta_1 < \theta_2 < \cdots$. A schematic interpretation of this result is shown in Fig. 5 for the two limiting cases: vacuum interferences remain in the development of the shower while the medium is not able to resolve the two emitters independently (Fig. 5, left); whenever the medium transverse resolution is able to probe quark and anti-quark separately, the destructive interferences are suppressed and a new angular regime becomes available, in which subsequent emissions are emitted at larger angles (Fig. 5, right).

When going beyond the eikonal limit, the color structure of the diagrams in Fig. 4 has to account for transverse Brownian motion, and so, there will be additional terms with respect to the previous case [6]. The new non-trivial color structures are represented in Fig. 6. As one can see, the direct terms (Fig. 6, left) will be proportional to a quadrupole that, within the considered limits (soft gluon radiation and multiple soft scattering approximation), can be approximated by the two independent broadenings of both quark and anti-quark:

$$\begin{aligned}\Delta'_{coh} &= \frac{1}{N2}Tr\left\langle G_1 G_{\bar{2}}^\dagger G_{\bar{1}}^\dagger G_{\bar{2}}\right\rangle = \frac{1}{N^2}Tr\left\langle G_1 G_{\bar{2}}^\dagger\right\rangle Tr\left\langle G_{\bar{1}}^\dagger G_{\bar{2}}\right\rangle \Delta_{coh}\\ &\simeq \frac{1}{N^2}Tr\left\langle G_1 G_{\bar{2}}^\dagger\right\rangle Tr\left\langle G_{\bar{1}}^\dagger G_{\bar{2}}\right\rangle,\end{aligned} \tag{19}$$

where $1(\bar{1})$, $2(\bar{2})$ refers to the quark and anti-quark in the (complex conjugate) amplitude. As for the interference term (Fig. 6, right), a similar Δ_{med} parameter that includes transverse momentum broadening can be defined such as:

$$1 - \Delta'_{med} = \frac{1}{N^2}Tr\left\langle G_1 G_{\bar{2}}^\dagger\right\rangle Tr\left\langle G_{\bar{1}}^\dagger G_{\bar{2}}\right\rangle. \tag{20}$$

The Dirac structure will be the same as Eq. (13). The spectrum in Eq. (12) (or Eq. (15)) is modified such that:

$$\frac{dI}{d\Omega_q d\Omega_{\bar{q}} d\Omega_k} = \Delta'_{coh}(R_q + R_{\bar{q}}) - 2(1-\Delta'_{med})J = \Delta'_{coh}R_{coh} + 2\overline{\Delta_{med}}J \tag{21}$$

where it was defined $(1-\Delta'_{med}) = \Delta'_{coh} - \overline{\Delta_{med}}$ to have the same diagrammatic interpretation as in Fig. 5. There is a correction due to the transverse Brownian motion such that the angular-ordering *cone* opens by:

$$\Delta'_{coh} \sim \frac{4\pi}{(\hat{q}\Delta L_+)^2}\left[(k_\perp + q_{1,\perp} - q_{2,\perp})^2 + \hat{q}\Delta L_+\right]\exp\left\{-\frac{(q_{1,\perp}+q_{2,\perp})^2}{\hat{q}\Delta L_+}\right\}, \tag{22}$$

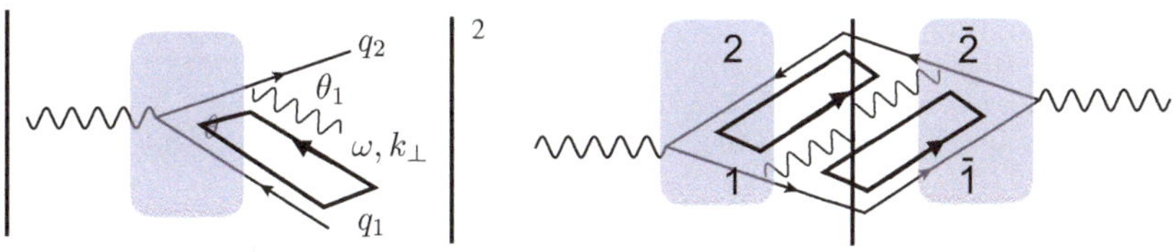

Fig. 6 Schematic representation of the *color* structure beyond the eikonal approximation of the diagrams in Fig. 4. The traces in Eqs. (19) and (20) are represented as *closed arrow* loops

where $\Delta L_+ = L_+ - x_{0+}$. As for the anti-angular component, instead of straight trajectories, the two emitters will oscillate in the transverse plane with:

$$1 - \Delta'_{med} \sim \exp\left\{-\frac{i\tan(\Omega\Delta L_+)}{\Omega}\left[(1-z)(q_{1,\perp}+k_\perp) - zq_{2,\perp}\right]^2 + \frac{i\tan(\Omega'\Delta L_+)}{\Omega'}\left[(1-z)q_{1,\perp} - z(q_{2,\perp}+k_\perp)\right]^2\right\}, \tag{23}$$

where $\Omega^2 \sim \hat{q}/(z(1-z)p_+)$, being p_+ the photon longitudinal energy, z the energy fraction carried by the quark and $\Omega'^2 = -\Omega^2$.

As such, even when considering non-eikonal corrections, one can understood a medium modified jet as composed by angular ordered mini vacuum-like jets. Each of these hard and collimated structures will behave as a different emitter inside of the jet that will lose energy independently. Therefore, the hard structure of the jet should be unaffected by the medium while the amount of soft fragments should be enhanced with respect to pp collisions. This expectation seems to be in agreement with the latest experimental results from both ATLAS [1] and CMS [17] collaborations.

4 Conclusions

Significant progress been made to understand the pQCD evolution of the parton shower in the presence of a hot and dense medium. In this work, we calculated the medium-induced gluon radiation from a non-eikonal QCD antenna in a finite medium. The results show that it is possible to generalize the previous results derived in a strict eikonal approximation to account for transverse momentum broadening. Nonetheless, the qualitative picture of a medium-modified jet remains to be valid, even when including such non-eikonal corrections. The resulting picture seems to be also in agreement with experimental observations.

Continuously efforts are still necessary to improve the current limitations of the qualitative picture obtained so far. Although it is clear that there is a strong interplay between medium- and vacuum-like jets, it is still unclear how to derive analytical evolution equations for both as the ordering variable by which one is able to resum the dominant contributions is not the same ($t \sim lnQ^2$ for vacuum-like jets while $t \sim L$ for medium-like jets). Moreover, the generalisation to multiple branchings is necessary to accurately evaluate the level of approximation when considering a complete factorised picture during the parton shower evolution.

Acknowledgements The author thanks N. Armesto, G. Milhano and C. Salgado for carefully reading this manuscript. This work was funded by the portuguese Fundação para a Ciência e Tecnologia, Grants SFRH/ BPD/103196/2014 and CERN/FIS-NUC/0049/2015.

References

1. G. Aad et al., Measurement of inclusive jet charged-particle fragmentation functions in Pb + Pb collisions at $\sqrt{s_{NN}} = 2.76$ TeV with the ATLAS detector. Phys. Lett. B **739**, 320–342 (2014). doi:10.1016/j.physletb.2014.10.065
2. G. Abbiendi et al., Charged particle momentum spectra in e+ e− annihilation at s**(1/2) = 192-GeV to 209-GeV. Eur. Phys. J. **C27**, 467–481 (2003). doi:10.1140/epjc/s2002-01121-3
3. ALICE Collaboration, Centrality dependence of the nuclear modification factor of charged pions, kaons, and protons in Pb–Pb collisions at $\sqrt{s_{NN}}$ =2.76 TeV. Phys. Rev. C **93**(3):034913 (2016). doi:10.1103/PhysRevC.93.034913
4. L. Apolinario, N. Armesto, C.A. Salgado, Medium-induced emissions of hard gluons. Phys. Lett. B **718**, 160–168 (2012). doi:10.1016/j.physletb.2012.10.040

5. L. Apolinário, N. Armesto, J.G. Milhano, C.A. Salgado, Medium-induced gluon radiation and colour decoherence beyond the soft approximation. J. High Energy Phys. **02**(2), 119 (2015). doi:10.1007/JHEP02(2015)119
6. L. Apolinário, N. Armesto, J.G. Milhano, C.A. Salgado (2017) (in preparation)
7. P.B. Arnold, G.D. Moore, L.G. Yaffe, Photon and gluon emission in relativistic plasmas. JHEP **06**, 030 (2002). doi:10.1088/1126-6708/2002/06/030
8. R. Baier, Y. Dokshitzer, A. Mueller, S. Peigné, D. Schiff, Radiative energy loss and pT-broadening of high energy partons in nuclei. Nucl. Phys. B **484**, 265–282 (1997a). doi:10.1016/S0550-3213(96)00581-0
9. R. Baier, Y. Dokshitzer, A. Mueller, S. Peigné, D. Schiff, Radiative energy loss of high energy quarks and gluons in a finite-volume quark-gluon plasma. Nucl. Phys. B **483**(1), 291–320 (1997b). doi:10.1016/S0550-3213(96)00553-6
10. J.P. Blaizot, F. Dominguez, E. Iancu, Y. Mehtar-Tani, Medium-induced gluon branching. J. High Energy Phys. **01**(1), 143 (2013). doi:10.1007/JHEP01(2013)143
11. J.P. Blaizot, E. Iancu, Y. Mehtar-Tani, Medium-induced QCD cascade: democratic branching and wave turbulence. Phys. Rev. Lett. **111**(5), 3–6 (2013b). doi:10.1103/PhysRevLett.111.052001
12. J.P. Blaizot, F. Dominguez, E. Iancu, Y. Mehtar-Tani, Probabilistic picture for medium-induced jet evolution. J. High Energy Phys. **06**(6), 075 (2014). doi:10.1007/JHEP06(2014)075
13. J.P. Blaizot, Y. Mehtar-Tani, M.A.C. Torres, Angular structure of the in-medium QCD cascade. Phys. Rev. Lett. **114**(22), 1–4 (2015). doi:10.1103/PhysRevLett.114.222002
14. W. Braunschweig et al., Global jet properties at 14-GeV to 44-GeV center-of-mass energy in e^+e^- annihilation. Z. Phys. **C47**, 187–198 (1990). doi:10.1007/BF01552339
15. J. Casalderrey-Solana, E. Iancu, Interference effects in medium-induced gluon radiation. J. High Energy Phys. (2011). doi:10.1007/JHEP08(2011)015
16. J. Casalderrey-Solana, Y. Mehtar-Tani, C.A. Salgado, K. Tywoniuk, New picture of jet quenching dictated by color coherence. Phys. Lett. B **725**(4–5), 357–360 (2013). doi:10.1016/j.physletb.2013.07.046
17. S. Chatrchyan et al., Measurement of jet fragmentation in PbPb and pp collisions at $\sqrt{s_{NN}} = 2.76$ TeV. Phys. Rev. **C90**(2), 024,908 (2014). doi:10.1103/PhysRevC.90.024908
18. Y.L. Dokshitzer, V.A. Khoze, A.H. Mueller, S.I. Troian, Basics of perturbative QCD (Edition Frontiers, 1991)
19. R.K. Ellis, W.J. Stirling, B.R. Webber, QCD and collider physics, (Cambridge University Press, 1996)
20. L. Fister, E. Iancu, Medium-induced jet evolution: wave turbulence and energy loss. J. High Energy Phys. **03**(3), 082 (2015). doi:10.1007/JHEP03(2015)082
21. M. Gyulassy, P. Levai, I. Vitev, Jet quenching in thin quark gluon plasmas. 1. Formalism. Nucl. Phys. **B571**, 197–233 (2000a). doi:10.1016/S0550-3213(99)00713-0
22. M. Gyulassy, P. Levai, I. Vitev, NonAbelian energy loss at finite opacity. Phys. Rev. Lett. **85**, 5535–5538 (2000b). doi:10.1103/PhysRevLett.85.5535
23. E. Iancu, B. Wu, Thermalization of mini-jets in a quark-gluon plasma. J. High Energy Phys. **10**(10), 155 (2015). doi:10.1007/JHEP10(2015)155
24. Y. Mehtar-Tani, C.A. Salgado, K. Tywoniuk, Anti-angular ordering of gluon radiation in QCD media. Phys. Rev. Lett. **106**, 122,002 (2011). doi:10.1103/PhysRevLett.106.122002
25. Y. Mehtar-Tani, C.A. Salgado, K. Tywoniuk, Jets in QCD media: from color coherence to decoherence. Phys. Lett. B **707**(1), 156–159 (2012). doi:10.1016/j.physletb.2011.12.042
26. U.A. Wiedemann, Jet quenching versus jet enhancement: a Quantitative study of the BDMPS-Z gluon radiation spectrum. Nucl. Phys. A **690**, 731–751 (2001). doi:10.1016/S0375-9474(01)00362-1
27. B.G. Zakharov, Fully quantum treatment of the Landau-Pomeranchuk-Migdal effect in QED and QCD. J. Exp. Theor. Phys. Lett. **63**(12), 952–957 (1996). doi:10.1134/1.567126
28. B.G. Zakharov, Radiative energy loss of high-energy quarks in finite-size nuclear matter and quark-gluon plasma. JETP Lett. **65**(8), 615–620 (1997). doi:10.1134/1.567389

Few-Body Syst (2017) 58:85
DOI 10.1007/s00601-017-1249-7

K. Tsushima · **J. P. B. C. de Melo**

In-Medium Pion Valence Distribution Amplitude

Received: 19 November 2016 / Accepted: 23 January 2017 / Published online: 16 February 2017

Abstract After a brief review of the quark-based model for nuclear matter, and some pion properties in medium presented in our previous works, we report new results for the pion valence wave function as well as the valence distribution amplitude in medium, which are presented in our recent article. We find that both the in-medium pion valence distribution and the in-medium pion valence wave function, are substantially modified at normal nuclear matter density, due to the reduction in the pion decay constant.

1 Introduction

One of the most exciting and challenging topics in hadronic physics is to study the modifications of hadron properties in nuclear medium, and to study how such modifications give impact on observables in medium. Since hadrons are composed of quarks, antiquarks and gluons, one can naturally expect that hadron internal structure would change when they are immersed in nuclear medium [1–5]. This question, becomes particularly interesting when it comes to that of pion. To be able to answer it, one first needs, simpler, effective quark–antiquark models of pion, which are successful in describing its properties in vacuum. Among such models, light-front constituent quark model has been very successful in describing the electromagnetic form factors, electromagnetic radii and decay constants of pion and kaon [6–11]. Recent advances in experiments, indeed suggest to make it possible to access to the pion properties in a nuclear medium [3–5, 12, 13].

Among the all hadrons, pion is the lightest, and it is believed as a Nambu–Goldstone boson, which is realized in nature emerged by the spontaneous breaking of chiral symmetry. This Nambu–Goldstone boson, pion, plays very important and special roles in hadronic and nuclear physics. However, because of its special properties, particularly the unusually light mass, it is not easy to describe the pion properties in medium as well as in vacuum based on naive quark models, even though such models can be successful in describing the other hadrons.

Recently, we studied the properties of pion in nuclear medium [12–14], namely, the electromagnetic form factor, charge radius and weak decay constant, distribution amplitude (DA) by using a light-front constituent quark model. There, the in-medium input was calculated by the quark–meson coupling (QMC) model [3, 15].

This article belongs to the Topical Collection "Light Cone 2016".

This work was partly supported by the Fundação de Amparo à Pesquisa do Estado de São Paulo (FAPESP) and Conselho Nacional de Desenvolvimento Científico e Tecnológico (CNPq) of Brazil.

K. Tsushima (✉)· J. P. B. C. de Melo
Laboratório de Física Teórica e Computacional, Universidade Cruzeiro do Sul, São Paulo 01506-000, Brazil
E-mail: kazuo.tsushima@gmail.com; kazuo.tsushima@cruzeirodosu.edu.br

J. P. B. C. de Melo
E-mail: joao.mello@cruzeirodosul.edu.br

The main purpose of this article is, to report our new results for the in-medium pion DA studied in Ref. [14], supplemented by the other pion properties already presented in Refs. [12,13].

Below, we first briefly review the QMC model, the quark-based model of nuclear matter, to study the pion properties in medium. The QMC model was invented by Guichon [15] to describe the nuclear matter based on the quark degrees of freedom. The self-consistent exchange of the scalar–isoscalar σ and vector–isoscalar ω mean fields coupled directly to the relativistic confined quarks, are the key and novelty for the new saturation mechanism of nuclear matter. The model was extended to finite nuclei [16–20], and has successfully been applied for various nuclear and hadronic phenomena [3].

The effective Lagrangian density for a uniform, spin-saturated, and isospin-symmetric nuclear system (symmetric nuclear matter) at the hadronic level is given by [3,15],

$$\mathcal{L} = \bar{\psi}\left[i\gamma\cdot\partial - m_N^*(\hat{\sigma}) - g_\omega\hat{\omega}^\mu\gamma_\mu\right]\psi + \mathcal{L}_{\rm meson}\,, \tag{1}$$

where ψ, $\hat{\sigma}$ and $\hat{\omega}$ are respectively the nucleon, Lorentz-scalar–isoscalar σ, and Lorentz-vector–isoscalar ω field operators with,

$$m_N^*(\hat{\sigma}) \equiv m_N - g_\sigma(\hat{\sigma})\hat{\sigma}. \tag{2}$$

Note that, in symmetric nuclear matter isospin-dependent ρ-meson mean filed is zero, and thus we have omitted it. Then the relevant free meson Lagrangian density is given by,

$$\mathcal{L}_{\rm meson} = \frac{1}{2}(\partial_\mu\hat{\sigma}\partial^\mu\hat{\sigma} - m_\sigma^2\hat{\sigma}^2) - \frac{1}{2}\partial_\mu\hat{\omega}_\nu(\partial^\mu\hat{\omega}^\nu - \partial^\nu\hat{\omega}^\mu) + \frac{1}{2}m_\omega^2\hat{\omega}^\mu\hat{\omega}_\mu. \tag{3}$$

Hereafter, we consider in the rest frame of symmetric nuclear matter. Then, within Hartree mean-field approximation, the nuclear (baryon) and scalar densities are respectively given by,

$$\rho = \frac{4}{(2\pi)^3}\int d\mathbf{k}\,\theta(k_F - |\mathbf{k}|) = \frac{2k_F^3}{3\pi^2}, \quad \rho_s = \frac{4}{(2\pi)^3}\int d\mathbf{k}\,\theta(k_F - |\mathbf{k}|)\frac{m_N^*(\sigma)}{\sqrt{m_N^{*2}(\sigma) + \mathbf{k}^2}}, \tag{4}$$

here, $m_N^*(\sigma)$ is the value (constant) of effective nucleon mass at given density [see also Eq. (2)]. In the standard QMC model [3,15–20] the MIT bag model is used, and the Dirac equations for the light quarks inside a nucleon (bag) composing nuclear matter, are given by,

$$\left[i\gamma\cdot\partial_x - (m_q - V_\sigma^q) \mp \gamma^0\left(V_\omega^q + \frac{1}{2}V_\rho^q\right)\right]\begin{pmatrix}\psi_u(x)\\ \psi_{\bar{u}}(x)\end{pmatrix} = 0, \tag{5}$$

$$\left[i\gamma\cdot\partial_x - (m_q - V_\sigma^q) \mp \gamma^0\left(V_\omega^q - \frac{1}{2}V_\rho^q\right)\right]\begin{pmatrix}\psi_d(x)\\ \psi_{\bar{d}}(x)\end{pmatrix} = 0\,. \tag{6}$$

As usual, Coulomb interaction is neglected, and SU(2) symmetry is assumed, $m_{u,\bar{u}} = m_{d,\bar{d}} \equiv m_{q,\bar{q}}$. The corresponding effective (constituent) quark masses are defined by, $m_{u,\bar{u}}^* = m_{d,\bar{d}}^* = m_{q,\bar{q}}^* \equiv m_{q,\bar{q}} - V_\sigma^q$, to be explained later.

As mentioned already, in symmetric nuclear matter within Hartree approximation, the ρ-meson mean field is zero, $V_\rho^q = 0$, in Eq. (6), and we ignore it. The constant mean-field potentials are defined as, $V_\sigma^q \equiv g_\sigma^q\sigma = g_\sigma^q < \sigma >$, and, $V_\omega^q \equiv g_\omega^q\omega = g_\omega^q < \omega >$, with g_σ^q, and g_ω^q, are the corresponding quark–meson coupling constants, where the quantities with the brackets stand for the expectation values in symmetric nuclear matter [3].

The same meson mean fields σ and ω for the quarks in Eqs. (5) and (6), satisfy self-consistently the following equations at the nucleon level [see Eq. (4)]:

$$\omega = \frac{g_\omega\rho}{m_\omega^2}, \quad \sigma = \frac{g_\sigma}{m_\sigma^2}C_N(\sigma)\frac{4}{(2\pi)^3}\int d\mathbf{k}\,\theta(k_F - |\mathbf{k}|)\frac{m_N^*(\sigma)}{\sqrt{m_N^{*2}(\sigma) + \mathbf{k}^2}} = \frac{g_\sigma}{m_\sigma^2}C_N(\sigma)\rho_s, \tag{7}$$

where $C_N(\sigma) \equiv \frac{-1}{g_\sigma(\sigma=0)}\left[\partial m_N^*(\sigma)/\partial\sigma\right]$ is the constant value of the scalar density ratio [3,15–20]. Because of the underlying quark structure of the nucleon used to calculate $M_N^*(\sigma)$ in nuclear medium [see Eq. (2)], $C_N(\sigma)$ gets nonlinear σ-dependence, whereas the usual point-like nucleon-based model yields unity, $C_N(\sigma) = 1$. It is this $C_N(\sigma)$ or $g_\sigma(\sigma)$ that gives a novel saturation mechanism in the QMC model, and contains the

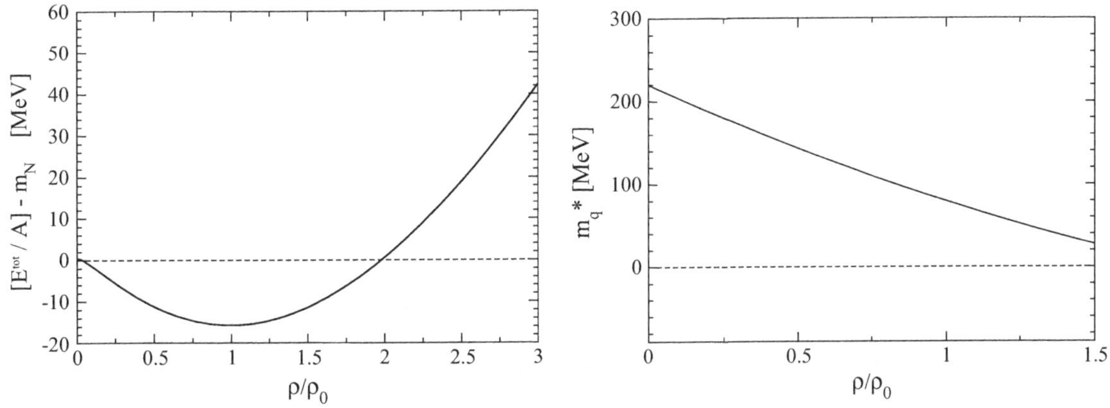

Fig. 1 Negative of the binding energy per nucleon for symmetric nuclear matter, $(E^{tot}/A) - m_N$, versus ρ ($\rho_0 = 0.15\,\text{fm}^{-3}$) with the vacuum quark mass $m_q = m_{\bar{q}} = 220\,\text{MeV}$ ($q = u, d$), calculated by the QMC model (*left panel*), and effective constituent quark mass m_q^* ($q = \bar{q} = u, \bar{u}, d, \bar{d}$) (*right panel*) (both figures are taken from Ref. [12]). The incompressibility K obtained is $K = 320.9\,\text{MeV}$

important dynamics which originates in the quark structure of the nucleon. Without an explicit introduction of the nonlinear couplings of the meson fields in the Lagrangian density at the nucleon and meson level, the standard QMC model yields the nuclear incompressibility of $K \simeq 280\,\text{MeV}$ with $m_q = 5\,\text{MeV}$, which is in contrast to a naive version of quantum hadrodynamics (QHD) [21] (the point-like nucleon model of nuclear matter), results in the much larger value, $K \simeq 500\,\text{MeV}$; the empirically extracted value falls in the range $K = 200$–$300\,\text{MeV}$.

Once the self-consistency equation for the σ, Eq. (7), has been solved, one can evaluate the total energy per nucleon:

$$E^{\text{tot}}/A = \frac{4}{(2\pi)^3\rho} \int d\mathbf{k}\, \theta(k_F - |\mathbf{k}|)\sqrt{m_N^{*2}(\sigma) + \mathbf{k}^2} + \frac{m_\sigma^2 \sigma^2}{2\rho} + \frac{g_\omega^2 \rho}{2m_\omega^2}. \tag{8}$$

We then determine the coupling constants, g_σ and g_ω, so as to fit the binding energy of 15.7 MeV at the saturation density $\rho_0 = 0.15\,\text{fm}^{-3}$ ($k_F^0 = 1.305\,\text{fm}^{-1}$) for symmetric nuclear matter.

In Refs. [11,12], the quark mass in vacuum was used $m_{q,\bar{q}} = 220$ MeV to study the pion properties in symmetric nuclear matter. With this value the model can reproduce the electromagnetic form factor and the decay constant well in vacuum [8]. So, we use the same value in this study. Then, all the nuclear matter saturation properties are generated by using this light-quark mass value. In other words, the different values of m_q in vacuum generate the corresponding different nuclear matter properties, except for the saturation point of the symmetric nuclear matter, $\rho = \rho_0$ (normal nuclear matter density, $0.15\,\text{fm}^{-3}$) with the empirically extracted binding energy of 15.7 MeV. This saturation point condition is generally used to constrain the models of nuclear matter. Thus, we have obtained the necessary properties of the light-flavor constituent quarks in symmetric nuclear matter with $m_q = 220$ MeV, namely, the density dependence of the effective mass (scalar potential) and vector potential. The same in-medium constituent quark properties which reproduce the nuclear saturation properties will be used as input to study the pion properties in symmetric nuclear matter.

In Fig. 1 we show negative of the binding energy per nucleon ($E^{\text{tot}}/A - m_N$) (left-panel), and effective constituent light-quark mass, m_q^* (right panel), in symmetric nuclear matter [12].

2 Pion Properties in Medium

The light-front constituent quark model used here [8], although simple, is quite successful in describing the properties of pion in vacuum.

As examples, we show in Fig. 2 pion charge form factor (left panel) and root mean-square pion charge radius in symmetric nuclear matter calculated in Ref. [12].

Here, we note that the pion mass up to normal nuclear matter density is expected to be modified only slightly, where the modification δm_π at nuclear density $\rho = 0.17\,\text{fm}^{-3}$, averaged over the pion isospin states, is estimated as $\delta m_\pi \simeq +3\,\text{MeV}$ [4,22–25]. Therefore, we approximate the effective pion mass in nuclear

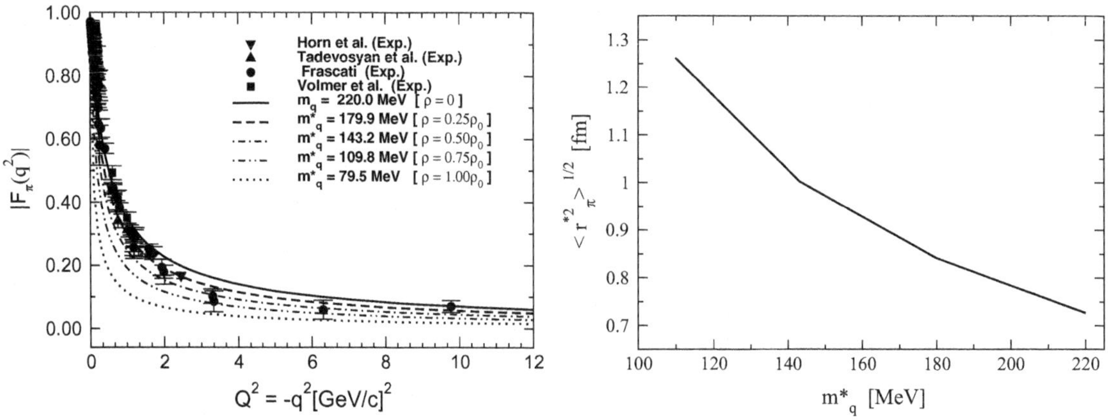

Fig. 2 Pion charge form factor (*left panel*), and root mean-square charge radius in symmetric nuclear matter (*right panel*). Both are taken from Ref. [12].

medium to be the same as in vacuum, $m_\pi^* = m_\pi$, up to $\rho = \rho_0 = 0.15\,\mathrm{fm}^{-3}$, the maximum nuclear matter density treated. Furthermore, since the present light-front constituent quark model is rather simple, the model cannot discuss the chiral limit of vanishing (effective) light-quark masses.

The effective interaction Lagrangian density for the quarks and pion in medium is given by,

$$\mathcal{L}_{\text{eff}} = -ig^* \, (\bar{q}\gamma^5 \boldsymbol{\tau} q \cdot \boldsymbol{\phi}) \, \Lambda^*, \tag{9}$$

where the coupling constant, $g^* = m_q^*/f_\pi^*$, is obtained by the "in-medium Goldberger–Treiman relation" at the quark level, with m_q^* and f_π^* being respectively the effective constituent quark mass and pion decay constant in medium, $\boldsymbol{\phi}$ the pion field [8,10], and Λ^* is the π-q-$\bar{q}$ vertex function in medium. Hereafter, the in-medium quantities are indicated with the asterisk, $*$.

The pion valence wave function used in this study is symmetric under the exchange of quark and antiquark momenta. This π-q-$\bar{q}$ vertex function, $\Lambda(k, P)$ in vacuum is the same as that used for studying the properties of pion [8] and kaon [10]. However, for the in-medium Λ^*, the arguments of the function are replaced by those of the in-medium [12]:

$$\Lambda^*(k+V, P) = \frac{C^*}{\left((k+V)^2 - m_R^2 + i\epsilon\right)} + \frac{C^*}{\left((P-k-V)^2 - m_R^2 + i\epsilon\right)}, \tag{10}$$

where $V^\mu = \delta_0^\mu V^0$ is the vector potential felt by the light quarks in the pion immersed in medium, and can be eliminated by the variable change in the k-integration, $k^\mu + \delta_0^\mu V^0 \to k^\mu$. The normalization factor associated with C^* is modified by the medium effects. The regulator mass m_R represents soft effects at short range of about the 1 GeV scale, and m_R may also be influenced by in-medium effects. However, we employ $m_R^* = m_R$ in Eq. (10), since there exists no established way of estimating this effect on the regulator mass. This can avoid introducing extra source of uncertainty.

The Bethe–Salpeter amplitude in medium, Ψ_π^*, with the vertex function in medium Λ^* is given by,

$$\Psi_\pi^*(k+V, P) = \frac{\not{k} + \not{V} + m_q^*}{(k+V)^2 - m_q^{*2} + i\epsilon}\gamma^5 \Lambda^*(k+V, P) \frac{\not{k} + \not{V} - \not{P} + m_q^*}{(k+V-P)^2 - m_q^{*2} + i\epsilon}. \tag{11}$$

By eliminating the instantaneous terms, or eliminating the terms with the matrix γ^+ in the numerators and k^+ and $(P^+ - k^+)$ in the denominators with the light-front convention $a^\pm \equiv a^0 \pm a^3$, and integrating over the light-front energy k^-, we obtain the in-medium pion valence wave function Φ_π^*,

$$\Phi_\pi^*(k^+, \mathbf{k}_\perp; P^+, \mathbf{P}_\perp) = \frac{P^+}{m_\pi^{*2} - M_0^2} \left[\frac{N^*}{(1-x)\left(m_\pi^{*2} - \mathcal{M}^2(m_q^{*2}, m_R^2)\right)} + \frac{N^*}{x\left(m_\pi^{*2} - \mathcal{M}^2(m_R^2, m_q^{*2})\right)} \right], \tag{12}$$

where $N^* = C^*(m_q^*/f_\pi^*)(N_c)^{\frac{1}{2}}$ is the normalization factor with the number of colors N_c [8,12], $x = k^+/P^+$ with $0 \leq x \leq 1$, $\mathcal{M}^2(m_a^2, m_b^2) \equiv \frac{\mathbf{k}_\perp^2+m_a^2}{x} + \frac{(\mathbf{P}-\mathbf{k})_\perp^2+m_b^2}{1-x} - \mathbf{P}_\perp^2$, the square of the mass M_0^2 is $M_0^2 = \mathcal{M}^2(m_q^{*2}, m_q^{*2})$, and m_R is the regulator mass with the value $m_R^* = m_R = 600\,\text{MeV}$ [8,12]. Note that the model used in Refs. [7,9] does not have the second term in Eq. (12). This means that the pion valence wave function in Refs. [7,9] is not symmetric under the exchange of quark and antiquark momenta.

The pion transverse momentum probability density in medium, $P_\pi^*(k_\perp)$, in the pion rest frame $P^+ = m_\pi^*$ is calculated by,

$$P_\pi^*(k_\perp) = \frac{1}{4\pi^3 m_\pi^*}\int_0^{2\pi} d\phi \int_0^{m_\pi^*} \frac{dk^+ M_0^{*2}}{k^+(m_\pi^* - k^+)}|\Phi_\pi^*(k^+, \mathbf{k}_\perp; m_\pi^*, \mathbf{0})|^2, \tag{13}$$

and the integration over $k_\perp$ for $P_\pi^*(k_\perp)$ leads to the in-medium probability of the valence component in the pion, η^* [8,12]:

$$\eta^* = \int_0^\infty dk_\perp k_\perp P_\pi^*(k_\perp). \tag{14}$$

The pion decay constant in medium, in terms of the pion valence component with $\Phi_\pi^*(k^+, \mathbf{k}_\perp; m_\pi^*, \mathbf{0})$, is calculate by [8,12]:

$$f_\pi^* = \frac{m_q^*(N_c)^{1/2}}{4\pi^3}\int \frac{d^2k_\perp dk^+}{k^+(m_\pi^* - k^+)}\Phi_\pi^*(k^+, \mathbf{k}_\perp; m_\pi^*, \mathbf{0}). \tag{15}$$

The result is shown in Fig. 4 (left panel) as a function of nuclear density (ρ/ρ_0). Note that, f_π^* above is calculated with the plus-component of the light front axial-vector current (light-front time component). Thus the f_π^* cannot be separated into the usual sense of the time and space components, where the corresponding two components of f_π^* decouple with the presence of background matter (in nuclear medium), and the time component is shown to be model independent and directly associated with the observables, and also with the Gell–Mann–Okes–Renner (GMOR) relation [26]. As nuclear density increases, the in-medium pion decay constant f_π^* decreases, although the amount of the reduction may be larger than that calculated by the NJL model [24,25]. However, in the treatment of the NJL model [24,25], the decoupling of the time and space components of f_π^* is not clear with the presence of background matter, and may not directly be compared with our result, as well as the empirical value extracted from the pionic-atom experiment [22].

In the following, we further discuss the in-medium "quark condensate" and the GMOR-like relation, just for an illustration. As we mentioned already, the present model is a light-front constituent quark model with the constituent quark mass of $m_q = 220$ MeV in vacuum, and thus we cannot discuss the chiral limit (model limitation). Keeping this point in mind, however, to get some idea, we simply write down the GMOR-like relation in vacuum and in medium, so that we try to compare the "quark condensate" value with that extracted experimentally [22]:

$$m_\pi^2 f_\pi^2 = -2m_q < \overline{q}q >, \tag{16}$$

$$m_\pi^{*2} f_\pi^{*2} = -2m_q^* < \overline{q}q >^* . \tag{17}$$

Then, the ratio of the in-medium to vacuum quark condensates in the present approach may be estimated as,

$$\frac{< \overline{q}q >^*}{< \overline{q}q >} = \frac{m_q}{m_q^*}\frac{m_\pi^{*2} f_\pi^{*2}}{m_\pi^2 f_\pi^2} \simeq \frac{m_q}{m_q^*}\frac{f_\pi^{*2}}{f_\pi^2}. \tag{18}$$

At normal nuclear matter density, ρ_0 ($0.15\,\text{fm}^{-3}$), the ratio gives $\simeq 0.52$ [12] (and also one can calculate by using the values listed in Table 1, to be given next). This implies a larger reduction in "quark condensate" compared to the value 0.67 ± 0.06 extracted in Ref. [22] at a density $0.17\,\text{fm}^{-3}$ (their value for the normal nuclear matter density). This feature may also be understood from the larger reduction in $(f_\pi^*/f_\pi)^2$ in our approach compared with that obtained in Ref. [22] (see also Ref. [12] for more details). We repeat that the

Table 1 Properties of pion in medium, taken from Ref. [12], with $\rho_0 = 0.15\,\text{fm}^{-3}$

ρ/ρ_0	m_q^* (MeV)	f_π^* (MeV)	$< r_\pi^{*2} >^{1/2}$ (fm)	η^*
0.00	220	93.1	0.73	0.782
0.25	179.9	80.6	0.84	0.812
0.50	143.2	68.0	1.00	0.843
0.75	109.8	55.1	1.26	0.878
1.00	79.5	40.2	1.96	0.930

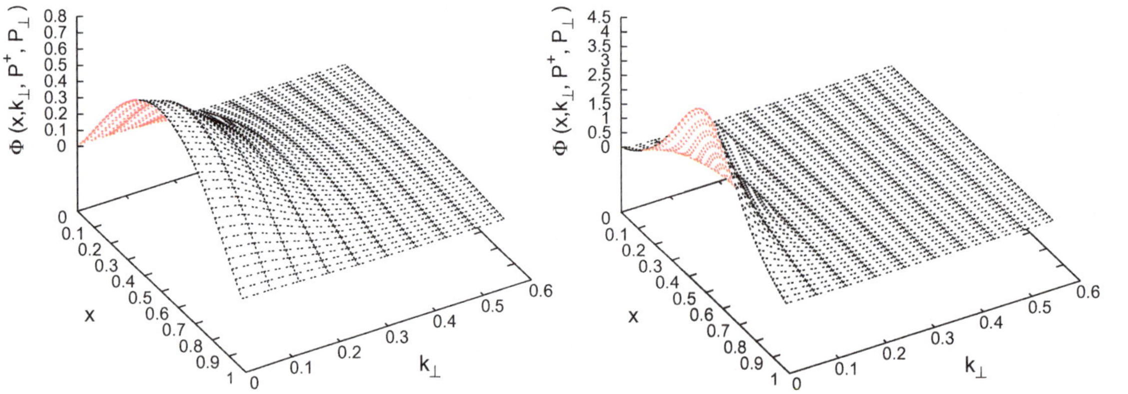

Fig. 3 Pion valence wave functions in vacuum ($\rho = 0$) (*left panel*) and in medium ($\rho = \rho_0$) (*right panel*) versus x and $k_\perp = |\mathbf{k}_\perp|$, where $P^+ = m_\pi = m_\pi^*$ and $P_\perp = |\mathbf{P}_\perp| = 0$. The wave functions are given in the units, $10^{-8}\times$ (GeV)$^{-1}$. Notice that the differences in the *vertical axis scales* for the *left* and *right panels* (taken from Ref. [14])

discussions above are not rigorous, but just for giving some intuition, since the vacuum structure in the light-front approach is usually considered as "trivial", and it is very difficult to study the quark condensate (or spontaneously broken chiral symmetry) in vacuum as well as in medium. This is a very interesting issue for the future elaboration within the light-front approach.

Some properties of the pion in symmetric nuclear matter obtained in Ref. [12], are summarized in Table 1. The results listed in Table 1 are summarized as follows [see also Fig. 1 (right panel) for m_q^*, Fig. 2 (right panel) for $< r_\pi^{*2} >^{1/2}$, and Fig. 4 (left panel) for f_π^*]. As the nuclear density increases, m_q^* and f_π^* decrease, while $< r_\pi^{*2} >^{1/2}$ and the probability of valence component in the pion, η^*, increase. This can be understood as follows. The reduction in mass, m_q^*, makes it easier to excite the valence quark component in the pion state, and resulting to increase the valence component probability η^* in the pion. Furthermore, the valence wave function spreads more in coordinate space by the decrease of m_q^*, and reduces the absolute value of the wave function at the origin ($f_\pi^* \propto |\Phi_\pi^*(\mathbf{r} = \mathbf{0})|$ reduction [27]), namely, increases $< r_\pi^{*2} >^{1/2}$.

Next we discuss the in-medium pion valence distribution amplitude. Pion DA provides information on the nonperturbative regime of the bound state nature of pion due to the quark and antiquark at higher momentum transfer. The pion valence wave function in vacuum is normalized by [28,29] (aside from the factor $\sqrt{2}$ difference):

$$\int_0^1 dx \int \frac{d^2k_\perp}{16\pi^3} \Phi_\pi(x, \mathbf{k}_\perp) = \frac{f_\pi}{2\sqrt{6}}. \tag{19}$$

This is an important constraint on the normalization of the $q\bar{q}$ wave function [28,29], associated with a probability of finding a pure $q\bar{q}$ state in the pion state. According to this normalization, the in-medium pion valence wave function is normalized by replacing $f_\pi \to f_\pi^*$ in the above. Since the pion decay constant in nuclear medium is modified, the pion valence wave function in nuclear medium is also modified via this normalization.

In order to examine more in detail as to how the change in f_π^* impacts on the in-medium pion valence wave function, we show in Fig. 3 the pion valence wave functions in vacuum (left panel) and $\rho = \rho_0$ (right panel). One can notice that the in-medium pion valence wave function in momentum space has a sharper peak and localized in narrower regions both in x and $k_\perp$ than those in vacuum. Of course, the total volume, the quantity

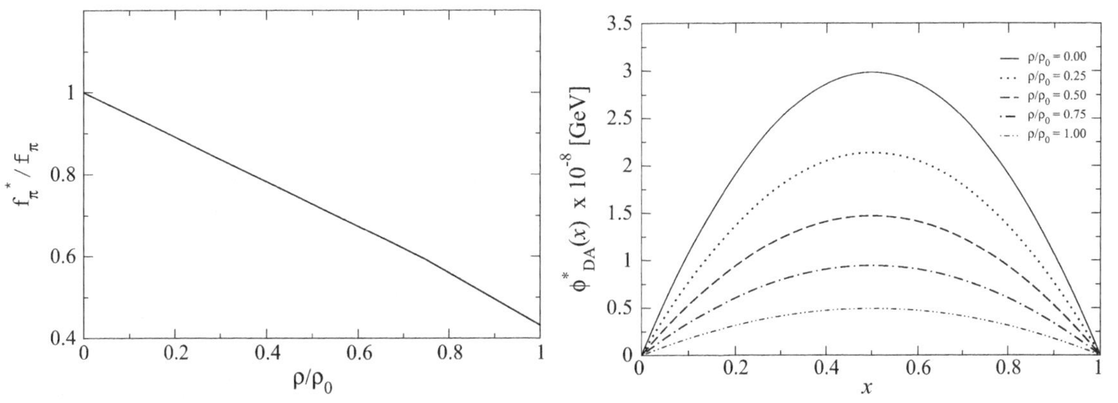

Fig. 4 Pion decay constant calculated in symmetric nuclear matter (*left panel*) taken from Ref. [12], and pion valence distribution amplitudes (*right panel*), taken from Ref. [14]

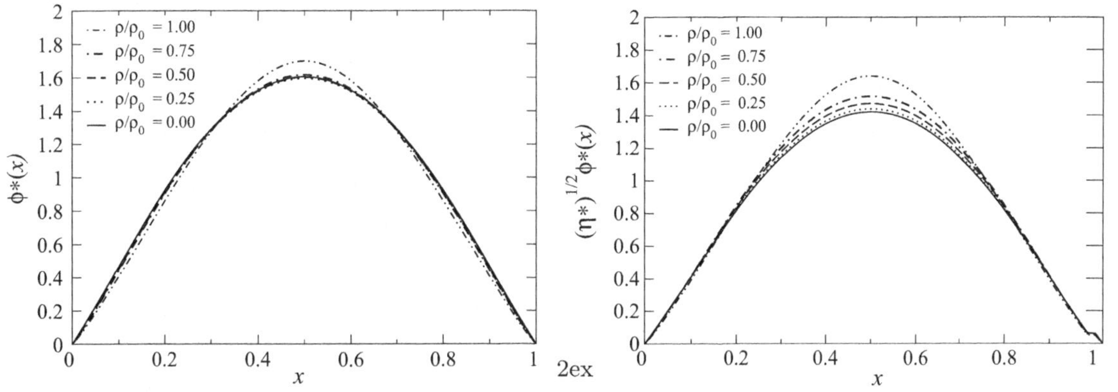

Fig. 5 Effective pion valence distribution amplitudes in vacuum and in medium, respectively multiplied by $\sqrt{\eta}$ and $\sqrt{\eta^*}$

integrated over x and $\mathbf{k}_\perp$, is reduced in medium, corresponding to the reduced f_π^*. This fact is reflected in the wave function in coordinate space, that it becomes spread wider, and generally its hight is reduced.

The corresponding pion valence DA in medium, denoted by $\phi^*_{DA}(x)$ (not normalized to unity), is calculated as

$$\phi^*_{DA}(x) = \int \frac{d^2k_\perp}{16\pi^3} \Phi^*_\pi(x, \mathbf{k}_\perp). \tag{20}$$

Note that, Eq. (20) holds also for the other pseudoscalar mesons M_{ps} such as kaon and D-meson, by replacing $\Phi^*_\pi(x, \mathbf{k}_\perp) \to \Phi^*_{M_{ps}}(x, \mathbf{k}_\perp)$ in the above.

We show in Fig. 4 (right panel) the obtained pion DAs, $\phi^*_{DA}(x)$, in vacuum ($\rho/\rho_0 = 0$) and in medium for several nuclear densities. The significant reduction of the in-medium pion valence DA is obvious, that reflects the reduction of f_π^* (left panel).

Next, we study pion valence DA normalized to unity, or normalized pion valence DAs in vacuum and in medium. By this, we can study the change in shape due to the medium effects. We show in Fig. 5 (left panel) the normalized pion valence DAs, $\phi^*(x)$, both in vacuum ($\rho/\rho_0 = 0$) and in medium. The in-medium change in shape is moderate when the nuclear matter densities are small, but it becomes evident when the density becomes ρ_0.

Furthermore, it may be useful to define *effective pion valence DA* using the valence probability in vacuum η and in medium η^* [see Eq. (14) and Table 1]. The pion states in vacuum, $|\pi>$, and in medium, $|\pi>^*$, can respectively be written as,

$$|\pi> = \sqrt{\eta}|q\bar{q}> + a|q\bar{q}q\bar{q}> + b|q\bar{q}g> + \cdots, \tag{21}$$

$$|\pi>^* = \sqrt{\eta^*}|q\bar{q}>^* + c|q\bar{q}q\bar{q}>^* + d|q\bar{q}g>^* + \cdots, \tag{22}$$

where a, b, c and d are constants, and g denotes a gluon, and $+\cdots$ stands for the higher Fock components in the pion states. The quantity η^* in Table 1 indicates that the valence $q\bar{q}$ component in the pion state increases in medium as nuclear density increases. The effective pion valence DAs, $\sqrt{\eta^*}\phi(x)^*$, in vacuum ($\rho/\rho_0 = 0$) and in medium, are shown in Fig. 5. They may respectively correspond to the first terms of Eqs. (21) and (22).

Since η^*/η is enhanced in medium, effective pion valence DA in medium is also enhanced, on the top of the corresponding medium-(shape)modified normalized pion valence DA. The obvious enhancement of the effective pion valence DA in medium can be seen around $x = 0.5$. This quantity may be useful when one studies some reactions in medium (in a nucleus) involving a pion, based on a constituent quark picture of pion.

3 Summary

We have studied the impact of in-medium effects on the pion valence distribution amplitudes using a light-front constituent quark model, combined with the in-medium input calculated by the quark–meson coupling model. The in-medium constituent light-quark properties inside the pion are consistently constrained by the saturation properties of symmetric nuclear matter.

The in-medium pion mass is assumed to be the same as that in vacuum, based on the extracted information from the pionic-atom experiment, and some theoretical studies. This information extracted is valid up to around the normal nuclear matter density. Thus, the results obtained in this study, combined with the light-front constituent quark model, are valid up to around the normal nuclear matter density, but cannot discuss reliably the chiral limit, the vanishing limit of the (effective constituent) light-quark masses. We need to rely on more sophisticated models of pion to be able to discuss the chiral limit in medium, as well as in vacuum.

Due to the reduction in the pion decay constant in medium, the pion distribution amplitude in medium normalized with the pion decay constant, is substantially reduced at relatively higher nuclear densities. Because the valence component probability in medium increases as nuclear density increases, we have defined an effective pion distribution amplitude normalized to the square root of the valence probability in the pion state. This may give some information for the effectiveness of the valence quark picture of pion in nuclear medium. Within the present light-front constituent quark model approach, the effectiveness of the valence quark picture of the pion in medium, becomes more enhanced as nuclear density increases, due to the increase of the valence component in the pion state.

Although the present study is based on a simple, light-front constituent quark model, this is a first step to understand the impact of the medium effects on the internal structure of the pion immersed in nuclear medium.

Acknowledgements The authors would like to thank the organizers of Light Cone 2016 for invitation, and warm hospitality, that motivated them and made them enjoy during the workshop.

References

1. G.E. Brown, M. Rho, Scaling effective Lagrangians in a dense medium. Phys. Rev. Lett. **66**, 2720 (1991)
2. H. Hatsuda, S.H. Lee, QCD sum rules for vector mesons in the nuclear medium. Phys. Rev. C **46**(1), R34 (1992)
3. K. Saito, K. Tsushima, A.W. Thomas, Nucleon and hadron structure changes in the nuclear medium and impact on observables. Prog. Part. Nucl. Phys. **58**, 1 (2007)
4. R.S. Hayano, T. Hatsuda, Hadron properties in the nuclear medium. Rev. Mod. Phys. **82**, 2949 (2010)
5. W.K. Brooks, S. Strauch, K. Tsushima, Medium modifications of hadron properties and partonic processes. J. Phys. Conf. Ser. **299**, 012011 (2011)
6. J.P.B.C. de Melo, T. Frederico, L. Tomio, A.E. Dorokhov, Quark–anti-quark correlation in the pion. Nucl. Phys. A **623**, 456 (1997)
7. J.P.C.B. de Melo, H.W.L. Naus, T. Frederico, Pion electromagnetic current in the light cone formalism. Phys. Rev. C **59**, 2278 (1999)
8. J.P.B.C. de Melo, T. Frederico, E. Pace, G. Salmè, Pair term in the electromagnetic current within the front form dynamics: spin-0 case. Nucl. Phys. A **707**, 399 (2002)
9. E.O. da Silva, J.P.B.C. de Melo, B. El-Bennich, V.S. Filho, Pion and kaon elastic form factors in a refined light-front model. Phys. Rev. C **86**, 038202 (2012)
10. G.H.S. Yabusaki, I. Ahmed, M.A. Paracha, J.P.B.C. de Melo, B. Bennichi, Pseudoscalar mesons with symmetric bound state vertex functions on the light front. Phys. Rev. D **92**, 034017 (2015)
11. J.P.B.C. de Melo, I. Ahmed, K. Tsushima, Parton distribution in pseudoscalar mesons with a light-front constituent quark model. AIP Conf. Proc. **1735**, 080012 (2016)
12. J.P.B.C. de Melo, K. Tsushima, B. El-Bennich, E. Rojas, T. Frederico, Pion structure in the nuclear medium. Phys. Rev. C **90**, 035201 (2014)

13. J.P.B.C. de Melo, K. Tsushima, T. Frederico, Pion in the medium with a light-front model. AIP Conf. Proc. **1735**, 080006 (2016)
14. J.P.B.C. de Melo, K. Tsushima, I. Ahmed, In-medium pion valence distributions in a light-front model. Phys. Lett. B **676**, 125 (2017)
15. P.A.M. Guichon, A possible quark mechanism for the saturation of nuclear matter. Phys. Lett. B **200**, 235 (1988)
16. P.A.M. Guichon, K. Saito, E.N. Rodionov, A.W. Thomas, The role of nucleon structure in finite nuclei. Nucl. Phys. A **601**, 349 (1996)
17. K. Saito, K. Tsushima, A.W. Thomas, Selfconsistent description of finite nuclei based on a relativistic quark model. Nucl. Phys. A **609**, 339 (1996)
18. K. Saito, K. Tsushima, A.W. Thomas, Variation of hadron masses in finite nuclei. Phys. Rev. C **55**, 2637 (1997)
19. K. Tushima, K. Saito, J. Haidenbauer, A.W. Thomas, The quark–meson coupling model for lambda, sigma and xi hypernuclei. Nucl. Phys. A **630**, 691 (1998)
20. P.A.M. Guichon, A.W. Thomas, K. Tsushima, Binding of hypernuclei in the latest quark–meson coupling model. Nucl. Phys. A **814**, 66 (2008)
21. B.D. Serot, J.D. Walecka, The relativistic nuclear many body problem. Adv. Nucl. Phys. **16**, 1 (1986)
22. R. Kienle, T. Yamazaki, Pions in nuclei, a probe of chiral symmetry restoration. Prog. Part. Nucl. Phys. **52**, 85 (2004)
23. U.G. Meissner, J.A. Oller, A. Wirzba, In-medium chiral perturbation theory beyond the mean field approximation. Ann. Phys. **297**, 27 (2002)
24. W. Weise, U. Vogl, The Nambu and Jona Lasinio model: its implications for hadrons and nuclei. Prog. Part. Nucl. Phys. **27**, 195 (1991)
25. M. Lutz, S. Klimt, W. Weise, Meson properties at finite temperature and baryon density. Nucl. Phys. A **542**, 521 (1992)
26. M. Kirchbach, A. Wirzba, In-medium chiral perturbation theory and pion weak decay in the presence of background matter. Nucl. Phys. A **616**, 648 (1997)
27. R. Van Royen, V.F. Weisskopf, Hadron decay processes and the quark model. Nuovo Cim. A **50**, 617 (1967). Erratum: [Nuovo Cim. A **51**, 583 (1967)]
28. G.P. Lepage, S.J. Brodsky, T. Huang, P.B. Mackenzie, hadronic wave functions in QCD. CLNS-82-522, FERMILAB-CONF-81-114-T
29. S.J. Brodsky, G.P. Lepage, Exclusive processes in quantum chromodynamics. Adv. Ser. Direct. High Energy Phys. **5**, 93 (1989)

Few-Body Syst
DOI 10.1007/s00601-017-1233-2

J. P. B. C. de Melo · K. Tsushima

In-Medium ρ-Meson Properties in a Light-Front Approach

Received: 18 November 2016 / Accepted: 12 January 2017

Abstract Properties of ρ-meson in symmetric nuclear matter are investigated within a light-front constituent quark model (LFCQM), using the in-medium input calculated by the quark-meson coupling (QMC) model. The LFCQM used here was previously applied in vacuum to calculate the ρ-meson electromagnetic properties, namely, charge G_0, magnetic G_1, and quadrupole G_2 form factors, as well as the electromagnetic radius and decay constant. We predict the in-medium modifications of the ρ-meson electromagnetic form factors in symmetric nuclear matter.

1 Introduction

One of the main objectives in hadronic physics is to understand the hadron structure in terms of the fundamental degrees of freedom in quantum chromodynamics (QCD), i.e, quarks and gluons. The Standard Model (SM) of elementary particles includes QCD as the strong interaction sector. Although solving QCD is an important part for understanding the SM of the particles physics, it is still a very difficult task (see Refs. [1,2] for QCD details).

On the other hand, light-front approach is an alternative to calculate observables with the ingredients from QCD [3]. With light-front quantum field theory, it is possible to describe the lower Fock components of the hadronic bound-state wave functions, i.e, mesons and baryons in terms of quarks and gluons [3–9].

In the present work the previous model for the ρ-meson in vacuum [10–12] is used to calculate the in-medium electromagnetic observables such as the electromagnetic form factors, charge G_0, magnetic G_2, and quadrupole G_2, as well as the electromagnetic radius and decay constant, using the plus-component of the current J^μ applied in symmetric nuclear matter [13]. (For a review of the in-medium properties of hadrons, see e.g., Refs. [14,15].) Next, we present the light-front model of ρ-meson, and briefly review the main ingredients of the model.

This article belongs to the Topical Collection “Light Cone 2016”.

This work was supported by the Fundação de Amparo à Pesquisa do Estado de São Paulo (FAPESP), Brazil, and Conselho Nacional de Desenvolvimento Científico e Tecnológico of Brazil (CNPq).

J. P. B. C. de Melo (✉) · K. Tsushima
Laboratório de Física Teórica e Computacional - LFTC, Universidade Cruzeiro do Sul, São Paulo 01506-000, Brazil
E-mail: joao.mello@Cruzeirodosul.edu.br

K. Tsushima
E-mail: kazuo.tsushima@gmail.com; kazuo.tsushima@cruzeirodosul.edu.br

2 The Light-Front Model

The electromagnetic current in the impulse approximation for the spin-1 bound state systems, with the plus component of the electromagnetic current J^+, is given by,

$$J_{ji}^{+} = \imath \int \frac{d^4k}{(2\pi)^4} \frac{Tr[\epsilon_j^{'\alpha} \Gamma_\alpha(k, k-p_f)(\not k - \not p_f + m_q)\gamma^+(\not k - \not p_i + m_q)\epsilon_i^\beta \Gamma_\beta(k, k-p_i)(\not k + m_q)]}{((k-p_i)^2 - m_q^2 + \imath\epsilon)(k^2 - m_q^2 + \imath\epsilon)((k-p_f)^2 - m_q^2 + \imath\epsilon)} \times \Lambda(k, p_f)\Lambda(k, p_i)\,, \tag{1}$$

where m is the quark mass, and ϵ_i^β and $\epsilon_j^{'\alpha}$ $(i, j = x, y, z)$ are the four-vector polarizations, respectively, for the initial state ρ-meson,

$$\epsilon_x^\beta = (-\sqrt{\eta}, \sqrt{1+\eta}, 0, 0),\ \epsilon_y^\beta = (0, 0, 1, 0),\ \epsilon_z^\beta = (0, 0, 0, 1), \tag{2}$$

and for the final state ρ-meson,

$$\epsilon_x^{'\alpha} = (\sqrt{\eta}, \sqrt{1+\eta}, 0, 0),\ \epsilon_y^{'\alpha} = (0, 0, 1, 0),\ \epsilon_z^{'\alpha} = (0, 0, 0, 1), \tag{3}$$

with $\eta \equiv -q^2/4m_\rho^2 = Q^2/4m_\rho^2 > 0$ (q; four-momentum transfer).

The electromagnetic current, Eq. (1), is divergent, and in order to turn the Feynman amplitude finite, the regulator function $\Lambda(k, p)$ is utilized [10],

$$\Lambda(k, p) = \frac{1}{((k-p)^2 - m_R^2 + \imath\epsilon)^2}, \tag{4}$$

here, the regulator mass m_R, is chosen to reproduce the experimental value of the ρ-meson decay constant in vacuum, 156 ± 8 MeV [16].

The vertex with the spinor structure for ρ-$q\bar{q}$ is modeled by,

$$\Gamma^\mu(k, k') = \gamma^\mu - \frac{m_\rho}{2} \frac{k^\mu + k'^\mu}{p.k + m_\rho m_q - \imath\epsilon}, \tag{5}$$

where, the ρ-meson is on-mass-shell, and its four momentum is $p^\mu = k^\mu - k'^\mu$, the quark momenta are k^μ and k'^μ, and their masses by m_q [10,17].

The light-front ρ-meson wave function is obtained after the substitution of the on-mass-shell condition, $k^- = (k_\perp^2 + m_q^2)/k^+$ in the propagator of the quark that absorbs the photon and in the corresponding regulator,

$$\Phi_i(x, \vec{k}_\perp) = \frac{N^2}{(1-x)^2(m_\rho^2 - M_0^2)(m_\rho^2 - M_R^2)^2} \vec{\epsilon}_i \cdot \left[\vec{\gamma} - \frac{\vec{k}}{\frac{M_0}{2} + m}\right], \tag{6}$$

where the polarization state is given by $\boldsymbol{\epsilon}_i$. The wave-function corresponds to an S-wave state [4,5]. The free quark-antiquark mass operator, and the M_R function are given below,

$$M_0^2 = \frac{k_\perp^2 + m^2}{x} + \frac{(\vec{p} - \vec{k})_\perp^2 + m^2}{1-x} - p_\perp^2,$$
$$M_R^2 = \frac{k_\perp^2 + m^2}{x} + \frac{(\vec{p} - \vec{k})_\perp^2 + m_R^2}{1-x} - p_\perp^2, \tag{7}$$

with $x = k^+/p^+$.

In the case of spin-1 particles with the light-front approach, the matrix elements of the plus-component of the electromagnetic current, J^+, is constrained by the angular condition [10,18,19]:

$$\Delta(q^2) = (1 + 2\eta)I_{11}^+ + I_{1-1}^+ + -\sqrt{8\eta}I_{10}^+ - I_{00}^+ = 0. \tag{8}$$

With the equation above, it is possible to express the electromagnetic form factors with different linear combinations [10,18], and eliminate some matrix elements $I_{m'm}^+$, $(m' = 0, 1; m = \pm 1, 0)$. But, some linear

Table 1 The MIT bag model (see Ref. [32] for details) quantities and coupling constants, the parameter Z_N, bag constant B (in $B^{1/4}$), and the properties for symmetric nuclear matter at normal nuclear matter density $\rho_0 = 0.15\ \text{fm}^{-3}$, for $m_q = 5,\ 220$ and $m_q = 430$ MeV)

m_q (MeV)	$g_\sigma^2/4\pi$	$g_\omega^2/4\pi$	m_N^*	K	Z_N	$B^{1/4}$ (MeV)
5	5.39	5.30	754.6	279.3	3.295	170
220	6.40	7.57	698.6	320.9	4.327	148
430	8.73	11.93	565.3	361.4	5.497	69.8

The effective nucleon mass, m_N^*, and the nuclear incompressibility, K, are quoted in MeV (the free nucleon bag radius used is $R_N = 0.8$ fm, the standard value in the QMC model [32])

combinations of the electromagnetic matrix elements, J_{ji}^+, break the covariance as well as the rotational symmetry, because the zero mode contributions or the pair terms [17,20,21]. In order to restore the covariance, it is necessary to add the pair-term contributions or zero modes [10,17,20]. The light-front and instant form spin bases are connected by the Melosh rotation matrix (see Ref. [10] for details).

However, in the combinations of the matrix elements of the electromagnetic current from Grach et al. [22,23], the zero modes are canceled out. This was demonstrated numerically in Ref. [10] and analytically in Ref. [17], since the electromagnetic matrix element of the current, I_{00}^+, was eliminated by the angular condition [10,18,22,23].

The electromagnetic form factors with the prescription above, are given below in both the light-front spin basis, $I_{mm'}^+$, and the instant form spin basis, J_{ji}^+:

$$
\begin{aligned}
G_0^{GK} &= \frac{1}{3}\left[(3-2\eta)\, I_{11}^+ + 2\sqrt{2\eta} I_{11}^+ + I_{1-1}^+\right] \\
&= \frac{1}{3}\left[J_{xx}^+ + 2J_{yy}^+ - \eta J_{yy}^+ + \eta J_{zz}^+\right], \\
G_1^{GK} &= 2\left[I_{11}^+ - \frac{1}{\sqrt{2\eta}} I_{10}^+\right] \\
&= J_{yy}^+ - J_{zz}^+ - \frac{J_{zx}^+}{\sqrt{\eta}}, \\
G_2^{GK} &= \frac{2\sqrt{2}}{3}\left[-\eta I_{11}^+ + \sqrt{2\eta}.I_{10}^0 - I_{1-1}^+\right] \\
&= \frac{\sqrt{2}}{3}\left[J_{xx}^+ + J_{yy}^+(-1-\eta) + \eta J_{zz}^+\right].
\end{aligned} \tag{9}
$$

In next section, the light-front constituent quark model described above is applied to study the ρ-meson form factors in symmetric nuclear matter.

3 Rho-Meson In-Medium

The light-front model for the vector ρ-meson was developed in Ref. [10] in order to calculate the observables in vacuum. In the present work, we explore that model, to calculate the ρ-meson observables in symmetric nuclear matter (see Refs. [14,15] for a review of in-medium quark and hadron properties). For modeling nuclear matter, "Quark-Meson Coupling" (QMC) model is utilized here, and combined with the light-front approach, analogous to that was done for the pion case with symmetric vertex for the pion-quark coupling [13,24].

The QMC model, which describes nuclear matter based on the quark degrees of freedom, was introduced using MIT bag model by Guichon [25], and Frederico et al. [26] with a confining harmonic potential. The model was successfully applied for finite nuclei [27–31], and also meson properties in medium [32]. In the QMC model, because of the surrounding medium, the bound states of mesons and baryons are modified, compared with those in vacuum. For some properties calculated for symmetric nuclear matter, we show in Fig. 1 (negative of the binding energy), and summarize in Table 1.

Next, the main ingredients of the QMC model are presented in order to calculate the ρ-meson properties in medium. For the system of a uniform, spin- and isospin-saturated symmetric nuclear matter, the effective Lagrangian density is given by [32],

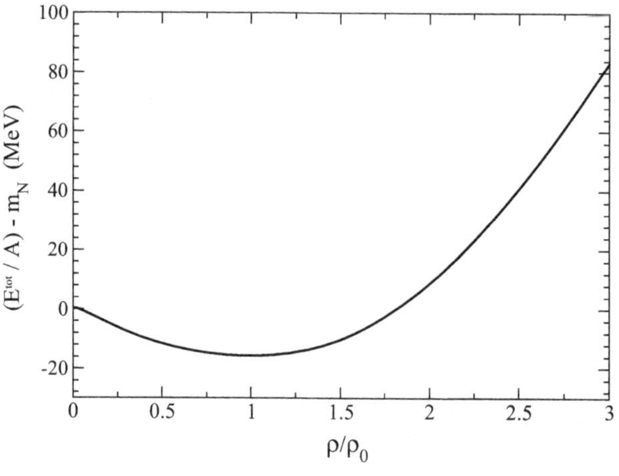

Fig. 1 Negative of the binding energy per nucleon for symmetric nuclear matter, calculated with the vacuum up and down quarks masses, $m_q = 430\,\text{MeV}$, with the QMC model [32]

$$\mathcal{L} = \bar{\psi}[i\gamma \cdot \partial - m_N^*(\hat{\sigma}) - g_\omega \hat{\omega}^\mu \gamma_\mu]\psi + \mathcal{L}_{\text{meson}}, \tag{10}$$

where $\mathcal{L}_{\text{meson}}$ is the free meson Lagrangian,

$$\mathcal{L}_{\text{meson}} = \frac{1}{2}(\partial_\mu \hat{\sigma} \partial^\mu \hat{\sigma} - m_\sigma^2 \hat{\sigma}^2) - \frac{1}{2}\partial_\mu \hat{\omega}_\nu (\partial^\mu \hat{\omega}^\nu - \partial^\nu \hat{\omega}^\mu) + \frac{1}{2} m_\omega^2 \hat{\omega}^\mu \hat{\omega}_\mu \ ,$$

and, ψ, $\hat{\sigma}$ and $\hat{\omega}$ are respectively the nucleon, Lorentz-scalar-isoscalar σ, and Lorentz-vector-isoscalar ω field operators, with the effective nucleon mass defined by,

$$m_N^*(\hat{\sigma}) \equiv m_N - g_\sigma(\hat{\sigma})\hat{\sigma}. \tag{11}$$

In the present work, the nuclear matter is in its rest frame. For symmetric nuclear matter with the Hartree mean-field approximation, the baryon (ρ) and scalar (ρ_s) densities, are calculated as,

$$\rho = \frac{4}{(2\pi)^3} \int d\mathbf{k}\, \theta(k_F - |\mathbf{k}|) = \frac{2k_F^3}{3\pi^2}, \tag{12}$$

$$\rho_s = \frac{4}{(2\pi)^3} \int d\mathbf{k}\, \theta(k_F - |\mathbf{k}|) \frac{m_N^*(\sigma)}{\sqrt{m_N^{*2}(\sigma) + \mathbf{k}^2}}, \tag{13}$$

In the equations above, k_F is the Fermi momentum, and $m_N^*(\sigma)$ is the value of the effective nucleon mass at a given density, self-consistently calculated with the QMC model [27–32]. The Dirac equations for the light quark and antiquark, are given by,

$$\left[i\gamma \cdot \partial_x - (m_q - V_\sigma^q) \mp \gamma^0 \left(V_\omega^q + \frac{1}{2}V_\rho^q\right)\right] \begin{pmatrix} \psi_u(x) \\ \psi_{\bar{u}}(x) \end{pmatrix} = 0,$$

$$\left[i\gamma \cdot \partial_x - (m_q - V_\sigma^q) \mp \gamma^0 \left(V_\omega^q - \frac{1}{2}V_\rho^q\right)\right] \begin{pmatrix} \psi_d(x) \\ \psi_{\bar{d}}(x) \end{pmatrix} = 0, \tag{14}$$

here, Coulomb interaction is neglected, because the nuclear matter has the properties adequately described by the strong interactions, also, the SU(2) symmetry for the light quarks ($m_q = m_u = m_d$) is assumed. Because of the symmetric nuclear matter in Hartree approximation, V_ρ^q is zero.

The bag radius in medium for the ρ-meson, R_ρ^*, is determined by the stability condition for the mass of the hadron against the ρ-meson against the variation of the bag radius [32] [see Eq. (16)], and, the eigenenergies in units of $1/R_\rho^*$ are given by,

$$\begin{pmatrix} \epsilon_u \\ \epsilon_{\bar{u}} \end{pmatrix} = \Omega_q^* \pm R_\rho^* \left(V_\omega^q + \frac{1}{2}V_\rho^q\right),$$

$$\begin{pmatrix} \epsilon_d \\ \epsilon_{\bar{d}} \end{pmatrix} = \Omega_q^* \pm R_\rho^* \left(V_\omega^q - \frac{1}{2}V_\rho^q\right). \tag{15}$$

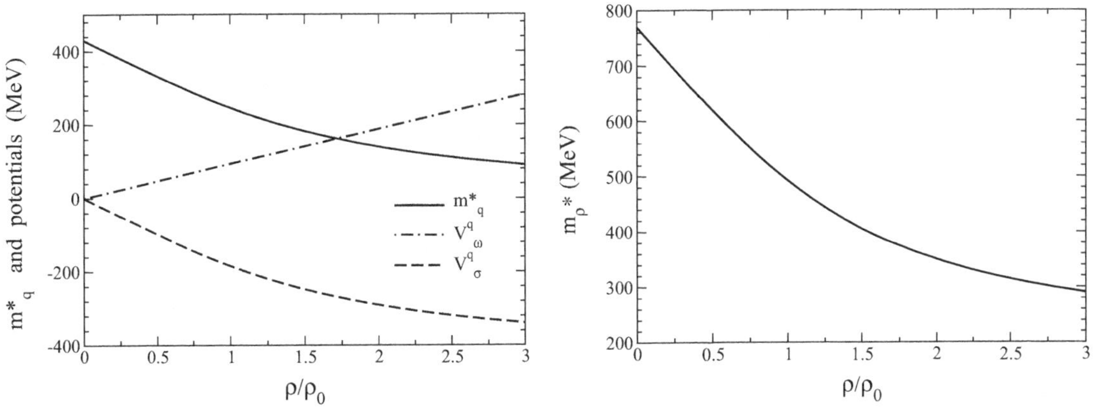

Fig. 2 Effective constituent light-quark masses and the potentials felt by the quarks (*left panel*), and ρ-meson effective mass, m_ρ^* (*right panel*), in symmetric nuclear matter calculated by the QMC model [32]

The mass of the ρ-meson in medium is calculated with the expression below,

$$m_\rho^* = \sum_{j=q,\bar{q}} \frac{n_j \Omega_j^* - z_\rho}{R_\rho^*} + \frac{4}{3}\pi R_\rho^{*3} B, \quad \left.\frac{\partial m_\rho^*}{\partial R_\rho}\right|_{R_\rho = R_\rho^*} = 0, \tag{16}$$

here, $\Omega_q^* = \Omega_{\bar{q}}^* = [x_q^2 + (R_\rho^* m_q^*)^2]^{1/2}$, with $m_q^* = m_q - g_\sigma^q \sigma$, and x_q being the lowest bag eigenfrequencies; and n_q ($n_{\bar{q}}$), is the light-quark (light-antiquark) number. We show in Fig. 2 the calculated in-medium light-quark effective mass and the potentials (left panel), and the effective mass of the ρ-meson (right panel) in symmetric nuclear matter. In this study, the in-medium input calculated by the QMC model [32], namely, the effective constituent quark mass m_q^* and effective ρ-meson mass m_ρ^*, and the light-front model for the ρ-meson [10], are utilized in order to calculate the ρ-meson properties in symmetric nuclear matter.

For the constituent quark model with the light-front approach utilized here, the sum of the constituent quark masses forming the bound state ρ-meson should be larger than the bound ρ-meson mass for the vacuum case, as well as for the case of the ρ-meson in medium, since the constituent quark and antiquark masses as well as the effective ρ-meson mass decrease with increasing nuclear density (see Fig. 2, and expected from the QMC model, similar to that found in the study of the pion case [13]).

4 Results and Conclusions

The LFCQM model scale is obtained by adjusting the value of the experimental ρ-meson decay constant in vacuum [16,21], with the quark and antiquark mass values $m_q = m_{\bar{q}} = 430\,\text{MeV}$ and the regulator mass $m_R = 3.0\,\text{GeV}$. We show in Figs. 3 and 4 the calculated ρ-meson electromagnetic form factors, G_0^*, G_1, and G_2^* in symmetric nuclear matter.

The electromagnetic form factors, G_0^*, G_1^* and G_2^*, are appreciably modified by the medium effects compared with the vacuum case in Figs. 3 and 4, the case $\rho/\rho_0 = 0$ (see also Ref. [10]).

The charge electromagnetic form factor, G_0^*, has a zero both in vacuum and in medium in the present model, which is about $q_{zero}^2 \simeq 3.0\ GeV^3$ in vacuum [10]. It is very interesting to notice in Fig. 3 (left) that the zero of the charge form factor changes the positions for different nuclear densities due to the medium effects. As increasing nuclear density, the position of the zero shifts to the smaller Q^2.

The effects of nuclear medium for the magnetic moment of the ρ-meson is not so strong at $Q^2 \approx 0$. In vacuum, the ρ-meson magnetic moment μ is about 2.20 $[e/2m_\rho]$, and the value agrees with some recent works [33,34]. At the maximum nuclear density considered here, the magnetic moment has the value of about 2.10, which shows the decrease of 4.5 % relative to the value in vacuum. However, the magnetic form factor decreases as increasing nuclear density and momentum transfer (Q^2), as shown in Fig. 3 (right), for six different nuclear densities.

Above the nuclear density $\rho/\rho_0 = 0.75$, the quadrupole form factor G_2^* changes the behavior, and increases (see Fig. 4). The non-zero values for the quadrupole moment is a consequence of the relativistic

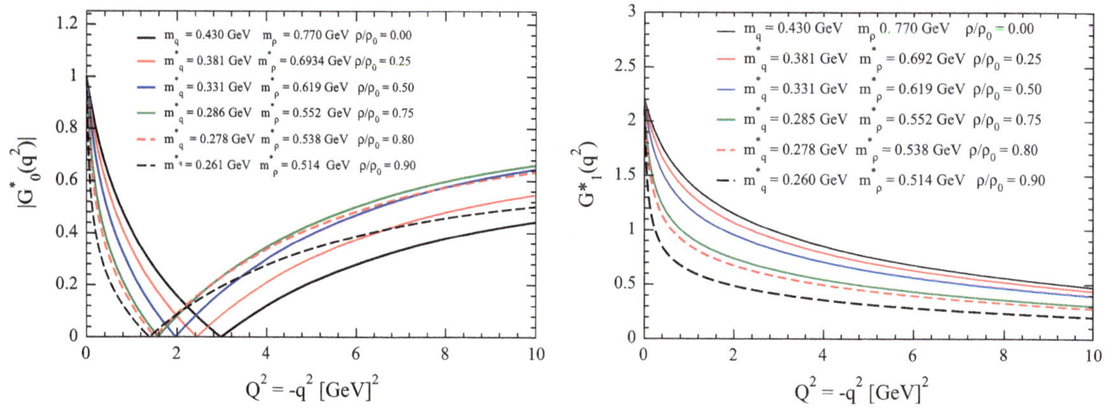

Fig. 3 Electromagnetic charge form factor, G^*_0 (*left*), and magnetic form factor, G^*_1 (*right*), both in symmetric nuclear matter, calculate by LFCQM [10] using the in-medium input obtained by the QMC model [32]

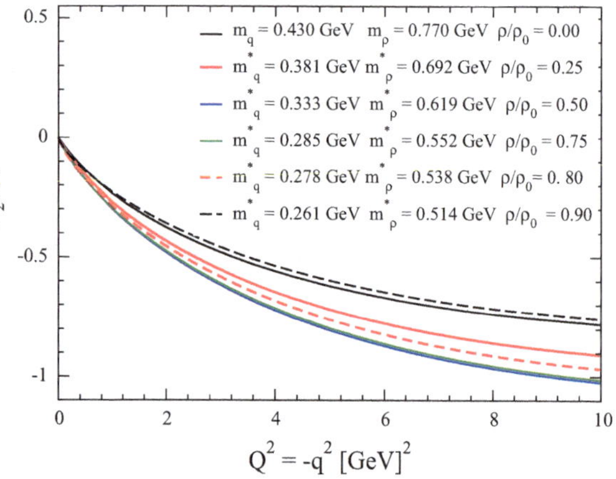

Fig. 4 In-medium ρ-meson quadrupole electromagnetic form factor, G^*_2, in symmetric nuclear matter, calculate with QMC model [32] and LFCQM [10]

character of LFCQM. In addition, the reduction rate as increasing Q^*_2 is modified differently because of the medium effects.

In conclusion, we have studied the ρ-meson electromagnetic form factors in symmetric nuclear matter, charge, magnetic and quadrupole form factors up to 10 GeV^2, with a relativistic constituent quark model in the light-front and using the in-medium input obtained by the QMC model. The modifications of the form factors due the medium effects are quite appreciable. The main point of this study is to explore the effects of the medium with well-established model in vacuum case [10], and compare with the results with those in symmetric nuclear matter.

Acknowledgements The authors would like to thank the organizers of Light Cone 2016 for invitation, and warm hospitality, that motivated them and made them enjoy during the workshop.

References

1. F.J. Yndurâin, *The Theory of Quark and Gluon Interactions* (Springer Verlag, Berlin, 1983)
2. P.Z. Skand, *Introduction to QCD*. Lecture Notes from a Course Given at TASI 2012. CERN-PH-TH-2012-196, arXiv:1207.2389
3. S.J. Brodsky, Pauli, H.C. Pauli, S.S. Pinsky, Quantum chromodynamics and other field theories on the light cone. Phys.Rep. **301**, 299 (1998)
4. W. Jaus, Semileptonic decays of B and d mesons in the light front formalism. Phys. Rev. D **41**, 3394 (1990)
5. W. Jaus, D. Wyler, Rare decays B $\to K l\bar{l}$ and B $\to K^* l\bar{l}$. Phys. Rev. D **41**, 3405 (1990)
6. W. Jaus, Relativistic constituent-quark model of eletroweak propreties of light mesons. Phys. Rev. D **44**, 2851 (1991)

7. J.P.B.C. de Melo, T. Frederico, E. Pace, G. Salme, Space-like and time-like pion electromagnetic form-factor and Fock state components within the light-front dynamics. Phys. Rev. D **73**, 074013 (2006)
8. F.M. Lev, E. Pace, G. Salmé, Deuteron magnetic and quadrupole moments with a Poincare covariant current operator in the front form dynamics. Phys. Rev. Lett. **83**, 5250 (1999)
9. J.P. Vary et al., Hamiltonian light-front field theory in a basis function approach. Phys. Rev. C **81**, 035205 (2010)
10. J.P.B.C. de Melo, T. Frederico, Covariant and light front approaches to the rho meson electromagnetic form-factors. Phys. Rev. C **55**, 2043 (1997)
11. J.P.B.C. de Melo, T. Frederico, Eletroweak form factors in the light-front for spin-1 particles. Few Body Syst. **52**, 403 (2012)
12. J.P.B.C. de Melo, A.N. da Silva, S. Clayton, T. Frederico, Spin-1 particles with light-front approach. EPJ Web Conf. **73**, 03017 (2014)
13. J.P.B.C. de Melo, K. Tsushima, B. El-Bennich, E. Rojas, T. Frederico, Pion structure in the nuclear medium. Phys. Rev. C **90**, 035201 (2014)
14. R.S. Hayano, T. Hatsuda, Hadron properties in the nuclear medium. Mod. Phys. **82**, 2949 (2010)
15. W.K. Brooks, S. Strauch, T. Tsushima, Medium modifications of hadron properties and partonic processes. J. Phys. Conf. Ser. **299**, 012011 (2011)
16. K.A. Olive, Olive et al., Particle data group; The review of particle physics. Chin. Phys. C **38**, 090001 (2014)
17. J.P.B.C. de Melo, T. Frederico, Light-front projection of spin-1 electromagnetic current and zero-modes. Phys. Lett. B **708**, 87 (2012)
18. F. Cardarelli, I.L. Grach, I.M. Narodetskii, E. Pace, G. Salmé, S. Simula, Electromagnetic form factors of the ρ meson in a light-front constituent quark model. Phys. Lett. B **349**, 393 (1995)
19. P.L. Chung, W.N. Polyzou, F. Coester, B.D. Keister, Hamiltonian light front dynamics of elastic electron deuteron scattering. Phys. Rev. C **37**, 2000 (1988)
20. B.L.G. Bakker, H.M. Choi, C.R. Ji, The vector meson form-factor analysis in light front dynamics. Phys. Rev. D **65**, 116001 (2002)
21. Clayton S. Mello, A.N. da Silva, J.P.B.C. de Melo, Light-front spin-1 model: parameters dependence. Few Body Syst. **56**, 509 (2015)
22. L. Grach, L.A. Kondratyuk, The electromagnetic form factor of the deuteton in relativistic dynamics. The two nucleon and the six components. Sov. J. Nucl. Phys. **39**, 198 (1984)
23. L.L. Frankfurt, I.L. Grach, L.A. Kondratyuk, M. Strikman, Is the structure in the deuteron magnetic form factor at $Q^2 \approx$ new evidence for nuclear core? Phys. Rev. Lett. **62**, 387 (1989)
24. J.P.B.C. de Melo, T. Frederico, E. Pace, G. Salmé, Pair term in the electromagnetic current within the front form dynamics: spin-0 case. Nucl. Phys. A **707**, 399 (2002)
25. P.A.M. Guichon, A possible quark mechanism for the saturation of nuclear matter. Phys. Lett. B **200**, 235 (1988)
26. T. Frederico, B.V. Carlson, R.A. Rego, M.S. Hussein, Quark structure of the nucleon and quantum hadrodynamics. J. Phys. G **15**, 297 (1989)
27. P.A.M. Guichon, K. Saito, E.N. Rodionov, A.W. Thomas, The role of nucleon structure in finite nuclei. Nucl. Phys. A **601**, 349 (1996)
28. K. Saito, K. Tsushima, A.W. Thomas, Variation of hadron masses in finite nuclei. Phys. Rev. C **55**, 2637 (1997)
29. K. Saito, K. Tsushima, A.W. Thomas, Selfconsistent description of finite nuclei based on a relativistic quark model. Phys. Rev. C **55**, 2637 (1997)
30. K. Tsushima, K. Saito, J. Haidenbauer, A.W. Thomas, The quark—meson coupling model for lambda, sigma and xi hypernuclei. Nucl. Phys. A **630**, 691 (1998)
31. P.A.M. Guichon, A.W. Thomas, K. Tsushima, Binding of hypernuclei in the latest quark-meson coupling model. Nucl. Phys. A. **814**, 66 (2008)
32. K. Saito, K. Tsushima, A.W. Thomas, Nucleon and hadron structure changes in the nuclear medium and impact on observables. Prog. Part. Nucl. Phys. **58**, 1 (2007)
33. D.García Guidi no, G.Toledo Sánchez, Determination of the magnetic dipole moment of the rho meson using four-pion electroproduction data. Int. J. Mod. Phys. A **30**(18n19), 1550114 (2015)
34. V. Simonis, Magnetic properties of ground-state mesons. Eur. Phys. J. A **52**(4), 90 (2016)

Few-Body Syst (2017) 58:129
DOI 10.1007/s00601-017-1279-1

Yuichi Hoshino

Phase Transition in Topologically Massive QED

Received: 10 January 2017 / Accepted: 27 February 2017 / Published online: 4 April 2017

Abstract We study the effects of Chern–Simons term on chiral condensate in (2+1)-dimensional QED. In solving the Dyson–Schwinger equation for the fermion self-energy, we show that there exists a critical value of the topological mass above which the chiral condensate is washed away at zero temperature.

1 Introduction

We examine the effects of Chern–Simons term on chiral condensate in (2+1)-dimensional QED. It has been known that this model has a vortex solution [1]. In Kosterlitz–Thouless type transition vortex washes away order parameter at finite temperature in statistical physics [2]. One may expect that the same kind of phenomenon take place in the relativistic model as QED with Chern–Simons term. We evaluate the fermion propagator by solving Dyson–Schwinger equation. After that we show the first order phase transition where chiral condensate is washed away by parity odd mass induced by Chern–Simons term.

2 Topologically Massive QED

2.1 Lagrangian

The Lagrangian density of Topologically Massive QED_3 is written [1]

$$\mathcal{L} = \overline{\psi}(i\gamma \cdot (\partial - ieA) - m)\psi - \frac{1}{4}F_{\mu\nu}F^{\mu\nu} - \frac{\mu}{4}\epsilon^{\mu\nu\rho}F_{\mu\nu}A_\rho - \frac{1}{2\xi}(\partial \cdot A)^2. \quad (1)$$

This system is characterized by equations of motion

$$(\gamma \cdot (i\partial + eA) - m)\psi = 0, \quad (2)$$

$$\partial_\mu F^{\mu\nu} + \frac{\mu}{2}\epsilon^{\nu\alpha\beta}F_{\alpha\beta} = J^\nu. \quad (3)$$

Under the gauge transformation

$$\psi \to e^{i\Omega}\psi, \quad (4)$$

$$A_\mu \to A_\mu + \frac{1}{e}\partial_\mu \Omega, \quad (5)$$

This article belongs to the Topical Collection "Light Cone 2016".

Y. Hoshino (✉)
Kushiro National College of Technology, Otanoshike-nishi 2-32-1, Kushiro City, Hokkaido 084, Japan
E-mail: hoshino@kushiro-ct.ac.jp
Tel.: +81-154-41-5177

The Lagrangian is invariant but the gauge fixing term changes by a total derivative

$$\mathcal{L}_g \to \mathcal{L}_g + \partial_\alpha \left(\frac{\mu}{4e} \epsilon^{\alpha\mu\nu} F_{\mu\nu} \Omega \right). \tag{6}$$

However, the action $\int d^3x \mathcal{L}(x)$ is gauge invariant. It is easily seen by partial integration.

2.2 Symmetry

Here we consider the dynamical effects as mass generation in the presence of Chern–Simons term. If the fermion is massless, $m = 0$, $\mathcal{L}$ has a $U(2)$ symmetry [3] generated by $\{I, \gamma_3, \gamma_5, \tau\}$, where $\{\gamma_\mu, \gamma_\nu\} = 2g_{\mu\nu}$ for $\mu, \nu = (0, 1, 2)$, $\gamma_0 = \begin{pmatrix} \sigma_3 & 0 \\ 0 & -\sigma_3 \end{pmatrix}$, $\gamma_{1,2} = -i \begin{pmatrix} \sigma_{1,2} & 0 \\ 0 & -\sigma_{1,2} \end{pmatrix}$, $\gamma_3 = \begin{pmatrix} 0 & I \\ I & 0 \end{pmatrix}$, $\gamma_5 = \begin{pmatrix} 0 & -iI \\ iI & 0 \end{pmatrix}$, $\tau = -i[\gamma_3, \gamma_5]/2 = \begin{pmatrix} I & 0 \\ 0 & -I \end{pmatrix}$. γ_3 and γ_5 act as chiral transformation

$$\begin{aligned} \psi(x) &\to e^{i\gamma_3\theta_3}\psi(x), \\ \psi(x) &\to e^{i\gamma_5\theta_5}\psi(x). \end{aligned} \tag{7}$$

The scalar density is mixed by the above transformation

$$\begin{aligned} \overline{\psi}(x)\psi(x) &\to \cos(2\theta_{3,5})\overline{\psi}(x)\psi(x) + i\sin(2\theta_{3,5})\overline{\psi}(x)\gamma_{3,5}\psi(x), \\ \overline{\psi}(x)\tau\psi(x) &\to \overline{\psi}(x)\tau\psi(x). \end{aligned} \tag{8}$$

Dynamical mass generation breaks $U(2)$ symmetry down to $U(1)_I \times U(1)_\tau$ which is generated by $\{I, \tau\}$. There exists a discrete parity transformation

$$P\psi(t, x, y)P^{-1} \to i\gamma^1\gamma^3\psi(t, -x, y). \tag{9}$$

Parity even and odd mass are transformed

$$Pm_e\overline{\psi}\psi P^{-1} \to m_e\overline{\psi}\psi, \tag{10}$$

$$Pm_o\overline{\psi}\tau\psi P^{-1} \to -m_o\overline{\psi}\tau\psi. \tag{11}$$

Both of these mass terms are invariant under $U(1)_I \times U(1)_\tau$ transformation

$$\begin{aligned} \psi(x) &\to e^{i\theta}\psi(x), \\ \psi(x) &\to e^{i\tau\theta_\tau}\psi(x). \end{aligned} \tag{12}$$

We find the eigenvalue of the free particle Hamiltonian

$$H = \gamma^0(\gamma^i p^i + m_e I + m_o\tau) \tag{13}$$

as $E^2 = p^2 + m_\pm^2$, $m_\pm = m_e \pm m_o$, where $i = 1, 2$, I is a 4×4 unit matrix and τ is an operator defined above. Two kinds of mass are written by

$$m_e I + m_o\tau = \begin{pmatrix} m_+ & 0 \\ 0 & m_- \end{pmatrix} = \begin{pmatrix} m_e + m_o & 0 \\ 0 & m_e - m_o \end{pmatrix}. \tag{14}$$

So that we may split 4-component spinors into upper and lower components by the projection operators

$$\psi_\pm = \chi_\pm\psi = \chi_\pm \begin{pmatrix} \psi_+ \\ \psi_- \end{pmatrix}, \chi_\pm = \frac{1 \pm \tau}{2}, \chi_+ = \begin{pmatrix} 1 & 0 \\ 0 & 0 \end{pmatrix}, \chi_- = \begin{pmatrix} 0 & 0 \\ 0 & 1 \end{pmatrix}. \tag{15}$$

From the Lagrangian

$$\mathcal{L} = \overline{\psi}_+(i\gamma \cdot \partial - m_+)\psi_+ + \overline{\psi}_-(i\gamma \cdot \partial - m_-)\psi_- \tag{16}$$

we have the free propagator

$$S_0(p) = -\frac{\gamma \cdot p + m_+}{p^2 - m_+^2}\chi_+ - \frac{\gamma \cdot p + m_-}{p^2 - m_-^2}\chi_-. \tag{17}$$

2.3 Ward–Takahashi Relation

In our 4-component spinor representations the parity violating mass $m_o\overline{\psi}\tau\psi$ is a pseudoscalar, and the current $J^{3,5}_\mu = \overline{\psi}\gamma_{3,5}\gamma_\mu\psi$ transforms as a vector, which mix with each other under parity and charge conjugation transformation with an arbitrary phase [3]. Here we start from the Dirac equation

$$(\gamma \cdot \partial + m_0)\psi = ie\gamma \cdot A\psi, \overline{\psi}(\gamma \cdot \overleftarrow{\partial} - m_0) = -ie\overline{\psi}\gamma \cdot A. \tag{18}$$

From a combination of the above two equations, we obtain

$$\partial_\lambda J^{3,5}_\lambda = 2m_0 J^{3,5} = D, \tag{19}$$

where

$$J^{3,5}_\lambda = \overline{\psi}\gamma_{3,5}\gamma_\lambda\psi, J^{3,5} = \overline{\psi}\gamma_{3,5}\psi. \tag{20}$$

The general Ward–Takahashi-relation of the vector current is known [4],

$$\partial^x_\mu T\{J^{3,5}_\mu(x)\Pi\phi_i(y_i)\} = T\{D(x)\Pi\phi_i(y_i)\} + \sum T\{\delta^{(3)}(x - y_i)\phi_1(y_1)...\delta\phi(y_i)...\phi(y_n)\}. \tag{21}$$

For the vertex function it is given as

$$\begin{aligned}\partial^x_\mu \left\langle 0|T(J^{3,5}_\mu(x)\psi(y)\overline{\psi}(z))|0\right\rangle &= \left\langle 0|T(D\psi(y)\overline{\psi}(z))|0\right\rangle \\ &+ \delta^{(3)}(x - y)\gamma_{3,5}S_F(y - z) + \delta^{(3)}(x - z)S_F(y - z)\gamma_{3,5},\end{aligned} \tag{22}$$

provided the operator of the symmetry transformation

$$\delta(x_0 - y_0)[J^{3,5}_0(x), \psi(y)] = \delta\psi(y)\delta^{(3)}(x - y) = \gamma_{3,5}\psi(y)\delta^{(3)}(x - y), \tag{23}$$

$$\delta(x_0 - y_0)[J^{3,5}_0(x), \overline{\psi}(y)] = \delta\overline{\psi}(y)\delta^{(3)}(x - y) = \overline{\psi}(y)\gamma_{3,5}\delta^{(3)}(x - y). \tag{24}$$

The above equation can be written as

$$\begin{aligned}&\int d^3x'd^3y'S_F(x - x')\partial^x_\mu \Gamma^{3,5}_\mu(y', z'; x)S_F(y' - y) \\ &= 2m_0 \int d^3x'd^3y'S_F(x - x')\Gamma^{3,5}(y', z'; x)S_F(y - y') \\ &+ \delta^{(3)}(x - y)\gamma_{3,5}S_F(y - z) + \delta^{(3)}(x - z)S_F(y - z)\gamma_{3,5}.\end{aligned} \tag{25}$$

In terms of the vertex function Γ,

$$\Gamma^{3,5}_\mu(y, z; x) = \frac{1}{(2\pi)^6}\int d^3pd^3qe^{ip(y-x)+iq(x-z)}, \tag{26}$$

it is rewritten in momentum space

$$\begin{aligned}-i(p - q)_\mu S_F(p)\Gamma^{3,5}_\mu(p, q)S_F(q) &= 2m_0 S_F(p)\Gamma^{3,5}(p, q)S_F(q) \\ &+\gamma_{3,5}S_F(q) + S_F(p)\gamma_{3,5}.\end{aligned} \tag{27}$$

For the proof of the Ward–Takahashi relation in terms of integral equation for $\Gamma^{3,5}$, $\Gamma^{3,5}_\mu$ see [5]. For vanishing bare mass m_0, current conservation is spontaneously broken by the parity even mass generation

$$\lim_{p\to q} i(p - q)_\mu \Gamma^{3,5}_\mu(p, q) = \{\gamma_{3,5}, -S_F^{-1}(p)\} \neq 0. \tag{28}$$

In other form it is well known

$$\partial^x_\mu T(J^{3,5}_\mu(x)\overline{\psi}(y)\gamma_{3,5}\psi(y)) = -2\overline{\psi}(x)\psi(x). \tag{29}$$

Taking the vacuum expectation value of both sides of the above equation, non-vanishing of the r.h.s indicates the existence of a massless scalar NG boson, in the vertex $\Gamma^{3,5}_\mu(p,q)$, of the form

$$\frac{if_\pi (p-q)_\mu X_\pi(P=0,p-q)}{(p-q)^2}, \tag{30}$$

where X_π is a Bethe–Salpeter amplitude for total momentum $P = 0$. Here we have two order parameters $\langle\bar\psi\psi\rangle, \langle\bar\psi\tau\psi\rangle$ for chiral symmetry breaking and explicit parity violation by Chern–Simons term. With the Dyson–Schwinger equation it is familiar how to obtain the non-perturbative propagator and dynamical mass. Therefore we set up the Dyson–Schwinger equation for the electron propagator in the chiral representation.

3 Dyson–Schwinger Equation

The quenched Schwinger–Dyson equation for the self-energy $\Sigma(p)$ is written

$$-i\Sigma(p) = (-ie)^2 \int \frac{d^3q}{(2\pi)^3}\gamma_\mu S(q)\Gamma_\nu(p,q)D_{\mu\nu}(k), \tag{31}$$

where $k = q - p$. $D_{\mu\nu}(k)$ is a free propagator of gauge boson with covariant ξ gauge

$$iD_{\mu\nu}(k) = \frac{g_{\mu\nu} - k_\mu k_\nu/k^2 - i\mu\epsilon_{\mu\nu\rho}k^\rho/k^2}{k^2-\mu^2+i\epsilon} + \xi\frac{k_\mu k_\nu}{k^4}. \tag{32}$$

$S(p)$ is a full fermion propagator

$$S(p) = \frac{i}{A(p)\gamma\cdot p - B(p)} \to i\frac{(\gamma\cdot pA_+(p)+B_+(p))}{p^2A_+^2(p)-B_+^2(p)}\chi_+ + i\frac{(\gamma\cdot pA_-(p)+B_-(p))}{p^2A_-^2(p)-B_-^2(p)}\chi_-. \tag{33}$$

In this normalization Σ is real in the region $p^2 < 0$. We separate the vector and scalar part of the propagator by taking the trace of Σ . Trace formulae for (4×4) γ matrices are reduced to (2×2) ones with the projection operator $\chi_\pm$

$$tr(\chi_\pm) = 2, tr(\gamma_\gamma\gamma_\nu\chi_\pm) = 2g_{\mu\nu}, tr(\gamma_\mu\gamma_\nu\gamma_\rho\chi_\pm) = \mp 2i\epsilon_{\mu\nu\rho}.$$

For the bare vertex we set $\Gamma_\mu(p,q) = \gamma_\mu$. To keep the gauge invariance of the action we adopt the gauge covariant approximation by introducing the Ball–Chiu vertex ansatz for the longitudinal part [6].

$$\begin{aligned}\Gamma^{BC}_\mu(p,q) &= \Gamma^T_\mu(p,q) + \frac{A(p)+A(q)}{2}\gamma_\mu + \frac{A(p)-A(q)}{2(p^2-q^2)}\gamma\cdot(p+q)(p+q)_\mu \\ &\quad - \frac{B(p)-B(q)}{p^2-q^2}(p+q)_\mu.\end{aligned} \tag{34}$$

For the bare vertex in the Landau gauge we have, in Euclidean momentum space,

$$B(p)_\pm = \frac{e^2}{4\pi^2}\int \frac{d^3q}{(q^2A(q)_\pm^2 + B(q)_\pm^2)(k^2+\mu^2)}\left[B(q)_\pm \mp \mu A(q)_\pm\left(\frac{1}{2} - \frac{p^2-q^2}{2k^2}\right)\right], \tag{35}$$

$$\begin{aligned}A(p)_\pm &= 1 + \frac{e^2}{4\pi^2p^2}\int \frac{d^3q}{(q^2A(q)_\pm^2 + B(q)_\pm^2)(k^2+\mu^2)}\left[\pm\mu B(q)_\pm\left(\frac{1}{2}+\frac{p^2-q^2}{2k^2}\right)\right. \\ &\quad \left. + A(q)_\pm\left(\frac{(p^2-q^2)^2}{4k^2} - \frac{k^2}{4}\right)\right].\end{aligned} \tag{36}$$

After angular integration we have the following coupled integral equation

$$B(p)_\pm = \frac{e^2}{2\pi^2}\int_0^\infty \frac{dqq^2}{q^2A(q)_\pm^2 + B(q)_\pm^2}(B(q)_\pm I_0[p,q] \mp \mu A(q)_\pm I_2(p,q)_-), \tag{37}$$

$$A(p)_\pm = 1 + \frac{e^2}{2\pi^2p^2}\int_0^\infty \frac{dqq^2}{q^2A(q)_\pm^2 + B(q)_\pm^2}(\pm\mu B(q)_\pm I_2(p,q)_+ + A(q)_\pm I_3(p,q)). \tag{38}$$

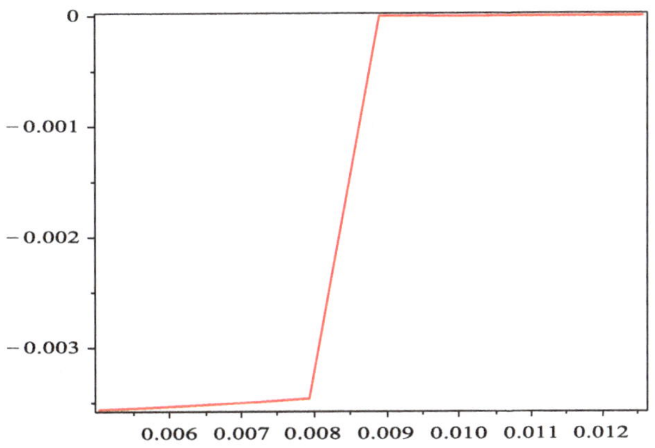

Fig. 1 $\langle\bar{\psi}\psi\rangle$ with BC vertex as a function of μ in unit of $e^2 = 1$

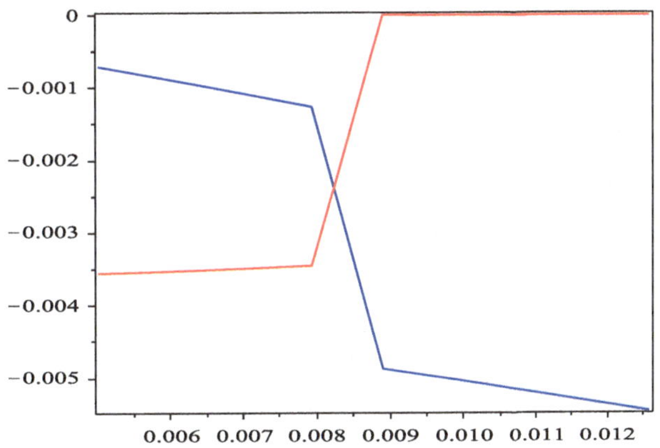

Fig. 2 Parity even and odd order parameters $\langle\bar{\psi}\psi\rangle$(red),$\langle\bar{\psi}\tau\psi\rangle$ (blue) as a function of μ

Angular integral formulae are given as

$$I_0(p,q) = \int_{-1}^{1} \frac{d\cos\theta}{k^2+\mu^2} = \frac{-1}{2pq}\ln(\frac{(p-q)^2+\mu^2}{(p+q)^2+\mu^2}), \tag{39}$$

$$\begin{aligned} I_2(p,q)_\pm &= \int_{-1}^{1} \frac{d\cos\theta}{k^2+\mu^2}\left(\frac{1}{2} \pm \frac{p^2-q^2}{2k^2}\right) \\ &= \frac{-1}{4pq}\ln\left(\frac{(p-q)^2+\mu^2}{(p+q)^2+\mu^2}\right) \pm \frac{p^2-q^2}{4\mu^2 pq}\ln\left(\frac{1+\mu^2/(p-q)^2}{1+\mu^2/(p+q)^2}\right), \end{aligned} \tag{40}$$

$$\begin{aligned} I_3(p,q) &= \int_{-1}^{1} \frac{d\cos\theta}{k^2+\mu^2}\left(\frac{(p^2-q^2)^2}{4k^2} - \frac{k^2}{4}\right) \\ &= \frac{(p^2-q^2)^2}{8\mu^2 pq}\ln\left(\frac{1+\mu^2/(p-q)^2}{1+\mu^2/(p+q)^2}\right) - \frac{1}{2} - \frac{\mu^2}{8pq}\ln\left(\frac{(p-q)^2+\mu^2}{(p+q)^2+\mu^2}\right). \end{aligned} \tag{41}$$

4 Numerical Results

Here we show the numerical results for the vacuum expectation value $\langle\bar{\psi}\psi\rangle$, $\langle\bar{\psi}\tau\psi\rangle$ and the vector and scalar part of the propagator in the chiral representation. Here we refer to earlier works related to the present work [7–9].

In Figs. 1 and 2 we see the vacuum expectation value as a function of the topological mass. There exists a critical value of the topological mass above which the chiral condensate is washed away. In Figs. 3 and 4 we see the vector and scalar part of the propagator which are the solutions of the Dyson–Schwinger equation with

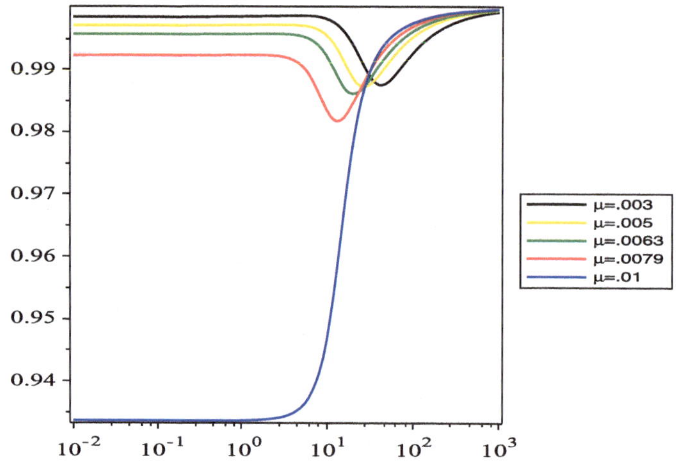

Fig. 3 $A_{+}(p)$ for p in unit of $e^2 = 1$

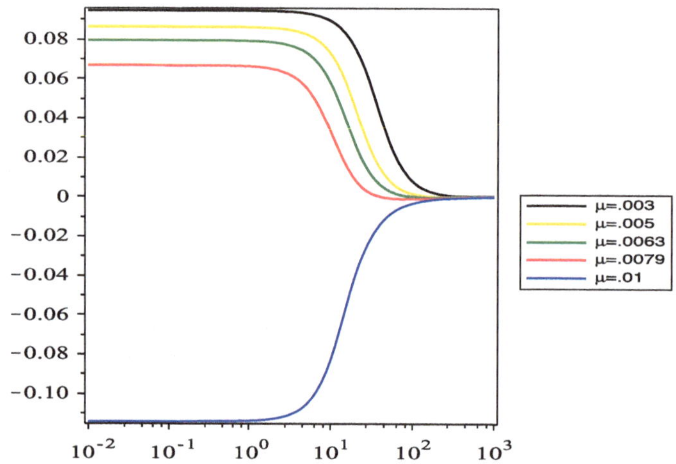

Fig. 4 $B_{+}(p)$ for p in unit of $e^2 = 1$

a bare vertex for various values of the topological mass. At the critical point of the topological mass, $B_{+}(p)$ changes its sign and $A_{+}(p)$ shows its effect. At least at one loop level, parity even and odd mass both behave as $1/p$ at large p. At the critical value of μ, these cancel each other, $B_{+}(p)$ vanishes and only the parity odd mass survives in the next iteration in eq (37) that leads to a vanishing chiral condensate. The critical value of the topological mass in the BC vertex is the same as that in the bare vertex case in our approximation.

5 Summary

In this work we studied the dynamics of Kosterlitz–Thouless type transition in Topologically Massive Abelian Gauge Theory in three dimensional space-time. In solving the Dyson–Schwinger equation in the chiral representation, which is useful in the presence of parity odd mass term, we find that the induced parity violating fermion mass is destructive to parity even mass at short distance and $B_{+}(p)$ vanishes. However we have not found a clear way to determine the critical value of topological mass. We simply compared the vacuum energy of parity odd and even one as shown in Fig. 2. Below the critical value of the topological mass there exist two kinds of mass. However if they coincide at the critical value of the topological mass, the parity violating vacuum is favored above the critical value of the topological mass and a first order phase transition take places. We saw that the chiral condensate is washed away at that point. This phenomenon turned out to be gauge invariant by the choice of the BC vertex in solving the Dyson–Schwinger equation for the fermion self-energy. In the next step we would like to study the more realistic case at finite temperature.

References

1. S. Deser, R. Jackiw, S. Templeton, Topologically massive gauge theories. Ann. Phys. **281**, 409–449 (2000)
2. J.M. Kosterlitz, The critical properties of the two-dimensional xy model. J. Phys. C **7**, 1046–1060 (1974)
3. C.J. Burden, Bound states from Bethe–Salpeter equation in QED3. Nucl. Phys. **B387**, 419–446 (1992)
4. D.J. Gross, *in Method in Field Theory*, ed. by R. Balian, J. Zinn-Justin. Application of the renormalization group to high-energy physics (Les Houches, North Holland, 1975), pp. 141–250
5. T. Maskawa, H. Nakajima, Spontaneous breaking of chiral symmetry in a Vector–Gluon model. Prog. Theo. Phys. **52**(4), 1326–1354 (1974). doi:10.1143/PTP.52.1326
6. C.S. Fisher, R. Alkofer, T. Dham, P. Maris, Dynamical breaking of chiral symmetry in unquenched QED_3. Phys. Rev. D **70**, 073007 (2004)
7. K.I. Kondo, P. Maris, Spontaneous chiral symmetry breaking in three-dimensional QED with a Chern-Simons term. Phys. Rev. D **52**, 1212 (1995)
8. S.S. Madrigal, C.P. Hofmann, A. Raya, Dynamical mass generation and confinement in Maxwell–Chern–Simons planar electrodynamics. J. Phys. Conf. Ser. **287**, 012028 (2011)
9. Y. Hoshino, T. Inagaki, Y. Mizutani, Gauge covariant solution for Schwinger–Dyson equation in 3D QED with a Chern–Simons term. PTEP. 023B03 (2015)

Few-Body Syst (2017) 58:128
DOI 10.1007/s00601-017-1291-5

Paul Hoyer

Born Level Bound States

Received: 13 December 2016 / Accepted: 21 March 2017 / Published online: 3 April 2017

Abstract Bound state poles in the S-matrix of perturbative QED are generated by the *divergence* of the expansion in α. The perturbative corrections are necessarily singular when expanding around free, $\mathcal{O}\left(\alpha^0\right)$ *in* and *out* states that have no overlap with finite-sized atomic wave functions. Nevertheless, measurables such as binding energies do have well-behaved expansions in powers of α (and $\log\alpha$). It is desirable to formulate the concept of "lowest order" for gauge theory bound states such that higher order corrections vanish in the $\alpha \to 0$ limit. This may allow to determine a lowest order term for QCD hadrons which incorporates essential features such as confinement and chiral symmetry breaking, and thus can serve as the starting point of a useful perturbative expansion. I discuss a "Born" (no loop, lowest order in $\hbar$) approximation. Born level states are bound by gauge fields which satisfy the classical field equations. Gauss' law determines a distinct field $A^0(\boldsymbol{x})$ for each instantaneous position of the charges. A Poincaré covariant boundary condition for the gluon field leads to a confining potential for $q\bar{q}$ and qqq states. In frames where the bound state is in motion the classical gauge field is obtained by a Lorentz boost of the rest frame field.

1 Introduction

The standard perturbative expansion of the S-matrix, defined by the master formula (1) below, is the basis for analytic approaches to the Standard Model. However, Eq. (1) is not valid for *bound* states in i and f, since they have zero overlap with the free *in* and *out* states. Even a first approximation to a bound state (such as that given by the Schrödinger equation for atoms) necessarily involves all powers of the coupling. This raises the question whether there is a physically motivated first approximation for bound states, around which a convergent power series in the coupling may be developed. In this talk I discuss $\mathcal{O}\left(\hbar^0\right)$ "Born states". In analogy to standard tree diagrams the bound state Born terms have no loop contributions. I refer to my lecture notes [1] and [2] for a more detailed treatment.

Section 2 concerns the $\hbar$ expansion and the need to modify (1) so that it applies to bound states. Section 3 verifies that the Born term concept gives the standard Schrödinger approximation for QED atoms. The (equal time) Hamiltonian formulation furthermore allows to consider atomic states with arbitrary CM momenta. In Sect. 4, I discuss the states of an electron that is relativistically bound by an external potential. The Dirac wave functions determine not only the valence electron distribution, but also those of the e^+e^- pairs. In Sect. 5, I study whether QCD Born states can serve as a first approximation for hadrons. Confinement requires to consider homogeneous solutions of Gauss' law for which the instantaneous gluon field $A_a^0(\boldsymbol{x} \to \infty) \neq 0$. In solutions which maintain Poincaré symmetry each quark color component of the hadron is bound by a linear

This article belongs to the Topical Collection "Light Cone 2016".

P. Hoyer (✉)
Department of Physics, University of Helsinki, POB 64, 00014 Helsinki, Finland
E-mail: paul.hoyer@helsinki.fi

potential. An external observer does not see the confining A_a^0 field of color singlet hadrons since it vanishes when averaged over quark colors. I discuss some properties of the meson solutions in Sect. 6, and give a brief outlook in Sect. 7.

2 The Perturbative S-Matrix with Bound in and out States

Perturbation theory is our main analytic tool in studies of physical gauge theories. Its success in describing QED+EW and (hard) QCD scattering amplitudes has established the Standard Model. The expression for the S-matrix in the Interaction Picture (IP) is

$$S_{fi} = {}_{out}\langle f,\, t \to \infty | \Big\{ \mathrm{T} \exp\Big[-i \int_{-\infty}^{\infty} dt\, H_I(t) \Big] \Big\} |i,\, t \to -\infty\rangle_{in} \tag{1}$$

The in and out states are eigenstates of the free, $\mathcal{O}\left(g^0\right)$ Hamiltonian H_0. The expansion of the exponential in powers of the coupling g ($H_I \propto g$) defines the perturbative corrections. The eigenstates of H_0 evolve into eigenstates of the full Hamiltonian H during their evolution from the asymptotic times $t = \pm\infty$. The perturbative expansion is thus formally exact provided that there is a non-vanishing overlap between the in and out states and the physical i and f states of the scattering amplitude.

As it stands, the expression (1) for S_{fi} is inappropriate when the initial or final state involves a (stable) bound state. The infinite separations of free particles (wave packets) means that the in and out states have a vanishing overlap with finite-sized bound states B:

$$\langle B, t = -\infty | i, t = -\infty\rangle_{in} = \langle B, t = -\infty | U^\dagger(t) U(t)\, |i, t = -\infty\rangle_{in} = 0 \tag{2}$$

By definition, bound states are eigenstates of the full H and thus stationary in time. Hence the overlap of $|B, t\rangle$ will vanish with any state $U(t)\,|i, t = -\infty\rangle_{in}$ that the in state evolves into. Equivalently, no finite order Feynman diagram can have a bound state pole.

Inspection of the Feynman diagrams contributing to $e^+e^- \to e^+e^-$ shows (see, *e.g.*, sect. II of [1]) that diagrams with repeated single photon exchanges (the "ladder" diagrams) have a special role in bound state formation. When all momenta are scaled as in atoms ($|\boldsymbol{p}| \propto \alpha\, m_e$ in the rest frame) every ladder diagram contributes at the same power of α. On the other hand, non-ladders having crossed photons, loop corrections *etc*, are suppressed by one or more powers of α.

The sum of all ladders gives (through its divergence) rise to bound state poles. The pole residues show that the (rest frame) wave functions satisfy the Schrödinger equation with the classical potential $V(r) = -\alpha/r$. The Schrödinger atom thus appears as a first approximation of the physical atom, with perturbative corrections that vanish as $\alpha \to 0$. The ladder sum may be formulated as a Bethe-Salpeter equation for the wave function, valid in any frame and with a single photon exchange kernel. The perturbative corrections can then be systematically added through loop corrections to the kernel and propagators. In the rest frame the Non-Relativistic QED (NRQED) expansion in powers of $|\boldsymbol{p}|/m_e$ is very efficient [3]. In either case, the perturbative corrections are applied to a first approximation that is guessed, or obtained from a divergent expansion.

While the above approach is successful for QED atoms it fails to describe confinement and chiral symmetry breaking in QCD hadrons. This could be due to perturbation theory being inapplicable. However, some features of hadrons are strikingly similar to atoms, such as the spectra of heavy quarkonia. It therefore seems worthwhile to investigate the alternative possibility that $\alpha_s(0)$ does allow a perturbative expansion. The failure to describe color confinement would be due to the quarks and gluons being infinitely separated in the in and out states of (1).

As remarked previously, the Schrödinger atom is bound by a *classical* potential. This indicates that it is the lowest order term in an $\hbar$ expansion of the exact QED bound state. Recall that a Green function of a (bosonic) field φ has the functional integral expression

$$G(x_1, \ldots, x_n) = \int [d\varphi]\, \varphi(x_1) \ldots \varphi(x_n)\, e^{iS[\varphi]/\hbar} \tag{3}$$

In the limit $\hbar \to 0$ the (classical) field configurations for which the action $S[\varphi]$ is stationary give the leading contribution. Contributions of $\mathcal{O}(\hbar)$ correspond to a sum over field configurations, or equivalently, integrals over loop momenta in Feynman diagrams. By identifying the Schrödinger atom with a Born approximation we avoid the divergent sum of ladder diagrams.

It may seem surprising that an all-orders sum in α can correspond to lowest order in $\hbar$. In scattering amplitudes there is a one-to-one correspondence between the powers of α and $\hbar$ (*i.e.*, the number of loops). However, $\mathcal{O}\left(\hbar^0\right)$ (tree) diagrams with n vertices are of $\mathcal{O}\left(g^n\right)$: The power of g, but not of $\hbar$, increases with the number of external legs. Ladder diagrams are (approximated as) convolutions of $\mathcal{O}\left(\alpha\,\hbar^0\right)$ single photon exchange amplitudes. Their geometric sum is non-polynomial in α while remaining of lowest order in $\hbar$.

The Interaction Picture of the S-matrix (1) is based on separating the full Hamiltonian H into its free H_0 and interacting H_{int} parts,

$$H = H_0 + H_{int} \tag{4}$$

The usefulness of the IP rests on having analytic (plane wave) solutions of the eigenstates of H_0, and on their similarity with the asymptotic scattering states. Infrared singularities arise because the scattering amplitudes of charged particles are not gauge invariant. In QED the IR singularities can be removed by associating each charged particle with an infinite number of soft photons [4].

The constituents of (stable) bound states remain close to each other at all times. In order to allow physical Positronium atoms in the initial and final states of the S-matrix, their Born terms (Schrödinger atoms) should be included in the asymptotic *in* and *out* states. H_0 must then include the classical A^0 field generated by the electron (at $\boldsymbol{x}_1$) and positron (at $\boldsymbol{x}_2$),

$$-\boldsymbol{\nabla}_{\boldsymbol{x}}^2 A^0(\boldsymbol{x};\boldsymbol{x}_1,\boldsymbol{x}_2) = e\big[\delta(\boldsymbol{x}-\boldsymbol{x}_1) - \delta(\boldsymbol{x}-\boldsymbol{x}_2)\big] \tag{5}$$

$$A^0(\boldsymbol{x};\boldsymbol{x}_1,\boldsymbol{x}_2) = \frac{e}{4\pi}\left(\frac{1}{|\boldsymbol{x}-\boldsymbol{x}_1|} - \frac{1}{|\boldsymbol{x}-\boldsymbol{x}_2|}\right) \tag{6}$$

A^0 is determined instantaneously by the positions of the charges due to the absence of a $\partial_t A^0$ term in the QED Lagrangian. Each position of the charges corresponds to a distinct $A^0(\boldsymbol{x};\boldsymbol{x}_1,\boldsymbol{x}_2)$. In the rest frame of the atom the spatial, propagating components $\boldsymbol{A}$ of the gauge field give contributions of higher order in α. Loop effects of $\mathcal{O}(\hbar)$ are generated by H_I and treated perturbatively, with the charged particles propagating in the classical field. Infrared singularities are absent since (neutral) atoms decouple from soft photons. Because the corrections to the Schrödinger atom vanish in the $\alpha \to 0$ limit the perturbative expansion thus defined is expected to converge.

The modification of the Interaction Picture envisaged here requires a more precise definition. Here I summarize results obtained in [1] and [2] at the Born level, neglecting the effects of H_I.

3 Hamiltonian Formulation of Born Level (Schrödinger) Atoms

A Born level Positronium state in the rest frame may be expressed in terms of the electron and positron creation operators (evaluated at any given time t),

$$|n,t\rangle = \int \frac{d\boldsymbol{k}}{(2\pi)^3}\,\phi_n(\boldsymbol{k})\,b^\dagger(\boldsymbol{k},\lambda)\,d^\dagger(-\boldsymbol{k},\lambda')\,|0\rangle \tag{7}$$

where n labels the state and the Schrödinger wave function $\phi_n(\boldsymbol{k})$ is independent of the e^-, e^+ helicities λ, λ'. In terms of the 4×4 wave function

$$\Phi_n^{\alpha\beta}(\boldsymbol{k}) \equiv {}_{\alpha}\Big[\gamma^0 u(\boldsymbol{k},\lambda)\Big]\Big[v^\dagger(-\boldsymbol{k},\lambda')\Big]_{\beta}\,\phi_n(\boldsymbol{k}) \tag{8}$$

the state may in coordinate space be expressed in terms of the electron field $\psi(t,\boldsymbol{x})$,

$$|n,t\rangle = \int d\boldsymbol{x}_1 d\boldsymbol{x}_2\,\bar{\psi}_\alpha(t,\boldsymbol{x}_1)\Phi_n^{\alpha\beta}(\boldsymbol{x}_1-\boldsymbol{x}_2)\psi_\beta(t,\boldsymbol{x}_2)\,|0\rangle \tag{9}$$

The bound state should be an eigenstate of the QED Hamiltonian,

$$H(t) = \int d\boldsymbol{x}\,\Big\{\bar{\psi}(t,\boldsymbol{x})\big[-i\boldsymbol{\nabla}\cdot\boldsymbol{\gamma} + m_e + e\gamma^0 A^0(\boldsymbol{x})\big]\psi(t,\boldsymbol{x}) + \mathcal{H}_{field}\Big\} \tag{10}$$

$$H(t)\,|n,t\rangle = (2m_e + E_n^b)\,|n,t\rangle \tag{11}$$

where $E_n^b < 0$ is the binding energy. In the non-relativistic limit no pairs are produced, $b^\dagger d^\dagger |n, t\rangle \to 0$. The field energy $\mathcal{H}_{field}$ of the classical field (6) turns out (see Eq. (2.18) of [2]) to cancel half of the contribution from the fermion interaction term $\psi^\dagger(t, \boldsymbol{x})\, eA^0(\boldsymbol{x})\psi(t, \boldsymbol{x})$. Its effect can thus be taken into account by a factor $\frac{1}{2}$ in front of this term.

Having determined the Hamiltonian we may impose the stationarity condition (11). Using the canonical relations $\{\psi_\alpha^\dagger(t, \boldsymbol{x}), \psi_\beta(t, \boldsymbol{y})\} = \delta_{\alpha,\beta}\,\delta^3(\boldsymbol{x} - \boldsymbol{y})$ it gives the Schrödinger equation for the wave function ϕ_n in (7), with potential $V(\boldsymbol{x}_1 - \boldsymbol{x}_2) = -\alpha/|\boldsymbol{x}_1 - \boldsymbol{x}_2|$, as expected.

A similar analysis can be carried out for a Positronium state with any momentum $P = (P^0, \boldsymbol{P})$,

$$|n, \boldsymbol{P}, t\rangle = \int d\boldsymbol{x}_1 d\boldsymbol{x}_2\, e^{i\boldsymbol{P}\cdot(\boldsymbol{x}_1+\boldsymbol{x}_2)/2}\, \bar{\psi}(t, \boldsymbol{x}_1)\Phi_n^{(\boldsymbol{P})}(\boldsymbol{x}_1 - \boldsymbol{x}_2)\psi(t, \boldsymbol{x}_2)\,|0\rangle \tag{12}$$

Due to the Hamiltonian framework the fields are evaluated at equal time in any frame. Since the definition of time is frame-dependent, Lorentz covariance is not explicit, and emerges as a dynamical consequence of the Poincaré invariance of the action. It is thus non-trivial that

$$H(t)\,|n, \boldsymbol{P}, t\rangle = \sqrt{\boldsymbol{P}^2 + (2m + E_n^b)^2}\;|n, \boldsymbol{P}, t\rangle \tag{13}$$

This relation is verified (in the limit $\alpha \ll 1$) when the classical gauge field in $H(t)$ is related to the rest frame field (6) by a standard boost. The Positronium wave function Lorentz contracts similarly to rods in classical relativity, see sect. II.4 of [2].

4 Fock States of the Dirac Wave Function

An intriguing aspect of hadrons is that their quantum numbers are determined solely by their valence constituents ($q\bar{q}$ or qqq). The sea quarks and gluons seen in DIS do not enrich the spectrum. This is possible due to the relativistic binding, and as such can be studied in the Dirac framework of an electron bound in a strong external field $A^\mu(\boldsymbol{x})$. The Dirac wave function is characterized by the quantum numbers of a single electron, but the state it describes must, due to 'Z-diagrams' and to account for the Klein paradox, contain multiple e^+e^- pairs.

The positive and negative energy solutions of the Dirac equation ($E_n, \bar{E}_n > 0$),

$$\begin{aligned}
\big(-i\boldsymbol{\nabla}\cdot\boldsymbol{\gamma} + m + e\not{A}\big)\Phi_n(\boldsymbol{x}) &= E_n\gamma^0\Phi_n(\boldsymbol{x}) \\
\big(-i\boldsymbol{\nabla}\cdot\boldsymbol{\gamma} + m + e\not{A}\big)\bar{\Phi}_n(\boldsymbol{x}) &= -\bar{E}_n\gamma^0\bar{\Phi}_n(\boldsymbol{x})
\end{aligned} \tag{14}$$

determine positive energy eigenstates of the Dirac Hamiltonian operator H_D,

$$H_D = \int d\boldsymbol{x}\,\bar{\psi}(\boldsymbol{x})\big[-i\boldsymbol{\nabla}\cdot\boldsymbol{\gamma} + m_e + e\not{A}\big]\psi(\boldsymbol{x}) \tag{15}$$

The eigenstates $|n\rangle$ and $|\bar{n}\rangle$ of H_D with eigenvalues E_n and $\bar{E}_n$ are given in sect. III of [2],

$$|n\rangle = \int d\boldsymbol{x}\,\psi_\alpha^\dagger(\boldsymbol{x})\Phi_{n\,\alpha}(\boldsymbol{x})\,|\Omega\rangle \qquad\qquad |\bar{n}\rangle = \int d\boldsymbol{x}\,\bar{\Phi}_{n\,\alpha}^\dagger(\boldsymbol{x})\psi_\alpha(\boldsymbol{x})\,|\Omega\rangle \tag{16}$$

The ground state $|\Omega\rangle$ is a superposition of Fock states with any number of e^+e^- pairs,

$$|\Omega\rangle = N_0 \exp\Big[-b_q^\dagger\big(B^{-1}\big)_{qm}D_{mr}d_r^\dagger\Big]\,|0\rangle \tag{17}$$

where a sum over the 3-momenta and helicities q, r and over the state labels m is understood. The matrices B and D are given by the Dirac wave functions (14) and the free spinors u, v,

$$B_{mq} = \Phi_m^\dagger(\boldsymbol{q})u(\boldsymbol{q}, \lambda) \qquad\qquad D_{mr} = \Phi_m^\dagger(-\boldsymbol{r})v(\boldsymbol{r}, \lambda) \tag{18}$$

The vacuum state satisfies $H_D\,|\Omega\rangle = 0$. In the weak field limit $|n\rangle$ turns into a single electron and $|\bar{n}\rangle$ into a single positron state.

The Dirac states allow to understand a curious property of the Dirac wave functions that was noted already in 1932: The wave functions cannot be normalized and the energy spectrum is continuous for a linear A^0 potential [5]. A potential which confines electrons repulses positrons. Hence the states have both a confined and a deconfined (accelerating/decelerating) component. The latter is due to the positrons in the e^+e^- pairs, which (predominantly) appear in the region where $|eA^0(\boldsymbol{x})| \gtrsim 2m_e$ (see Fig. 14 of [1] for an example in $D = 1+1$ dimensions). The solutions are characterized by a continuous parameter which determines the ratio of e^+e^- pairs to valence e^-. The minimum ratio is given by the Schwinger pair production rate [6].

5 A Confining QCD Potential from a Boundary Condition

A primary motivation for the present study is to determine whether the concept of Born term for QED atoms is relevant for QCD hadrons. With no loop contributions the coupling is frozen at a value which may be perturbative, $\alpha_s^{\overline{MS}}/\pi \simeq 0.137$ [7]. A confining potential requires a parameter Λ_{QCD} with the dimension of mass. This parameter does not appear in the QCD Lagrangian, and can result from the classical gluon field equations only due to a boundary condition. The standard QED solution (6) of Gauss' law (5) is obtained with the boundary condition $\lim_{|\boldsymbol{x}|\to\infty} A^0(\boldsymbol{x}) = 0$.

A non-vanishing boundary condition at $|\boldsymbol{x}| \to \infty$ generally violates Poincaré invariance and generates long-distance effects. However, there is an (apparently unique) acceptable solution for mesons and baryons (see section VI B of [1] for an operator field formulation, and section V C of [2] for the equivalent classical gauge field). Translation invariance requires that the field strength have no spatial dependence and that the states be (global) color singlets. Rotational invariance is preserved when the field is correlated with the positions of the charges, analogously to (6). A non-vanishing boundary condition on the classical field can be imposed only on A_a^μ with $\mu = 0$ and $a = 3, 8$. $\mu = 0$ ensures instantaneity and the diagonal color matrices $T_{3,8}^{AB} \propto \delta_{AB}$ conserve the quark colors. This should hold in a gauge where the color structure of the meson wave functions is $\Phi^{AB} \propto \delta^{AB}$ and for baryons $\Phi^{ABC} \propto \epsilon^{ABC}$.

Gauss' law is imposed separately for each spatial position of the quarks as in (5), and for each quark color component. The boundary condition results in a confining potential which for mesons is exactly linear. Since the state is an overall color singlet the field vanishes when summed over colors. Thus an external observer does not see the hadron via its confining field, eliminating long range effects.

The $\mathcal{O}\left(\alpha_s^0 \hbar^0\right)$ confining gauge fields for the $|q_A(\boldsymbol{x}_1)\bar{q}_A(\boldsymbol{x}_2)\rangle$ Fock component of a meson are

$$\bar{\psi}_A(\boldsymbol{x}_1)\psi_A(\boldsymbol{x}_2)\,|0\rangle : \qquad \mathcal{A}_a^0(\boldsymbol{x}) = 6\Lambda^2 \frac{\boldsymbol{x}\cdot(\boldsymbol{x}_1-\boldsymbol{x}_2)}{|\boldsymbol{x}_1-\boldsymbol{x}_2|} T_a^{AA} \qquad (a=3,8) \tag{19}$$

where the constant Λ characterizes the boundary condition of this homogeneous solution ($\boldsymbol{\nabla}^2 \mathcal{A}_a^0(\boldsymbol{x}) = 0$) of Gauss' law. Since $\sum_A T_a^{AA} = 0$ a color singlet meson generates no overall color field.

The corresponding color field for the Fock component $|q_1(\boldsymbol{x}_1)q_2(\boldsymbol{x}_2)q_3(\boldsymbol{x}_3)\rangle$ of a baryon is

$$\psi_1^\dagger(\boldsymbol{x}_1)\psi_2^\dagger(\boldsymbol{x}_2)\psi_3^\dagger(\boldsymbol{x}_3)\,|0\rangle : \quad \mathcal{A}_3^0(\boldsymbol{x}) = 3\Lambda^2 \frac{\boldsymbol{x}\cdot(\boldsymbol{x}_1-\boldsymbol{x}_2)}{d(\boldsymbol{x}_1,\boldsymbol{x}_2,\boldsymbol{x}_3)} \qquad \mathcal{A}_8^0(\boldsymbol{x}) = \sqrt{3}\Lambda^2 \frac{\boldsymbol{x}\cdot(\boldsymbol{x}_1+\boldsymbol{x}_2-2\boldsymbol{x}_3)}{d(\boldsymbol{x}_1,\boldsymbol{x}_2,\boldsymbol{x}_3)} \tag{20}$$

where

$$d(\boldsymbol{x}_1,\boldsymbol{x}_2,\boldsymbol{x}_3) = \frac{1}{\sqrt{2}}\sqrt{(\boldsymbol{x}_1-\boldsymbol{x}_2)^2+(\boldsymbol{x}_2-\boldsymbol{x}_3)^2+(\boldsymbol{x}_3-\boldsymbol{x}_1)^2}$$

The total field energy $\int d\boldsymbol{x}\, \sum_{a=3,8}\left[\boldsymbol{\nabla}\mathcal{A}_a^0(\boldsymbol{x})\right]^2 = 12\Lambda^4 \int d\boldsymbol{x}$ must be a universal, the same for all Fock states of mesons and baryons. Hence Λ is a universal constant, equivalent to Λ_{QCD}.

Using the classical gauge fields (19), (20) in the QCD Hamiltonian the condition

$$H_{QCD}\,|n, \boldsymbol{P}=0\rangle = M_n\,|n, \boldsymbol{P}=0\rangle \tag{21}$$

determines the bound state equations for the meson and baryon wave functions in the rest frame. Quark pair production ("string breaking") is of higher order in $1/\sqrt{N_c}$ and can be included iteratively. The meson bound state equation is, with $\Phi^{AB}(\boldsymbol{x}_1-\boldsymbol{x}_2) = \delta^{AB}\Phi(\boldsymbol{x}_1-\boldsymbol{x}_2)/\sqrt{3}$,

$$i\boldsymbol{\nabla}\cdot\left\{\gamma^0\boldsymbol{\gamma}, \Phi(\boldsymbol{x})\right\} + m_1\gamma^0\Phi(\boldsymbol{x}) - m_2\Phi(\boldsymbol{x})\gamma^0 = \left[M - V_{\mathcal{M}}(\boldsymbol{x})\right]\Phi(\boldsymbol{x}) \tag{22}$$

where $m_{1,2}$ are the (current) quark masses. The potential $V_{\mathcal{M}}$ arises from the interaction term in H_{QCD}. For the $|q_A(\boldsymbol{x}_1)\bar{q}_A(\boldsymbol{x}_2)\rangle$ Fock component the color field (19) contributes at $\boldsymbol{x} = \boldsymbol{x}_1$ and $\boldsymbol{x} = \boldsymbol{x}_2$ (with opposite signs). The field energy subtracts half of the quark interaction energy as in QED (10) (see section V D of [2] for details),

$$V_{\mathcal{M}}(\boldsymbol{x}_1 - \boldsymbol{x}_2) = \tfrac{1}{2} g \sum_{a=3,8} T_a^{AA} \big[\mathcal{A}_a^0(\boldsymbol{x}_1) - \mathcal{A}_a^0(\boldsymbol{x}_2)\big] = g\Lambda^2\, |\boldsymbol{x}_1 - \boldsymbol{x}_2| \tag{23}$$

The potential is seen to be independent of the quark color A using the identity $\sum_a T_a^{AB} T_a^{CD} = \frac{1}{2}\delta^{AD}\delta^{BC} - \frac{1}{6}\delta^{AB}\delta^{CD}$.

The baryon bound state equation is analogous to (22), with the potential

$$V_{\mathcal{B}}(\boldsymbol{x}_1, \boldsymbol{x}_2, \boldsymbol{x}_3) = \frac{1}{\sqrt{2}}\, g\Lambda^2 \sqrt{(\boldsymbol{x}_1 - \boldsymbol{x}_2)^2 + (\boldsymbol{x}_2 - \boldsymbol{x}_3)^2 + (\boldsymbol{x}_3 - \boldsymbol{x}_1)^2} \tag{24}$$

The two gluon field components $a = 3, 8$ in (20) induce confinement in the two relative separations $\boldsymbol{x}_1 - \boldsymbol{x}_2$ and $\boldsymbol{x}_1 + \boldsymbol{x}_2 - 2\boldsymbol{x}_3$ of the three quarks in the baryon. No analogous solution is available for multiquark states such as $qq\bar{q}\bar{q}$ and $qqq\,q\bar{q}$. When two quarks coincide the baryon potential reduces to the meson one, $V_{\mathcal{B}}(\boldsymbol{x}_1, \boldsymbol{x}_2, \boldsymbol{x}_2) = V_{\mathcal{M}}(\boldsymbol{x}_1 - \boldsymbol{x}_2)$.

6 Properties of the Meson Solutions

The rotational invariance of the meson potential (23) allows to express the 4×4 wave function in terms of Dirac matrices, spherical harmonics and radial functions. For $q\bar{q}$ pairs of the same flavor ($m_1 = m_2 = m$) the solutions may be grouped into "trajectories" [8] on which the parity η_P and charge conjugation η_C depend on the total angular momentum J as $\eta_P = -\eta_C = (-1)^{J+1}$ ("π trajectory"), $\eta_P = \eta_C = (-1)^{J+1}$ ("a_1 trajectory") and $\eta_P = \eta_C = (-1)^J$ ("ρ trajectory"). Despite the relativistic dynamics there are no states with quantum numbers that would be exotic in the quark model.

The wave function of a state of mass M is in general singular at $M - V(r) = 0$. States with regular wave functions have a discrete spectrum. For $m = 0$ the Regge trajectories are linear, with evenly spaced daughter trajectories as in dual models (Figs. 16 and 17 of [2]). Similarly to the Dirac case, the radial wave functions have constant norm at large radii r. This allows an interpretation of the $q\bar{q}$ components with large potential energy $V(r)$ as being dual to hadrons produced via string breaking, as in phenomenological models of quark hadronization. The string breaking amplitudes are of higher order in $1/\sqrt{N_c}$ and determined by the overlap of the hadron states obtained at leading order in $1/\sqrt{N_c}$. Thus $A \to B + C$ is given by $\langle BC|A\rangle$, which can be expressed in terms of the hadron wave functions Φ_A, Φ_B and Φ_C (Eq. (7.1) of [1]). The square of this amplitude gives a hadron loop correction to Φ_A, as required by unitarity at $\mathcal{O}\left(\hbar^0\right)$. Meson scattering amplitudes may be similarly evaluated.

Analogously to the weakly bound Positronium states, mesons with CM momentum $\boldsymbol{P}$ are bound by the potential obtained by boosting (19). They have energy eigenvalue $E = \sqrt{M^2 + \boldsymbol{P}^2}$ and their wave functions are Lorentz contracted compared to the rest frame ones.

The states based on a chiral invariant ground state appear in parity degenerate pairs as expected. There is a massless meson solution with $J^{PC} = 0^{++}$. Since it has vanishing 4-momentum it may mix with the ground state without breaking Poincaré invariance. This would cause spontaneous chiral symmetry breaking.

7 Outlook

The application of perturbation theory to bound states is subtle: the expansion must be developed around an interacting state (see, *e.g.*, section II C of [1]). Since the Schrödinger atom has no loop contributions it serves as a Born term for bound states, analogously to tree diagrams of scattering amplitudes. The $\mathcal{O}\left(\hbar^0\right)$ approximation is not limited to non-relativistic dynamics – Schrödinger atoms can be viewed in any reference frame. Loop corrections can be added once the master formula (1) for the S-matrix is generalized to include bound state Born terms in the *in* and *out* states.

Conceivably, Born bound states could serve as a first approximation also for QCD hadrons. If radiative gluon contributions are suppressed at low scales Q the coupling $\alpha_s(Q^2)$ freezes, enabling a perturbative expansion.

The novel features of hadrons, including confinement and chiral symmetry breaking, must then appear already at Born level.

The QCD confinement scale Λ_{QCD} can arise from the Born level field equations only via a boundary condition. The instantaneity of A_a^0 allows to consider non-trivial homogeneous solutions of Gauss' law. Physical requirements restrict the solutions to color singlet states bound by a linear potential. The spectrum found this way includes massless states, which may mix with the vacuum, causing chiral symmetry breaking.

The perturbative expansion is a central tool in analytic studies of the Standard Model. This, together with features of the data, motivate careful studies of its applicability even to the hadron spectrum and dynamics.

Acknowledgements I thank the organizers of Light Cone 2016 in Lisbon for this opportunity to present my work. I have benefited from discussions with Matti Järvinen. I am grateful to the Department of Physics at Helsinki University for my privileges as Professor Emeritus and to the Magnus Ehrnrooth Foundation for a travel grant.

References

1. P. Hoyer, *Lecture Notes: "Bound States—from QED to QCD"*. arXiv:1402.5005 [hep-ph]
2. P. Hoyer, *"Lectures on Bound States"*. arXiv:1605.01532 [hep-ph]
3. T. Kinoshita, G.P. Lepage, in *Quantum Electrodynamics*, ed. by T. Kinoshita (World Scientific, Singapore, 1990), p. 81
4. P.P. Kulish, L.D. Faddeev, Theor. Math. Phys. **4**, 745 (1970) [Teor. Mat. Fiz. **4**, 153 (1970)]
5. M.S. Plesset, Phys. Rev. **41**, 278 (1932)
6. J.S. Schwinger, Phys. Rev. **82**, 664 (1951)
7. Y.L. Dokshitzer, *Proceedings of ICHEP 1998*, p. 305–324 [hep-ph/9812252] and hep-ph/0306287
8. D.A. Geffen, H. Suura, Phys. Rev. D **16**, 3305 (1977)

Printed by Printforce, the Netherlands